T0252615

A Revised Handbook
to the
FLORA OF CEYLON

VOLUME X

Araliaceae

Callitrichaceae

Capparaceae

Caricaceae

Caryophyllaceae

Casuarinaceae

Celastraceae

Ceratophyllaceae

Dilleniaceae

Elaeagnaceae

Erythroxylaceae

Fabaceae (Leguminosae)

Flacourtiaceae

Hippocrateaceae

Icacinaceae

Loganiaceae

Monimiaceae

Nelumbonaceae

Nymphaeaceae

Olacaceae

Passifloraceae

Phytolaccaceae

Plantaginaceae

Podostemaceae

Polygalaceae

Portulacaceae

Ranunculaceae

Rhamnaceae

Stemonaceae

Theaceae

Tiliaceae

Violaceae

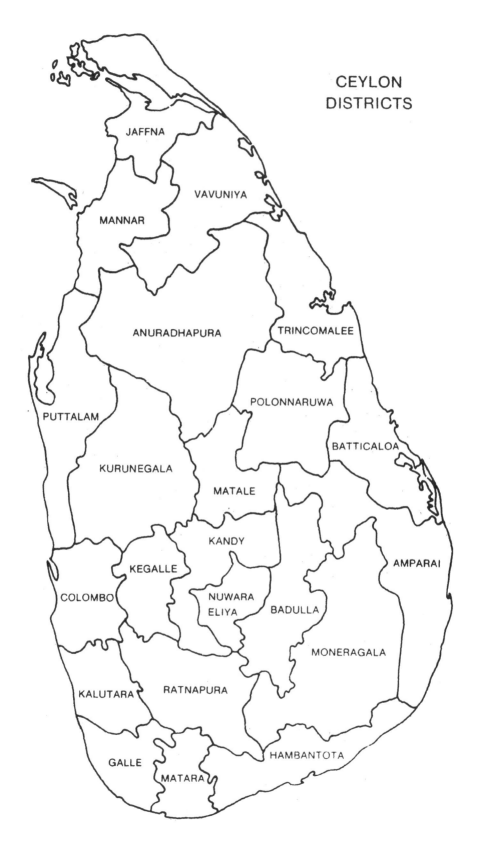

CEYLON
DISTRICTS

A Revised Handbook
to the
FLORA OF CEYLON

VOLUME X

Sponsored by the
University of Peradeniya,
Department of Agriculture, Sri Lanka,
and the Overseas Development Administration, U.K.

General Editor
M.D. DASSANAYAKE

Editorial Board
M.D. DASSANAYAKE
W.D. CLAYTON

A.A.BALKEMA/ROTTERDAM 1996

For the complete set of ten volumes, ISBN 90 6191 063 3

For volume I   ISBN 90 6191 064 1
For volume II   ISBN 90 6191 065 X
For volume III   ISBN 90 6191 066 8
For volume IV   ISBN 90 6191 067 6
For volume V   ISBN 90 6191 068 4
For volume VI   ISBN 90 6191 069 2
For volume VII   ISBN 90 6191 551 1
For volume VIII   ISBN 90 6191 552 X
For volume IX   ISBN 90 5410 267 5
For volume X   ISBN 90 5410 268 3

This edition is not for sale in North America and Sri Lanka

FOREWORD

In 1990 the Overseas Development Administration undertook to complete the Revised Handbook to the Flora of Ceylon by funding collaboration between Royal Botanic Gardens, Peradeniya and Royal Botanic Gardens, Kew, allowing a gradual withdrawal of the Smithsonian contribution. This volume is the first to be published wholly under the auspices of the new three-way partnership.

W.D. CLAYTON

CONTENTS

viii

A Revised Handbook
to the
FLORA OF CEYLON

VOLUME X

ARALIACEAE

(by David G. Frodin*)

Juss., Gen. Pl. 217. 1789 ("Araliae"). Type genus: *Aralia* L., Sp. Pl. 273. 1753.

Trees, shrubs, lianes, rhizomatous herbs or woody hemi-epiphytes, sometimes unbranched and palmlike and often armed, hermaphrodite, polygamous or dioecious; indumentum, when present, of simple or stellate hairs. Stems and branches mostly orthotropic, usually aromatic when cut or bruised and often quite pithy; shoots all alike or differentiated into long and short states (only the latter then being fertile); buds covered by the stipular sheaths of the leaves or by cataphylls. Foliage of juvenile plants or reversion shoots sometimes differing from that of adults, in certain species markedly so. Leaves often large, spirally arranged or alternate, rarely opposite or in whorls, simple or compound; petiole usually more or less clasping the stem; stipules absent or when developed distinct or united into a sometimes prominent ligule; blades when single entire or pinnately or palmately lobed, when multiple digitately or pinnately arranged, sometimes to the second or third degree; margins entire or variously toothed or slashed. Inflorescence terminal or pseudolateral, rarely lateral (and usually then as part of a sometimes very reduced short shoot), simple or, when (as is more often the case) compound, usually paniculate but sometimes more or less umbelliform, usually distinct from the foliage, naked (the bracts quickly falling) or with more or less persistent internal or external and sometimes leaf-like bracts; ultimate parts mostly umbellate, capitulate, racemose or spicate. Pedicel, when developed, continuous with the flower or articulated at its base. Flowers bi- or unisexual, sometimes with both in a single inflorescence (the unisexual ones then being male), actinomorphic, at anthesis generally epigynous or nearly so with the top of the ovary usually forming a distinct, fleshy, more or less nectariferous disk. Calyx lobes small or reduced to an entire or undulating rim or (rarely) absent. Petals 3 to many (most often 5), valvate or imbricate in bud, sometimes fused into a wholly or partly sutured calyptra, the base broad but in a few genera more or less narrowed below. Stamens usually as many as the petals and alternating with them or more numerous; filaments inserted at the edge of the disk, in one

* Royal Botanic Gardens, Kew.

or more whorls or, when numerous, in an indeterminate fashion; anthers linear to oblong, dorsifixed, introrse, with 4 (rarely 8) pollen sacs opening by longitudinal slits. Ovary syncarpous, wholly or partly inferior or (very rarely) superior, 1 to 75 (but most commonly 2–5, sometimes to 10–12)-celled; styles and stigmas as many as the cells, either free or wholly or more or less partly united; ovules solitary, pendulous, anatropous, the raphe ventral. Fruit drupaceous or baccate or (rarely) a schizocarp with persistent carpophore, in transverse section round or more or less compressed, in outline round or more or less elongate or flattened or even somewhat winged, the exocarp usually fleshy, the endocarp usually distinctly cartilaginous or membraneous. Seeds one per pyrene, the endosperm oily, smooth or more or less ruminate (the ruminations sometimes inclusive of the pyrene wall).

A family of 50 or so genera and at least 1400 species. Three genera are found in Ceylon, with six native species; two of the genera have additional species in cultivation. Much of the family is associated with the land masses and fragments associated with or derived from Gondwanaland; few genera (*Eleutherococcus* and its allies, *Hedera, Panax*) are largely associated with northern lands. Of the native genera, *Schefflera* is the most prominently represented.

Karyotypes: $x = 12$ (based on the $n = 12$ or $n = 24$ of most reliable reports). Figures of 9, 10, 11 and 13 have also been deduced or recorded; these should be viewed with caution.

Scattered polyploidy occurs although too little information has been gathered for useful arguments. No counts have been made of the species in Ceylon, save *Aralia leschenaultii* where from north Indian material a figure of $n = 12$ has been obtained.

Pollen: primitively spheroidal, the sexine undifferentiated, the tectum imperforate, non-sculptured; in more advanced types the tectum is variously perforate to even reticulate and in shape the grains may be subprolate to prolate. Specific studies of those in Ceylon have yet to be made.

Morphology: The presence of more than 10–12 stamens and/or ovary cells (beyond that, for example, represented by *Schefflera actinophylla*) is likely to represent secondary increase. Likewise, heteromery in the floral parts, in Ceylon seen in *Schefflera heterobotrya* (with ovary-cells more numerous than the petals and stamens) and the cultivated *Polyscias fruticosa* (with 5 petals and stamens but only 2 cells in the ovary) is very likely an advanced feature.

The three genera are mutually relatively easily recognized. *Schefflera* is characterized by digitately compound (or, in one species, simple) leaves and flowers with valvate petals and no articulation at the top of the pedicel. *Polyscias* has pinnately compound (or, in one cultivated species, sometimes simple) leaves; in the native species the leaflets are entire while those in cultivation have the leaflets all more or less toothed or slashed. *Aralia* has pinnately compound leaves, an umbelliform inflorescence and flowers with imbricate petals.

KEY TO THE GENERA

1 Leaves palmately lobed, softly textured, white-hairy beneath. (Cultivated and possibly naturalized species.) .... **2. Tetrapanax**
1 Leaves simple, trifoliolate, palmately or pinnately compound; texture more or less firm, the undersurface glabrous
  2 Leaves simple or trifoliolate
    3 Climbers or scramblers. Pedicels of flowers inarticulate. (Native species.) ....
      .... **3. Schefflera**
    3 Erect shrubs or treelets. Pedicels of flowers articulate. (Cultivated species.) ....
      .... **4. Polyscias**
  2 Leaves with 5 or more leaflets
    4 Leaflets palmately compound. (Native and cultivated species.) .... **3. Schefflera**
    4 Leaflets pinnately compound, sometimes more than once so
      5 Leaflets entire. (Native species) .... **4. Polyscias**
      5 Leaflets toothed or slashed
        6 Leaves once-pinnate. Leaflets finely bristle-toothed. Petals in bud imbricate. (Native species.) .... **1. Aralia**
        6 Leaflets once- or twice-pinnate (or more finely divided). Leaflets repandly toothed or slashed or more or less pinnatifid. (Cultivated species.) .... **4. Polyscias**

1. ARALIA

L., Sp. Pl. 273. 1753; L., Gen. Pl. ed. 5: 134. 1754. Type: *A. racemosa* L. (Wen 1992 in Taxon 41(2): 69–75).

Pentapanax Seemann, J. Bot. 2: 294. 1864; Harms in Pflanzenfam. III, 8: 55. 1897. Type: *Pentapanax leschenaultii* (DC.) Seemann.

Trees, shrubs or rhizomatous perennial herbs (rarely lianas), armed or unarmed. Shoot growth sympodial, undifferentiated or differentiated, the stems when erect sometimes stout. Leaves spirally arranged or alternate, once- to four times pinnately compound with terminal leaflets always present; stipules generally developed. Inflorescence terminal, distinct from the leaves or mixed with them, paniculate, compound-umbellate or corymbose but occasionally comprising a solitary umbel, the main axis elongate or not. Flowers 5–8-merous, the pedicels articulated just below; calyx mostly united with the ovary but manifest as a 5–8-toothed rim surrounding the petals and disk; petals 5–8, imbricate in bud, free, more or less spreading at anthesis; stamens 5–8; ovary wholly inferior with 5–8 cells (these sometimes reduced to 3 by abortion), surmounted by the disk and stylopodium; styles free to connate at the base to completely united. Fruit drupaceous, fleshy, more or less globose, the endocarp when mature crustose or somewhat hardened; seeds in pyrenes, with smooth endosperm.

A genus of some 50 species, primarily in E and SE Asia and in Malesia with also additional centres in the Americas. The single species in Ceylon belongs to the former *Pentapanax*, distinguished by Seemann and Harms mainly with reference to the styles being wholly or mostly united (in *Aralia* proper they are free or united only at the base).

Aralia leschenaultii (DC.) J. Wen, Brittonia 45(1): 53. 1993.

Hedera fragrans D. Don, Prod. Fl. Nepal. 187. 1825.
Panax leschenaultii DC., Prod. 4: 254. 1830. Type: India (Nilghiri Hills), *Leschenault s.n.* (G, holotype).
Hedera leschenaultii (DC.) Wight & Arn., Prod. 1: 377. 1834.
Pentapanax leschenaultii (DC.) Seemann, J. Bot. 2: 296. 1864; Seemann, Rev. Hed. 22. 1868; Clarke in Hook. f., Fl. Br. Ind. 2: 724. 1879; Trimen, Handb. Fl. Ceylon 2: 282. 1894; Brandis, For. Fl. 248. 1906.
Pentapanax fragrans (D. Don) T.D. Ha, Novost. Sist. Vyssh. Rast. 11: 227. 1974; T.D. Ha, Vopr. Sravnit. Morfol. Semenn. Rast. 77. 1975.

A tree to 15 m tall or a scandent shrub. Leaves imparipinnately compound with 3–5 leaflets (in Ceylon 5); stipules scarcely developed; petioles (inclusive of the rhachis) to 15 cm long; leaflets ovate or elliptic-ovate, chartaceous to subcoriaceous, 5–12 cm long, 2–6 cm wide, the apex acuminate, the base rounded, the margins bristly-serrate; petiolules 0.3–1 cm long. Inflorescence compound, corymbose-umbellate, 8–15 cm long with 5–10 rays; main axis very short, to 2 cm long; primary rays puberulent to glabrous, ascending to erect, each bearing a terminal umbellule and a verticil of 0–4 subsidiary umbellules at about 2/3 the length from the base; umbellules 2–2.5 cm across; peduncles of lateral umbellules 1.5–3 cm long. Flowers puberulent to glabrous on pedicels 0.5–1 cm long, elongate in bud to 3 mm; calyx-rim with 5 small teeth; petals 5, 2 mm long, cohering into a calyptra and soon falling; stamens 5; ovary 5-celled, the styles connate nearly to the apex. Fruit ovoid, 3–4 mm long, becoming slightly 5-angled when dried; stylar column 1.5 mm long.

Distr. Northern India along the Himalaya to Burma and SW China; also in southern India and Ceylon. In Ceylon only one confirmed collection has been seen, recorded as being in thickets in the central highlands.

Note. The presence of this species, and the genus, in Ceylon was long considered doubtful. Clarke included it for the island but 'I do not know on what authority. There are no specimens.' This statement is repeated by Trimen. The species is omitted from the checklists of Trimen, Willis, Abeyesundere and de Rosayro, and Abeywickrama, appearing only in that of Gunawardena. It can only be assumed that it is very local or rare; further records should be sought.

If *Pentapanax* is retained as a distinct genus, then the name of the species must be *P. fragrans*. If, however, a narrower view of taxa (for which there is ·in this instance not a good case) is once more adopted, that name is applicable only to the populations north of the Ganges (including *Pentapanax umbellatus* Seemann). *P. leschenaultii* would then be available for those in Ceylon and southern India.

Specimen Examined. NUWARA ELIYA DISTRICT: Horton Plains, Agrapatana Road, altitude not given, *Nowicke & Jayasuriya 258* (PDA).

2. TETRAPANAX

K. Koch, Wochenschr. Gärtnerei Pflanzenk. 2: 371. 1859. Type: *T. papyriferus* (Hook.) K. Koch.

Unarmed clonal evergreen shrubs or small trees, with successive initially rhizomatous stems (ramets) arising as underground branches. Shoot growth undifferentiated, sympodial; stems thick, pithy, rarely branching; young aerial parts covered with loose floss. Leaves arranged palm-like at the tops of stems, large, palmately lobed, toothed, softly textured, hairy; stipules conspicuous. Inflorescence terminal, paniculate, woolly, with conspicuous bracts. Flowers in small, round, racemosely arranged, pedunculate umbels, not jointed at the top of the pedicel and without a calyculus; calyx-rim almost obsolete; petals 4–5, valvate; stamens 4–5; ovary wholly inferior, 2-celled, styles free, initially erect and recurved in their upper parts but divergent after anthesis. Fruit almost globose, drupaceous, slightly laterally compressed; pyrenes 2; endosperm uniform.

A monotypic genus of east Asia; certainly native only in Taiwan, but widely introduced elsewhere, particularly in Asia, and here and there naturalised.

Tetrapanax papyriferus (Hook.) K. Koch, Wochenschr. Gärtnerei Pflanzenk. 2: 371. 1859; Seemann, J. Bot. 6: 58. 1868; Seemann, Rev. Hed. 88. 1868; Macmillan, Tropical Planting and Gardening, 6th edn. 236, 478. 1991.

Aralia (?) papyrifera Hook., Hooker's J. Bot. Kew Gard. Misc. 4: 53. 1852; Hook., Fl. Serres 8: pl. 806, 807. 1853; Hook., Bot. Mag. 82: pl. 4897. 1856.

Fatsia papyrifera (Hook.) Miquel ex Witte, Ann. Hort. Bot. 4: 93. 1861; Forbes and Hemsley, J. Linn. Soc. Bot. 23: 341. 1888 (ex Benth. and Hook. f.); Trimen, Hort. Zeyl. 40. 1888; Parsons. An alphabetical list of plants in the Royal Botanic Gardens, Peradeniya 70. 1926; Macmillan, Tropical Planting and Gardening, 5th edn. 200, 412. 1943.

Shrubs or trees to 7 m tall or so, the stems not or but few-branched, sympodial, to 2 cm in diameter. Leaves orbicular, to 50 cm or more across; stipules

2, awl-shaped; petiole to 60 cm, white-hairy when young, later rusty-hairy; primary segments 5–12, the central ones usually forming 2 secondary segments; under surface paler than the dull upper surface, the hairs persistent. Inflorescence three times compound, white, conspicuous, with age becoming crowded by renewed vegetative growth; main axis short; primary branches 3–4, elongate, radiating, with basal bracts and bearing numerous secondary branches each subtended by a conspicuous bract to 2 cm long; secondary branches each with 10–15 umbels about 12 mm in diameter on peduncles to 1.2 cm long, these latter subtended by bracts to 1.2 cm long. Umbels all alike; pedicels 4 mm long. Flowers yellowish-white; calyx wholly fused with the ovary, to 1 mm long; petals 2 mm long, tomentose outside, sometimes remaining together and falling as a calyptra; filaments 3 mm long. Fruit fleshy.

D i s t r. Taiwan. Introduced into Ceylon (1856, fide Trimen), initially grown in the Royal Botanic Gardens, Peradeniya, and now naturalized here and there in the highlands. In some countries a troublesome weed (Macmillan).

V e r n. Rice-paper plant, Chinese rice-paper plant (E); Tung-tsau (Ch).

U s e s. These plants, the pith of which has long been used in East Asia for the manufacture of rice paper and paper objects (particularly flowers), are also decorative and first were introduced to cultivation in Europe and elsewhere, including South and Southeast Asia, in the middle of the nineteenth century.

N o t e. In Ceylon *Tetrapanax papyriferus* performs best in 'up country', and by the second quarter of the twentieth century had become occasionally naturalized therein by roadsides. Balakrishnan's collection suggests that this spread has continued. More information is needed.

Variegated forms are also known.

S p e c i m e n s   E x a m i n e d. NUWARA ELIYA DISTRICT: Foot trail from Ohiya to Pattipola, in forest edge, *Balakrishnan 481* (PDA). LOCALITY UNKNOWN: *s. coll. s.n.* (PDA).

3. SCHEFFLERA

J.R. & G. Forst., Char. Gen. 45, pl. 23. 1775, nom. cons. Type: *S. digitata* J.R. & G. Forst. (Harms 1894 in Pflanzenfam. III, 8: 39).

Heptapleurum Gaertn., Fruct. 2: 472, pl. 178. 1791. Type: *Heptapleurum stellatum* Gaertn. (= *Schefflera stellata* (Gaertn.) Baill.).

Brassaia Endl., Nov. Stirp., Dec. 1: 89. 1839. Type: *Brassaia actinophylla* Endl. (= *Schefflera actinophylla* (Endl.) Harms).

Agalma Miq., Bonplandia 4(9): 138. 1856 (1 May); Miq., Fl. Ind. Bat. 1(1): 751. 1856 (10 July). Type: *Agalma rugosum* (Blume) Miq. (= *Schefflera rugosa* (Blume) Harms).

Tupidanthus Hook. f. and Thoms., Bot. Mag. 82: pl. 4908. 1856. Type: *Tupidanthus calyptratus* Hook. f. and Thoms. (=*Schefflera pueckleri* (K. Koch) Frodin).

Unarmed glabrous or pubescent trees, shrubs, subshrubs or vines, very often epiphytic, occasionally also strangling host trees and becoming self-supporting. Shoot growth sympodial, almost always undifferentiated, rarely divided into long- and short shoots; stems sometimes pithy. Leaves generally once-palmately compound (rarely more than once palmately compound or unifoliolate or simple), petiolate; base more or less forming a sheath around the twig and commonly developing 2 stipules, these generally fused into a ligule sometimes up to 10 cm long. Leaflets 3–20 but in a few species reduced to 1 or further subdivided, developing in a single plane or occasionally fascicled, usually distinctly stalked. Juvenile leaves sometimes differing markedly from adults. Inflorescences compound, paniculate or umbellate, usually leafless, terminal or becoming falsely lateral but never on short shoots, generally solitary on any one branch but occasionally more numerous. Flowers in small umbels, heads, racemes or spikes; calyx-rim conspicuous to obscure, lobed, toothed, wavy or uniform; petals (4–) 5 or more, valvate, sometimes coherent and falling as a cap or even more or less fused; stamens as many as the petals or more numerous (up to 500); ovary (1–) 2–30(–75)-locular, wholly or largely inferior at anthesis, the styles free or variously united or entirely reduced. Fruit drupaceous, elongate, round or more or less compressed, inferior or up to 1/2 or more superior at maturity, the exocarp usually fleshy, the endocarp cartilaginous; seeds in pyrenes, the endosperm generally smooth.

A genus of 900 or more species, throughout warmer and more humid parts of the world (but not reaching the Mascarenes, the Society Islands, Hawaii, the Marianas and the Carolines); best represented in the Americas and in Malesia. Apart from the characteristically digitately compound leaves (manifest in most species, though not, for example, in *S. emarginata* where the leaves are most often unifoliolate), it is distinguished by the presence of some comparatively unspecialized features which set it apart from related genera such as *Cussonia* (Africa), *Fatsia* (Japan and other East Asian islands), *Macropanax* (East Asia to West Malesia and the Himalaya), and *Hedera* (temperate parts of Eurasia). Four species are native to Ceylon; three are endemic though closely related to those in southern India while *S. stellata* is presently considered to be in both Ceylon and southern India.

The Queensland Umbrella-tree, *S. actinophylla* (Endl.) Harms, an Australian and New Guinean species, was introduced to Ceylon in 1873 and is sometimes cultivated. It is distinguished from the native species by its leaves, which have more numerous leaflets, and particularly by the distinctive terminal inflorescences which comprise laterally radiating primary branches bearing numerous heads of large red flowers. It is listed and illustrated in Macmillan, Tropical Planting and Gardening, 6th edn. (1991) and, as *Brassaia actinophylla*, in its predecessors.

KEY TO THE SPECIES

1 Leaves unifoliolate or trifoliolate. Plants differentiated into primary and secondary branches, the latter fertile ....**4. S. emarginata**
1 Leaves with 5 or more leaflets
  2 Leaflets with the widest point usually above the middle, obovate or narrowly obovate. Veins on under surface well-defined when blade dry. Climbers or scramblers, usually of 'low-country' .... **3. S. stellata**
  2 Leaflets with the widest point usually at or below the middle, more or less oblong-elliptic. Veins on under surface flush with mesophyll when blade dry. Trees or shrubs (sometimes epiphytic) of 'up-country'.
    3 Small trees to 10 m or shrubs, sparsely-branched, sometimes epiphytic. Leaflets 5–10, 7.5–20 × 2.5–8 cm, 2–2.5 times as long as broad, usually narrowly oblong to oblong-ovate. Primary branches of inflorescence longer than its central axis. Stylopodium raised, but stigmata flush and stylar column absent. Juvenile leaves little different from adults, the leaflets without gashes ....**2. S. exaltata**
    3 Medium to large trees, 15–30 m and much-branched when fully grown. Leaflets (3–) 5–9, 4.5–13.5 × 2.5–6.5 cm, 1.5–2 times as long as broad, usually elliptic to oblong or oblong-ovate. Primary branches of inflorescence shorter than its central axis. Stylopodium remaining more or less flat, the stigmata at the tip of a stylar column to 1.5 mm long. Juvenile leaves distinct from adults, the leaflets commonly with 4 deep gashes ..... **1. S. heterobotrya**

1. Schefflera heterobotrya Frodin, sp. nov.

Hedera racemosa auct., non Wight; Thw., Enum. Pl. Zeyl. 132. 1864.

Heptapleurum racemosum auct., non (Wight) Bedd.; Bedd., Fl. Sylv. pl. 214. 1872; Clarke in Hook. f., Fl. Br. Ind. 2: 729. 1879; Trimen, Handb. Fl. Ceylon 2: 283. 1894.

Schefflera racemosa auct., non (Wight) Harms; Alston in Trimen, Handb. Fl. Ceylon 6: 139. 1931; Abeyesundere and de Rosayro, Draft of First Descriptive Checklist 8. 1939.

Schefflera wallichiana auct., non (Wight & Arn.) Harms; Worthington, Ceylon Trees 288, illus. 1959.

Arbor sempervirens magna vel mediocris *Schefflerae racemosae* peninsulae Indiae non dissimilis, foliolis juvenalibus vulgo profunde lobatis vel pinnatifidis, floribus interdum umbellatis interdum racemosis, petalis et staminibus plerumque 5–7, ovariis (5–) 6–8(–9)-locularis, fructibus in statu sicco obtuse (5–)6–8(9)-costulatis. Partes juniores ultro plus minusve paullo rubiginose-puberulentes et venae laterales in foliolis plus numerosae.

Habitat in Zeylona in sylvis regionum superarum de 610 ad 1900 metros supra mare. Typus: *Thwaites C.P. 549*, "Pupillawa" (? Pussellawa) Regionis Centralis (K, holotypus; BM, CGE, Fl, GH, MEL, PDA, W, isotypi).

A shrub or tree to 30 m tall, much-branched when well-grown, the diameter at breast height 0.5–0.9 m. Twigs in first-year growth 6–8 mm in diameter, with leaves aggregated towards their ends; young parts closely though thinly covered with rusty or furfuraceous scurfy hairs, later wholly glabrous or nearly so. Outer bark light brown, thin, smooth; inner bark to 15 mm thick, pale brown with a resinous smell when cut. Foliage dimorphic; leaflets on juvenile or sucker shoots commonly larger than in adults, sometimes appearing fascicled on the verticil, the blades membraneous, up to 23 × 12 cm on petiolules 2–10 cm long, often with 1–3 (most commonly 2) deep gashes on each side with the resulting segments unequal, very acute and sometimes slightly curved; petioles up to 42 cm, the free portion of the stipular ligule up to 1 cm long. Adult leaves (3–)5–9-foliolate, the leaflets spreading in a single whorl; petioles 9.5–21 cm long, the free portion of the stipular ligule only 2–3 mm long; petiolules 1.5–3.4 cm long; blades elliptic to oblong-elliptic, oblong-ovate or oblong, 4.5–13.5 × 2.5–6.5 cm, widest at or a little below the middle, entire, moderately thick, the under surface distinctly paler than the upper; apex acute to obtuse, the tip to 10 mm, acuminate, sometimes funnel-like; base obtuse to rounded; main veins 7–9, somewhat brochidodromous, the venation slightly impressed above when dry and slightly but distinctly raised below; reticulation sometimes relatively conspicuous. Inflorescence developing from apex of previous year's wood and maturing in one season, terminal or pseudolateral, narrow in outline, erect or ascending at first but later drooping, usually ferrugineous-tomentose and also bearing small but persistent axillary bracts to 0.5 mm long on the main axis and primary branches; main axis elongate, to 32 cm long, with 18 or so primary branches arising throughout its length and at least sometimes ending in a single, early-developing and -maturing umbel. Primary branches 3.5–7.2 cm long, radiating in all directions. Flowers racemosely arranged, solitary or more or less in clusters or pseudowhorls or at branch ends in small umbels, those nearer the main axis often replaced by short, later-developing secondary peduncles bearing umbellules of male or fertile tertiary-level flowers; pedicels 3–7 mm (shorter in tertiary-level flowers). Buds ovoid, green. Calyx pale green, furfuraceous, the rim (limb) very narrow with 5–7 small teeth; petals 5–7, green to greenish-white to white, opening from apex and spreading before falling. Stamens as numerous as the petals. Ovary (5–)6–8(–9)–locular, 3–5 × 2.5 mm; disk flat, topped by a conical stylar column up to 1.5 mm long. Fruit remaining wholly inferior, initially green, ultimately black, globose or depressed-globose, to 5 × 5.5 mm, the pericarp shrinking around the (usually) 6–8 pyrenes on drying; disk 2–3 mm across, flat, sloping only into the stylar column; stylar column (except in the top terminal umbel) usually about 1.5 mm long, the stigmata forming a small boss at the apex.

Distr. Endemic.

Ecol. Forest in 'up-country' in central parts, sometimes by streams, 610–1900 m; reported at times as rather common. Fl. and fr. after monsoon, with full fruiting from October.

Vern. Itha (S).

Note. Beddome was the first to call attention to differences between these trees and *Schefflera racemosa* in south India. These distinctions have been observed in subsequent collections, and in addition the juvenile foliage of the Ceylon trees is distinguished by pinnatifid leaflets. There are also differences in inflorescence morphology and flower arrangement, notably the far greater incidence of umbellules in *S. heterobotrya*. With more abundant material now available, their recognition as a distinct species is justified which, in some respects, is more closely related to *S. micrantha* (Clarke) Gamble than *S. racemosa.*

New shoots begin development following the approximately mid-year emergence of the inflorescence from the wood of the previous cycle. Growth continues through the summer wet season with gradual maturation across the following winter and spring, the differentiation of new buds, and transformation of the shoot apex. The inflorescence develops over the next summer wet and begins flowering afterwards; fruit development and set follow during the latter part of the second season, tailing off by February/March. In *S. racemosa*, with greater seasonality in shoot development, the whole cycle is more prolonged.

Specimens Examined. MATALE DISTRICT: Laggala Estate of Brae Group, abandoned cardamom plantation, *Jayasuriya & Balasubramaniam 3474* (PDA). KANDY DISTRICT: ca. 2 miles N of Hunasgiriya near mile post 23, cardamom plantation, *Davidse 8563* (PDA); 3 mi. east beyond Madugoda, near road marker 29/17, *Jayasuriya 369* (K, POM, US); 'Kandy', *Moon s.n.* (BM); Rajamallay, forest patch in Moray Estate by stream, *Sohmer & Sumithraarachchi 9877* (GH, PDA); Imboolpitiya, Nawalapitiya, *Worthington 110* (K); Nawalapitiya, *Worthington 385* (BM, K); Kandy-Patiagama Road, road marker 19/20, *Worthington 1764* (K); Deltota Estate, Galaha, *Worthington 4971* (K); Moragolla, Patahewaheta, *Worthington 2887* (BM, K); Deltota, *Worthington 6590* (K, PDA). BADULLA DISTRICT: Madawelagama—at 6th mile post behind the pine nursery, along stream under shade, *Balakrishnan & Jayasuriya 823* (PDA, juvenile); Adisham forest near Haputale, *Kostermans 23425* (BO, K, L); Ohiya-Boralanda road, *Sohmer et al. 8550* (BISH, BM, GH, NY, PDA); Palugama, *s. coll. s.n.* (PDA); Craig Estate, Bandarawela, *A.M. Smith s.n.* (PDA, juvenile); Haputale, *A.M. Smith s.n.* (PDA); Glenanore Estate near Haputale, *Stone 11220* (BISH, KLU, L, PH, RSA). RATNAPURA DISTRICT: Adams Peak, from Carney, *Bernardi*

16085, 16087(?) (both PDA); Adams Peak trail, NE of Carney, *Davidse & Sumithraarachchi 8763* (PDA); Indikatupana, S slope of Peak Wilderness Sanctuary above Palabaddala, *Jayasuriya & Gunatilleke 3169* (BISH, GH, NY, PDA). NUWARA ELIYA DISTRICT: Moon Plains, Parawella, *St. John 24112* (BISH); portion of Sita-Eliya Forest Reserve, N side of Nuwara Eliya-Hakgala Rd., 52-mile, *Sohmer et al., 8461* (BISH, GH, NY); jungle margin, Hakgala, *N.D. Simpson 9036* (BM); Sita Eliya, along stream near highway A–5, *Theobald & Krahulik 2845* (PDA, US); Pidurutalagala, *Waas 172* (PDA); Hanguranketa, roadside near a stream, *Wirawan & Corner 772* (PDA); Hakgala Botanic Garden, Cultivated, *Meijer et al. C-610* (K), *Corner s.n.* (CGE, K, L). LOCALITY UNKNOWN: Central Province, without more precise locality, *Thwaites C.P. 0549* (BM, CGE, FI, GH, K, MEL, PDA,W); "Pupillawa", July 1846, *Thwaites C.P. 0549, 0559* (both K); *Macrae s.n.* (CGE); before 1844, *Walker ss.nn.* (E, E–GL, K, OXF).

2. Schefflera exaltata (Thw.) Frodin, comb. nov. Type: based on *Hedera exaltata* Thw.

Hedera exaltata Thw., Enum. Pl. Zeyl. 132. 1864. Type: Ceylon, without
   precise locality, *Thwaites C.P. 1633* (PDA, lectotype; BM, FI, K, W,
   isotypes).
Heptapleurum exaltatum (Thw.) Seemann, J. Bot. 3: 80. 1865; Seemann, Rev.
   Hed. 44. 1868.
Heptapleurum wallichianum auct., non (Wight & Arn.) Seemann; Clarke in
   Hook. f., Fl. Br. Ind. 2: 730. 1879.
Schefflera wallichiana auct., non (Wight & Arn.) Harms; Alston in Trimen,
   Handb. Fl. Ceylon 6: 139. 1931; Abeyesundere and de Rosayro, Draft of
   First Descriptive Check-List 8. 1939; Holmes, Imperial Forestry Institute
   Paper 28: 53. 1951.

A glabrous tree (recorded as 'large' by Thwaites and Trimen but more recent collections do not give a height of more than 10 m), sometimes a liana or an epiphyte. Twigs hollow, along with the leaves aromatic when cut or crushed (once reported with a mangiferous smell). Leaves palmately compound, the leaflets 5–10 in a single plane; petioles 12–23 cm, stout, cylindrical, the free part of the basal ligule small, to 6 mm or so with a broad base and acute tip; petiolules 1.3–3.5 cm long. Blades oblong-lanceolate, 7.5–20 × 2.5–8 cm (in young plants or reversion shoots somewhat larger), thinly coriaceous, entire, abruptly rounded or occasionally slightly cordate at base, the surfaces contrasting, the apex acute, the tip caudate-acuminate up to an extra 1.7 cm in length; midrib slightly raised below and recessed above when dry; lateral venation inconspicuous, the primary veins numerous, curving up near margin. Inflorescence terminal but by maturity becoming pseudolateral, paniculate, the branches relatively stiff, any bracts early caducous; main axis to at least 12 cm long; primary branches 7–9, to 36 cm long but more usually not over 20 cm, the umbellules racemosely disposed along them; peduncles 1.3–3.2 cm (occasionally to 3.9 cm) long. Flowers in bud to 5 mm long or more, when expanded to 10 mm across, 12–20 on pedicels from 5–7 mm in bud to 8–12 mm at maturity; buds nearly globose; calyx-rim truncate; petals 7–10, linear-oblong, acute, thick; stamens as many as the petals, spreading at anthesis, the filaments flat, narrow; ovary 7–10-celled, half-inferior at anthesis; disk and stylopodium conspicuous, blunt, conical, the stigmata flush at the apex. Fruit initially green, then yellow orange, at maturity about half-inferior, globose, to 5(–7) mm across with when dry 7–10 blunt vertical ribs, the upper half above the calyx-rim and surmounted by the persistent, slightly raised stigmata; pyrenes 7–10.

Distr. Endemic.

Ecol. In moist closed forests or scrubland or in *Eucalyptus* plantations in the up-country, 1100–1600 m. Trimen noted that it was relatively rare.

Uses. None recorded.

Note. The union of this species with *S. wallichiana* was first made by Clarke in 1879. Several small but apparently constant differences, however, separate this from *S. wallichiana*, which thus is limited to South India. In *S. exaltata*, the leaflets tend to be narrowly oblong, the stigmata range from 7 to 10 (most commonly 8–9) in contrast to 5–6 in *S. wallichiana*, there are fewer umbellules per primary branch with their peduncles being somewhat longer. The leaves are moreover in the dry state exceptionally clearly lined on the under surface when compared with *S. wallichiana*. This feature, along with the comparatively small leaves in adult plants, has led to confusion with what is now *S. heterobotrya*. The leaflets in *S. exaltata* are, however, usually more than twice as long as broad; those in *S. heterobotrya*, less than twice as long as broad.

The distribution of the species appears well to correlate with desirable areas for cardamom cultivation. In addition, it reportedly is a colonizer of *Eucalyptus robusta* plantings originally established in dry patana, e.g. in Kinigama Reserve in the Uva basin (Holmes 1951); no voucher specimens have, however, so far been seen.

The *Smith* collection from Maturata has exceptionally large leaflets, inflorescence branches and fruits (to 7 mm across) when compared to most others. Some transitions to these states exist in *Thwaites C.P. 1633*. More collections are needed.

Specimens Examined. MATALE DISTRICT: Rattota-Ilukkumbura Road, mile 35, *Huber 0744* (PDA). KANDY DISTRICT: Hunasgiriya, *Alston s.n.* (PDA); Loolecondara Estate, 2 mi S of Deltota, [at] upper edge of plantation, *Fosberg 57922* (NY); Midcar Estate, Knuckles, *Greller 810329–50* (BKL); Looloowatte near Knuckles, *Jayasuriya 945* (K, PDA, US); Knuckles, from Rangala to Looloowatte, *Nooteboom 3078* (KEP, L, PDA, US); locality not given (Hunasgiriya added on PDA sheet), *Thwaites C.P. 1633* (BM, FI, K, P, PDA, W), Nawanagalla Estate, Madugoda, *Worthington 1586* (K). BADULLA DISTRICT: Haputale, *Kostermans 23191* (L); Bandarawela, *Worthington 0584* (BM). NUWARA ELIYA DISTRICT: Maturata, high forest, *A. M. Smith s.n.* (PDA). RATNAPURA DISTRICT: Road Kalawana to Rakwana, mile 24/8, *S. Lucas 1141* (US). LOCALITY UNKNOWN: "Central Province", *Beckett 590* (MEL). Additional localities cited by Trimen are Pussellawe (Kandy District) and Ramboda (Nuwara Eliya District).

3. Schefflera stellata (Gaertn.) Baill., Hist. Pl. 7: 161. 1879; Harms in Pflanzenfam. III, 8: 39. 1894; Alston in Trimen, Handb. Fl. Ceylon 6: 139. 1931; Alsẓon, Kandy Flora 44, fig. 234. 1938.

[Ittawael: arbor zeylanica *Itta* dicta, resinam Terebinthinae similem fundens Hermann, Mus. Zeyl. 50. 1717.]
[Arbor *Itta* dicta (etc.) Burm., Thes. Zeyl. 28. 1737.]
[Ittawel L., Fl. Zeyl. 249. 1747.]
Heptapleurum stellatum Gaertn., Fruct. 2: pl. 178, fig. 3. 1791; Seemann, J. Bot. 3: 80. 1865; Seemann, Rev. Heder. 45. 1868; Clarke in Hook. f., Fl. Br. Ind. 2: 730. 1879; Trimen, Handb. Fl. Ceylon 2: 283. 1894; Pearson, J Linn. Soc. Bot. 34: 343. 1899. Type: not yet established.
Heptapleurum acutangulum Gaertn., Frucht. 2: 472. 1791.
Hedera terebinthacea Vahl, Symb. Bot. 3: 42. 1794; Moon, Cat. 18. 1828; DC., Prod. 4: 265. 1830; Wall., Cat. no. 4920 in part, *s. str.* 1831–32. Type: Ceylon, c. 1777–1781, *Koenig s.n.* (BM, lectotype).
Paratropia terebinthacea Arn., Pugill. 20. 1836; Arn., Nova Acta Phys.-Med. Acad. Caes. Leop.-Carol. Nat. Cur. 18: 338. 1837; Walp., Rep. 2: 433. 1843. Type: Ceylon, 'Highlands', *Col. Walker 14* (K, lectotype).
Hedera vahlii Thw., Enum. Pl. Zeyl. 132. 1859. Type: based on *Hedera terebinthacea.*

A large, wholly glabrous shrub, climber or epiphyte, or a tree to 8 × 0.70 m branching from the base, or the whole plant sprawling among rocks; branches becoming stout and sometimes pendant; twigs about as thick as a swan's quill, along with the leaves strongly aromatic when bruised or cut, the odour reported as 'mangiferous'. Leaves palmately compound, the leaflets (3–)5–7 in a single plane; petioles 5.3–15 cm long, cylindrical, the base amplexicaul, the ligule mostly fused with its free part to 1 mm long, rounded or bluntly acute; petiolules 1.1–5 cm long. Blades obovate to narrowly obovate, nearly always widest distinctly above the middle, 4.5–12 × 1.9–6.8 cm, coriaceous, entire, the base attenuate to acute, the apex broadly acute to obtuse with only a slightly developed tip; venation generally conspicuous when dry, the main lateral veins ascending, curving up near margin. Inflorescence terminal, paniculate, well distinct from leaves, bracteate when young but later naked, the bracts membraneous; main axis to at least 13.5 cm long with the branches diverging within the upper 7.5 cm or less; primary branches (–3)5–12, to 17 cm long, the umbellules racemosely and usually somewhat repandly disposed along them; peduncles 6–12 but sometimes to 21 cm long. Buds rather blunt, 2–3 × 1.5–2 mm, glabrous; flowers when expanded to 5–6 mm across, (3–) 5–8 on slender pedicels 2–9 mm long; calyx-rim truncate, without a rim; petals 6–9 (most commonly 8), acute with an inflexed point; stamens as many as the petals but shorter, spreading at anthesis; ovary 6–9 (most commonly 8)-celled, mostly inferior; disk and stylopodium conspicuous, blunt, conical, the stigmata flush at the apex. Fruit initially green, later yellow to orange but ripening black with the glaucousness of the disk eventually fading, subglobose (Trimen: clavate-ovoid), to 7 × 6 mm, about 4/5 inferior at maturity and surmounted by the persistent stigmata, when dry with 6–9 blunt vertical ribs; pyrenes 6–9.

Distr. Ceylon, southern India.

Ecol. In forest or other vegetation, sometimes growing over rocks, in the moist or intermediate zone to 1200(–1525) m. It enters the patanas region in Badulla District but there remains below 1220 m in sheltered areas (Pearson). Fl. and fr. May–July(–October).

Uses. The leaves are used as a cattle medicine (Trimen).

Vern. Itha, Itta, Ittaw[a]el, Maha-itta-waela (S). Gaertner listed 'Bukera' but without indication of a language.

Note. The Indian plants, including *Hedera obovata* Wight and probably also *Paratropia rotundifolia* Tenore, differ from those in Ceylon by having somewhat thinner leaflets which moreover are broader relative to their length; in addition, the ovary is 5–8-celled. Most records are from 600–1800 m, and in some at least the leaves are semi- or strongly deciduous. *S. venulosa* (Wight & Arn.) Harms, of the lower wet parts of the Western Ghats of India, differs in having more elliptic and ± acuminate leaflets; the ovary moreover is 5–6-celled. Basic relationships, however, are with those plants as well as *S. roxburghii* Gamble and *S. clarkeana* Craib.

The fruits studied by Gaertner had mostly 7 pyrenes, with a few 6-pyrened. This contrasts with the usual number of 8–9 as described by Arnott and most commonly observed in available Ceylon specimens. *Matthew RHT.49866* (K) from Dindigul in the Palni Hills (Tamilnadu, India) and identified as *S. stellata* is, however, 6–7-pyrened, a state also seen in a number of other collections from southern India. It is thus possible that Gaertner's material came not from Ceylon, as stated by him, but from India. Further studies, and additional material from India, may show that the Indian plants are truly distinct from those in Ceylon. If *S. stellata* is then applied to Indian plants, the name *Hedera terebinthacea* would have to be revived, necessitating a new combination in Schefflera.

The illustration in De Wildeman, Ic. Sel. Hort. Then. 4: pl. 133. 1903, offered as this species, represents a form of *S. roxburghii*.

Specimens Examined. ANURADHAPURA DISTRICT: Andiyagala, *Amaratunga 931* (PDA); Ritigala S.N.R., summit, *Jayasuriya 0904* (PDA), *Willis s.n.* (PDA). KURUNEGALA DISTRICT: Dolukande, *Cramer 3050* (PDA, US). MATALE DISTRICT: Hill SE of Dambulla Rock; on south slope, *Balakrishnan 502* (CAS, K, PDA); Illukkumbura, *Jayasuriya & Bandaranayake 1773* (K, US); E slope of Lenadora Hill, mi. 38/9 on Dambulla to Nalanda Road, route A–9, *Nowicke & Jayasuriya 378* (NY, PDA); Wiltshire Forest, Kandegedara, *Sumithraarachchi 0395* (BISH, PDA); Mile 24/5, Palapathwela—Pallepola Rd, *Sumithraarachchi et al., 0726* (K, PDA, POM). POLONNARUWA DISTRICT: Western slope of Gunner's Quoin, forest edge, *Huber 0438* (HBG, PDA, US). KEGALLE DISTRICT: Bible Rock, *Sumithraarachchi & Fernando 0135* (K, PDA, POM). KANDY DISTRICT: Peradeniya, moist up-country, *Balasubramaniam 222* (NA); Hunasgiriya, on Mahiyangana—Teldeniya Rd., *Cooray 690801–12R* (K, NY, POM); Gannoruwa Mtn., Peradeniya, *Greller 810114–6* (BKL); Madugoda Forest, SE of Knuckles, *Greller s.n.* (BKL); Peradeniya, Royal Botanic Gardens, *Pearson 976* (CGE), *Hosseus 11* (M); near Corbet's Gap, *Kostermans 23510* (L, US); Rangale to Corbet's Gap, *Kostermans 23510* (A); Hunasgiriya, near rivulet, *Kostermans 25274* (K, L, US); near railroad tunnel above downgrade on Hwy. A–1 E of Kadugannawa, *Theobald & Grupe 2369* (US); Kandy catchment, *Worthington 7208* (K). BADULLA DISTRICT: Road edge, from Ohiya to Haldummulla, *Balakrishnan 535* (BISH, K); Bandarawela, in patana, *Pearson 871* (CGE); patana, Fort Macdonald valley, SW of Kirklees, *A.M. Smith s.n.* (PDA); the same, on rock, *A.M. Smith s.n.* (PDA), Haputale, Thotulagala Estate, *Werner 292* (K). AMPARAI DISTRICT: Potuvil-Moneragala Road, c. 196-mile marker, *D.B. and D. Sumithraarachchi 0887* (PDA). KALUTARA DISTRICT: Ad templum Kalugala Ahran, circa Pelawatta, *Bernardi 15751* (G not seen, PDA). RATNAPURA DISTRICT: above Belihul Oya, 900 m, 25 May 1969, *Kostermans 23630* (L, US); roadside, Balangoda to Maratenna, at culvert 7/11, *Lucas 1116* (US); Lankaberiya, *Waas 1395* (BISH, PDA). MONARAGALA DISTRICT: mile 137/4, W of Wellawaya, route A–4, *Meijer C–202* (K, L, PDA, US); Nagala, Bibile-Nilgala rd., 3rd mile marker, *Waas 0650* (NY, PDA). GALLE DISTRICT: Galle, scandens, 100–200 m, Pierre

[Coll. Zeyl.] *55* (P). LOCALITY UNKNOWN: Labelled only as "Itta-wel", *s. coll. s.n.* (BM); 1838–47, *Champion s.n.* (CGE); before 1847, *Gardner 0325* (K); "in sylvis et nemorosis Zeylonae", 1777–81, *Koenig s.n.* (BM); before 1864, *Thwaites C.P. 1632* (BM, BO, BR, FI, K, MEL, P, W); "Highlands", before 1844, *Col. Walker 14* (K).

4. Schefflera emarginata (Moon) Harms in Pflanzenfam. III, 8: 38. 1894; Alston in Trimen, Handb. Fl. Ceylon 6: 139. 1931.

Hedera emarginata Moon, Cat. 18: 1824; Linden, Cat. 12: 18. 1857; Thw., Enum. Pl. Zeyl. 132. 1864. Type: Sabaragamuwa, Ceylon, *Moon s.n.* (not located).

Paratropia emarginata (Moon) Regel, Gartenfl. 13: 305. 1865. Type: as above.

Heptapleurum emarginatum (Moon) Seemann, J. Bot. 3: 80. 1865; Seemann, Rev. Heder. 44. 1868; Clarke in Hook. f., Fl. Br. Ind. 2: 729. 1879; Trimen, Handb. Fl. Ceylon 2: 284. 1894, Atlas, pl. 64. 1895. Type: as above.

A glabrous epiphytic or terrestrial shrub, sometimes climbing or creeping but often dwarf and bushy; bark silvery grey; branches dimorphic, with stout primary runners to 7 mm in diameter from which develop numerous lateral secondary shoots. Leaves numerous, usually closely spaced, 1(–3)-foliolate, the petiole slender, 7–20(–50) mm long, jointed at the top and with the base forming a sheath around the twig, the free portion of the stipular ligule to 1 mm long; leaflets pseudopetiolate, 22–50(–80) × 7–23(–38) mm, cuneate to oblong-cuneate, narrowing into a point about 3–8 mm above the base and then gradually tapering downwards, entire, the apex rounded to truncate and emarginate or shallowly 2-lobed at apex, thick when fresh but shrinking markedly on drying, the steeply sloping veins becoming prominent on the under surface. Inflorescences mostly developing on secondary shoots, terminal, small, quite lax, any bracts quickly falling during development; main axis very short or obscure; primary branches rather slender, 1–3, to 50 mm long but usually smaller, when multiple sometimes widely divaricating; peduncles up to 5 but often solitary or in pairs at branch ends, up to 23 mm long but usually short. Flowers very small, white with a red tinge, solitary or in clusters of 2–6(–7), the pedicels slender, 6–8 mm; calyx-rim absent; petals 5, spreading at anthesis but quickly falling; stamens as many as but longer than the petals; ovary largely inferior, with (4–) 5 cells and a prominent disk surmounted by the almost sessile stigmata; fruit fleshy, smooth, red to purple and finally black, broadly ovoid, to 3 × 3 mm at maturity, truncate, the pyrenes not becoming conspicuous on drying.

Distr. Endemic.

Ecol. In wet forest in moist country (generally with little or no seasonal water deficit) up to 1220 m, reportedly rather rare (Trimen) but best represented in the SW districts. Fl. May–June. As a shrub it has been said often to be dwarf and bushy, but also 'able to climb the largest trees'. It can also be a creeper spreading on the ground of wet primary forest or a climber or weak-stemmed treelet, and in the Knuckles area it can get into laurel-forest (Greller). In *Jayasuriya 2907* it is described as a large, hemi-epiphytic shrub climbing to 6 m with horizontal branches.

Uses. None recorded. It was, however, introduced into horticulture in Europe in the 1850s and there enjoyed a brief vogue, remaining in collections through the 1880s (e.g. at the Royal Botanic Gardens, Kew and the Botanic Garden, Liverpool, UK).

Vern. none recorded.

Note. A 'curious', delicate climbing species, with unifoliolate or rarely 3-foliolate leaves with obovate, emarginate leaflets. Most nearly related to *S. bourdillonii* (leaves 5-foliolate) and *S. chandrasekharanii* (leaves with 1–2 acute leaflets), both in southern India. The reportedly greenish-pink (Trimen) or white, pink-tipped (Thwaites) flowers suggest a further relationship with *S. pubigera* and its allies in northern India, the Himalayan states and adjacent areas. Pritzel, in the first edn. of his Thesaurus (1851), noted Moon's catalogue as being 'liber perrarus'. The species is missing from de Candolle's Prodromus and G. Don's General System.

Specimens Examined. KEGALLE DISTRICT: Allagalla Mt., the ridge, *Worthington 0297* (K). KANDY DISTRICT: Hantane Hill, *Burtt & Townsend [C]50* (K), in 1900s, *J.M. de Silva s.n.* (PDA), *Thwaites C.P. 0655* (K); Midcar Estate, Knuckles, *Greller 810329–51* (BKL). KALUTARA DISTRICT: Denihena, *Waas 1887* (E, GH, K, L, PDA, POM, US). RATNAPURA DISTRICT: Gilimale F.R., *Jayasuriya (apud Meijer) C–917* (PDA, RSA, US), trail to Gongala above Longford Div. of Hayes Group, *Jayasuriya 2907* (A, BISH, PDA not seen); Gilimale F.R., logging area, *Meijer C–406* (K, US); Sri Palabaddala, Mapulana ella, *Waas 0436* (PDA); Mannikkawatta Forest (S of Sinharaja primary forest) via Deniyaya, *Waas 1772* (K, PDA). GALLE DISTRICT: Haycock (Hinidumkande), near summit, *Kostermans 25514* (PDA, US). LOCALITY UNKNOWN: [1838–47], *Champion s.n.* (CGE); given only as 'Central Province' [before 1864], *Thwaites C.P. 0655* (CGE, FI, P, W); [before 1844], *Walker s.n.* (E, K, OXF); *Walker 158* (K).

Additional localities cited by Trimen are Hunasgiriya and Peradeniya (Kandy District), Kuruwita Korale (Ratnapura District), 'Sabaragamuwa' (from Moon), and Kukul Korale.

4. POLYSCIAS

J.R. & G. Forst., Char. Gen. 64, pl. 32. 1775. Type species: *P. pinnata* J.R. & G. Forst., Vanuatu (generally now united with *P. scutellaria* (Burm. f.) Fosberg).

Eupteron Miq., Pl. Jungh. 3: 420. 1855; Miq., Bonplandia 4(9): 139. 1856; Miq., Fl. Ind. Bat.1(1): 762. 1856. Type: *E. nodosum* (Blume) Miq., Indonesia (= *Polyscias nodosa* (Blume) Harms).

Unarmed shrubs or trees, glabrous or pubescent, sometimes strongly aromatic. Leaves 1–3 times compound and imparipinnate, rarely unifoliolate, with a short or elongated sheathing base; rhachis articulated, it and the petiole terete. Leaflets when present in pairs, entire, crenate or dentate. Inflorescence terminal, often large, usually compound and forming a panicle, corymb or umbel, rarely reduced to a simple umbel. Flowers racemosely arranged along primary inflorescence branches or more commonly in umbellules or capitula so disposed; pedicels articulated below the ovary. Calyx-rim sometimes well-developed, with a undulating or toothed margin. Petals valvate, 4–5(–8 or more). Stamens as many as the petals. Ovary inferior, 2–5(–8 or more)-celled, the disk fleshy; styles either free and recurved (at least in fruit) or joined into a beak-like stylopodium with the stigmata then flush. Fruit drupaceous, globose or ovoid, sometimes slightly or strongly compressed, topped by the persistent calyx-rim and styles or stylopodium; exocarp usually fleshy; endocarp chartaceous; endosperm ruminate, with an uneven surface or fissured, rarely smooth.

A genus of some 100 species, in Africa, SW Indian Ocean Islands, S Asia, Ceylon, and in Malesia from Java, NE Borneo and the Philippines eastwards to New Guinea; also in Australia and the Pacific Islands to the Marianas and the Societies. It is most highly developed in Madagascar, Papuasia and New Caledonia. The single native species is conventionally referred to sect. Eupteron (Miq.) Philipson; the commonly cultivated 'fancy' species and forms are all in sect. Polyscias. Miquel (1856) included *Polyscias acuminata* in his concept of *Eupteron* but further study may show that an alternative infrageneric disposition is necessary.

The classification of the many cultivated forms in sect. Polyscias is difficult and no satisfactory treatment has yet been published. They are not 'biological' species, being always associated with man. Any taxonomy which may be developed will naturally depend upon the proponent's species philosophy, but a conventional framework may not be suitable. The erect shrub or small tree forms have had a long history in the Indo-Pacific region, with two at least already described and figured by Rumphius in the seventeenth century. Another, in Vanuatu, was the basis for the Forsters' genus. Others, usually low-growing shrubs, may be chimaeras and, if allowed to grow up, become the same as or similar to the well-known shrubby or small tree forms. They have long been, and remain, fine useful and ornamental plants for all tropical homes and gardens. The more important are *Polyscias balfouriana* (Sander ex André) L.H. Bailey, *P. filicifolia* (C. Moore ex E. Fournier) L.H. Bailey, *P. fruticosa* (L.) Harms, *P. guilfoylei* (W. Bull) L.H. Bailey and *P. scutellaria* (Burm. f.) Fosberg. *P. fruticosa* was listed by Trimen (and by Abeywickrama), and all but *P. scutellaria* are listed, and the first three illustrated, in Macmillan, Tropical Planting and Gardening (6th edn. 1991) as well as in earlier editions under *Aralia* and *Panax*.

The ruminations of the endosperm are inclusive of the pyrene wall through inversion (in *P. ornifolia* and *P. australiana*; Harms 1894). *P. acuminata* in Ceylon, however, lacks ruminations (Trimen).

Polyscias acuminata (Wight) Seemann, J. Bot. 3: 181. 1865; Seemann, Rev. Hed. 56. 1868; Clarke in Hook. f., Fl. Br. Ind. 2: 727. 1879; Trimen, Handb. Fl. Ceylon 2: 282. 1894.

Hedera acuminata Wight, Ic. Pl. Ind. Or. 1062. 1850; Thw., Enum. Pl. Zeyl. 131. 1859, excl. synon. Type: India (Nilghiri Hills), *Wight s.n.* (K, lectotype).

Eupteron acuminatum (Wight) Miq., Bonplandia 4: 139. 1856; Miq., Fl. Ind. Bat. 1(1): 762. 1856.

A glabrous small tree or scandent shrub to 5 m. Leaves once pinnately compound, with 5–11 leaflets (in Ceylon 5–7), the rhachis 15–23 cm long; leaflets narrowly oblong to lanceolate, 9–12 cm long, entire, the apex acute, the tip caudate-acuminate, the base acute to rounded and abruptly recurving into the petiolule, the veins ascending, slender, fairly numerous, scarcely raised on drying, the midrib somewhat prominent on the under surface; petiolules 0–6 mm long. Inflorescence terminal, leaf-opposed, soon bypassed by subsequent growth and appearing pseudolateral; main axis thickened, 13–14 cm; primary branches 2–5 or more, usually all springing from apex but occasionally one arising from along the axis, to 23 cm long but usually less (in Ceylon only to 13 cm); umbellules developing in the upper part of the branches, racemosely arranged, the spreading peduncles 0.6–1 cm long. Flowers pale green, 3–5 in an umbellule, the pedicels thickened, 2.5–4 mm long, articulate at each end; bracts linear, setaceous; calyx-lobes acute; petals 5, very thick, acute; stamens 5, small; ovary wholly inferior, 4–5-celled, the disk broad, fleshy, the styles short, erect, blunt. Fruit initially green, later red before turning black, broadly ovoid, obscurely ribbed when dry, topped by the calyx-rim and the persistent, spreading styles.

Distr. South India and Ceylon.

Ec ol. In forest along streams, 1220–1525 m, evidently very local in the central highlands. Fl. Feb–May.

Vern. not recorded.

Note. Ceylon specimens are more depauperate than those from India, with fewer leaflets (5–7 as opposed to 7–11) and weaker inflorescences with fewer primary branches. In addition, in India the species is always erect. More material is required, however, to assess whether these distinctions are worthy of formal recognition.

Specimens Examined. KANDY DISTRICT: Adam's Peak, *Thwaites C.P. 4* (BM, K); 4 mi. SW of Maskeliya (straight line), fishing hut area on margin of Moray Group Tea Estate, at SE base of Adam's Peak, *Davidse & Sumithraachchi 8711* (PDA); Maskeliya, *Kostermans 24244A* (K).

Trimen has cited in addition Bogawantalawa (Nuwara Eliya District).

CALLITRICHACEAE
(by M.D. Dassanayake*)

Link, Enum. Hort. Berol. Alt. 1: 7. 1821 ("Callitrichinae"). Type genus: *Callitriche* L.

Small glabrous herbs, aquatic and submerged or emergent, or terrestrial in wet places, Stems slender and tufted, with small, decussate, entire, exstipulate leaves. Flowers unisexual, solitary in leaf axils (rarely two flowers, a staminate and a pistillate, in the same axil), very small, without perianth. Bracteoles two, opposite, caducous or persistent. Staminate flowers of usually a single stamen with a 2-celled anther dehiscing by lateral and longitudinal apically confluent slits. Pistillate flower of a single pistil. Ovary of two carpels, syncarpous, laterally 4-lobed, each carpel divided into 2 loculi by 'false' septa. One ovule in each loculus, pendulous, on axile placenta, unitegmic, anatropous, with ventral raphe. Styles 2, free, filiform, papillose. Fruit a schizocarp, splitting at maturity into four one-seeded mericarps; lobes winged or keeled. Seed minute, with straight embryo and oily endosperm.

One genus, *Callitriche*.

CALLITRICHE

L., Sp. Pl. 969. 1753; L., Gen. Pl. ed. 5. 5. 1754. Type species: *C. palustris* L.
Characters of the family.
About 17 species, almost all cosmopolitan.

Callitriche stagnalis Scop., Fl. Carniol. ed. 2, 2: 251. 1772; Trimen, Handb. Fl. Ceylon 2: 149. 1894. Type: from Europe.

Much-branched flaccid herb rooting at the nodes. Stems glabrous, 1–2 mm thick, green. Leaves broadly ovate- to very broadly ovate-spathulate, obtuse and rounded at the apex, tapering to the petiole, 2–15 mm long including the petiole, 1–9 mm broad, fleshy, glabrous, with 3 veins, dark green. Flowers mostly in only one axil of a pair of leaves. Bracts c. 1 mm long, falcate, white. Staminate flower: filament erect, filiform, 3–4 mm long, white. Anther c. 0.75 mm long, pale yellow. Pistillate flower: Ovary c.

* Flora of Ceylon Project, Peradeniya.

0.5 mm long, pale green. Styles 3–5 mm long, white, at first semierect, later divergent, minutely papillose, withering in fruit. Fruit on c. 1 mm long stalk, compressed, with rounded sides and depressed at the apex, c. 2 mm long and 2.5 mm broad, glabrous, each lobe with a wing c. 0.5 mm broad round the back. Seed ellipsoid, c. 1 mm long, 0.5 mm broad.

Distr. Throughout the old world: Indian mountains, the Himalayas to the Deccan, in Europe, northern Asia, tropical Africa, Australia, New Zealand.

Ecol. Upper montane zone. In stagnant or slow-moving water or on damp ground. Flowering throughout the year.

Specimens Examined. NUWARA ELIYA DISTRICT: Talawakelle, St. Coombes, 5 Nov 1940, *Eden s.n.* (PDA); Horton Plains, *Gardner in C.P. 303* (PDA); Maturata, Apr 1859, *s. coll. C.P. 3636* (PDA); stream on Albion Estate, 14 Mar 1906, *J.C. (Willis) s.n.* (PDA); Condegalla Estate, 18 Sept 1927, *Alston 1859* (PDA).

CAPPARACEAE
(by D. Philcox[*])

Juss., Gen. Pl. 242. 1789 ("Capparides"). Type genus: *Capparis* L.

Herbs, shrubs or trees, erect or sometimes scrambling or climbing. Leaves alternate, petiolate, simple or palmately divided, entire or rarely denticulate, pinnately veined. Stipules thorny, minute or wanting. Inflorescence usually terminal or lateral, racemose, corymbose or umbellate, but flowers occasionally solitary-axillary. Flowers bisexual, actinomorphic or zygomorphic. Sepals 4, free, rarely connate into tube, equal or in 2 series with outer pair slightly different from inner. Petals 4, rarely 2, free, usually clawed. Stamens 4 to indefinite, hypogynous or perigynous, or inserted at base of, or midway on, gynophore. Ovary on long gynophore, rarely sessile, 1-celled; ovules numerous, usually on 2, rarely more, parietal placentas. Fruit capsular, 2-valved, or berry. Seeds many, mostly curved-reniform, testa in seeds from dry fruits variously sculptured or smooth.

Family of about 45 genera containing approximately 675 species, from tropical and subtropical regions.

KEY TO THE GENERA

```
1 Herbs  ........................................................ 1. Cleome
1 Trees or shrubs
  2 Sepals connate at base  ............................................. 2. Maerua
  2 Sepals free
    3 Sepals biseriate
      4 Stamens 4–6, inserted midway on gynophore  ......................... 3. Cadaba
      4 Stamens indefinite, inserted at base of gynophore  .................. 4. Capparis
    3 Sepals equal, inserted on edge of large disk  ........................ 5. Crateva
```

1. CLEOME

L., Gen. Pl. ed. 5. 302. 1754. Lectotype species: *Cleome ornithopodioides* L. (see M.L. Green, Prop. Brit. Bot. 172. 1929.)

Annual herbs, glabrous to hairy, or at times glandular-hairy. Leaves simple or compound, petiolate. Flowers pedicellate, terminally racemose or

---

[*] Royal Botanic Gardens, Kew.

paniculate. Sepals 4, free. Petals usually 4, quite or nearly regular. Stamens 6—indefinite, inserted on receptacle or at base of gynophore. Ovary sessile or with short gynophore, 1-celled; ovules numerous on 2 parietal placentas. Capsule linear, 2-valved, beaked. Seeds reniform.

Genus of 150 species from the tropics, subtropics and warmer temperate regions.

KEY TO THE SPECIES

1 Leaves simple .... 1. C. monophylla
1 Leaves compound
  2 Stamens 20 or more
    3 Stamens 20–30, filaments not thickened at top; stems glandular-pubescent, viscid; flowers yellow .... 2. C. viscosa
    3 Stamens more than 30, filaments thickened at top; stems hirsute with pale, stiff, bulbous-based hairs; flowers rose to purple .... 3. C. chelidonii
  2 Stamens 6
    4 Leaflets linear .... 4. C. tenella
    4 Leaflets not linear
      5 Gynophore 1–4 mm long in flower, extending to 2–2.5 cm in fruit ..... 5. C. gynandra
      5 Gynophore minute or lacking
        6 Petals pink to pale purple; fruit narrowing into stalk-like base above pedicel; seeds with open cleft .... 6. C. rutidosperma
        6 Petals white to dull yellow; fruit terminating abruptly at base above pedicel; seeds with closed cleft .... 7. C. aspera

1. Cleome monophylla L., Sp. Pl. 672. 1753; DC., Prod. 1: 239. 1824; Wight & Arn., Prod. 21. 1834; Thw., Enum. Pl. Zeyl. 14. 1858; Hook. f. & Thoms. in Hook. f., Fl. Br. Ind. 1: 168. 1872; Trimen, Handb. Fl. Ceylon 1: 55. 1893. Type: from India.

Annual, 30–60 cm tall, erect, simple to much-branched above; stems strongly patent, glandular-pubescent. Leaves simple, slender petiolate, passing into bracts above; lamina 2–5 (–7.5) × 0.4–1.4 (–2) cm, lanceolate, acute, entire, rounded to cordate at base, finely glandular-pubescent on both surfaces; petiole 1–2.5 (–4) cm long. Flowers pale violet-pink, solitary in axils of leafy bracts, forming erect raceme 30–40 cm long; pedicels 0.7–1 cm long, slender, shorter than bracts. Sepals 3.5 × c. 1 mm, oblong, drying dull purplish, densely, shortly glandular-pubescent without. Petals c. 9 × 1.5 mm including 4 mm long slender claw. Stamens 6; filaments c. 9 mm long; anthers 1.5 × c. 0.2 mm, linear, greatly curved. Ovary 3.5 × 0.3 mm, narrowly ellipsoid, densely glandular. Capsule up to 8–9.5 × 0.2 cm, linear, straight, cylindrical, pubescent, prominently parallel veined. Seeds c. 1.5 mm diameter, laterally compressed-globose, strongly ribbed on back, drying dark brown to black, cleft more or less closed.

Distr. Ceylon and southern India; tropical and subtropical Africa, from the Somali Republic to South Africa.

Ecol. Weed of waste and cultivated ground in the lowlands; also in rocky areas in the dry districts.

Specimens Examined. ANURADHAPURA DISTRICT: Dambulla to Anuradhapura road, S of mile-marker 52/5, 4 Oct 1973, *Sohmer 8078* (PDA). POLONNARUWA DISTRICT: Polonnaruwa Rest House, 8 Jan 1970, *Fosberg 51908* (PDA). KURUNEGALA DISTRICT: Kurunegala, Mag Hill, 4 Sept 1962, *Amaratunga 312* (PDA). MATALE DISTRICT: Dambulla, 180 m, 2 Dec 1970, *Cramer 3283* (PDA), 335 m, 6 Oct 1970, *Dassanayake 529* (K); Dambulla Rock, 16 May 1931, *Simpson 8131* (BM); Dambulla Rock, 17 Nov 1961, *Amaratunga 22* (PDA), March 1868, *Thwaites s.n.* in *C.P. 2792*, p.p. (PDA), 20 Dec 1881, *s. coll. s.n.* (PDA). BADULLA DISTRICT: Badulla, Jan 1888, *s. coll. s.n.* (PDA), *s. coll.* in *C.P. 2792*, p.p. (HAK); Bintenna, 24 Apr 1923, *de Silva s.n.* (PDA). KANDY DISTRICT: Peradeniya, in 1860, *Thwaites s.n.* in *C.P. 2792*, p.p. (BM). KEGALLE DISTRICT: Galpitamade, 22 Aug 1966, *Amaratunga 1124* (PDA). LOCALITY UNKNOWN: *Walker 1007* (E), *s. coll. s.n.* (K).

2. Cleome viscosa L., Sp. Pl. 672. 1753; Hook. f. & Thoms. in Hook. f., Fl. Br. Ind. 1: 170. 1872; Trimen, Handb. Fl. Ceylon 1: 57. 1893. Type: Ceylon, Herb. *Hermann 241* (BM–SL, lectotype).

Polanisia viscosa (L.) DC., Prod. 1: 242. 1834.

Herb, annual, erect up to 0.4–1.5 m tall, widely branched, unarmed; stems yellowish hirsute, viscid, foetid. Leaves 3–5-foliolate; leaflets (0.8–) 2–3 × (0.25) 0.8–1.3 cm, oblanceolate-elliptic, acute or obtuse, cuneate to attenuate at base, densely glandular-pubescent, becoming glabrescent with age, but maintaining pubescence especially on major veins beneath, petiolule 0.5–1.5 mm long, densely glandular-pubescent; petiole c. 0.4–3.75 cm long, densely glandular-pubescent. Flowers pedicellate, various shades of yellow (white to pink or purple—*Wirawan 1220, 1222*), in corymbose racemes with flowers solitary in axils of reduced, leaf-like bracts. Sepals 0.5 × 0.15 cm, oblong, acute, obtuse or rounded, glandular-pubescent without, free. Petals 0.9–1 × 0.35 cm, oblong, rounded at apex, cuneate at base into short claw, glabrous. Stamens c. 20 or more, included; filaments 5–8 mm long; anthers 1.25 × 0.25 mm, linear, curved. Ovary 5 × 1.5 mm, densely glandular-hairy, included, style 1–1.5 mm long, glabrous. Fruit up to 8 × 0.35 cm, linear-cylindric, subdensely glandular-pubescent, prominently parallel-veined, beak 3–5 mm long; fruiting pedicel 1.2–2.2 cm long. Seeds c. 1 mm diameter, 0.6 mm thick, circular, prominently transversely ridged, cleft closed, orange-red.

Distr. Native of the Old World tropics, occurring from southern Arabia and tropical Africa, through tropical Asia including Malaysia, to Australia.

Ecol. Rocky and sandy places, roadsides and cultivated ground.

Vern. Wal-aba, Ran-manissa (S).

Specimens Examined. JAFFNA DISTRICT: Punkudutivu, 15 Nov 1970, *Kundu & Balakrishnan 668* (PDA). ANURADHAPURA DISTRICT: Anuradhapura, 17 Mar 1927, *Alston 1118* (PDA); Anuradhapura to Trincomalee road, near mile-marker 65/4, 5 Oct 1973, *Sohmer 8147* (PDA); Nuwara Wewa, 4 May 1931, *Simpson 8043* (BM); Vavuniya to Kebithigollewa road, 24 Jun 1975, *Sumithraarachchi & Sumithraarachchi 830* (PDA); Rajangana Reservoir, 1 Sept 1978, *Wirawan 1214* (PDA); margin of Hunuwillagama Wewa, 31 Aug 1978, *Wirawan 1210* (PDA). TRINCOMALEE DISTRICT: Harbour, York Island, 27 Aug 1860, *Dubuc s.n.* (E); Kantalai, S end of tank, 6 Oct 1973, *Sohmer 8185* (PDA), *8187* (PDA); c. 3 km N of Adampane, c. 16 km SW of Kuchchaveli, 1 Dec 1977, *Fosberg & Jayasinghe 57113* (K, PDA). KURUNEGALA DISTRICT: Athagala Rock, Kurunegala, 23 Sept 1978, *Wirawan 1231* (PDA); Nikaweratiya, 4 Sept 1962, *Amaratunga 322* (PDA). MATALE DISTRICT: Dambulla Rock, 29 Nov 1926, *Silva s.n.* (PDA); Sigiriya Wewa, 11 Nov 1973, *Sohmer 8643* (PDA). POLONNARUWA DISTRICT: Polonnaruwa, 1 Sept 1978, *Wirawan 1215* (PDA), Rest House, 7 Oct 1973, *Sohmer 8228* (PDA), Sacred Area, 61 m, 18 Dec 1970, *Ripley 308* (PDA); Minneriya Tank, 6 Oct 1973, *Sohmer 8194* (PDA), *8199* (PDA); c. 5 km SW of Elahera near mile-marker 12/5, c. 120–150 m, 10 Oct 1974, *Davidse 7326* (K, PDA). BATTICALOA DISTRICT: Tiruperümdurai, sea-level, 5 Dec 1976, *Cramer 4784* (K, PDA). COLOMBO DISTRICT: Colombo, Government Farm, 15 Dec 1926, *Alston 1119* (PDA); Minuwangoda, 16 Apr 1962, *Amaratunga 1272* (PDA); Negombo, 18 Nov 1969, *Amaratunga 1840* (PDA). KANDY DISTRICT: Kandy, Royal Palace Park, 16 Sept 1978, *Wirawan 1220* (K), *1221* (PDA), *1222* (K, PDA). BADULLA DISTRICT: Mahiyangana to Bibile road, c. 8 km W of Bibile, 11 Oct 1973, *Sohmer 8315* (PDA). AMPARAI DISTRICT: Gal Oya Reservoir, 270 m, 13 Nov 1967, *Comanor 564* (E, K); E side of Ulpasse Wewa, c. 3 km N of Panama, 26 Nov 1970, *Fosberg & Sachet 52974* (K, PDA). KALUTARA DISTRICT: Kalutara, Nov 1920, *de Alwis s.n.* (PDA). RATNAPURA DISTRICT: surrounds of sports pitch, Tunkama, c. 25 km NE of Ambalantota, 25 Feb 1994, *Philcox et al., 10676* (K, PDA). HAMBANTOTA DISTRICT: Ruhuna National Park, Sithul Pahuwa, 25 Nov 1970, *Fosberg & Sachet 52914* (PDA), Block I, Bambawa, near Guards' Hut, 17 Nov 1969, *Cooray 69111706R* (K, PDA). LOCALITY UNKNOWN: *Brodie 125* (E); in 1860, *Dubuc s.n.* (E); *Walker 1014* (E), *s. coll. s.n.* (E); *Thwaites s.n.* in *C.P. 1073*, p.p. (BM, PDA).

3. Cleome chelidonii L. f., Suppl. Pl. 300. 1782; Hook. f. & Thoms. in Hook. f., Fl. Br. Ind. 1: 170. 1872; Trimen, Handb. Fl. Ceylon 1: 56. 1893. Type: from India.

Polanisia chelidonii (L. f.) DC., Prod. 1: 242. 1824; Wight & Arn., Prod. 22. 1834.

Herb, annual, erect, to 0.8 m tall; stems angular with sparse, pale, stiff, bulbous-based hairs. Leaves (1–3–) 6–7-foliolate, with fewest leaflets towards top; leaflets 1–2.25 (–5.5) × 0.25–0.4 (–1.2) cm, obovate, usually densely appressed-hairy, apex rounded, subacuminate to acute, base cuneate, petiolules 0.1–0.25 cm long, appressed-hairy; petiole 1–5 (–10) cm long or almost wanting, indumentum as stem. Flowers rose to pinkish-purple, pedicellate, in corymbose racemes, bracteate with each flower subtended by small leaf-like bracts comprising 1–3 bractlets. Pedicels 1–2.5 cm long, sparsely hairy to subglabrous. Sepals 0.2–0.3 × 0.1–0.15 cm, elliptic to obovate, shortly acuminate, sparsely hairy without. Petals 4, 0.9–1.2 × 0.3–0.5 cm, obovate, rounded at apex, narrow at base. Stamens many, up to 65, 7–8 mm long, glabrous; filaments thickened at top; anthers c. 1 mm long. Ovary c. 7 × 1 mm, linear, glabrous. Fruit 8–9.5 × 0.2 cm including 0.5 cm long beak, cylindrical, glabrous, prominently parallel-veined. Seeds c. 1.5 mm diameter, 1 mm thick, somewhat reniform with open cleft, warty, not ribbed.

Distr. India and Ceylon; Burma and Thailand; Central and East Java.

Ecol. Dry low country.

Vern. Wal-aba (S).

Specimens Examined. MANNAR DISTRICT: Giant's Tank, 14 Mar 1932, *Simpson 9332* (BM, PDA), Murunkan, 12 Nov 1970, *Kundu & Balakrishnan 601* (PDA), Mannar, sea-level, 24 Mar 1970, *Cramer 2886* (PDA). ANURADHAPURA DISTRICT: between Dambulla and Habarana, 21 Oct 1973, *Jayasuriya 1341* (K, PDA). TRINCOMALEE DISTRICT: Kantalai Tank, 7 Oct 1978, *Wirawan 1244* (PDA). MATALE DISTRICT: Ereula Tank, c. 8 km ESE of Dambulla, c. 215 m, 11 Oct 1974, *Davidse 7389* (PDA). COLOMBO DISTRICT: Siduwa, 10 Sept 1967, *Jayasuriya 2642* (PDA). HAMBANTOTA DISTRICT: Tissamaharama, 12 Aug 1932, *Simpson 9933* (BM, PDA); Bundala Sanctuary, 9 Apr 1985, *Jayasuriya et al., 3301* (PDA); Ruhuna National Park, Andunoruwa Wewa, 16 Dec 1969, *Cooray 69121618R* (PDA), Block I, 1 Jun 1968, *Cooray 68060102R* (PDA). ANURADHAPURA OR KURUNEGALA DISTRICT: Migas Wewa, 4 Oct 1931, *Simpson 8708* (BM, PDA). POLONNARUWA DISTRICT: Mineri Lake, 6 Sept 1885, *s. coll.* (PDA).

4. Cleome tenella L. f., Suppl. Pl. 300. 1782; DC., Prod. 1: 240. 1824; Wight & Arn., Prod. 21. 1834; Hook. f. & Thoms. in Hook. f., Fl. Br. Ind. 1: 169. 1872; Trimen, Handb. Fl. Ceylon 1: 55. 1893. Type: from India.

Herb, annual, erect, 8–12 (–25) cm tall, much branched; branches spreading, very slender, glabrous. Leaves 3-foliolate; leaflets 0.6–2 × 0.02–0.04 cm, narrowly linear to filiform, glabrous, petiolule minute or lacking; petiole 0.5–1.2 cm long, appearing somewhat flattened, glabrous. Flowers white or yellowish with purple veined petals. Pedicel 1.5–5 mm long, slender, glabrous. Sepals c. 0.8 × 0.5 mm, oblong-ovate or elliptic, glabrous, solitary in axils of upper leaves. Petals 2.75–3 × 1 mm, including c. 0.5 mm long claw. Stamens 6; filaments 2 mm long, slender; anthers c. 0.5 mm long. Ovary 2–2.5 × 0.4 mm. Fruit 1.5–4.5 × 0.08–0.15 cm, linear-cylindric, faintly but prominently, parallel-veined, glabrous; fruiting pedicel 0.8–1 cm long. Seeds c. 0.5 mm diameter, suborbicular, black or blackish-brown, densely, minutely muricate, not ribbed, cleft not tightly closed, slightly open.

Distr. Southern India and Ceylon; Mali and Senegal, Ethiopia, Eritrea and the Sudan and Somali Republics.

Ecol. In Ceylon only recorded from coastal areas in Puttalam and Batticaloa Districts.

Specimens Examined. PUTTALAM DISTRICT: Chilaw, Nov 1881, *Ferguson s.n.* (BM, K, PDA). BATTICALOA DISTRICT: S of Batticaloa Lagoon, 28 Jun 1931, *Simpson 8303* (BM, PDA).

5. Cleome gynandra L., Sp. Pl. 671. 1753. Type: cultivated plant (BM-CLIFF, holotype).

Cleome pentaphylla L., Sp. Pl. ed. 2. 938. 1763, nom. illegit.
Gynandropsis pentaphylla (L.) DC., Prod. 1: 238. 1824.
Gynandropsis gynandra (L.) Briq., Annuaire Conserv. Jard. Bot. Geneve Annuaire 17: 382. 1914.

Herb, annual, 0.3–1 m tall, erect, widely branched; stems glandular-pubescent to long, patent, shaggy-pilose, or subglabrous, unarmed. Leaves long-petiolate, palmately (3–) 5 (–7)-foliolate; leaflets 1.2–7.5(–10) × 0.8–3 cm, terminal longest, shortly petiolulate to subsessile, thinly herbaceous, lanceolate to obovate-lanceolate or rhombic, acute, shortly acuminate, acute to subobtuse at base, ciliate to denticulate, minutely pubescent on both surfaces, glabrescent; petiolules 1–3 mm long, webbed at base, glandular-pubescent; petiole 2.5–10 cm long, densely glandular-pubescent, occasionally with short prickles. Flowers white, yellowish, pink or pale purplish, pedicellate in long corymbose racemes, bracteate with bracts composed of 3 palmately arranged bractlets. Pedicels 1.5–2 cm long, slender, glandular-pubescent. Sepals 2.5–3 (–5) mm long, c. 0.5 mm wide, linear, acute, glandular-pubescent. Petals 0.9–1.6 × 0.2–0.4 cm overall, rotund, slender-clawed. Stamens 6. Gynophore 1–4 mm long above point of insertion of sepals, slender, glabrous, extending to 2–2.5 cm long in fruit, slightly curved upwards; ovary c. 4 × 0.5 mm, linear-oblong-cylindric, glandular-pubescent. Fruit 6.5–8 × 0.3–0.35 cm, linear-cylindric, somewhat compressed, slightly curved. Seeds c. 1.3 mm diameter, helicoid-reniform, black, rough.

Distr. Most of Africa to the Seychelles and Madagascar; tropical Asia and America.

Ecol. Disturbed, waste or cultivated ground.

Vern. Wela (S), Tayirvalai (T).

Note. The closely allied *Cleome speciosa* Raf., a native of the New World, is frequently encountered in Ceylon, where it is cultivated as an ornamental. The two species are easily distinguished by way of the gynophore, which in *C. speciosa* reaches about 6 cm in length, and by the pink petals which are almost twice as large in that species, measuring 25–35 mm long.

Specimens Examined. MANNAR DISTRICT: Vankalai, 12 Nov 1970, *Kundu & Balakrishnan 625* (PDA). ANURADHAPURA DISTRICT: Anuradhapura, 25 Sept 1969, *Beusekom & Beusekon 1600* (PDA). TRINCOMALEE DISTRICT: Trincomalee, 30 Aug 1860, *Dubuc s.n.* (E). KURUNEGALA DISTRICT: near Nikaweratiya Rest House, 12 Oct 1978, *Wirawan 1256* (K, PDA). POLONNARUWA DISTRICT: Minneriya, roadside from Polonnaruwa, 2 Sept 1978, *Wirawan 1217* (K, PDA). MATARA DISTRICT: Polhena, 8 Oct 1971, sea-level, *Cramer 3444* (PDA). BATTICALOA DISTRICT: Kennedy Estate, Kalkudah, sea-level, *Cramer 2758* (PDA). HAMBANTOTA DISTRICT: Hambantota, 31 Dec 1926, *Alston 1120* (PDA); Hambantota to Tissamaharama road, mile marker 152/4, 19 Jun 1973, *Sohmer et al., 8844* (BM); Ruhuna National Park, Block I, Patanagala, 2–3 m, 6 Apr 1968, *Fosberg 50349* (BM, K, PDA), 24 May 1968, *Cooray 68052405R* (PDA), 22 Oct 1968, *Cooray 68102210R* (K, PDA), 3 Dec 1969, *Cooray 69120301R* (K, PDA), 25 Aug 1967, *Mueller-Dombois 67082508* (PDA). LOCALITY UNKNOWN: *s. coll.* in *C.P. 2460* (PDA).

6. Cleome rutidosperma DC., Prod. 1: 241. 1824; Iltis, Brittonia 12: 290. 1960. Type: presumably from West Africa, by an unknown collector, possibly *H. Smeathman* (G–DEL).

Cleome burmannii Wight & Arn., Prod. 22. 1834; Hook. f. & Thoms. in Hook. f., Fl. Br. Ind. 1: 170. 1872. Type: from India.
Cleome aspera sensu Trimen, Handb. Fl. Ceylon 1: 56. 1893, p.p., non Koen., 1824.

Herb, annual, erect or spreading, up to 50 cm tall; stems weak, sparsely hairy with simple mixed with stipitate glandular hairs, otherwise subglabrous. Leaves 3-foliolate; leaflets 0.8–2.5 × 0.4–1.4 cm, ovate, ovate-elliptic to rhombic, acute to acuminate, occasionally rounded, minutely mucronate, attenuate to cuneate at base, serrulate, minutely ciliolate, sparsely glabrous to glandular-pilose, petiolules 0–0.5 mm long; petiole 0.5–2.25 cm long, slender, indumentum as for stem. Flowers pink to pale purple, or white, pedicellate, solitary in axils of slightly reduced leaves. Pedicels 1–2 cm long, filiform. Sepals 2 × 0.5 mm, ovate or acute, shortly ciliate, obtuse, slightly pubescent. Petals c. 6 mm long, including 2.5 mm long claw. Stamens 6; filaments 7–8 mm long; anthers c. 2 mm long, linear, strongly curved. Ovary 5–8(–12) mm long, including 1.5 mm long style, linear-cylindric, glabrous. Fruit 3–5 × 0.3 cm, linear-cylindric, narrowing into stalk-like base before junction with pedicel, prominently longitudinally parallel-veined; fruiting pedicel up to 2.5 cm long. Seeds 1.5–1.75 mm diameter, deeply laterally ribbed, dark brown, cleft semi-open.

Distr. West tropical Africa; introduced into the New World and recorded from Florida, through Honduras and Panama to one collection in Brazil and then recorded widely from the Caribbean. Again widely reported from tropical Asia from Burma, Thailand and Ceylon, through to Sumatra and the Philippines.

Ecol. Waste land and sites of old cultivation; roadsides, gardens etc.

Note. Great confusion has been caused by the fact that De Candolle (1834) considered *C. rutidosperma* to possibly have originated from "Tabago" [? Tobago]. This confusion has been studied in great detail by Iltis and is well documented by him. (in Brittonia 12: 290–294. 1960.) As is to be seen above, the type of the name is a specimen housed in Geneva and presumably collected from Sierra Leone and then probably by H. Smeathman.

Several collections exist in various herbaria under the name *Cleome graveolens* Raf. On study these have been found to be more correctly placed here under *Cleome rutidosperma*. The basis of Rafinesque's name was from his earlier *Polanisia graveolens*, a plant described as being glandular-pilose and having 8–12 stamens and further localized from North America. A. De Candolle cited this latter name in his treatment of the family for his Prodromus, keeping it under *Polanisia*. However, in this same work he described *C. rutidosperma*, a plant which he considered to be glabrous and having six stamens. The present author has not seen the type of this name in Geneva, but seriously doubts that it would be totally glabrous. Every single specimen of this taxon seen for this work, which otherwise fits the original description, is glandular-pilose, albeit, in some cases, sparsely so. There is no hesitation about including the '*C. graveolens*' material from our area here, but the possible inclusion of the name in synonymy is left to later authors who should have access to a wider range of material for reference.

Specimens Examined. MANNAR DISTRICT: Illupaikadavai, Feb 1890, *s. coll. s.n.* (PDA). PUTTALAM DISTRICT: Lunuwila, Bandirippuwa Estate, 21 Jan 1949, *Child s.n.* (PDA). ANURADHAPURA DISTRICT: Ritigala, 10 Jan 1974, *Waas 322* (PDA). KURUNEGALA DISTRICT: Kurunegala, 23 Sept 1978, *Wirawan 1234* (K); Athagala Rock, 23 Sept 1978, *Wirawan 1230* (K); Bakmigalle, c. 9 km from Kurunegala on road to Dambulla, 7 Oct 1978, *Wirawan 1242* (K). POLONNARUWA DISTRICT: Polonnaruwa Rest House, 2 Sept 1978, *Wirawan 1216* (K). COLOMBO DISTRICT: Colombo, 22 Oct 1928, *de Alwis s.n.* (PDA), near Lake Lodge Hotel, 28 Mar 1973, *Bremer et al., 60* (K); Urapola, 10 Dec 1968, *Amaratunga 1684* (PDA); Pasyala, 18 Apr 1968, *Amaratunga 1597* (PDA); Veyangoda, 9 Oct 1969, *Amaratunga 1878* (PDA). KEGALLE DISTRICT: Noori Estate, Deraniyagala, 10 Aug 1940, *Estate Superintendent s.n.* (PDA). KANDY DISTRICT: Royal Palace Park, Kandy, 16 Sept 1978, *Wirawan 1221* (PDA), *1222* (PDA); Peradeniya, University campus, 10 Nov 1970, *Fosberg 52706* (K, PDA), 5 July 1975, *Jayasuriya & Premadasa 2234* (PDA); Peradeniya Botanic Gardens, April 1938, *de Silva 731* (PDA); Akkarawatta, Galagedera, 180 m, 18 Feb 1994, *Philcox et al., 10613* (K, PDA). BADULLA DISTRICT: Alawatugoda, 27 May 1967, *Amaratunga 1307* (PDA). AMPARAI DISTRICT: Pottuvil Circuit Bungalow, S point of Arugam Bay, 26 Nov 1970, *Fosberg & Sachet 53025* (PDA). HAMBANTOTA DISTRICT: Ruhuna National Park, Patanagala, 9 Nov 1969, *Cooray 69110901R* (K), below Palatupana Wewa,˙5 Dec 1973, *Sohmer 9013* (PDA), Block I, Rakinawala, 7 Dec 1969, *Cooray 69120713R* p.p. (PDA), 22 Oct 1968, *Mueller-Dombois 68102208* (PDA), mile 11, Yala road, 18 Oct 1968, *Mueller-Dombois 68101827* (PDA); Kirinda, December 1882, *s. coll.* (PDA). LOCALITY UNKNOWN: *s. coll.* in *C.P. 1068*, p.p. (PDA).

7. Cleome aspera Koen. ex DC., Prod. 1: 241. 1824; Wight & Arn., Prod. 22. 1834; Thw., Enum. Pl. Zeyl. 14. 1858; Hook. f. & Thoms. in Hook. f., Fl. Br. Ind. 1: 169. 1872; Trimen, Handb. Fl. Ceylon 1: 56. 1893, p.p.; Alston in Trimen, Handb. Fl. Ceylon 6: 12. 1931. Type: from India.

Herb, annual, 15–40 cm tall, erect, diffusely branched; branches glabrous except for very occasional short, pale, patent or antrorse, soft prickles. Leaves 3 (–5)-foliolate, short petiolate; leaflets 0.5 × 0.1 × 1.75–0.5 cm, oblong to obovate, sparsely short, appressed, hispid-hairy, entire, ciliolate, shortly petiolulate with petiolules up to 0.8 mm long; petiole 0.15–0.8 cm long. Flowers dull-yellow to white, slender-pedicellate, solitary in axils of small leaf-like bracts, racemose. Pedicels 0.5–1.5 cm long, filiform. Sepals 2–3 ×

0.25–0.5 mm, acute or obtuse. Petals 3 × 1 mm, obovate, narrowed at base but not clearly clawed. Stamens 6; filaments 2.5–2.75 mm long, shorter than petals; anthers c. 0.5 mm long. Ovary c. 3 × 0.5 mm, including 0.5 mm long style, linear, glabrous. Gynophore c. 1 mm long or much less. Fruit 3 × 0.2 cm, including 2.5 mm long beak, cylindrical, narrowing abruptly at base at junction with pedicel, valves prominently parallel-veined. Seeds c. 1.3–1.5 mm diameter, compressed-suborbicular, finely longitudinally ridged, not markedly transversely so, cleft closed, dark orange-brown.

Distr. India and Ceylon; Malesia.

Ecol. Low country and dry districts, especially near the sea.

Specimens Examined. JAFFNA DISTRICT: Jaffna, in 1846, *Gardner s.n.* in *C.P. 1068*, p.p. (PDA). ANURADHAPURA DISTRICT: Puttalam road, 10 Oct 1970, *Kundu 301* (PDA). BATTICALOA DISTRICT: Batticaloa, Mar 1865 (?), *Thwaites s.n.* in *C.P. 1068*, p.p. (PDA). COLOMBO DISTRICT: Colombo, 6 Jun 1926, *Ball s.n.* (PDA). AMPARAI DISTRICT: E side of Ulpasse Wewa, c. 3 km N of Panama, 26 Nov 1970, *Fosberg & Sachet 52961* (K, PDA). HAMBANTOTA DISTRICT: Ruhuna National Park, Patanagala, 30 Nov 1969, *Cooray 69113002R* (PDA), Block I, Buttawa Plain, 2 Dec 1969, *Cooray 69120218R* (PDA), Rakinawala, 7 Dec 1969, *Cooray 69120713R*, p.p. (PDA). LOCALITY UNKNOWN: 1853, *Thwaites s.n.* in *C.P. 1068*, p.p. (BM, K); *s. coll.* in *C.P. 1068* p.p. (PDA).

2. MAERUA

Forssk., Fl. Aegypt.-Arab. 104. 1775. Type species: *Maerua crassifolia* Forssk.

Shrubs, straggling or climbing. Leaves simple, alternate. Flowers usually in terminal or lateral corymbose racemes or at times solitary in axils of upper leaves. Sepals 4, connate at base into slender tube half length of lobes or less. Petals 4, smaller than calyx lobes, ovate, inserted on rim of disk. Stamens numerous, free. Ovary 1-celled, ovules many. Fruit elongate, moniliform, soft. Seeds large.

Genus of about 50 species, from tropical and South Africa to India and Ceylon.

Maerua arenaria Hook. f. & Thoms. in Hook. f., Fl. Br. Ind. 1: 171. 1872; Trimen, Handb. Fl. Ceylon 1: 58. 1893. Type: from India.

Maerua oblongifolia sensu Thw., Enum. Pl. Zeyl. 15. 1858, non A. Rich. 1847.

Shrub, scrambling to 2 m or more; stems slender. Leaves: lamina 2–5 × 1.25–3 cm, elliptic to broadly ovate-elliptic, entire, apex obtuse, retuse or minutely apiculate, base obtuse, glabrous; petiole 0.4–1 cm long, glabrous. Flowers pedicellate in terminal corymbose racemes or solitary axillary, white becoming pale greenish-white, sweet-scented. Calyx-tube 2–2.5 mm long, lobes 10–13 × 3.5–4.5 mm, ovate to oblong-ovate, acute, mucronate, glabrous except minutely pubescent on margins. Petals c. 6 × 3 mm, ovate, markedly veined. Stamens up to c. 30 or more, spreading; filaments 2–2.5 cm long, white; anthers c. 1.5–2 mm long, green. Gynophore 2–4 mm long. Ovary 2–4 mm long, cylindric, glabrous. Fruit 3–5 cm long, up to 12-seeded, irregularly constricted between seeds, smooth, obscurely verrucose, greenish-yellow when mature.

Distr. India and Ceylon.

Ecol. Dry country, especially dry, stony, sandy areas near the coast.

Note. Our species differs from the African *Maerua oblongifolia* (Forssk.) A. Rich. only in the relative size of the calyx tube compared with its lobes. Both Hook. f. & Thoms. and Ellfers doubted whether this justified their continued separation. The limited material available for the present study was insufficient to decide the issue, and *M. arenaria* is provisionally maintained as distinct.

Specimens Examined. JAFFNA DISTRICT: Pallavarayankaddu, Feb 1890, *s. coll. s.n.* (PDA); Elephant Pass, Jaffna Lagoon, 10 Dec 1970, *Fosberg et al., 53564* (K, PDA); Vaddukoddai, 12 Apr 1971, *Balasubramaniam 188* (K). PUTTALAM DISTRICT: Puttalam, July 1883, *s. coll. s.n.* (PDA); near Chilaw, 11 Sept 1931, *Simpson 8570* (BM); Wilpattu National Park, Maduru Odai, 30 Jun 1969, *Wirawan et al., 925* (PDA). ANURADHAPURA DISTRICT: near Kudagama, between Ratmalagahawewa and Medawachchiya, 8 Oct 1978, *Wirawan 1246* (K, PDA); road to Talava, 15 Jun 1969, *Read 2173* (PDA). TRINCOMALEE DISTRICT: Trincomalee, in 1846, *Gardner s.n.* in *C.P. 1064,* p.p. (PDA). KURUNEGALA DISTRICT: 49 mile-marker on Kurunegala to Puttalam road, just after Deduru Oya bridge, 29 Aug 1978, *Wirawan 1200* (E, K, PDA), *1202* (K, PDA). BATTICALOA DISTRICT: Batticaloa, in 1856, *Glenie s.n.* in *C.P. 1064,* p.p. (PDA). RATNAPURA DISTRICT: Madampe road from south, 16 Sept 1931, *Simpson 8641* (BM). HAMBANTOTA DISTRICT: Hambantota, Dec 1882, *s. coll.* (PDA); Mandagala Tank, between Nonagama and Hungama, 16 Sept 1978, *Wirawan 1261* (K, PDA); between Tissamaharama and Ruhunu National Park, 22 Apr 1969, *Cooray 69042202R* (PDA). LOCALITY UNKNOWN: *Thwaites s.n.* in *C.P. 1064,* p.p. (BM, PDA).

3. CADABA

Forssk., Fl. Aegypt. Arab. 67. 1775. Type species not designated.

Shrubs or small trees, erect or scrambling, branches unarmed. Leaves simple or trifoliolate. Flowers terminal-corymbose or racemose, or solitary axillary. Sepals 4, unequal, free, caducous, in 2 whorls, outer pair usually enclosing bud, inner smaller. Petals 2 or 4, equal, clawed. Disk about equalling claw, tubular, trumpet-shaped. Stamens 4–6, inserted about midway on gynophore; filaments filiform, spreading, exserted; anthers somewhat large. Ovary on long gynophore, 1-celled; ovules many on 2–4 parietal placentas. Fruits slender, cylindric, dehiscent with 2 valves. Seeds numerous, reniform to subglobose.

Genus of 30 species from the Old World tropics, mainly Africa.

KEY TO THE SPECIES

1 Leaves trifoliolate; petals 2; stamens 6 .... **1. C. trifoliata**
1 Leaves simple: petals 4; stamens 4 .... **2. C. fruticosa**

1. Cadaba trifoliata (Roxb.) Wight & Arn., Prod. 24. 1834; Hook. f. & Thoms. in Hook. f., Fl. Br. Ind. 1: 172. 1872; Thw., Enum. Pl. Zeyl. 15. 1858; Trimen, Handb. Fl. Ceylon 1: 59. 1893.

Stroemeria trifoliata Roxb., Fl. Ind. ed. Carey 2: 79. 1832. Type: from India. *Cadaba triphylla* Wight in Hook., Bot. Misc. 3: 296, t. 37. 1833. Type: from India.

Shrub or small tree to 6 m tall, much branched; young branches shortly patent hirsute to subglabrous. Leaves subcoriaceous, palmately trifoliolate, petiolate; petiole 1.25–3.5 cm long, shortly hirsute, glabrescent; leaflets 2–5.75 × 0.9–2 cm, ovate-lanceolate to oblanceolate, apex obtuse to broadly rounded, rarely acute, base acute or obtuse, entire, glabrous, somewhat glossy above, venation subprominent especially above; petiolule 2–4 mm long, glabrous. Flowers few, white, short terminal-corymbose. Sepals 1.25–2.5 × 0.5–1 cm, broadly ovate, outer pair larger, minutely pubescent without. Petals 2, c. 3.5 cm long, limb up to 2 cm wide, tapering into claw up to 2 cm long. Stamens 6; filaments c. 2.5 cm long; anthers c. 6–7 mm long. Disk 1.5–2 cm long, tubular, very dilated at apex, entire. Gynophore 2.5 cm long; ovary 6–7 × 1 mm, linear-oblong, glabrous. Fruit capsular, 10 × 0.8 cm, cylindric, coarsely muricate, on fruiting gynophore up to 6.5 cm long. Seeds 3 × 2 mm, reniform, muricate.

Distr. Southern India and Ceylon.

Ecol. Dry areas at low altitude, up to 150 metres.

Vern. Oothi perali (T, fide Wight).

Specimens Examined. MANNAR DISTRICT: Illupaikaduvai, Feb 1890, *s. coll. s.n.* (PDA). PUTTALAM DISTRICT: Wilpattu National Park,

N of Marai Villu, 1 Mar 1969, *Mueller-Dombois 69030101* (K), 30 Jun 1969, *Wirawan et al., 880* (PDA), 25 Sept 1969, *Cooray 69092501R* (PDA), 4 km S of Marai Villu, 30 Apr 1969, *Mueller-Dombois et al., 69043018* (PDA), Maduru Odai, Mannar to Puttalam road at junction to Periya Naga Villu, 30 Jun 1969, *Wirawan et al., 916* (K, PDA); 6.5 km S of Kalpitiya, 14 Nov 1970, *Fosberg & Jayasuriya 52763* (K, PDA); near Maraitivu, 10 July 1932, *Simpson 9845* (BM, PDA). ANURADHAPURA DISTRICT: Wilpattu National Park, 30–150 m, 21 Mar 1969, *Robyns 6978* (E, K). TRINCOMALEE DISTRICT: Elephant Island, Trincomalee, 1 m, *Worthington 1215* (BM), 2 Aug 1943, *Worthington 1296* (BM). LOCALITY UNKNOWN: 26 Sept 1969, *Beusekom & Beusekom 1623* (PDA); *Gardner s.n.* in *C.P. 1066*, p.p. (PDA); *Macrae 670* (BM); in 1853, *Thwaites* in *C.P. 1066*, p.p. (BM, K, PDA).

2. Cadaba fruticosa (L.) Druce, Rep. Bot. Soc. Exch. Club 1913: 415. 1914; Alston in Trimen, Handb. Fl. Ceylon 6: 13. 1931.

Cleome fruticosa L., Sp. Pl. 671. 1753. Type: from India.
Cadaba indica Lam., Enc. 1: 544. 1785; Hook. f. & Thoms. in Hook. f., Fl. Br. Ind. 1: 172. 1872; Trimen, Handb. Fl. Ceylon 1: 60. 1893. Type: from India.

Straggling shrub or small tree up to 3 m tall, much-branched; branches glabrous. Leaves thin, membranous, shortly petiolate; lamina 2.25–3 (–5) × 0.9–1.8 (–3) cm, ovate-oblong, obtuse-mucronate at apex, base obtuse, glabrous, venation well defined beneath, less so above; petiole 0.3–0.5 (–0.9) cm long. Flowers few, white, in terminal corymbose inflorescences. Sepals 0.8–1.2 × 0.4–0.5 cm, ovate, acute, equal, outer concave. Petals 4, 0.8–1 cm long, limb c. 0.4 × 0.25 mm, ovate, claw c. 0.4 cm long, slender. Stamens 4, c. 1.2 cm long. Disk tubular, c. 0.7 cm long, dilated at apex, toothed, sited between petals. Gynophore 1.75–2 cm long, exserted; ovary c. 3 × 0.75 mm long, cylindrical, slightly curved. Fruit capsular, 3.5–4 cm long, linear-oblong, cylindric, muricate. Seeds c. 2 mm diameter, subreniform, when dry with helicoid black and white stripes.

Distr. Southern India and Ceylon.

Ecol. Lowland dry areas.

Vern. Vili (T).

Specimens Examined. JAFFNA DISTRICT: Analativu, Kayts, 22 Sept 1897, *s. coll. s.n.* (PDA); Vaddukoddai, 22 Apr 1954, *Koshy s.n.* (PDA). MANNAR DISTRICT: Illupaikaduvai, Feb 1890, *s. coll. s.n.* (PDA); 8 km SE of Murunkan, 10 Apr 1973, *Bremer et al., 65* (K), *69* (K); Giant's Tank, 14 Mar 1932, *Simpson 9325* (BM). PUTTALAM DISTRICT: Puttalam, 2 Dec 1910, *Willis s.n.* (PDA); Wilpattu National Park, 4 Apr 1970, *Cooray*

70040402R (K, PDA), Maduru Odai, Mannar to Puttalam road, junction to Periya Naga Villu, 30 Jun 1969, *Wirawan et al., 925* (K), W end of Mail Villu, 30 Apr 1969, *Mueller-Dombois 69043027* (K, PDA). LOCALITY UNKNOWN: *Gardner s.n.* in *C.P. 1070*, p.p. (PDA), in 1853, *Thwaites s.n.* in *C.P. 1070*, p.p. (BM, K).

4. CAPPARIS

L., Sp. Pl. 503. 1753; L., Gen. Pl. ed. 5. 222. 1754; DC., Prod. 1: 245. 1824. Lectotype species: *Capparis spinosa* L. (see Britton & Millspaugh, Bahama Fl. 150. 1920).

Trees or shrubs, erect, decumbent or climbing, unarmed or with stipular thorns; stems mostly pubescent, glabrescent, at times base of shoots surrounded by cataphylls, i.e. much reduced leaves appearing like narrow budscales. Leaves simple, membranous or coriaceous, petiolate, with or without pair of stipular thorns. Flowers pedicellate, often showy, variously arranged. Sepals 4, free, biseriate, inserted at base of gynophore, outer pair subconcave and enveloping bud, inner pair flattish. Petals 4, imbricate, free. Receptacle or torus flattish or subconical. Stamens indefinite, longer than petals; anthers small, basifixed. Gynophore about equalling stamens or may be longer, extending or not in fruit. Fruit a berry with leathery pericarp, usually globose, 1-celled. Seeds 1–many, embedded in pulp, subreniform.

Genus of about 250 species from the warmer temperate, subtropical and tropical areas of the world.

KEY TO THE SPECIES

1 All sepals free in bud
  2 Flowers with 2–10 arranged in supra-axillary rows, or sometimes only small bundles of cataphylls in their place
    3 Young stem glabrous; flowers small, c. 1.25 cm diam., petals 0.5–0.7 cm long . . . .**1. C. tenera**
    3 Young stem rufous-scurfy; flowers larger, c. 3.5 cm diam., petals 2.5–3.5 cm long . . . . **2. C. zeylanica**
  2 Flowers arranged otherwise, in racemes or terminal corymbs or spikes or subumbels or solitary-axillary
    4 Young shoots and inflorescences surrounded at base by cataphylls . . . . .**3. C. brevispina**
    4 Young shoots and inflorescences not surrounded by cataphylls
      5 Gynophore 3.5 cm long or longer
        6 Stamens more than 100 . . . . **4. C. moonii**
        6 Stamens up to c. 50 . . . . **5. C. roxburghii**
      5 Gynophore shorter than 3.5 cm
        7 Leaves with cordate base, 4.5 cm or shorter with mucronate apex; thorns often long and acicular . . . . **6. C. rotundifolia**
        7 Leaves with base rarely subcordate, mostly larger than 4 cm; thorns if present, recurved
          8 Gynophore hairy at base during anthesis . . . . **7. C. grandis**
          8 Gynophore glabrous during anthesis

9 Leaves glabrous throughout; sepals 2–3.5 × 1.5–2 mm; stamens 7–10; umbels of
       15–25 or more flowers .... **8. C. floribunda**
9 Leaves pubescent, persistently beneath; sepals 5–7 × 3–5 mm; stamens 30–45;
       umbels of 8–18 flowers .... **9. C. sepiaria**
1 Outer pair of sepals connate in bud .... **10. C. divaricata**

1. Capparis tenera Dalz., Hooker's J. Bot. Kew Gard. Misc. 2: 41. 1850;
Hook. f. & Thoms. in Hook. f., Fl. Br. Ind. 1: 179. 1872, as for var. *dalzellii*;
Trimen, Handb. Fl. Ceylon 1: 65. 1893; Jacobs, Blumea 12: 497. 1965. Type:
from India.

Capparis tetrasperma Thw., Enum. Pl. Zeyl. 15. 1858. Type: Ceylon,
    Thwaites s.n. in *C.P. 614* (BM, K, PDA).
Capparis tenera var. *zeylanica* Hook. f. & Thoms. in Hook. f., Fl. Br. Ind.
    1: 179. 1872. Type: Ceylon, *Thwaites s.n.* in *C.P. 614* (BM, K, PDA).

Shrub, climbing or scandent up to 3 m; stems slender, glabrous. Thorns
1.5–2 mm long, recurved. Leaves membranous to subcoriaceous; lamina
3.5–5.75 (–9) × 1.5–2.75 (–4.25) cm, ovate, obovate or oblong, acute or
shortly acuminate, base obtuse to rounded, glabrous; petiole 6–8 mm long,
glabrous. Flowers white, solitary or in axillary clusters of 2–4. Pedicels
1–1.8 cm long, slender, glabrous. Sepals 4–5 × 2 mm, ovate-elliptic, outer
pair navicular, glabrous without, shortly crisped-ciliate, inner pair slightly
shorter, flatter. Petals 5.5–7 × 1.5–2 mm, ovate-oblong, rounded-obtuse,
pubescent on both surfaces. Torus c. 1 mm diameter or less. Stamens 10–20.
Gynophore 10–18 mm long, filamentous, glabrous; ovary c. 1.25 × 0.6 mm,
ovate-ellipsoid or pyriform, glabrous. Fruit 0.7 cm diameter (immature), globose, short-apiculate, gynophore not enlarged or thickened, nor pedicel. Seeds
not seen.

Distr. India, Burma, Ceylon and Andaman Islands.

Ecol. Low country areas.

Specimens Examined. MATALE DISTRICT: Matale, May 1846,
Thwaites s.n. in *C.P. 614*, p.p. (K, PDA). NUWARA ELIYA DISTRICT:
Hanguranketa, July & Sept 1851, *Thwaites s.n.* in *C.P. 614*, p.p. (PDA).
MONERAGALA DISTRICT: Bulupitiya, E of Bibile, 1 May 1975,
Jayasuriya 1947 (E, K, PDA); Illukkepatana, near Bibile, 1 Jun 1901, *Trimen
s.n.* (PDA); Nelliyadda, W of Mullegama, 4 May 1975, *Jayasuriya 2049* (K,
PDA). PUTTALAM DISTRICT: Halaba, 26 Jun 1931, *Simpson 8241* (BM).
LOCALITY UNKNOWN: *Macrae 827* (BM); 1 Feb 1819, *Moon 84* (BM);
in 1860, *Thwaites s.n.* in *C.P. 614*, p.p. (BM, K); *Walker 136* (K); June 1863,
s. coll. in *C.P. 614* bis, p.p. (K, PDA s. dat.).

2. Capparis zeylanica L., Sp. Pl. ed 2. 720. 1762; DC., Prod. 1: 247. 1824,
as for Ceylon material.

Capparis horrida L. f., Suppl. Pl. 264. 1781; DC., Prod. 1: 246. 1824, as for Ceylon material; Thw., Enum. Pl. Zeyl. 15. 1858; Hook. f. & Thoms. in Hook. f., Fl. Br. Ind.1: 178. 1872; Trimen, Handb. Fl. Ceylon 1: 64. 1893; Jacobs, Blumea 12: 505. 1965. Type: Ceylon and India, *Koenig s.n.*

Shrub, 0.75–10 m tall, erect or straggling, much-branched; branches greyish- or brownish-tomentose, soon glabrescent, then branches somewhat shiny. Thorns 2–4 mm long, recurved to occasionally patent. Leaves subcoriaceous; lamina 3.75–11 × (1.25–) 1.75–4.5 cm, ovate-elliptic or ovate-lanceolate, apex acute or rounded-obtuse, often with stiff mucro up to 3 mm long, base rounded to at times obscurely subcordate, pubescent, soon glabrescent above, becoming glossy, at times pubescence persisting beneath, otherwise glabrescent, not glossy, venation prominent on both surfaces; petiole 3–8 mm long, glabrous. Flowers white, upper pair of petals yellow at base, becoming reddish-violet with age, solitary or paired-axillary, spread serially along branches at times for distances up to 30 cm. Pedicels (0.5–) 2.5–3.5 cm long, minutely pubescent to subglabrous. Sepals 7.5–10 × 3 mm, ovate, densely pubescent without. Petals 2–2.5 × 0.8–1.2 cm, rounded-oblong, subglabrous. Torus c. 1 mm diameter. Stamens 25–62, 1.5–3.5 cm long; anthers c. 2 × 1 mm. Gynophore 2.5–4.5 cm long; ovary 4–5 × 1–1.5 mm, narrowly pyriform to ellipsoid, minutely pubescent, placentas 4. Fruit c. 5 × 3 cm, ripening red, orange or purple (Jacobs). Seeds 12–15 embedded in white, somewhat gelatinous, pulp, 0.8 × 0.6 mm, irregularly reniform, blackish-mottled-brown, totally enclosed by white aril c. 0.5 mm thick.

Distr. India and Ceylon; Burma, Thailand, Indo-China, Malaysia (Java, Lesser Sunda Islands, Celebes), Philippines.

Ecol. Low altitude, scrubland of dry country, sandy beaches and other well drained soils.

Vern. Sudu-welangiriya (S).

Specimens Examined. JAFFNA DISTRICT: 1846, *Gardner s.n.* in *C.P. 1058*, p.p. (PDA), *C.P. 1060*, p.p. (PDA), *C.P. 1062*, p.p. (PDA); Elephant Pass, 10 Mar 1912, *s. coll.* (PDA). VAVUNIYA DISTRICT: Mullaitivu, near lagoon, 13 Mar 1932, *Simpson 9305* (BM); Mankulam, 10 Oct 1978, *Wirawan 1254* (PDA). PUTTALAM DISTRICT: Chilaw, March 1882, *Nevill s.n.* (PDA); Wilpattu National Park, Atha Villu, 28 Sept 1969, *Cooray 69092812R* (K, PDA); Pallugahaturai beach, 26 Sept 1969, *Cooray 69092609R* (PDA); N of Marai Villu, 30 Jun 1969, *Wirawan et al., 885* (PDA). ANURADHAPURA DISTRICT: Anuradhapura, 25 Mar 1971, *Amaratunga 2255* (PDA), 30 Mar 1883, *s. coll. s.n.* (PDA); Ritigala Strict Natural Reserve, 290 m, 28 Sept 1972, *Jayasuriya 847* (PDA). TRINCOMALEE DISTRICT: Trincomalee, *Gardner s.n.* in *C.P. 1059* p.p. (PDA); Kantalai, Aug 1885, *s. coll.* (PDA). KEGALLE DISTRICT:

Galpitamada, 27 July 1968, *Amaratunga 1621* (PDA). KURUNEGALA
DISTRICT: Nikaweratiya Rest House, 11 Oct 1978, *Wirawan 1255* (K,
PDA); Wariapola, 8 Mar 1963, *Amaratunga 534* (PDA); mile-marker 49
on Kurunegala to Puttalam road, just after Deduru Oya bridge, 29 Aug
1978, *Wirawan 1201* (PDA). MATALE DISTRICT: Habarane, 3 May 1927,
Alston 505 (PDA); Alawatugoda, 18 Feb 1971, *Balakrishnan 598* (PDA);
Dambulla, 30 Apr 1965, *Amaratunga 864* (PDA); Naula, 340 m, 25 Mar
1971, *Dassanayake 346* (K); road from Naula to Dambulla, at marker 36/4,
14 July 1973, *Nowicke et al., 323* (K, PDA); Amban Ganga, 19 Jun 1932,
Simpson 9821 (PDA). KANDY DISTRICT: Kandy (planted), 27 Oct 1978,
Wirawan 1268 (E, K); Peradeniya, Royal Botanic Gardens, 23 Feb 1964,
Amaratunga 789 (PDA); Haragama, July 1881, *s. coll.* (PDA); Weragantota,
18 July 1927, *Alston 802* (PDA). BADULLA DISTRICT: Maturata July
1837, *s. coll. s.n.* (PDA); Mahiyangana, 30 Apr 1975, *Jayasuriya 1927* (K,
PDA), road to Hunasgiriya, 28 July 1974, *Kostermans 25283* (K). AMPARA
DISTRICT: Okanda to Kumana road, 30 July 1969, *Cooray 69073020R*
(PDA); Uma Oya, April 1883, *s. coll.* (PDA). NUWARA ELIYA DISTRICT:
Hangurangketa, July 1851, *s. coll.* in *C.P. 1058*, p.p. (PDA). HAMBANTOTA
DISTRICT: near Ranna, April 1900, ? *Brown s.n.* (PDA); near Andala,
Tissamaharama, 13 Aug 1932, *Simpson 9959* (PDA); between Wirawila and
Tissamaharama, 17 Oct 1978, *Wirawan 1263* (K, PDA); Ruhuna National
Park, between Sithulpahuwa and Katagamuwa, c. 8 m, 4 Apr 1969, *Grierson
1142* (PDA), Kumana, near well, 31 July 1969, *Cooray 69073125R* (PDA),
75 m, 2 Mar 1993, *Philcox, Weerasooriya & Abeysiri 10555* (K, PDA),
Buttawa Modera, 18 Oct 1968, *Wirawan 1266* (PDA); Block I, Rakinawala,
22 Oct 1968, *Cooray 68102217R* (PDA); Buttawa dunes, 17 Jan 1969,
Cooray 69011706R (PDA), N of Buttawa, 28 Feb 1968, *Mueller-Dombois &
Cooray 68022804* (PDA), Buttawa Plain, 10 Mar 1970, *Cooray 70031917R*
(PDA), between Tissamaharama and Yala main entrance, c. 15 m, 3 Mar
1993, *Philcox et al., 10563* (K, PDA), near Yala bungalow, 2 m, 26 Aug
1967, *Mueller-Dombois 67082607* (PDA), Yala, Jamburalagala Rock, 19
July 1973, *Nowicke & Jayasuriya 418* (PDA); Block II, NW of Walaskema
Rocks, 2 m, 1 Oct 1967, *Mueller-Dombois 67100110* (PDA), *67100110A*
(PDA); Palatupana, 3 m, 4 Mar 1952, *Worthington 5690* (K). DISTRICT
UNCERTAIN: Wilpattu National Park, 26 Sept 1969, *Beusekom & Beusekom
1628* (PDA). LOCALITY UNKNOWN: in 1851, *Brodie s.n.* in *C.P. 1060*,
p.p. (PDA); in 1853, *Thwaites s.n.* in *C.P. 1058*, p.p. (BM, K); *Walker s.n.* (K).

3. Capparis brevispina DC., Prod. 1: 246. 1824; Thw., Enum. Pl. Zeyl. 15.
1858, incl. var. *rheedii* but excl. var. *rotundifolia.* Type: from India.

Capparis zeylanica sensu Hook. f. & Thoms. in Hook. f., Fl. Br. Ind. 1: 174. 1872; Trimen, Handb. Fl. Ceylon 1: 61. 1893, non L. 1753.

Shrub or slender tree; stems densely fawn, stellate-pubescent, soon glabrescent, minutely verrucose. Thorns 1–3 mm long, patent to antrorse. Leaves coriaceous, very varied in size and shape, petiolate; lamina 3–8.5 × 1–4.5 cm, lanceolate- to ovate-elliptic or lanceolate to obovate, apex acute or obtuse, often with stiff mucro up to 1.5 mm long, base obtuse or rounded, entire, glabrous, major veins prominent on both surfaces, especially beneath; petiole 0.3–0.7 cm long, deeply channelled above, verrucose. Flowers white, yellow turning maroon at base, solitary, axillary, at times on short lateral branchlets. Pedicels 1.5–3.5 cm long, shortly pubescent, soon glabrescent, or subglabrous. Sepals 5–10 × 3–5 mm, ovate, outer pair glabrous without, except for finely pubescent margin, tomentose to villous within, inner pair slightly longer than outer, pubescent without, subglabrous within. Petals c. 1.5–3 × 0.5–1.25 cm, obovate. Torus c. 1.5–2 mm diameter. Stamens up to 40–50, c. 2.5 cm long, slender. Gynophore 1.5–3 cm long, glabrous. Ovary c. 5 × 2 mm, ovoid, beaked, placentas 4. Seeds 6–7 × 4–5 × 2.5–3.5 mm, globose-reniform, densely tuberculate, yellow, enclosed in thin, smooth, white pericarp.

Distr. Southern India and Ceylon.

Ecol. Dry country.

Vern. Wal dehi (S).

Note. Of the specimens cited below, *Davidse 8077* and *Mueller-Dombois 69042201* have hitherto been considered to represent the Indian *Capparis nilgiriensis* Subba Rao, Kumari & Chandrasekaran of which only one specimen exists at Kew. The present author is not assured of the wisdom in upholding this differentiation, as the poor state of both of the Ceylon collections gives no clear reason for keeping them separate. Until more material becomes available, they will be kept under the closely allied species, *C. brevispina*. In addition to the two collections cited above, there are six collections made by Wirawan from Hambantota and Anuradhapura Districts, and housed at Peradeniya. These are unfortunately sterile and along with the specimen representing the Mueller-Dombois collection, are considered to relate to *Capparis heyneana* Wall. Decisions of accurate identification of these seven collections are presently outside the scope of this work.

Specimens Examined. JAFFNA DISTRICT: Elephant Pass, 11 Mar 1932, *Simpson 9251* (BM). TRINCOMALEE DISTRICT: Norway Island, 2 Apr 1943, *Worthington 1248* (BM). POLONNARUWA DISTRICT: bank of Ambanganga near Matuveyaya, 19 Jun 1932, *Simpson 9821* (BM). MATALE DISTRICT: Erawalagala Mts, E of Kandalama Tank, c. 9.5 km E of Dambulla,

29 Oct 1974, *Davidse 8077* (K, PDA). KANDY DISTRICT: Botanic Gardens, Peradeniya, 12 Oct 1926, *Alston 393* (K); Weragantota, 18 July 1927, *Alston 802* (K). HAMBANTOTA DISTRICT: Talgasmankada ford, 22 Apr 1969, *Mueller-Dombois 69042201* (K). LOCALITY UNKNOWN: *Macrae 714*, p.p. (BM); 6 Mar 1819, *Moon 322* (BM); *Thwaites s.n.* in *C.P. 1060*, (K), *1062* (BM, K), *1853, Thwaites s.n.* in *C.P. 2509*, p.p. (BM, K).

4. Capparis moonii Wight, Ill. Ind. Bot. 35. 1840; Thw., Enum. Pl. Zeyl. 16. 1858; Hook. f. & Thoms. in Hook. f., Fl. Br. Ind. 1: 175. 1872; Trimen, Handb. Fl. Ceylon 1: 62. 1893; Jacobs, Blumea 12: 472. 1965. Types: India and Ceylon: March 1836, *Wight s.n.* (K).

Climber, very large, woody, up to 3 m tall, glabrous after initial pubescence; stems much branched, smooth, reputedly up to 18 cm diameter. Thorns up to 4.5 mm long, usually recurved. Leaves subcoriaceous; lamina 5–12 × 2.5–5.75 cm, ovate-oblong to oblong-lanceolate, obtuse or acute, apiculate, rounded to subacute, cuneate at base, midrib shallowly channelled above, otherwise venation slightly prominent on both surfaces; petiole 1–1.5 cm long. Flowers white, fragrant, up to 8 cm diameter, 2–6 aggregated towards end of branchlet with lowest frequently appearing solitary-axillary; pedicels 3.5–6.5 cm long, stout, angular, glabrous. Sepals: outer pair 2 × 1.5 cm, subcircular, deeply concave, minutely pubescent to subglabrous, inner pair flatter. Petals up to 3.5 × 1–2 cm, obovate. Torus 4–5 mm wide. Stamens c. 130–140 (Jacobs), 4–5 cm long, white, turning red. Gynophore 6–9.5 cm long, glabrous. Ovary 5 × 3 mm, ovoid to ellipsoid, glabrous; placentas 4. Fruit 5–6 cm diameter, globose, pericarp 8 mm thick, fruiting pedicel enlarged to c. 8 cm long, this with gynophore much thickened, woody. Seeds c. 17 × 13 × 11 mm.

Distr. India and Ceylon.

Ecol. Moist forests up to 950 metres.

Vern. Rudanti (S).

Specimens Examined. ANURADHAPURA DISTRICT: Ritigala, 24 Mar 1905, *s. coll. s.n.* (PDA); Ritigala Strict Natural Reserve, Andikanda, 575 m, 21 Jan 1973, *Jayasuriya 1074* (PDA), S edge of Weweltenna, c. 600 m, 2 Jun 1974, *Jayasuriya 1750* (PDA). KURUNEGALA DISTRICT: Wariyapola, 13 Nov 1965, *Amaratunga 1045* (PDA). MATALE DISTRICT: Laggala, c. 900 m, 10 Nov 1978, *Cramer 5235* (K, PDA). KANDY DISTRICT: Hantane, 915 m, *Gardner 1197* (K), 27 Jan 1957, *Appuhamy s.n.* (PDA); c. 3.5 km W of double-cut junction, 700 m, 26 Mar 1974, *Jayasuriya & Sumithraarachchi 1557* (K, PDA); Hunnasgiriya, April 1851, *Moon s.n.* in *C.P. 2415* (PDA); near Dehigama Tea Estate, Murutalawa, 22 Sept 1978, *Wirawan 1227* (K, PDA); Murutalawa, Dec 1977, *Tennakoon s.n.* (PDA). NUWARA

ELIYA DISTRICT: near Welimada, 24 Mar 1906, *Silva s.n.* (HAK); above Welimada, 1 May 1932, *Simpson 9631* (BM). LOCALITY UNKNOWN: March 1836, *Wight s.n.* (K); in 1853, *Thwaites s.n.* in *C.P. 2415*, p.p. (BM, PDA).

5. Capparis roxburghii DC., Prod. 1: 247. 1824; Thw., Enum. Pl. Zeyl. 15. 1858; Hook. f. & Thoms. in Hook. f., Fl. Br. Ind. 1: 175. 1872, p.p.; Trimen, Handb. Fl. Ceylon 1: 62. 1893; Jacobs, Blumea 12: 486. 1965. Type: from India.

Shrub, climbing or straggling to 5 m tall; stems woody, young branches finely pubescent, glabrescent at length. Thorns where present, 2–2.5 mm long, recurved, but mostly wanting. Leaves subcoriaceous; lamina 3–7 × 1.25–3.5 cm, obovate-oblong, obtuse or acute, base obtuse, glabrous, venation slightly impressed above, prominent beneath; petiole 0.9–1.7 cm long, slender, minutely pubescent to subglabrous. Flowers white, pedicellate, 3–7 aggregated towards ends of, or spaced along, short branchlets, or occasionally similar numbers arranged in pseudo-umbels. Pedicels 1.2–4 cm long, slender. Sepals 0.8–1 cm diameter, subcircular, concave, glabrous. Petals 1.3–1.5 × 0.5–1 cm, obovate. Torus 3.5–4 mm wide. Stamens up to c. 50. Gynophore 5–5.5 cm. Ovary c. 3.5 × 2 mm, ellipsoid, glabrous; placentas 4. Fruit up to 5 cm diameter (7.5 cm, *Balakrishnan 1070*), globose, shortly apiculate, glossy, scarlet. Seeds 1–1.3 cm long, c. 0.8 cm deep, reniform, brown.

Distr. India and Ceylon.

Ecol. Open scrubland and thickets.

Vern. Kalu illangedi (S), Punai-virandi, Velungiriya (T).

Specimens Examined. MANNAR DISTRICT: in 1890, *Crawford 134* (PDA). PUTTALAM DISTRICT: Puttalam, 23 Apr 1796, *s. coll. 942* (K); Wilpattu National Park, Occaypu Kallu, 1 Oct 1969, *Cooray 69100107R* (PDA); crossing of Modaragam Aru, below Marichchukkladdi on road to Mannar, 15 m, *Davidse & Sumithraarachchi 8232* (PDA); Chilaw, 30 May 1931, *Simpson 8204b* (BM). ANURADHAPURA DISTRICT: Anuradhapura, 8 Sept 1965, *Amaratunga 1005* (PDA); Lulnewa, SW of Kahatagasdigiliya, 105 m, 2 Nov 1974, *Davidse & Sumithraarachchi 8238* (PDA); Dambulla to Kekirawa road, mile post 51, 6 July 1971, *Meijer 714* (PDA); Kekirawa Tank, July 1887, *s. coll.* (PDA); Wilpattu National Park, close to entrance, 5 July 1969, *Wirawan et al., 963* (K, PDA), Galge Vihare, 16 Sept 1968, *Mueller-Dombois 68091604* (PDA), between Magul Illaima and Malimaduwa, 13 July 1969, *Wirawan 963A* (PDA); S of Wilpattu National Park, gate c. 3 km N of junction on road to Anuradhapura, 29 Apr 1969, *Mueller-Dombois et al., 69042901* (K, PDA); mile marker 4/5 on road between Nekkatunuwewa and Seepukulamewewa, 16 Nov 1971, *Balakrishnan &*

Jayasuriya 1070 (K, PDA); along Noochiyagama to Galewewa Otappuwa road, 1 Sept 1978, *Wirawan 1211* (PDA); Mihintale to Horowapatana, 28 Mar 1963, *Amaratunga 565* (PDA). TRINCOMALEE DISTRICT: Trincomalee, 1846, *Gardner s.n.* in *C.P. 1065,* p.p. (PDA); Norway Island, 6 m, 2 Apr 1943, *Worthington 1249* (BM). KURUNEGALA DISTRICT: Tumbulla, c. 5 km beyond Nikaweratiya along Kurunegala to Puttalam road, 29 Aug 1978, *Wirawan 1203* (K, PDA). MATALE DISTRICT: Dambulla, near Tank, 15 May 1931, *Simpson 8086* (BM); Kurunegala road near Dambulla, 16 May 1931, *Simpson 8104* (BM). NUWARA ELIYA DISTRICT: Hanguranketa, Aug 1852, *s. coll.* in *C.P. 1065,* p.p. (PDA). HAMBANTOTA DISTRICT: Ruhuna National Park, Okanda to Kumana road, 30 July 1969, *Cooray 69073020R* (PDA). LOCALITY UNKNOWN: *Thwaites s.n.* in *C.P. 2480* (K); *s. coll. s.n.* (K); *s. coll* in *C.P. 1065,* p.p. (BM, PDA).

6. Capparis rotundifolia Rottl., Neue Schriften Ges. Naturf. Freunde Berlin 4: 185. 1803; DC., Prod. 1: 245. 1824; Jacobs, Blumea 12: 485. 1965. Type: from India.

Capparis pedunculosa Wall. ex Wight & Arn. Prod. 27. 1834; Thw., Enum. Pl. Zeyl. 16. 1858; Hook. f. & Thoms. in Hook. f., Fl. Br. Ind. 1: 176. 1872; Trimen, Handb. Fl. Ceylon 1: 63. 1893. Type: from India.
Capparis brevispina var. *rotundifolia* (Rottl.) Thw., Enum. Pl. Zeyl. 15. 1858.
Capparis longispina Hook. f. & Thoms. in Hook. f., Fl. Br. Ind. 1: 176. 1872. Type: from India.
Capparis pedunculosa var. *longispina* (Hook. f. & Thoms.) Trimen, Handb. Fl. Ceylon 1: 63. 1893.

Shrub, spreading or straggling, much-branched; branches densely to subdensely minutely pubescent, eventually glabrescent. Thorns recurved, 1.5–5 mm long, or 2.5–12 mm long, slender, straight, acicular. Leaves numerous, subcoriaceous; lamina 1–2.5 × 1–2 cm, subcircular to very broadly ovate, rounded, obtuse or retuse apex, base cordate, somewhat glossy above, less so beneath, margin slightly revolute or not, major venation subprominent above, obscure beneath; petiole up to 1 mm long, rarely more. Flowers white, 5–12 together in umbelliform cluster at ends of short axillary peduncles, often peduncles very short with flowers appearing clustered-axillary; peduncles (0.05–)0.5–4 cm long, slender, glabrous; pedicels 0.8–1.5 cm long, filamentous. Sepals c. 4–5 × 2–2.5 mm, glabrous, thin, membranous margin. Petals c. 5 × 3 mm, obovate. Stamens c. 20–36. Torus c. 1 mm diameter. Gynophore 7–17 mm, filiform, glabrous; ovary 1–1.5 mm long, glabrous. Fruit c. 0.9–1.5 cm diameter, globose, often with short point. Seeds not seen.

Distr. India, Ceylon and Burma; one collection doubtfully from Cambodia.

Ecol. Dry low country.

Vern. Pichchuvilatti (T).

Specimens Examined. VAVUNIYA DISTRICT: just S of Mullaitivu, 9 Dec 1970, *Fosberg & Balakrishnan 53504* (K, PDA). PUTTALAM DISTRICT: Chilaw, March 1882, *Nevill s.n.* (PDA); Wilpattu National Park, 26 Sept 1969, *van Beusekom & van Beusekom 1626* (PDSA), Kurutu Pandi Villu, 8 July 1969, *Wirawan et al., 1055* (K, PDA); between Kokkare and Kurutu Pandi Villu, 29 Jun 1969, *Wirawan et al., 858* (K, PDA). ANURADHAPURA DISTRICT: Na Ela, W of Ritigala, 20 Mar 1951, c. 150 m, *Worthington 5173* (BM, PDA). KANDY DISTRICT: Udawatte, off Hanguranketa, 950 m, 14 Dec 1971, *Jayasuriya & Balasubramaniam 465* (PDA); Haragama, July 1851, *Thwaites s.n.* in *C.P. 1069* p.p. (PDA). MATARA DISTRICT: E. of Kirinda, 12 Aug 1932, *Simpson 9938* (BM, PDA). HAMBANTOTA DISTRICT: Ruhunu National Park, Block I, N of Buttawa, 28 Feb 1968, *Mueller-Dombois & Cooray 68022805* (PDA), 10 Dec 1967, *Mueller-Dombois & Cooray 67121026* (PDA), next to Yala Camp Site, 30 Jan 1968, *Mueller-Dombois & Cooray 68013003* (PDA), Block II, 8 m, 25 Feb 1968, *Comanor 1038* (E, K, PDA). LOCALITY UNKNOWN: *Macrae 714*, p.p. (BM); 6 Mar 1819, *Moon 323* (BM); in 1853, *Thwaites s.n.* in *C.P. 1069* p.p. (BM, K); *s. coll.* in *C.P. 1069* (PDA).

7. Capparis grandis L. f., Suppl. Pl. 263. 1781; DC., Prod. 1: 248. 1824; Moon, Cat. 41. 1824; Thw., Enum. Pl. Zeyl. 16. 1858; Hook. f. & Thoms. in Hook. f., Fl. Br. Ind. 1: 176. 1872; Trimen, Handb. Fl. Ceylon 1: 63. 1893; Jacobs, Blumea 12: 456. 1965. Type: Ceylon, *Koenig s.n.*

Tree, 3.5–7 m tall, young branchlets finely, pale tomentose. Thorns c. 2 mm long, patent, more usually wanting. Leaves thickened-membranous, not coriaceous; lamina 3–5.5 (–7.5) × 2.5–4.5 (–5.5) cm, rhombic to broadly ovate, acute, rarely obtuse or slightly emarginate, base obtuse or subcuneate, somewhat glabrescent above but remaining pubescent on midrib, densely pubescent beneath, venation prominent on both surfaces with well defined reticulation; petiole 0.8–1.2 cm long, densely pubescent, obscurely channelled above. Flowers white, pedicellate, up to 10 aggregated at or towards ends of branches, appearing pseudoumbellate; pedicels (0.5–)1.5–3.75 cm long, minutely pubescent, glabrescent. Sepals 0.6–1.2 × 0.4–0.7 cm, outer pair navicular, inner slightly longer, obovate, flat, glabrous with shortly ciliate margins. Petals 0.8–1.1 × 0.3–0.4 cm, obovate. Stamens numerous, numbers and length uncertain. Gynophore 2–3 cm long, glabrous but minutely pubescent at base. Ovary c. 1.5 × 1 mm, glabrous. Fruit c. 2 cm diameter (? immature), globose, green when immature, ripening purple. Seeds few, ? c. 4, 10–12 × 6–10 mm.

Distr. India and Ceylon; Burma, northern Thailand and Indo-China.

Ecol. Dry lowland areas; open scrub with or without light tree cover.

Vern. Mudkondai (T).

Specimens Examined. PUTTALAM DISTRICT: Elluvankulam, on Puttalam to Mannar road, near border of Wilpattu National Park, 30 Aug 1978, *Wirawan 1208* (K, PDA); Wilpattu National Park, Kattakandal Kulam, 1 Dec 1969, *Cooray 69100109R* (K, PDA), between Sengapada Villu and Kattankandal, 6 July 1969, *Cooray & Balakrishnan 964* (K). ANURADHA-PURA DISTRICT: Wilpattu National Park, Erige Ela, 15 Sept 1968, *Wirawan & Cooray 68091501* (K, PDA), Huniwilagama, c. 200 m, from gate to Park, 30 Sept 1969, *Cooray 69093004R* (K, PDA). HAMBANTOTA DISTRICT: Mandagala Tank, between Nonagama and Hungama, 16 Oct 1978, *Wirawan 1260* (K, PDA); Ruhuna National Park, Palugaswala, 2–4 m, 29 Sept 1967, *Comanor 430* (E, K), Block I, Andunoruwa Tank, 11 Nov 1968, *Wirawan 710* (K, PDA); Komawa Wewa, 25 Aug 1967, *Mueller-Dombois 67082501* (PDA); Block II, Ketagalwala water-hole, 2 m, 27 Aug 1967, *Mueller-Dombois 67082701* (PDA). MONERAGALA DISTRICT: Kataragama, 22 Aug 1974, *Kostermans 25450* (K). LOCALITY UNKNOWN: *Forster s.n.* (K); *Mackenzie s.n.* (K); *Thwaites s.n.* in *C.P. 1071* (BM, K).

8. Capparis floribunda Wight, Ic. Pl. Ind. Or. 2: 35, t. 14. 1840; Hook. f. & Thoms. in Hook. f., Fl. Br. Ind. 1: 177. 1872; Trimen, Handb. Fl. Ceylon 1: 64. 1893; Jacobs, Blumea 12: 453. 1965. Type: from India.

Woody climber (Trimen), glabrous or rarely fulvous-pubescent. Thorns c. 2 mm long, recurved, frequently wanting. Leaves submembranous; lamina 4.5–9 × 1.5–3.5 cm, ovate-elliptic, rounded or obtuse at base, apex obtuse to subacute, glabrous, margin entire, often slightly revolute, midrib channelled above, prominent beneath, other venation obscure; petiole 4–8 mm long, glabrous. Flowers white, numerous in small umbels arranged in terminal or smaller axillary panicles; panicles up to 12 cm or more long; peduncles 1.5–5 cm long; umbels composed of up to c. 15–25 or more slender-pedicellate flowers; pedicels 0.8–1.7 cm long; bracts 2–5 mm long, linear, early caducous. Sepals 2–3.5 × 1.5–2 mm, ovate, minutely pubescent at base without. Petals slightly longer than sepals, narrowly ovate to oblong. Stamens c. 7–10. Gynophore 8–10 mm long. Fruit and seeds not seen.

Distr. India and Ceylon, Burma, Thailand, Indo-China; Philippines, Moluccas and Java.

Ecol. Lowland dry areas.

Specimens Examined. KANDY DISTRICT: Haragama, in 1862, *Thwaites s.n.* in *C.P. 3766*, p.p. (PDA). LOCALITY UNKNOWN: *Thwaites s.n.* in *C.P. 3766*, p.p. (BM, K).

9. Capparis sepiaria L., Syst. Nat. ed. 10, 2: 1071. 1759; DC., Prod. 1: 247. 1824; Hook. f. & Thoms. in Hook. f., Fl. Br. Ind. 1: 177. 1872, as for var.

vulgaris; Trimen, Handb. Fl. Ceylon 1: 65. 1893; Jacobs, Blumea 12: 489. 1965. Type: from India.

Capparis retusella Thw., Enum. Pl. Zeyl. 16. 1858. Type: Ceylon, *Thwaites s.n.* in *C.P. 2550* (BM, K, PDA).
Capparis sepiaria var. *retusella* (Thw.) Thw., Enum. Pl. Zeyl., Addenda 400. 1864; Hook. f. & Thoms. in Hook. f., Fl. Br. Ind. 1: 177. 1872.

Shrub, 2.5–5 m tall, woody, much branched, sometimes straggling or climbing; branches densely, finely fulvous or grey pubescent, especially when young, glabrescent when older. Thorns 1–3 (–5) mm long, recurved. Leaves herbaceous to subcoriaceous; lamina 2.25–6.5 × 0.75–3.5 cm, ovate, oblong-ovate or obovate, apex rounded-obtuse or retuse, at times narrowly lanceolate-acuminate, usually in juvenile state, base acute, obtuse or rounded, pubescent on both surfaces, upper surface early glabrescent except on midrib, becoming glossy, pubescence persistent beneath; petiole 2–3.5 mm long, densely pubescent. Flowers white, 8–18 in subumbels at ends of short, lateral branchlets; pedicels 1–1.8 (–3) cm long, filiform, glabrous. Sepals 5–7 × 3–5 mm, ovate, outer pair densely short-pubescent, inner smaller than outer, pubescence less dense. Petals c. 7 × 3 mm, pubescent, especially without. Stamens 30–45 (Jacobs). Gynophore 3–10 (–25) mm long, if pubescent, then towards or at base; ovary c. 1–2.5 × 1–1.75 mm, ovoid, glossy; placentas 2 (–3). Fruit 1–1.25 cm, subglobose, drying pale brown, verrucose. Seeds not seen.

Distr. India and Ceylon; Burma to southern China, southwards through the Philippines, Celebes and New Guinea to Australia.

Ecol. Dry country, open scrubland, often near coast.

Vern. Karunchurai (T).

Specimens Examined. JAFFNA DISTRICT: Jaffna, in 1846, *Gardner s.n.* in *C.P. 1063*, p.p. (PDA). MANNAR DISTRICT: Mannar Island, Feb 1890, *s. coll. s.n.* (PDA). VAVUNIYA DISTRICT: Cheddikulam along Medawachchiya to Mannar road, 8 Oct 1978, *Wirawan 1247* (K, PDA), *1248* (K, PDA), *1249* (E, K, PDA), *1250* (E, K, PDA), *1251* (K, PDA); Parayanalankulam to Vavuniya, 24 Jun 1973, *Kostermans 25138* (PDA). PUTTALAM DISTRICT: Maduru Odai, Mannar to Puttalam road, 30 Jun 1969, *Wirawan et al. 924* (K, PDA); Mannativu, sea-level, 19 May 1976, *Cramer 4648* (E, K, PDA). ANURADHAPURA DISTRICT: Anuradhapura, *Trimen s.n.* (PDA); along Nochchiyagama to Galewela to Otappuwa road, 1 Sept 1978, *Wirawan 1212* (K, PDA); near Kekirawa, July 1887, *s. coll. s.n.* (PDA). KURUNEGALA DISTRICT: Tumbulla c. 4.5 km from Nikaweratiya on Kurunegala to Puttalam road, 29 Aug 1968, *Wirawan 1204* (E, PDA). MATALE DISTRICT: Matale Estate, June 1853, *Thwaites s.n.* in *C.P. 2550*, p.p. (PDA). KANDY DISTRICT: Haragama, July 1851, *Thwaites s.n.*

in *C.P. 2550*, p.p. (PDA). AMPARAI DISTRICT: Arugam Bay, sea-level, 1 Jun 1972, *Maxwell & Jayasuriya 754* (PDA). NUWARA ELIYA DISTRICT: Maturata, 5 Nov 1823, *Moon s.n.* in *C.P. 2550*, p.p. (PDA). MONERAGALA DISTRICT: 17 km W of Tanamalwila, 125 m, 24 Nov 1974, *Davidse & Sumithraarachchi 8817* (K, PDA). HAMBANTOTA DISTRICT: milepost 152, NE of Hambantota, 1 July 1970, *Meijer 247* (PDA); along Maha Ela, near Mandagala Tank, between Nonagama and Hungama, 16 Oct 1978, *Wirawan 1258* (PDA), *1259* (PDA); Bundala, 20 July 1972, *Hepper & de Silva 4749* (K, PDA); Ruhuna National Park, Buttawa Modera, 5–8 m, 5 May 1968, *Fosberg 50304* (PDA), Uda Potana, c. 100 m from ocean, 3–4 m, 28 Sept 1967, *Comanor 421* (E, K, PDA), near Yala Bungalow, 26 Aug 1967, *Mueller-Dombois 67082605* (PDA), Derasgala Tank, 30 Jun 1970, *Meijer 240* (K, PDA), N of Yala Camp, 20 m, 25 Jun 1967, *Mueller-Dombois & Comanor 67062507* (PDA); Block I, 0–10 m, 4 Apr 1969, *Robyns 6993* (E, PDA), Yala Camp, 11 Nov 1968, *Wirawan 705* (K, PDA), Karaugaswala, 30 May 1968, *Cooray 68053005R* (PDA), Andunoruwa Wewa, 26 Aug 1967, *Mueller-Dombois 67082610* (PDA). LOCALITY UNKNOWN: in 1853, *Thwaites s.n.* in *C.P. 2550*, p.p. (BM, K); *Thwaites s.n.* in *C.P. 1063*, p.p. (BM, K); 15 Dec 1881, *s. coll.* (PDA); *s. coll.* in *C.P. 1063*, p.p. (PDA).

10. Capparis divaricata Lam., Enc. 1: 606. 1785; DC., Prod. 1: 252. 1824; Thw., Enum. Pl. Zeyl. 15. 1858; Hook. f. & Thoms. in Hook. f., Fl. Br. Ind. 1: 174. 1872; Trimen, Handb. Fl. Ceylon 1: 61. 1893; Jacobs, Blumea 12: 515. 1965. Type: from India.

Shrub or small tree, 2–6 m tall, much-branched, spreading, glabrous, shiny, becoming somewhat hoary with age. Thorns (1.5–) 2–4 mm long, straight, patent or forwardly directed, not recurved. Leaves subcoriaceous, very variable; lamina 3.25–7.5 × (0.5–) 1.5–4 cm, linear-oblong or ovate, apex acute or obtuse, at times mucronulate, base obtuse or subacute, venation pronounced throughout, (3–) 5–7-veined from base, pinnately veined above; petiole (3–) 5–8 mm long. Flowers white to cream, solitary, axillary, up to 4.5 cm diameter in bloom. Pedicels 0.8–2 cm long. Sepals c. 1–1.6 × 0.4–0.8 cm, elliptic, acute-acuminate with 1–2 mm long acumen, navicular, outer pair glabrous, verrucose, inner pair larger, subtomentose without. Petals 1.8–3.5 × 0.5–1 cm, narrowly oblong-spathulate. Stamens 24–50 or more. Gynophore 2–2.5 cm long; ovary 7–8.5 × 3 mm, glabrous, strongly, longitudinally 5–10-ribbed, beak 1–2 mm long. Fruit 5–5.5 cm long, globose to broadly ellipsoid, ripening red, pericarp 4–5 mm thick, up to 10 longitudinal ridges, coarsely largely verrucose. Seeds up to 7–8 × 4 mm, purplish-brown, white arillate.

Distr. India and Ceylon.

Ecol. Open forests and scrub vegetation.

Vern. Torikei (T).

Specimens Examined. MANNAR DISTRICT: Madhu Road, 28 Mar 1932, *Simpson 9389* (BM); S of Murunkan, 15 Mar 1932, *Simpson 9380* (PDA), *9381* (PDA); Parayanalankulam, 28 Jan 1950, *Worthington 4503* (BM, PDA); Kumbusanyawadi, near Murunkan, 8 Oct 1978, *Wirawan 1252* (K, PDA). PUTTALAM DISTRICT: "Pomparappowa" (?=Pomparripu), in 1848, *Gardner s.n.* in *C.P. 1072*, p.p. (PDA); Kattakandalkulam, *Cooray 69100207* (PDA). TRINCOMALEE DISTRICT: Fort Frederick ramparts, 40 m, 22 Sept 1971, *Ripley 467* (PDA). AMPARA DISTRICT: Kumbukkan Aru (Oya), October 1884, *Nevill s.n.* (PDA). HAMBANTOTA DISTRICT: near Hambantota salterns, 26 Nov 1974, *Sumithraarachchi & Davidse 576* (PDA); Ruhuna National Park, towards Warahena, 2 m, 22 Jun 1967, *Mueller-Dombois & Comanor 67062214* (PDA), Andunoruwa Wewa, 3–5 m, 3 Apr 1968, *Fosberg & Mueller-Dombois 50153* (K, PDA), road from Andunoruwa to Komawa Wewa, 14 Nov 1969, *Cooray 69111401R* (K, PDA), 25 Mar 1970, *Cooray 70032518R* (PDA); Block I, 0–10 m, 4 Apr 1969, *Robyns 6993* (PDA), Ruhuna National Park, 10 m, 9 Mar 1973, *Bernardi 14202* (PDA); Yala to Debbara Wewa, 11 Mar 1969, *Ripley 122* (PDA); Jamburagala, sea-level, 27 Apr 1973, *Cramer 4131* (PDA), Talgasmakanda, sea-level, 26 Apr 1973, *Cramer 4122* (PDA), Gonagala, sea-level, 5 Jan 1971, *Cramer 3322* (PDA), c. 3.5 km N of Yala Bungalow, 2 m, *Mueller-Dombois et al., 67093014* (PDA), Komawa Wewa, 4 m, 28 Mar 1968, *Comanor 1154* (K), 29 May 1968, *Cooray 68052904* (PDA), between Korawakwewa and Warahena, 21 Oct 1968, *Cooray 68102107R* (K, PDA), Kotabendu Wewa, 5–9 m, 25 Aug 1967, *Comanor 412* (E, K); Patanagala, 3 m, 1 Mar 1952, *Worthington 5687* (BM). LOCALITY UNKNOWN: in 1853, *Thwaites s.n.* in *C.P. 1072*, p.p. (BM, K).

5. CRATEVA

L., Sp. Pl. 444. 1753; L., Gen. Pl. ed. 5: 203. 1754. Type species: *Crateva tapia* L. (see Correa, Trans. Linn. Soc. London 5: 222. 1800.)

Trees, small to medium height, glabrous, shortly deciduous, flowering when leafless; branches smooth, distinctly leaf-scarred. Leaves stipular, 3-foliolate; stipules small, deciduous; top of petiole at times glandular; leaflets sessile to shortly petiolulate, lateral asymmetrical. Flowers large, white to yellow or purplish, pedicellate, in terminal corymbose racemes. Sepals 4, equal, free, ovate-spathulate, inserted on edge of large disk. Petals 4, long-clawed. Stamens up to 24 or more; filaments long filiform, spreading, connate at base with gynophore, white initially, becoming lilac to purple. Gynophore about equalling stamens. Ovary on slender stalk, 1-locular, ovules many on 2 parietal placentas. Fruit globose, fleshy. Seeds embedded in pulp.

Tropical and subtropical genus of 6 species.

Crateva adansonii DC., Prod. 1: 243. 1824.

ssp. **odora** (Buch.-Ham.) Jacobs, Blumea 12, 2: 198. 1964.

Crateva odora Buch.-Ham., Trans. Linn. Soc. London 15: 118. 1827. Type: from India.

Crateva roxburghii sensu Wight & Arn., Prod. 23. 1834; Thw., Enum. Pl. Zeyl. 14. 1858; Trimen, Handb. Fl. Ceylon 1: 59. 1893, non R. Br., 1826.

Crateva religiosa var. *roxburghii* (R. Br.) Hook. f. & Thoms. in Hook. f., Fl. Br. Ind. 1: 172. 1872.

Tree, 3–12 (–20) m tall. Leaves long-petiolate. Lamina 5–8 (–12) × 6.5–17 (–20) cm overall, terminal leaflet 5–12 × 2–6 cm, broadly elliptic to lanceolate, acute, acuminate, laterals 3.5–7 × 1–10 cm, asymmetric, acute, acuminate, widest below middle; petiolules 0.4–1 cm long. Petiole 3–7 (–c. 13) cm long, usually with small glands above at top. Flowers white or cream to yellow, sweetly scented. Inflorescence on small twigs, growing through or not, floriferous portion up to 10 cm long, 17–30-flowered; pedicels 1.5–4 cm long. Torus 4–5 mm wide. Sepals 3.5–6 × 2–2.5 mm, elliptic. Petals: limb 1–2.5 × 0.6–1.2 cm, broadly elliptic; claw 0.2–0.5 (–0.7) cm long. Stamens 18–22 or more, up to 5 cm long. Gynophore 3–5 cm long. Ovary 2.5–3 × c. 2 mm, ellipsoid. Fruit 1.8–2.5 cm diameter (immature), globose, fruit stalk 2.75–5 cm long, up to 2 mm thick. Seeds c. 5 × 2 mm, deeply reniform or somewhat horseshoe-shaped, smooth.

Distr. India, Ceylon and Burma.

Ecol. Low altitude dry woodland and scrub.

Vern. Lunu-warana (S), Navala, Navilankai (T).

Note. According to Jacobs (1964), this species is comprised, in addition to the typical subspecies, of four others. The geographical range of the species reaches from Africa, through much of Asia into Malesia thence to the Philippines, Taiwan and to the Ryukyu Islands in the north Pacific. As mentioned above, our subspecies is restricted to India and Burma in addition to Ceylon.

Specimens Examined. JAFFNA DISTRICT: Jaffna, *Gardner s.n.* in *C.P. 1067*, p.p. (PDA); Tikuvil, 13 July 1970, *Dassanayake 439* (K); 16 km S of Pooneryn, sea-level, 9 July 1971, *Meijer 795* (PDA). MANNAR DISTRICT: Mantai, 8.5 km from Mannar on Mannar to Pooneryn road, 9 Oct 1978, *Wirawan 1253* (PDA); 8 km SW of Murunkan, 10 Apr 1973, *Bremer et al., 69* (K). VAVUNIYA DISTRICT: Mullaitivu, near lagoon, 13 Mar 1932, *Simpson 3291* (BM). PUTTALAM DISTRICT: near Mampuri, 16 Feb 1975, *Sumithraarachchi 705* (PDA); Chilaw, 30 May 1931, *Simpson*

9876 (BM). ANURADHAPURA DISTRICT: Awkana, 9 Sept 1975, *Bernardi 15263* (PDA); Habarana, beside Habarana to Kantalai road, c. 100 m, 28 Mar 1977, *Cramer 4941* (K, PDA); Wilpattu National Park, 30–150 m, 21 Mar 1969, *Robyns 6978* (PDA), between Kokkari Villu and Kurutu Pandi Villu, 29 Jun 1969, *Wirawan et al., 873* (K, PDA), between Magul Illaima and Malimaduwa, 13 July 1969, *Wirawan 873A* (K, PDA). TRINCOMALEE DISTRICT: Trincomalee, in 1862, *Glenie s.n.* in *C.P. 1067*, p.p. (PDA), 3 m, 12 Mar 1940, *Worthington 843* (BM); Norway Island, 5 m, 27 Nov 1939, *Worthington 667* (BM); Kuchchaveli, seaside, 24 July 1939, *Worthington 525* (BM). KURUNEGALA DISTRICT: Bakmigalle, 8.5 km from Kurunegala on Kurunegala to Dambulla road, planted as fence plant, 7 Oct 1978, *Wirawan 1241* (PDA). MATALE DISTRICT: Dambulla, *s. coll.* in *C.P. 1067*, p.p. (PDA). POLONNARUWA DISTRICT: Uradiwetti, Kandakaduwa, 8 Jun 1974, *Waas 621* (K). KANDY DISTRICT: Roseneath Reservoir, Kandy, 26 Apr 1960, *Worthington 6924* (PDA); Peradeniya, Royal Botanical Gardens, 12 Mar 1910, *Silva s.n.* (PDA). BADULLA DISTRICT: Bibile to Alutnuwara road, 30 Apr 1978, *Soejarto & Balasubramaniam 4873* (K). AMPARAI DISTRICT: near Panakala Lagoon, S. of Panama, 3 May 1975, *Jayasuriya 2002* (K, PDA). HAMBANTOTA DISTRICT: Hambantota, December 1882, *s. coll. s.n.* (PDA), 20 July 1972, *Hepper & de Silva 4759* (K, PDA); Hambantota to Tissamaharama road mile-marker 152/4, 3 Oct 1973, *Waas 65* (K), 19 Nov 1973, *Sohmer et al., 8842* (PDA); Ruhunu National Park, Patanagala Rocks, 3–5 m, 3 Apr 1968, *Fosberg & Mueller-Dombois 50133* (BM, K, PDA), Jamburagala, 16 July 1975, *Jayasuriya & Austin 2248* (K, PDA), Block I, Komawa Wewa, 4–8 m, 25 Aug 1967, *Comanor 408* (E, K, PDA), 28 Mar 1968, *Comanor 1155* (K, PDA), near Gonagala, 25 Jan 1968, *Comanor 853* (K), between Situlpawa and Katagamuwa, c. 3 m, 4 Apr 1969, *Grierson 1141* (E, PDA), Palatupana, 25 Oct 1968, *Wirawan 666* (K, PDA), 200 m W of Karaugaswala at Yala Road, 19 Oct 1968, *Mueller-Dombois 68101907* (K, PDA); Ruhunu National Park, Block II, Kumbukkan Oya, c. 20 m from river, 28 Sept 1967, *Comanor 426* (E, K), near Maha Gajabawa, 2 m, 2 Oct 1967, *Cooray 67100205* (PDA), Mahaseelawa, 4 Dec 1973, *Sohmer 8981* (PDA); Bundala Sanctuary, 10 m, 8 Mar 1973, *Bernardi 14168* (PDA). RATNAPURA DISTRICT: by Chandrika wewa, Tunkama, 25 Feb 1994, *Philcox et al., 10677* (K, PDA, US). LOCALITY UNKNOWN: in 1853, *Thwaites* in *C.P. 1067*, p.p. (BM, K); *s. coll.* in *C.P. 1067*, p.p. (PDA).

CARICACEAE
(by M.D. Dassanayake*)

Dumort., Anal. Fam. Pl. 37, 42. 1829. Type genus: *Carica* L.

Small pachycaul trees or shrubs with sparsely branched or unbranched, mostly soft-wooded and unarmed stems bearing often large, palmately lobed to palmate or digitate, spirally arranged leaves with long petioles. Stipules absent or spine-like. Milky latex present in all parts, in anastomosing articulated laticifers. Inflorescence axillary, cymose, sometimes flowers solitary in leaf-axils. Flowers regular, 5-merous, unisexual or bisexual (plants dioecious, monoecious or polygamous). Calyx 5, very small, gamosepalous, 5-toothed or lobed. Corolla of 5 mostly united petals twisted in bud; in staminate flowers with a long corolla tube, in pistillate flowers with petals united only at the base, lobes contorted or valvate. Perfect flowers with petals united into a short corolla tube or free. Stamens 5 + 5 or reduced to 5 and alternate with corolla lobes. Staminodes absent in pistillate flowers. Filaments epipetalous or arising from the receptacle, free or basally united into a short tube. Anthers dorsifixed, with longitudinal slits. Connective often shortly prolonged. Ovary superior, of 5 united carpels, unilocular with deeply intruded parietal placentas, or plurilocular. Ovules many, anatropous, bitegmic, with enlarged funicle. Styles free, very short. Stigmas 5, simple or lobed or branched and antler-like. Rudimentary gynoecium present and thread-like in staminate flowers, or absent. Fruit a large fleshy berry. Seeds many, with a succulent outer layer (sarcotesta) and a harder warty or tuberculate inner layer (sclerotesta). Embryo straight, endosperm fleshy, oily.

Four genera, with c. 30 species, mainly in tropical and subtropical America and tropical Africa.

CARICA

L., Sp. Pl. 1036. 1753; L., Gen. Pl. ed. 5. 458. 1754. Lectotype species: *C. papaya* L.

Small trees with typically unbranched, soft-wooded stems. Leaves simple or palmately lobed, borne in a terminal crown. Plants dioecious, polygamous

* Flora of Ceylon Project, Peradeniya.

or monoecious. In plants with mostly staminate flowers inflorescences of long, axillary, often pendulous, many-flowered panicles; in plants with mostly pistillate or bisexual flowers, flowers in erect, few-flowered cymes or solitary in leaf-axils. Pedicels short. Sepals very small. Stamens 10 or 5; filaments free, pilose. Anthers introrse, with connectives sometimes enlarged. In staminate and some bisexual flowers the outer whorl of stamens with longer filaments and shorter anthers than the inner. Stigmas variously lobed. Fruit a smooth or more or less 5-ridged berry with a short or long stalk.

c. 22 species, native to parts of Mexico and central and South America.

Carica papaya L., Sp. Pl. 1036. 1753; Willis, Cat. 135. 1911. Type: from "Habitat in Indiis".

Small tree to c. 5(–7) m tall. Stem usually unbranched, 10–60 cm broad, hollow, marked with large prominent leaf-scars. Wood soft and succulent. Leaf lamina to c. 75 cm long and broad, palmately (3) 7–11 (–13)-lobed to varying extents, lobes pinnatifid or pinnatipartite, prominently veined beneath. Petiole hollow, to c. 100 cm long, abscising at the base after the leaf dries up. Inflorescences or single flowers appearing in the axil of each leaf after their first formation. Staminate inflorescences to c. 75 cm long, with the first branch c. 35 cm from the base of the peduncle, profusely branched thereafter, ending in dichasia. Pistillate flowers solitary, in few-flowered inflorescences, or single pistillate flowers with a staminate flower on either side. Bracts to c. 20 mm long, light green, caducous. Staminate flower: pedicel solid. Calyx 5-toothed, with segments c. 1 mm long. Corolla tube 2.5–3 cm long, 3–5 mm broad, sparsely hairy within, glabrous outside. Lobes 12–20 × c. 4 mm, narrowly elliptic, spreading, reflexed, contorted, greenish yellow. Filaments of outer stamens 2–3 mm long, inner stamens almost sessile. Anthers yellow, of inner stamens 2–3 mm long, confluent at their tips, of outer c. 2 mm. Filaments and anthers with white hairs. Flowers fragrant. Pistillate flower: pedicel c. 7 mm long, jointed at apex. Calyx lobes 3–5 mm long, light green. Petals free, greenish yellow, 3.5–5 cm long, lanceolate to linear, acuminate at apex, reflexed, more or less thick and leathery, valvate or contorted, caducous. Staminodes absent. Stigmas subsessile, spreading, c. 15 mm long, branching into c. 5 segments, pale yellow. Bisexual flower: calyx lobes c. 3 mm long. Corolla tube c. 2 cm long, lobes 2.5–3 cm long, oblong or linear, reflexed, contorted, greenish below, yellowish above. Fruit a fleshy berry, varying in shape from almost globose to ellipsoid and pyriform to cylindrical, smooth or with 5 furrows, usually yellow-green outside when ripe, with yellow, orange or red pulp. Central cavity round or with 5 ridges, c. 200–750 seeds in 5 strips on the fruit wall. Seeds yellowish brown to black, ovoid, c. 7 mm long, 5 mm broad.

Distr. Native to tropical America. Cultivated in all parts of the country (occasionally an escape from cultivation) and throughout the tropics.

Ecol. Flowering throughout the year.

Uses. The ripe fruit and the cooked unripe fruit are eaten. Unripe fruits are the source of papain, a proteolytic enzyme with many uses. The laticifers are particularly well distributed in the outer layers of the fruit wall.

Vern. Papol, Gaslabu (S); pappali (T), papaw, papaya, pawpaw (E).

Note. There is much variation in the structure of the flowers. The above description is of three basic flower types: more than 40 intergrading morphological types have been described by Storey (Hort. Advance 2: 49–60. 1958). Chromosome number 2n=18.

Specimens Examined. COLOMBO DISTRICT: Kirillawela, *Amaratunga 2295* (PDA); Pugoda, *Amaratunga 2220* (PDA).

Carica pubescens Lenne & K. Koch (*C. candinamarcensis* Hook. f.), the "mountain papaw" of the Andes, is sometimes cultivated in the montane zone, and occurs sporadically as an escape. It has smaller fruits which are made into preserves.

CARYOPHYLLACEAE
(by B.M. Wadhwa*)

Juss., Gen. Pl. 299.1789 "Caryophylleae". Type genus: *Caryophyllus* Mill., non L., nom. illegit. (= *Dianthus* L.).

Herbs, rarely undershrubs; branches usually jointed or thickened at the joints. Leaves usually opposite, less commonly alternate or whorled, simple, entire; stipules present or 0. Flowers actinomorphic, usually bisexual, sometimes unisexual or the plant gynodioecious, often in bracteate, dichasial cymes. Sepals 4–5, free or united, often united by commissures alternating with the calyx-teeth, persistent. Petals 4–5, sometimes 0, free. Stamens 5–10, rarely fewer, inserted with the petals and sometimes slightly connate to them; anthers 2-celled. Disk small or annular, sometimes segmented. Ovary superior, unilocular or imperfectly 3–5-locular; ovules 1–many on a basal or free-central placenta; styles 2–5, free or connate. Fruit a capsule dehiscing by as many or twice the number of valves or apical teeth as styles, rarely a berry or achene. Seeds few-many, rarely single, globose, pyriform or reniform, smooth or ornamented on testa; embryo usually curved round the copious, starchy perisperm.

A large family of 66 genera and about 1,750 species, cosmopolitan but chiefly in the temperate and alpine areas of Europe and W. Asia, 9 genera and 15 species in Sri Lanka.

KEY TO THE GENERA

1 Sepals free
  2 Stipules 0; styles free
    3 Leaves linear-filiform; petals entire, fruits dehiscent by 4–5-valves down to base . . . .
    . . . . **1. Sagina**
    3 Leaves wider, ovate to oblong; petals emarginate or more deeply incised or 0; fruits dehiscent by 3–10-valves
      4 Capsules short, straight, 3–6-valved, if 3-valved, then 1-seeded; petals split to the base or 0; styles 3 . . . . **2. Stellaria**
      4 Capsules long, often curved, dehiscent by 8–10 valves, many-seeded; petals shortly notched or more incised, rarely lacking; styles 4–5 . . . . **3. Cerastium**
  2 Stipules present, setaceous, scarious or membranous; styles united below (except *Spergula*)

---

* Royal Botanic Gardens, Kew.

5 Stipules scarious or membranous; petals entire or toothed; styles 3
   6 Sepals keeled, green, membranous along margins ....**4. Polycarpon**
   6 Sepals not keeled, scarious or membranous
      7 Axillary leaf-fascicles present; stamens 5–10; seeds keeled or winged; stipules scarious ....**5. Spergula**
      7 Axillary leaf-fascicles absent; stamens 5; seeds neither keeled nor winged; stipules membranous ....**6. Polycarpaea**
5 Stipules setaceous, filiform, forming an interpetiolar fringe, style 1, 3-partite as far as middle or a little more; petals 2–6-fid ....**7. Drymaria**
1 Sepals distinctly connate ....
   8 Styles 2; capsules shortly 4-valved; leaves clasping the stem ....**8. Vaccaria**
   8 Styles 3; capsules 3–6-valved; leaves narrowed below, not clasping the stem .... **9. Silene**

1. SAGINA

L., Sp. Pl. 128. 1753; L., Gen. Pl. ed. 5. 176. 1754; Benth. & Hook. f., Gen. Pl. 1: 151. 1862; Edgew. & Hook. f. in Hook. f., Fl. Br. Ind. 1: 242. 1874; Gamble, Fl. Pres. Madras 1: 63. 1915; Alston in Trimen, Handb. Fl. Ceylon 6: 18. 1931; Clapham & Jardine in Tutin et al., Fl. Europaea 1: 146. 1964, ed. 2, 176. 1993. Lectotype species: *S. procumbens* L.

Small annual or perennial herbs, often caespitose; stems procumbent or ascending. Leaves opposite, linear-lanceolate or subulate, slightly connate at base, exstipulate. Flowers small, 4–5-merous, solitary or in few-flowered cymes. Sepals 4–5, free. Petals 4–5, sometimes 0, free, white, minute, entire. Stamens as many or twice as many as the sepals. Ovary 1-celled; styles 4–5, alternating with the sepals; ovules numerous. Fruit a capsule, dehiscing to the base by 4–5 valves. Seeds minute, numerous.

A north temperate genus of c. 30 species, also in East and South Africa, India, Pakistan (Himalaya), New Guinea and the Andes; in Sri Lanka 1 species.

Sagina saginoides (L.) Karst., Deutsch. Fl. Pharm.—Med. Bot. 539. 1882; Alston in Trimen, Handb. Fl. Ceylon 6: 18. 1931; Steinb. in Kom., Fl. URSS. 6: 471. 1936; Clapham & Jardine in Tutin et al., Fl. Europaea 1: 147. 1964; Ramamurthy in Nair & Henry, Fl. Tamilnadu 23. 1983; Ghazanfar in Nasir & Ali, Fl. Pakistan 175: 43. 1986.

Spergula saginoides L., Sp. Pl. 441. 1753. Type: Herb. Linn. 604/6 (LINN).
Sagina linnaei Presl, Rel. Haenk. 2: 14. 1831. Type: from Mexico.
Spergula procumbens Auct.: Edgew. & Hook. f. in Hook. f., Fl. Br. Ind. 1: 242. 1874; Gamble, Fl. Pres. Madras 1: 63. 1915, non L., 1753.

Small, caespitose, perennial herb; stems procumbent to ascending, 4–9 cm, slender, glabrous. Leaves linear, rosette ones to 15 mm long, cauline leaves 4–8 mm, glabrous, sometimes ciliate, shortly mucronate, base connate, margin scarious. Flowers usually solitary, sometimes 2, 2–3 mm diam., on slender pedicels; pedicels glabrous, rarely glandular-puberulent. Sepals ovate-oblong, 3–4.5 mm long, obtuse, glabrous, spreading in fruits, margin hyaline. Petals slightly shorter than the sepals, sometimes 0, white. Stamens usually 10, rarely 5. Styles 5. Capsules 4–5 mm long, pale straw-coloured, shining, splitting by 5 valves. Seeds numerous, subglobose to triangular, smooth.

Distr. Scandinavia, C. & W. Asia, India (Himalaya & Nilgiris), Pakistan (temperate and alpine Himalaya), Sri Lanka, W. China, N. Japan, N. America and Mexico.

Ecol. Amongst boulders and rocky slopes, in montane wet zone, rare.

Notes. The species is included on the authority of Alston (l. c.). I have not seen any specimen from Sri Lanka and the description is based on specimens from India.

2. STELLARIA

L., Sp. Pl. 421. 1753; L., Gen. Pl. ed. 5, 193. 1754; Thw., Enum. Pl. Zeyl. 24. 1858; Edgew. & Hook. f. in Hook. f., Fl. Br. Ind. 1: 229. 1874; Trimen, Handb. Fl. Ceylon 1: 86. 1893; Pax & Hoffmann in Pflanzenfam. ed. 2, 16C: 322. 1934. McNeil, Notes Roy. Bot. Gard. Edinburgh 32(3): 389–395. 1973. Lectotype species: *S. holostea* L.

Annual or perennial herbs; stems prostrate to erect, glabrous or hairy. Leaves variable, sessile or petiolate, distinctly longer than wide, glabrous or sparingly hairy, exstipulate. Inflorescences usually few to many-flowered dichasial cymes, rarely flowers solitary or 2 together. Bracts scarious or herbaceous. Sepals 4–5, free. Petals 4–5 or 0, white, rarely greenish, usually deeply bilobed. Stamens 2–10, nectaries usually present. Ovary 1-celled, rarely 3-celled; styles usually 3, rarely 2 or 5, free; ovules usually numerous. Fruit a capsule, globose, ovoid or cylindrical, dehiscing with 3 or 6 teeth, usually to about the middle. Seeds 1–many, compressed, usually tuberculate, sometimes smooth.

A large genus of 120 species, cosmopolitan, in cold and temperate regions; in Sri Lanka 2 species.

KEY TO THE SPECIES

1 Leaves ovate-elliptic, 0.6–3.0 × 0.4–1.5 cm; petioles of lower leaves 0.5–2.0 cm, those of upper ones very short or leaves almost sessile; petals bifid, almost to base; fruit 6-valved, 5-many-seeded .... **1. S. media**

1 Leaves ovate, 1.8–5.5 × 1.0–3.0 cm, petioles of lower leaves 1.0–4.0 cm, those of upper ones 0.5–1.0 cm long; petals bifid to ± half way down or slightly less; fruit 3-valved or ultimately 6-valved by longitudinal splitting of valves, 1-seeded .... **2. S. pauciflora**

1. **Stellaria media** (L.) Vill., Hist. Pl. Dauph. 3: 615. 1789; Thw., Enum. Pl. Zeyl. 24. 1858; Edgew. & Hook. f. in Hook. f., Fl. Br. Ind. 1: 230. 1874; Alston in Trimen, Handb. Fl. Ceylon 6: 17. 1931; Schischkin in Kom., Fl. URSS 6: 395. 1936; Chater & Heywood in Tutin et al., Fl. Europaea 1: 134. 1964, ed. 2, 1: 161. 1993; Coode in Davis, Fl. Turkey 2: 69. 1967; Matthew, Fl. Tamilnadu Carnatic 1: 79. 1983; Ghazanfar in Nasir & Ali, Fl. Pakistan 175: 22. 1986.

Alsine media L., Sp. Pl. 272. 1753. Lectotype: Described from Europe, Herb. Linn. 388. 1 (LINN).

subsp. **media**

Annual herb, 10–25 cm long, striate, stems terete, rather weak, prostrate to ascending or decumbent with 1 or 2 lines of hairs. Leaves ovate-elliptic, 0.6–3.0 × 0.4–1.5 cm, acute or shortly acuminate, glabrous or ciliate near the base, lower petiolate, upper ones almost sessile, commonly green; petioles of the lower leaves 0.5–2.0 cm long. Inflorescence terminal and axillary, few or many-flowered dichasial cymes, sometimes flowers solitary; peduncles up to 5 cm. Flowers 5 mm across; pedicels slender, 0.8–1.0 cm long, with simple, fine hairs. Sepals ovate-oblong or lanceolate, 3–5 mm long, with scarious margins. Petals linear, shorter than the calyx, 2–2.5 mm, obtuse, deeply bifid, almost to base. Stamens 3–5; anthers globose. Ovary ovoid, 1-locular; ovules many; styles 3-fid, c. 1 mm long. Capsules globose, exceeding the calyx, splitting by 6 valves, nearly to the base. Seeds subreniform, tuberculate.

Distr. Cosmopolitan.

Ecol. A weed of cultivated ground, common locally; fls Dec–Feb.

Notes. Charter & Heywood (l. c.) have recorded two other subspecies, viz. subsp. *postii* Holmboe and subsp. *cupaniana* (Jordon & Fourr.) Nyman, from S. Balkan and Mediterranean regions respectively.

Thwaites (l. c.) has recorded specimen No. *C.P. 3090* from Nuwara Eliya, but I have not been able to locate this specimen in K, BM or PDA herbaria.

Specimens Examined. KANDY DISTRICT: Bogawantalawa, ± 1000 m, 22 Dec 1972, *Cramer, Balasubramaniam & Tirvengadum 3984* (K, PDA). NUWARA ELIYA DISTRICT: Nuwara Eliya, *Thomson s.n.* (K), *s. coll. C.P. 3090* (HAK).

2. **Stellaria pauciflora** Zoll. & Moritzi in Moritzi, Syst. Verz. 30. 1845–46; Briq., Ann. Conserv. Jard. Bot. Geneve 13–14: 378. 1909–1911; Alston in Trimen, Handb. Fl. Ceylon 6: 17. 1931; Backer & Bakh., Fl. Java 1: 208. 1963. Type: Java, *Zollinger s.n.*

Stellaria drymarioides Thw., Enum. Pl. Zeyl. 24. 1858; Edgew. & Hook. f. in Hook. f., Fl. Br. Ind. 1: 229. 1874; Trimen, Handb. Fl. Ceylon 1: 86. 1893. Type: Sri Lanka, Haputale, 915–1220 m, Apr 1856, *Thwaites C.P. 400* (PDA).

Annual herbs; stems decumbent at base, top erect, flaccid, 4-angled, glabrous below, glandular-pubescent above. Leaves broadly ovate, 1.8–5.5 × 1.0–3.0 cm, with intra-marginal vein distinctly remote from the margin, apex acute or apiculate, with a few scattered cilia on margin; petioles of lower leaves 1.0–4.0 cm long, those of upper ones 0.5–1.0 cm long. Inflorescence usually of 3–9-flowered dichasial cymes; peduncles 3.5–5.0 cm long, densely viscid-pubescent. Flowers small, densely pubescent; pedicels 4–5 mm long, thickened at top, viscid-pubescent. Sepals oblong, 3.5–6 mm long, obtuse, glandular-pubescent. Petals white, wedge-shaped, 1.5–3 mm long, 2-fid to halfway down or slightly less. Stamens 5–10. Ovary subglobose, 3-celled; ovules 3; styles c. 2 mm long, filiform. Capsules included, 3-valved becoming 6-valved by longitudinal splitting of valves, 1-seeded.

Distr. Java, Mauritius and Sri Lanka.

Ecol. In shaded, moist places in lower montane zone, 915–1220 m, rare; fls April.

Specimens Examined. BADULLA DISTRICT: Haputale ("Hapootelle") Pass, 915–1220 m, Apr 1856, *Thwaites C.P. 400* (Type of *S. drymarioides* Thw., PDA, holotype; BM, K, isotypes). LOCALITY UNKNOWN: *Walker 83* (K); *s. coll. s.n.* (ACC. No. 000074, HAK).

3. CERASTIUM

L., Sp. Pl. 437, 1753; L., Gen. Pl. 585. 1754; Thw., Enum. Pl. Zeyl. 24. 1958; Benth. & Hook. f., Gen. Pl. 1: 148. 1862; Edgew. & Hook. f. in Hook. f., Fl. Br. Ind. 1: 227. 1874; Trimen, Handb. Fl. Ceylon 1: 85. 1893; Sell & Whitehead in Fedde, Repert. 69: 14–24. 1964. Lectotype species: *C. arvense* L.

Annual or perennial herbs, pubescent, often with glandular hairs. Leaves ovate to lanceolate, elliptic, obovate to oblong, exstipulate. Flowers in terminal cymes, 4–5-merous. Sepals free, margin scarious. Petals emarginate or bilobed, white, rarely absent. Stamens 8 or 10. Ovary 1-celled; styles 3–5. Capsules cylindrical, straight or slightly curved, with 8–10 teeth; teeth flat, revolute or with revolute margins. Seeds compressed, tuberculate.

A large cosmopolitan genus with 60 species; in Sri Lanka 3 species.

KEY TO THE SPECIES

1 Stems flaccid, ascending, viscid; intermediate leaves always more than 4 times longer than wide; cymes glandular-pubescent; fruiting pedicels 15–30 mm long; petals subemarginate, slightly longer than the calyx; capsule teeth erect-recurved with incurved or flat edges . . . .
. . . . **1. C. indicum**
1 Stems not flaccid, erect, sometimes ascending; inter-mediate leaves $1\frac{1}{2}$–4 times longer than wide; cymes with or without glandular hairs; fruiting pedicels 3–15 mm long; petals with 1–1.5 mm deep incision; capsule-teeth straight, with recurved edges
   2 Flowers in terminal heads or clusters; sepals with or without a membranous margin, green central part pubescent, the uppermost hairs protruding beyond tip of sepals; pedicels shorter than the calyx . . . .**2. C. glomeratum**
   2 Flowers in lax cymes; sepals with apical portion entirely membranous, glabrous, the green central part villous with hairs not protruding beyond tip of sepals; pedicels usually longer than the calyx . . . . **3. C. fontanum** subsp. **vulgare**

1. Cerastium indicum Wight & Arn., Prod. 43. 1834; Wight, Ill. Ind. Bot. 1: t. 26. 1840; Thw., Enum. Pl. Zeyl. 24. 1858; Edgew. & Hook. f. in Hook. f., Fl. Br. Ind. 1: 227. 1874; Trimen, Handb. Fl. Ceylon 1: 85. 1893; Gamble, Fl. Pres. Madras 1: 61. 1915; Backer & Bakh., Fl. Java 1: 208. 1963; Ramamurthy in Nair & Henry, Fl. Tamilnadu 1: 22. 1983. Type: Peninsular India, *Wight 149* (K).

Perennial herbs; stems erect-ascending, slender, 30–60 cm, flaccid, viscid, glandular-pubescent. Leaves ± sessile, narrowly-lanceolate, 15–50 × 3–10 mm, acute, covered with viscid pubescence or hairs on both surfaces, middle leaves more than 4 times longer than wide. Flowers few in terminal, dichotomous, glandular pubscent cymes, 4–5-merous, white; bracts without membranous margins; pedicels densely glandular-pubescent, 10–15 mm long, in fruits 15–30 mm long. Sepals oblong-lanceolate, 3–4.5 mm long, acute, margin narrowly membranous or at least membranous at the extreme top. Petals linear-oblong or spathulate, slightly longer than the sepals, subemarginate. Ovary ovoid-oblong; styles 3–5. Capsules ovoid, 4–6 mm long; teeth 10, erect-recurved, with incurved or flat edges. Seeds numerous, c. 1.5 mm diam., dark-brown, muriculate.

Distr. India (S. India), Sri Lanka, and Java.

Ecol. In shaded situations in open forest or grassy slopes in upper montane wet zone, 1800–2400 m, common; fls & frts Mar–Sept.

Specimens Examined. NUWARA ELIYA DISTRICT: Horton Plains, at entrance to Fog Intercept Station, 2135 m, 16 May 1968, *Mueller Dombois 68051601* (PDA), near Farr Inn, 2135 m, 19 Sept 1969, *C.F.& R.J. van Beusekom 1486* (PDA), 2400 m, 4 Nov 1971, *Balakrishnan 1049* (K, PDA), along roadside at Old Farm, c. 2100 m, 14 March 1971, *Balakrishnan 469* (K, PDA); Hakgala, Apr 1920, *Alston s.n.* (PDA).

LOCALITY UNKNOWN: 1830–2135 m, *s. coll. C.P. 2957* (BM, HAK, K, PDA); *Col. Walker 182* (PDA).

2. Cerastium glomeratum Thuill., Fl. Par. ed. 2, 226. 1799; Gamble, Fl. Pres. Madras 1: 61. 1915; Alston in Trimen, Handb. Fl. Ceylon 6: 17. 1931; Sell & Whitehead in Tutin et al., Fl. Europaea 1: 144. 1964; E. Rao in Saldanha, Fl. Karnataka 1: 151. 1984; Ghazanfar in Nasir & Ali, Fl. Pakistan 175: 38. 1986. Type: France: Paris, *Thuiller s.n.* (P?).

Cerastium viscosum L., Sp. Pl. 437. 1753, p. p., nom. ambig.
Cerastium vulgatum L. var. *glomeratum* (Thuill.) Edgew. & Hook. f. in Hook. f., Fl. Br. Ind. 1: 228. 1874; Trimen, Handb. Fl. Ceylon 1: 85. 1893.

Annual, 10–35 cm; stem simple or branched, erect to ascending, with eglandular to glandular hairs. Leaves (lower) obovate to spathulate, upper ones elliptic to elliptic-ovate, 8–20 × 3–10 mm, sessile, with sparse to dense eglandular and glandular hairs, apex obtuse to subacute. Bracts herbaceous or green. Flowers in compact cymose heads or clusters; pedicels shorter than the sepals, 2–3 mm long. Sepals lanceolate, 4–5 mm, acute, with a narrow scarious margin, with glandular hairs and eglandular hairs exceeding the apex. Petals membranous, bifid to 1/3 of the way down, as long as or slightly shorter than the sepals, with a few cilia at base. Stamens 10, free. Styles 5. Capsules 6–10 mm; teeth flat with revolute margins. Seeds small, pale brown, finely tuberculate.

Distr. Cosmopolitan.

Ecol. A weed in the montane wet zone, 1800–2400 m, common; fls & frts Mar–Jan.

Specimens Examined. NUWARA ELIYA DISTRICT: Horton Plains, along Jeep track, 2400 m, 12 Dec 1971, *Balakrishnan 409* (K, PDA); Ohiya Station, c. 2100 m, 15 March 1971, *Balakrishnan 473* (PDA); Nuwara Eliya, 15 Sept 1920, *J.M. Silva s.n.* (PDA); Hakgala, Aug 1881, *s. coll. s.n.* (PDA).

3. Cerastium fontanum Baumg., Enum. Strip. Transs. 1: 425. 1816.

subsp. **vulgare** (Hartm.) Greuter & Burdet, Willdenowia 12: 37. 1982; Jackson in Tutin et al., Fl. Europaea ed. 2, 1: 171. 1993.
Cerastium viscosum L., Sp. Pl. 437. 1753, nom. ambig.
Cerastium vulgare Hartman, Handb. Skand. Fl. ed. 1, 182. 1820. Type: Scandanavia, *Hartman*.
Cerastium triviale Link, Enum. Hort. Berol. 1: 433. 1821, nom. superfl. illegit. pro *C. viscosum* L.
Cerastium vulgatum L. var. *triviale* (Link) Edgew. & Hook. f. in Hook. f., Fl. Br. Ind. 1: 228. 1874. (excl. basionym, quoad pl. ex India).

Cerastium vulgare Hartman subsp. *triviale* Murb., Bot. Not. 1898: 252. 1893. Type: *Fries* Herb. fasc. 10. n. 40 (excl. B).

Cerastium fontanum Baumg. var. *triviale* (Link) Jalas, Arch. Soc. Zool. Bot. Fenn. "Vanamo" 18(1): 63. 1963; Ghazanfar in Nasir & Ali, Fl. Pakistan 175. 39. 1986.

Perennial, 10–50 cm, laxly caespitose with leafy basal shoots; stems erect to ascending, patently hairy and also glandular hairy. Leaves 10–25 × 3–10 mm, lower obovate-spathulate, upper ones elliptic or oblong, apex obtuse or acute, sparsely villous, grassgreen. Bracts herbaceous, upper ones with scarious margins. Flowers in lax cymes; pedicels as long as or slightly longer than the sepals, 3–8 mm long, somewhat longer in fruits, hairy. Sepals 5.5–7 mm, lanceolate, apical portion entirely membranous and glabrous, the green parts villous, with hairs not protruding beyond the tip of the sepal. Petals white, bifid near the apex, shorter than the sepals, 3.5–4.5 mm long, not ciliate at base. Capsules 9–12 mm long; teeth flat, with revolute margins. Seeds small, rounded, tuberculate.

Distr. Widespread in mountainous areas of Europe, Malaysia, India, Pakistan, China, Japan, Korea, Taiwan and Sri Lanka.

Ecol. On grassy, shady areas in submontane wet zone, 900 m, rare; fls & frts Apr–June.

Notes. Jackson (l. c.) has referred to 2 more subspecies, viz. subsp. *lucorum* (Schur) Soo and subsp. *scoticum* Jalas & Sell, from Europe.

Specimens Examined. NUWARA ELIYA DISTRICT: Pidurutala-gala, c. 900 m, 21 May 1971, *Jayasuriya 196* (K); Nuwara Eliya, May 1880, *s. coll. s.n.* (PDA). BADULLA DISTRICT: Palugama, Jan 1888, *s. coll. s.n.* (HAK, PDA).

4. POLYCARPON

Loefl. ex L., Iter. Hisp. 7. 1758; L., Syst. ed. 10, 881. 1759; Benth. & Hook. f., Gen. Pl. 1: 152. 1862; Edgew. & Hook. f. in Hook. f., Fl. Br. Ind. 1: 244. 1874; Trimen, Handb. Fl. Ceylon 1: 87. 1893; Cullen in Davis, Fl. Turkey 2: 95. 1967; Chater & Akeroyd in Tutin et al., Fl. Europaea ed. 2, 1: 184. 1993. Type species: *P. tetraphylla* (L.) L.(= *Mollugo tetraphylla* L.).

Hapalosia Wall. ex Wight & Arn., Prod. 358. 1834; Thw., Enum. Pl. Zeyl. 25.1858. Type species: *H. loeflingiae* Wall. ex Wight & Arn.

Small annual to perennial herbs; stems ascending or erect and dichotomously branched, rough at the angles. Leaves ± petiolate, opposite, often apparently whorled, from the presence of axillary fascicles of leaves; stipules scarious. Flowers small, in cymose clusters, with scarious bracts. Sepals 5, keeled. Petals 5, hyaline, shorter than the sepals, entire or toothed. Stamens 3–5, rarely 1, filaments ± united at the base. Ovary 1-locular; styles short, 3-lobed. Capsules dehiscing by 3 valves almost to the base. Seeds numerous; embryo nearly straight. A genus with 16 species, cosmopolitan; 2 species in Sri Lanka.

KEY TO THE SPECIES

1 Annual or biennial herbs, without woody rootstocks; leaves mostly in whorls of 4, obovate to spathulate, shortly petiolate; inflorescences lax, much branched cymes ....
.... 1. P. tetraphyllum
1 Perennial herbs, with woody rootstocks; leaves apparently in whorls, linear-lanceolate to obovate, sessile, inflorescences congested cymes .... 2. P. prostratum

1. Polycarpon tetraphyllum (L.) L., Syst. ed. 10, 881. 1759; Gamble, Fl. Pres. Madras 1: 64. 1915; Fyson, Fl. Ind. Hill Stns 2: 44. t. 28. 1932; Cullen in Davis, Fl. Turkey 2: 96. 1967; Ramamurthy in Nair & Henry, Fl. Tamilnadu 1: 23. 1983; Chater & Akeroyd in Tutin et al., Fl. Europaea ed. 2, 1: 185. 1993.

Mollugo tetraphyllum L., Sp. Pl. 89. 1753. Type: Italy, Herb. Linn.

subsp. **tetraphyllum.**

Annual herbs, sometimes biennial, without woody rootstock, 15–20 cm high; stems much-branched, ascending or decumbent. Leaves usually in whorls of 4, obovate to spathulate, 1.0–1.7 × 0.4–06 cm, shortly petiolate, glabrous. Inflorescences terminal, spreading, much-branched cymes. Flowers c. 3 mm across, white; pedicels 3–4 mm long, pilose. Sepals lanceolate, 2 mm long, scarious on margin. Petals suborbicular, usually emarginate. Stamens 3–5. Ovary 1-locular; styles 3-lobed. Capsules subglobose, c. 2 mm, dehiscing by 3 valves. Seeds numerous, very small, punctulate.

Distr. In temperate and subtropical regions in Europe, India (Nilgiris) and Sri Lanka.

Ecol. Weed along railway track and roadside, in montane wet zone, abundant; fls & frts Oct–May.

Notes. Chater & Akeroyd (l. c.) have mentioned 2 more subspecies from Europe, viz. subsp. *diphyllum* (Cav.) Bolos & Font and subsp. *alsinifolium* (Biv.) Ball.

Specimens Examined. NUWARA ELIYA DISTRICT: At Ohiya Station, 2100 m, 15 March 1971, *Balakrishnan 472* (K); Ohiya Station by railway siding, 27 Apr 1932, *Simpson 9561* (BM); Hakgala, 19 Feb 1927, *Alston 869* (PDA); Hakgala, on Patanas by roadside, Apr 1920, *Alston s.n.* (PDA); Vicinity of Farr Inn, 11 May 1970, *Gould & Cooray 13829* (PDA). BADULLA DISTRICT: Soraborawewa, Bintenna, 11 Oct 1908, *J.M. Silva s.n.* (PDA).

2. Polycarpon prostratum (Forssk.) Aschers & Schweinf., Oesterr. Bot. Z. 39: 128. 1889; Matthew, Ill. Fl. Tamilnadu Carnatic t. 47. 1982; Matthew,

Fl. Tamilnadu Carnatic 1: 78. 1983; Ghazanfar in Nasir & Ali, Fl. Pakistan 175: 53. 1986.

Alsine prostrata Forssk., Fl. Aegypt-Arab. 207. 1775. Type: Egypt, near Cairo, *Forsskal s.n.* (C).
Hapalosia loeflingiae Wall. ex Wight & Arn., Prod. 358. 1834; Thw., Enum. Pl. Zeyl. 25. 1858. Type: India, *Wallich 6962* (K).
Polycarpon loeflingiae (Wight & Arn.) Benth. & Hook. f., Gen. Pl. 1: 153. 1862, in nota; Edgew. & Hook. f. in Hook. f., Fl. Br. Ind. 1: 245. 1874, nom. illegit.; Trimen, Handb. Fl. Ceylon 1: 87. 1893; Gamble, Fl. Pres. Madras 1: 64. 1915.

Perennial herbs, with many stems arising from a woody rootstock; stems glabrous or pubescent. Leaves apparently in whorls, sessile, lanceolate to obovate, 4–6 × 2 mm, chartaceous, puberulous, base attenuate, margin entire; stipules scarious. Inflorescence axillary, congested cymes. Flowers c. 4 mm across. Bracts ovate, scarious. Sepals 5, ovate, 2.5–3 mm long, thick, keeled, green, with scarious margin. Petals 5, suborbicular, c. 1 mm. Stamens 3; anthers ovoid. Ovary 1-locular; styles 3-fid. Capsules ovoid or subglobose, c. 2 mm, dehiscing by 3 valves. Seeds numerous, oblong, pale-brown.

D i s t r. Africa, tropical Asia, India, Pakistan, Nepal and Sri Lanka.

E c o l. In dry low country in exposed places, fairly common; fls & frts Aug–Jan.

S p e c i m e n s   E x a m i n e d. ANURADHAPURA DISTRICT: Tissawewa, alt. low, 9 Jan 1971, *Jayasuriya & Theobald 40* (PDA), *Jayasuriya 41* (K). TRINCOMALEE DISTRICT: Kantale, "Kantalai", 29 Aug 1885, *s. coll. s.n.* (PDA). POLONNARUWA DISTRICT: Minneriya tank area, 6 Oct 1973, *Sohmer 8197* (PDA); 12 miles E. of Habarane, along road to Polonnaruwa, 100 m, 12 Oct 1974, *Davidse 7481* (PDA). LOCALITY UNKNOWN: *Thwaites C.P. 1090* (BM, K, PDA); in 1839, *Mackenzie s.n.* Herb. Soc. Hort. (K).

5. SPERGULA

L., Sp. Pl. 440. 1753; L., Gen. Pl. ed. 5, 189. 1754; Edgew. & Hook. f. in Hook. f., Fl. Br. Ind. 1: 243. 1874; Alston in Trimen, Handb. Fl. Ceylon 6: 18. 1931; Ratter in Davis, Fl. Turkey 2: 92. 1967; Ratter & Akeroyd in Tutin et al., Fl. Europaea, ed. 2, 1: 185. 1993. Lectotype species: *S. arvensis* L.

Annual or perennial herbs; stems decumbent to ascending, often profusely branching at the base. Leaves with scarious-margined stipules, linear-filiform, bearing fascicles of leaves in axil or at each node, apparently verticillate. Flowers in lax, terminal, paniculate cymes, 5-merous; pedicels recurved after flowering. Sepals free, obtuse, green, with scarious margins. Petals entire, white. Stamens 5–10. Ovary 1-celled. Styles (3–)5; ovules many.

Capsules ovoid to subglobose, dehiscing by 3–5 valves; valves alternating with the sepals. Seeds compressed, often winged.

A small genus with 5 species centred in the temperate region; represented in Sri Lanka by 1 species.

Spergula arvensis L., Sp. Pl. 440. 1753; Boiss., Fl. Orient. 1: 731. 1867; Edgew. & Hook. f. in Hook. f., Fl. Br. Ind. 1: 243. 1874; Gamble, Fl. Pres. Madras 1: 63. 1915; Alston in Trimen, Handb. Fl. Ceylon 6: 18. 1931; Ratter in Davis, Fl. Turkey 2: 93. 1967; Ghazanfar in Nasir & Ali, Fl. Pakistan 175: 47. 1986. Lectotype: Europe, Herb. Linn. 604/1 (LINN).

Annual herbs, 12–30 cm; stems ascending, branched from base upwards, glabrous, ± glandular-pubescent towards the top. Leaves in false whorls, linear-subulate, 10–35 mm, ± fleshy, longitudinally grooved beneath, glabrous or glandular-pubescent; stipules triangular-ovate. Flowers in glandular-pubescent, few to many-flowered cymes; pedicels 6–25 mm, slender, spreading or deflexed, often pubescent. Sepals ovate, 2.5–3.5 mm long (in fruits 3–4.5 mm long), sparsely glandular-hairy. Petals ovalobovate, obtuse, as long as or slightly longer than the calyx, white. Stamens 5–10. Capsules ovoid, slightly longer than the sepals, c. 5 mm long. Seeds subglobose, 1–2 mm, keeled or with narrow wing, densely verrucose.

Distr. Cosmopolitan.

Ecol. A weed of cool climates on loose soil, rare; fls Apr–May.

Notes. Alston (l. c.) has stated that it occurred at Ambawela (Nuwara Eliya District).

Specimens Examined. NUWARA ELIYA DISTRICT: Nuwara Eliya, *Thwaites C.P. 564* (BM, HAK, K).

6. POLYCARPAEA

Lam., J. Hist. Nat. 2: 3. t. 25. 1792, nom. cons.; Thw., Enum. Pl. Zeyl. 25. 1858; Benth. & Hook. f., Gen. Pl. 1: 154. 1862; Edgew. & Hook. f. in Hook. f., Fl. Br. Ind. 1: 245. 1874; Trimen, Handb. Fl. Ceylon 1: 88. 1893. Type species: *P. teneriffae* Lam. (typ. cons.).

Annual or perennial herbs; stems ascending to erect, branched. Leaves opposite or apparently whorled, subulate, linear or spathulate; stipules scarious. Flowers numerous, in effuse or capitate cymes, bisexual. Sepals 5, scarious, often coloured. Petals 5, shorter than the sepals, entire, 2-toothed or with the margins erose. Stamens 5, free or basally cohering with the petals. Ovary 1-locular; style slender, 3-lobed; ovules numerous. Capsules included in the sepals, dehiscing by 3-valves. Seeds obovoid or compressed; embryo usually curved.

A genus of 50 tropical and subtropical species; represented in Sri Lanka by 3 species.

KEY TO THE SPECIES

1 Leaves obovate-spathulate, the basal forming a rosette; stems glabrous; flowers in spikes at the summits of subumbellate cymes .... **1. P. spicata**
1 Leaves linear-lanceolate or linear-subulate, basal ones not rosette-forming; stems woolly; flowers in branched corymbose cymes
  2 Leaves linear-subulate, 12 ×1 mm, less than 6 per node, margin inrolled or convolute; sepals ovate-elliptic, coloured .... **2. P. aurea**
  2 Leaves linear-lanceolate, 25 × 3 mm, more than 8 per node, margin flat; sepals elliptic-lanceolate, white .... **3. P. corymbosa**

1. Polycarpaea spicata Wight & Arn., Ann. Mag. Nat. Hist. ser. 1. 3: 91. 1831; Wight, Ic. Pl. Ind. Or. t. 510. 1840–43; Edgew. & Hook. f. in Hook. f., Fl. Br. Ind. 1: 246. 1874; Willis, Ann. Roy. Bot. Gard. (Peradeniya) 5(3): 167. 1911; Gamble, Fl. Pres. Madras 1: 65. 1915; Alston in Trimen, Handb. Fl. Ceylon 6: 18. 1931. Type: Not designated.

Polycarpaea staticaeformis Hochst. & Steud. ex Fenzl in Endl., Gen. Pl. 961. 1840. Type: Egypt, Herb. Steud. & Hochst. No. 940.

Annual herb with woody tap root 5–10 cm long; stems many, slender, erecto-patent, purplish-brown, glabrous. Leaves obovate-spathulate, somewhat thick, fleshy, 5–15 × 3–5 mm, radical ones forming rosette, cauline leaves apparently fascicled or whorled at the nodes, at the point of branching; stipules lanceolate, lacerate, acuminate, scarious. Flowers sessile, in spikes at the summits of subumbellate cymes; peduncles 1.0–2.5 cm long. Sepals 5, scarious, lanceolate, 2.5–3 mm long, acute, with a broad, brown midrib at the back. Petals oblong, obtuse, much smaller than the sepals. Stamens 5, basally cohering with the ovary forming a ring. Ovary 1-locular; style 3-lobed. Capsules ovoid, c. 2 mm long, included. Seeds small, subtrigonous, shining.

Distr. India, Pakistan, Saudi Arabia, Egypt and N. Australia.

Ecol. On sandy beaches, rare; fls & frts Nov–Jan.

Specimens Examined. JAFFNA DISTRICT: Island off the coast of Jaffna, *Lewis s.n.* (PDA). LOCALITY UNKNOWN: May 1848, *Gay s.n.* (K).

2. Polycarpaea aurea Wight & Arn., Ann. Mag. Nat. Hist. ser. 1, 3: 91. 1839; Gamble, Fl. Pres. Madras 1: 47. 1915; E. Rao in Saldanha, Fl. Karnatic 1: 152. 1984; Singh, Fl. E. Karnatic 1: 146. 1988. Type: India, Bellary (Karnataka State), Herb. *Wight 114* (K).

Polycarpaea corymbosa (L.) Lam. var. *aurea* Wight, Ill. Ind. Bot. 2: 44. t. 110. 1850; Edgew. & Hook. f. in Hook. f., Fl. Br. Ind. 1: 245. 1874; Cooke, Fl. Pres. Bombay 1: 66. 1901.

Perennial herbs 12–16 cm long; stems branched, from the woody rootstocks; densely white-tomentose. Leaves linear-subulate, 4–6 per node, 12 × 1 mm, margin inrolled or convolute. Flowers numerous, small, in dense, much-branched terminal cymes, coloured; pedicels short, densely white-tomentose. Bracts 2. Sepals 5, free, ovate-elliptic, c. 2 mm long, scarious, brown, shining. Petals 5, free, orange or orange-yellowish, emarginate, shorter than the sepals. Capsules included in the sepals, dark-brown. Seeds 4–5, light-brown.

Distr. India (Peninsular India) and Sri Lanka.

Ecol. In rocky crevices on bare slopes in dry region, rare; fls & frts January.

Specimens Examined. POLONNARUWA DISTRICT: S.E. of Giritale wewa, near Circuit Bungalow, 7 miles N.W. of Polonnaruwa, 7 Jan 1970, *Fosberg & Ripley 51952* (PDA).

3. Polycarpaea corymbosa (L.) Lam., Tabl. Enc. 2: 129. 1797; Wight & Arn., Prod. 350. 1834; Wight, Ic. Pl. Ind. Or. t. 712. 1840; Thw., Enum. Pl. Zeyl. 25. 1858; Edgew. & Hook. f. in Hook. f., Fl. Br. Ind. 1: 245. 1874; Trimen, Handb. Fl. Ceylon 1: 88. 1893; Gamble, Fl. Pres. Madras 1: 65. 1915; E. Rao in Saldanha, Fl. Karnatic 1: 153. 1984; Ghazanfar in Nasir & Ali, Fl. Pakistan 175: 52. 1986.

Achyranthes corymbosa L., Sp. Pl. 205. 1753. Lectotype: Ceylon, Herb. *Herman* (BM).

Annual herbs, 10–30 cm, erect to ascending, branching from the base; branchlets white-tomentose. Leaves subsessile, opposite or in false whorls, linear-lanceolate, 10–25×3 mm, 1-nerved, villous or glabrescent; stipules lanceolate, c. 3–4 mm, acuminate, scarious. Flowers numerous, 4 mm across, in compact, terminal, much branched cymes; peduncles to 5 cm long; pedicels to 2 mm long, woolly. Bracts scarious, silvery, bristle-shaped. Sepals 5, lanceolate, 3.5–4 mm long, acuminate, glabrous, entirely scarious, silvery-white. Petals 5, ovate-suborbicular, emarginate, shorter than the sepals. Stamens 5; filaments c. 1 mm. Ovary globose, 1-celled; ovules many. Capsules oblong, dark-brown, included in the sepals, 3-valved; valves with a narrow thickened yellow margin. Seeds small, subreniform, smooth.

Distr. Widely distributed in the tropics of both E. & W. Hemispheres.

Ecol. In the plains in exposed places on gravelly, often poor soil, sea level to 1400 m, very common; fls & frts Dec–May.

Specimens Examined. JAFFNA DISTRICT: About 30 miles S.E. of Jaffna, 20 Feb 1973, *Townsend 73/78* (PDA). MANNAR DISTRICT: Kovitkulum, in 1800, *Jonville s.n.* (BM). VAVUNIYA DISTRICT: South of Mullaittivu, northeast coast, 9 Dec 1970, *Fosberg & Balakrishnan 53513* (K, PDA). PUTTALAM DISTRICT: Wilpattu National Park at Kali Villu, 1 May 1969, *Cooray 69050107* (K, PDA); Tatperiyakulam, 18 March 1927, *Alston 1126* (PDA). TRINCOMALEE DISTRICT: Trincomalee, 27 Aug 1846, *Reynaud* (Herb. J. Gay) *s.n.* (K); Near Foul Point, 20 Nov 1932, *Simpson 9671* (K). COLOMBO DISTRICT: Jaela, Ekkala, 21 Aug 1971, *Cramer 3416* (PDA); Colombo, 12 Apr 1927, *A. D Silva s.n.* (PDA). MATALE DISTRICT: Dambulla, ± 130 m, 18 Feb 1977, *Cramer 4847* (K, PDA); Dambulla Rock, 7 March 1963, *Amaratunga 528* (PDA). POLONNARUWA DISTRICT: Polonnaruwa Govt. Farm, 22 June 1943, *Senarathna 3496* (PDA). BATTIÇALOA DISTRICT: Batticaloa, alt. low, 7 May 1975, *Jayasuriya 2107* (K, PDA). AMPARAI DISTRICT: Arugam Bay, 4 May 1975, *Jayasuriya 2035* (K, PDA); Near Kalmunai, 30 May 1971, *Kostermans 24370* (PDA); Talaryadi sand dunes, 3 May 1931, *Simpson 8013* (PDA). KANDY DISTRICT: Kandy, in 1919, *Moon s.n.* (BM). KALUTARA DISTRICT: Kalutara seashore ('Caltura'), *Macrae 145* (K). NUWARA ELIYA DISTRICT: Elephant Plains, *s. coll. C.P. 2383* (BM, HAK, PDA). MONARAGALA DISTRICT: Galoya National Park, Nilgala, alt. low, 1 May 1975, *Jayasuriya 1959* (K, PDA). HAMBANTOTA DISTRICT: Uraniya near Buttawa, 28 Jan 1968, *D. Mueller-Dombois & Cooray 68012823* (PDA); Ruhuna National Park, Block I, at Kohombagaswala, Plot R39, 21 Jan 1969, *Cooray & Balakrishnan 69012115* (K, PDA), Patanagala area, 8 Dec 1967, *Mueller-Dombois 67120827* (PDA); Yala, Patanagala, sea level, 4 Jan 1971, *Cramer 3301* (PDA); Hambantota, near shore, 30 Dec 1926, *Alston 1127* (PDA). BADULLA DISTRICT: Patana in Fort Macdonald valley, *A. Silva s.n.* (PDA). LOCALITY UNKNOWN: in 1849, *Fraser 185* (K); Apr 1848, *Gay s.n.* (K); *Macrae 66* (BM).

7. DRYMARIA

Willd. ex Schult. in Roem. & Schult., Syst. Veg. 5: 406. 1819; Seringe in DC., Prod. 1: 395. 1824; Thw., Enum. Pl. Zeyl. 25. 1858; Benth. & Hook. f., Gen. Pl. 1: 152. 1862; Edgew. & Hook. f. in Hook. f., Fl. Br. Ind. 1: 244. 1874; Trimen, Handb. Fl. Ceylon 1: 87. 1893; Pax & Hoffmann in Pflanzenfam. ed. 2, 16C: 306. 1934; Mizushima, Jap. J. Bot. 32: 78. 1957; Duke, Ann. Missouri Bot. Gard. 48: 173. 1961; Majumdar, Bull. Bot. Surv. India 10: 293. 1968. Lectotype species: *D. arenarioides* Humboldt & Bonpland apud Willd. ex Roem. et Schult.

Pinosia Urban, Arkiv. Bot. 23A (5): 70. t. 2. 1930. Type species: *P. ortegioides* (Griseb.) Urban (= *Drymaria ortagiodes* Griseb.).

Mollugophytum M.E. Jones, Contr. W. Bot. 18: 35. 1933. Type species: Not designated.

Annual or perennial herbs; stems usually suberect or diffuse. Leaves opposite, sessile to long-petiolate, glabrous or glandular to villose; stipules small, of several bristles. Flowers axillary or terminal, solitary or in dichasial cymes. Sepals 5, free, margin scarious. Petals 5, white, 2–6-fid, often auriculate. Stamens 2–5; filaments slightly connate at base; anthers versatile, 2-celled. Ovary shortly stipitate, 1-locular; style 3-fid; ovules 2-many, on free-central plancentae. Capsules dehiscing by 3 valves to the base. Seeds orbicular, reniform or compressed, usually tuberculate.

A genus of 48 species in the tropical and subtropical regions of the world, chiefly American; represented by 1 species in Sri Lanka.

Drymaria cordata (L.) Roem. & Schult., Syst. Veg. 5: 406. 1819; Mizushima, Jap. J. Bot. 32: 78. 1957; Duke, Ann. Missouri Bot. Gard. 48: 251. f. 18, A–C. 1961.

Holosteum cordatum L., Sp. Pl. 88. 1753. Lectotype: Linn. Herb. No. 109–1.

subsp. **diandra** (Blume) Duke, Ann. Missouri Bot. Gard. 48: 253. f. 18, D-E. 1961; Mizushima, Jap. J. Bot. 38: 150. 1963; Majumdar, Bull. Bot. Surv. India 10: 294. 1983; Matthew, Fl. Tamilnadu Carnatic 1: 76. 1983; Grierson in Grierson & Long, Fl. Bhutan 1(2): 215. 1984.

Drymaria diandra Blume, Bijdr. 62. 1825. Lectotype: after Mizushima, sect. 99, 143–199 (L).
Drymaria cordata Willd. ex Schult. in Roem. & Schult., Syst. Veg. 5: 406. 1819; Thw., Enum. Pl. Zeyl. 25. 1850; Edgew. & Hook. f. in Hook. f., Fl. Br. Ind. 1: 244. 1874, excl. syn.
Cerastium cordifolium Roxb.; Trimen, Handb. Fl. Ceylon 1: 87. 1893.
Alsine nervosum Moon, Cat. n. 23. 1824, nom. nud.
Drymaria extensa Wall., Cat. n. 647. 1829, nom. nud.

Annual herbs; stems prostrate or ascending, striate, internodes longer than the leaves, glabrous to stipitate glandular. Leaves opposite, deltoid-ovate, suborbicular or cordate, 5–25 × 3–20 mm, often apiculate at apex, subtruncate to obtuse at base, 3–7-nerved; petioles 2–6 mm long, usually exceeding the stipules; stipules lacerate, segments filiform, 1–2 mm long. Inflorescences terminal 3-many-flowered cymes; peduncles to 8.5 cm long, glabrous or glandular-puberulent. Flowers 4 mm across; pedicels finely glandular-papillose, up to 8 mm long. Bracts lanceolate, 2–4 mm long. Sepals 5, oblong or elliptic-ovate, 2.5–4 mm long, strongly carinate, 3-veined, inflexed, glandular-papillose on veins. Petals 2.5–3 mm long, bifid up to middle or more; lobes oblong, obtuse, emarginate, base narrow to a linear claw; stamens 2–3, filaments c. 1.5 mm, connate at base; anthers suborbicular. Ovary ovoid to subglobose, 2 mm, 1-locular; style short, 3-fid far beyond the middle; stigma simple. Capsules oblong, 3-valved, c. 2 mm long, 1-2-seeded. Seeds orbicular or reniform, c. 2 mm across, densely tuberculate.

Distr. India, Sri Lanka, Malaysia, Indo-China, Formosa, China, Australia and Tropical Africa.

Ecol. Weed of gardens, plantations and ditches in shady places, common; fls & frts throughout the year.

Vern. Kukulupala (S).

Specimens Examined. KANDY DISTRICT: Pussellawa, Peacock Estate, ± 1050 m, 4 Feb 1975, *Cramer 4582* (PDA); Moray Estate, near trail to Adam's Peak, 15 Nov 1973, *Sohmer & Ariyaratne 8716* (PDA); Peradeniya, 10 Nov 1970, *Fosberg 52703* (K); Near Rangala, 19 July 1931, *Simpson 8359* (BM). BADULLA DISTRICT: Thotulagalla Ridge, 4 Jan 1970, *Fosberg 51828* (PDA). RATNAPURA DISTRICT: Depedene, S. of Ratnapura, 29 Oct 1976, *Fosberg 56622* (K). NUWARA ELIYA DISTRICT: Kotmale, 915 m, 23 June 1972, *Hepper, Maxwell & Jayasuriya 4484* (K, PDA); Rothchild Estate, mile 25 above Pussellawa on Kandy-Nuwara Eliya road, 2 Dec 1970, *Fosberg & Sachet 53210* (PDA); Ramboda, *Thwaites C.P. 1091* (BM, HAK, K, PDA). LOCALITY UNKNOWN: *Col. Walker 181* (K).

8. VACCARIA

Medikus, Philos. Bot. 1: 96. 1789; Gorshkova in Kom., Fl. URSS. 6: 802. 1936; Chater in Tutin et al., Fl. Europaea 1: 186. 1964, ed. 2, 1: 224. 1993; Cullen in Davis, Fl. Turkey 2: 177. 1967; Ghazanfar in Nasir & Ali, Fl. Pakistan 175: 109. 1986. Lectotype species: *V. pyramidata* Medikus.

Annual herbs; stems erect, simple or branched, glabrous. Leaves sessile. Epicalyx absent. Inflorescence a lax panicle, or flowers in dichasia. Calyx-tube inflated below, white, with 5 green wings at the angles, without scarious commissures. Petals 5, long-clawed; coronal scales absent. Stamens 10. Ovary 1-locular; styles 2. Capsules ovoid, with thick, papery exocarp, dehiscing by 4 teeth. Seeds subglobose. A small genus with 4 species in the Mediterranean region and S.W. Asia, India and Pakistan; represented by 1 species in Sri Lanka.

Vaccaria hispanica (Mill.) Rauschert in Fedde, Repert. 73: 52. 1966; Ghazanfar in Nasir & Ali, Fl. Pakistan 175: 109. 1986; Chater in Tutin et al., Fl. Europaea ed. 2, 1: 224. 1993.

Saponaria hispanica Mill., Gard. Dict. ed. 8, in err. 1768. Type: Specimen cultivated in Chelsea Garden.

Saponaria vaccaria L., Sp. Pl. 409. 1753; Edgew. & Hook. f. in Hook. f., Fl. Br. Ind. 1: 217. 1874; Gamble, Fl. Pres. Madras 1: 61. 1915; Alston in Trimen, Handb. Fl. Ceylon 6: 17. 1931. Type: from England.

Annual herbs up to 50 cm high, simple or sparingly branched above, glabrous. Leaves sessile, lanceolate to ovate-lanceolate, 25–60 × 5–20 mm, acute, clasping the stem above, glaucous. Flowers erect, in much-branched dichasia; pedicels slender. Bracts scarious, with green midrib. Calyx 10–16 mm long, cylindric; teeth triangular or ovate, with a scarious margin, apex with 5 prominent green veins and yellowish-green wings between them. Petals pink; limbs 5–10 mm, sometimes exserted, entire or bifid or irregularly dentate above. Capsules subglobose, 8–10 mm, included in the calyx. Seeds subglobose, c. 2 mm diam., black.

Distr. Cosmopolitan.

Ecol. A weed of cultivation, rare.

Notes. This species has been included on the authority of Alston (l. c.) who has recorded that it occurred in Peradeniya as a casual.

9. SILENE

L., Sp. Pl. 416. 1753; L., Gen. Pl. ed. 5, 194. 1754; Rohrb., Monogr. de Gatt. *Silene* 1867; Edgew. & Hook. f. in Hook. f., Fl. Br. Ind. 1: 217. 1874; Schischkin in Kom., Fl. URSS 6: 577. 1936; Chowdhuri, Notes Roy. Bot. Gard. Edinburgh 22: 221–278. 1957; Chater, Walters & Akeroyd in Tutin et al., Fl. Europaea ed. 2, 1: 191. 1993. Type species: *S. anglica* L.

Annual or perennial herbs, usually with woody rootstock. Leaves radical and cauline, of varied shapes. Flowers solitary or in dichasial or paniculate cymes. Epicalyx absent. Calyx tubular, campanular or inflated, 5-toothed, (5–) 10–30-nerved. Petals 5, with narrow claws; limbs entire, bifid or laciniate, usually with 2 basal scales. Stamens 10. Disk usually produced into a long carpophore. Ovary incompletely 3-celled, rarely 1-celled; styles usually 3; ovules many. Capsules with or without a basal septum, opening by 6 teeth, twice the number of styles. Seeds reniform or subreniform, smooth or variously sculptured, sometimes winged or papillose.

A large genus represented by 600 species in the temperate regions of the world, with main concentration in Europe, Asia and N. Africa; represented by 1 species in Sri Lanka.

Silene armeria L., Sp. Pl. 420. 1753; Alston in Trimen, Handb. Fl. Ceylon 6: 17. 1931; Backer & Bakh., Fl. Java 1: 210. 1963; Chater, Walters & Akeroyd in Tutin et al., Fl. Europaea ed. 2, 1: 211. 1993. Type: from England.

Annual or biennial herbs, up to 40 cm; stems simple or branched, glabrous, often viscid below the higher nodes. Basal leaves spathulate, withering early; cauline leaves ovate-cordate to oblong-lanceolate, amplexicaul, decreasing gradually upwards, with ± sinuate margin. Inflorescence usually corymbiform cymes, often multiflorous. Calyx slender, cylindrical-clavate, 12–16 mm long; teeth obtuse. Petal-limbs narrowly obovate-cuneate, 6–8 mm long, emarginate, usually pink; coronal scales lanceolate, acute, of the same colour as the corolla. Stamens 10. Ovary incompletely 3-celled; styles 3. Capsules oblong, 7–10 mm, equalling or somewhat longer than the carpophore.

Distr. Western & Central Europe, Java and Sri Lanka.

Ecol. On moist, loose soil in montane areas, occasional.

Notes. Alston (l. c.) has reported that this species has occurred as a casual at Bandarawela, but I have not seen any specimen from Ceylon, and the description is based on material from Java.

This species is variable in habit, leaf-shape and inflorescence.

CASUARINACEAE

(by B.A. Abeywickrama*)

R. Br. in Flinders, Voy. Terra Austr. 2: 571. 1814 ("Casuarineae"). Type genus: *Casuarina* [Rumph. ex] L.

Conifer-like trees and shrubs. Young shoots and branchlets green or greenish, cylindrical, longitudinally grooved, and jointed. True leaves much reduced and forming a whorl of small scales at each node. Plants monoecious or dioecious. Flowers minute, each with 1 or 2 bracteoles; perianth absent. Male flowers in terminal spikes formed of short, superposed, toothed cups; each flower with a single stamen which is inflexed in bud; anther basifixed. Female flowers in ovoid or globose heads; ovary superior, initially bicarpellary and bilocular but later becoming unilocular through suppression of posterior loculus; style with two long filiform stigmas; ovules 2, collateral, bitegmic. Fructification (fruit-body) a cone with achenes enclosed in the enlarged, coriaceous or woody bracteoles. Seed with a terminal wing; embryo straight, cotyledons flat, endosperm absent.

A family with a single genus distributed mainly in the Australasian region.

CASUARINA

[Rumphius ex] L., Amoen. Acad. 4: 143. 1753.

Characters same as for the family.

About 50 species, of which one is commonly cultivated in Sri Lanka.

Fig. 1. *Casuarina equisetifolia* L. Vegetative features: 1, a portion of stem (×5). 2, Shoot tip (×10). 3, a joint disconnected showing scale leaves, and joint above as seen from below and from one side (×10). Male shoot: 4, male reproductive shoot (×1). 5, male inflorescence (×3). 6, male flowers at different stages of development, a bract and stamen (×10). 7, male inflorescence, entire and with bracts removed (×10). Female Shoot: 8, female reproductive shoot (×1); with base of stem enlarged (×2). 9, a single mature cone (×1). 10, a young female inflorescence with numerous protruding styles (×10). 11, a single female flower with the two bracteoles, and bract separated (enlarged). Mature cone: 12, a single cone (×3). 13, a bract (×5). 14, bracteoles closed (×5). 15, bracteoles slightly opened (×5). 16, fruit (samara) entire (×5). 17, same in L.S. (×5). 18, seed with part of testa removed showing 2 cotyledons (×5).

* Dept. of Botany, University of Colombo.

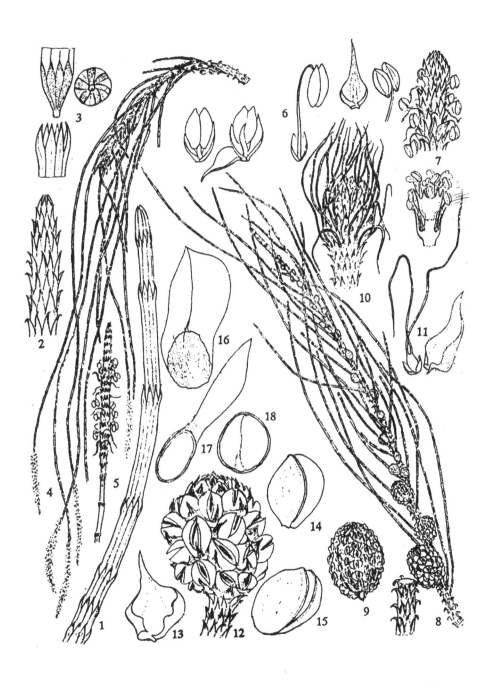

Casuarina equisetifolia L., Amoen. Acad. 4: 143. 1759 as *'equisefolia'* sphalm.; Hook. f., Fl. Br. Ind. 5: 598. 1885; Hook. f. in Trimen, Handb. Fl. Ceylon 4: 120. 1898.—**Fig. 1.**

A large fast-growing tree with straight boles and drooping branches. Stem apices 1–2 mm thick; joints of branchlets about 1 mm or less thick and 5–10 mm long with 6–8 striae on each. Bark brown, rough, and peeling off in vertical strips. Leaves reduced to subulate scales, about 6–8 per whorl, and connate at base to form short sheaths at the nodes. Male flowers in 7–25 mm long spikes which are terminal on branchlets; each flower with a single stamen which is inflexed in bud; anther basifixed. Female flowers in globose or cylindrical heads on short lateral branchlets. Ovary minute; ovules 2, collateral; stigmas 2, long, filiform. Infructescence (fruit-body) cone-like, with the fruits (samaras) enclosed in the enlarged and hardened bracteoles. Seed flattened and with a terminal wing. Endosperm absent. Embryo straight.

Distr. South East Asia from Chittagong southwards and in Australia. Widely cultivated elsewhere in the Tropics.

Ecol. Much cultivated especially in coastal areas and other sandy places.

Uses. A good sand-binder, it is useful for reclamation of sand dunes; its wood has a high calorific value. Sometimes pruned and trained as a hedge-plant.

Vern. Kassa (S), Chavukku (T), Whip Tree (E). ·

Notes. It thrives best on well drained soils on coasts where sand accretion is taking place. It has root nodules with nitrogen-fixing bacteria. In favourable locations it grows to a height of about 30 m and reaches a diameter of about 60 cm at breast height.

Specimens Examined. COLOMBO DISTRICT: Colombo fort, *Amaratunga 2580* (PDA), male, *Abeywickrama s.n.* (PDA); female, *Abeywickrama s.n.* (PDA). NUWARA ELIYA DISTRICT: Nuwara Eliya, *Worthington 5617* (PDA); Hakgala, 4 Nov 1925, *s. coll. s.n.* (PDA), 2 Feb 1926, *s. coll. 138* (PDA).

CELASTRACEAE
(by B.M. Wadhwa*)

R.Br. in Flinders, Voy. Terra Austr. 2: 554. 1814 ("Celastrinae"). Type genus: *Celastrus* L.

Trees, straggling shrubs or climbers. Leaves alternate or opposite, simple, coriaceous, usually entire; stipules absent or present, then minute and caducous. Inflorescence axillary or terminal dichasial cymes, sometimes paniculate or racemose, usually bracteate. Flowers regular, 4– or 5–merous, usually bisexual, sometimes polygamous. Sepals 4 or 5, imbricate or valvate, persistent. Petals 4 or 5, spreading, free, contorted, imbricate, rarely valvate, usually caducous. Disk conspicuous, fleshy, flat or cupular, sinuous or lobed. Stamens 4 or 5, opposite sepals; filaments free, inserted on the margin of the disk or slightly below it. Ovary superior, 2–5-locular, rarely 1-locular, usually wholly or partly immersed in the disk; ovules often 2 in each locule, usually basal; style short, simple; stigma usually lobed. Fruit a loculicidal, globose capsule or an indehiscent, obovoid drupe. Seeds usually arillate.

A large tropical and subtropical family of about 50 genera and about 800 species; in Sri Lanka 9 genera and 16 species.

KEY TO THE GENERA

1 Leaves alternate, fruit a dehiscent capsule
  2 Leaves with cross-bar veins between the nerves; petals contorted; styles 2, free or slightly united at the base ....**1. Bhesa**
  2 Leaves with reticulate venation; petals usually imbricate; styles simple
    3 Scandent dioecious shrubs, always unarmed; ovary usually free from the disk; ovules with a cupular aril at base .... **2. Celastrus**
    3 Small trees or erect shrubs with bisexual flowers, often armed; ovary usually completely immersed in the disk; ovules without cupular aril at the base .... **3. Maytenus**
1 Leaves opposite; fruit a dry or pulpy drupe or a dehiscent capsule
  4 Fruit drupaceous
    5 Disk more or less flat; anthers subglobose, connective not distinct on the dorsal side; ovary usually 2-locular; fruit with terminal, persistent style or its scar ......**4. Cassine**
    5 Disk cupular; anthers ovoid, short apiculate with thick and broad connectives on the dorsal side; ovary usually 1-celled by abortion; fruits with lateral, persistent style ....
    ....**5. Pleurostylia**

* Royal Botanic Gardens, Kew.

4 Fruits capsular
6 Petals slightly connate at the base; disk ± absent; filaments united at the base in a ring, usually united with the petals; fruits usually 1-celled, splitting on one side, usually 1-seeded . . . . 6. Microtropis
6 Petals completely free; disk conspicuous, fleshy, flat or cupular, free from the petals; fruits 3–5-celled, loculicidally dehiscent, 3–many-seeded
7 Ovary 3-celled; ovules 4–8–(16) in each cell; fruits 3-angled or lobed; seeds winged at the apical end, without aril . . . . 9. Kokoona
7 Ovary usually 4–5-celled; ovules 1 or 2 in each cell; fruits 4–5-angled or lobed; seed not winged, ± arillate
8 Petals bifoveolate; ovules 1 in each cell, pendulous . . . . 7. Glyptopetalum
8 Petals efoveolate; ovules 2 in each cell . . . . 8. Euonymus

1. BHESA

Buch.-Ham. ex Arn., Edinburgh New. Philos. J. 16. 315. 1834; Endl., Gen. pl. 1089. 1840; Walp., Rep. 1: 538. 1842 as *"Rhesa"* sphalm.; Ding Hou, Blumea Suppl. 4: 150. 1958; Ding Hou in Fl. Males. ser. 1, 6: 280. 1963. Lectotype species: *B. paniculata* Arn.

Kurrimia Wall. [Cat. n. 4334–4336, 7200. 1831, nom. nud. pro maj. part.; Arn., Nov. Actorum Acad. Caes. Leop.-Carol. Nat. Cur. 18: 328. 1836, nom. nud.] ex Thw., Enum. Pl. Zeyl. 72. 1858, homonym illegit.; Lawson in Hook. f., Fl. Br. Ind. 1: 621. 1875; Loes. in Pflanzenfam. 3, 5: 120. 1892 & ed. 2, 20b: 158. 1942, non *Kurrimia* Wall. ex Meisn., 1837. Type species: *Kurrimia ceylanica* Arn. ex Thw.

Trochisandra Bedd., Fl. Sylv. 1: t. 120. 1871. Type species: *T. indica* Bedd.

Tall, buttressed trees; branchlets terete, their tips enclosed by stipules. Stipules veined lengthwise, usually with colletors at the base on inner side, caducous, convolute in buds. Leaves ovate or ovate-lanceolate, glabrous, coriaceous, entire; midrib & nerves prominent with distinct crossbar veins; petioles long, slightly furrowed above, with knee-like thickening on the upper end. Flowers bisexual in axillary, simple or paniculate racemes, usually 5-merous, bracteate. Bracts small, caducous. Calyx deeply lobed; lobes imbricate. Petals contorted. Disk fleshy, thick, sub-entire or lobed. Stamens inserted on the disk or just beneath the outer margin; filaments subulate; anthers ovate, lengthwise dehiscent, introrse, latrorse or extrorse, basifixed. Ovary ovoid, seated on the disk, 2-celled, usually with a tuft of hairs at the top; styles 2, filiform, free or slightly united at the base; stigma small, capitate. Fruit a leathery, deeply 2-lobed capsule, 1–2-celled, lengthwise dehiscing loculicidally by 2-valves, mostly 1–2-seeded. Seeds large, completely or partly covered by the fleshy aril; albumen abundant, fleshy; cotyledons linear-oblong.

About 6 species in the tropical areas of India (s. & e. India), Bangladesh, Burma, Thailand, Sri Lanka, Malesia (Sumatra, Malay Peninsula, Borneo), Philippines and New Guinea; 2 species in Sri Lanka.

KEY TO THE SPECIES

1 Inflorescences red; fruits cherry red; Leaves when dry yellowish-brown, at sapling stage broadly ovate with subcordate or rounded base, dull grey beneath .... **1. B. ceylanica**
1 Inflorescences white; fruits yellowish; leaves when dry dark brown; at sapling stage narrow oblong, with acute base, glossy green beneath .... **2. B. nitidissima**

1. Bhesa ceylanica (Arn. ex Thw.) Ding Hou, Blumea Suppl. 4: 151. 1958.

Kurrimia ceylanica Arn. [Nov. Actorum Acad. Caes. Leop.-Carol. Nat. Cur. 18: 328. 1836, nom. nud.] ex Thw., Enum. Pl. Zeyl. 72: 1858; Bedd., Fl. Sylv. 1: t. 147. 1871; Lawson in Hook. f., Fl. Br. Ind. 1: 622. 1875, incl. var. *montana*; Loes. in Pflanzenfam. 3, 5: 210. f. 124. 1892 & ed. 2, 20b: 158. f. 43. 1942; Trimen, Handb. Fl. Ceylon 1: 274. 1893 *"zeylanica"*; Worthington, Ceylon Trees 136. 1959 *"zeylanica"*. Type: Sri Lanka: *s. coll. C.P. 1225* (PDA).
Kurrimia ceylanica var. *montana* Thw., Enum. Pl. Zeyl. 72. 1858. Type: Sri Lanka, *s. coll. C.P. 409* (PDA).

Large trees, 25–35 m tall; branchlets terete, glabrous, their buds acute, enveloped in large stipules; bark smooth, rather thick, grey. Leaves ovate-lanceolate or ovate, 10.2–14.4 × 4.1–7.8 cm, glabrous, rounded or subcordate at base, in vivo dull-grey beneath, apex acute, often twisted, margin entire; veins conspicuous, 12–15 pairs on each side; petioles 2.5–3.5 (–4.2) cm long; stipules large, oblong, up to 2.4 cm long, convolute, deciduous. Flowers sessile in red paniculate racemes, 5-merous, bracteate, 5–6 mm across; panicles usually longer than the leaves, sometimes contracted and shorter. Calyx lobes elliptic, 1.5 × 1.0 mm, acute. Petals obovate-oblong, 2.5–3.0 × 1.0–1.5 mm, concave. Disk fleshy, thick, lobed. Stamens inserted in the sinuation of the disk; filaments subulate; anthers ovate, basifixed, extrorse, dehiscing lengthwise. Ovary ± globose, 1 mm in diam., 2-celled with a dense tuft of hairs on top; styles entirely free; stigma capitate. Capsules broadly ellipsoid to suborbicular, up to 3.5 cm long, 1-celled, 1-seeded, fleshy, glabrous, cherry-red. Seed erect, oblong-ovoid, brownish; aril thick, transparent-white, acidic.

Distr. Endemic.

Ecol. In forests of wet zone, chiefly in lower areas, extending to 1750 m, fairly common; fls Feb–May, frts June–Sept.

Vern. Pelang, Uruhonda, Palen, Et-heraliya (S); Konnai (T).

Note. On the sheets *C.P. 1225* in PDA, Kalutara "Caltura", *Gardner*, Sept 1852, Galagamuwa, May 1852; Hantane, Ambagamuwa, *Gardner* and Maturata, July 1851 are annotated.

On the sheets *C.P. 409* in K, Dambulla, Mar 1946, and in PDA, Maturata, Sept 1851 & Sept 1853 are annotated. Sheet *C.P. 2477* in K is labelled as Peninsula India Orientalis.

The panicles in mountain forms are contracted and shorter than the leaves.

Plate No. 422 in PDA, based on specimen *C.P. 409*, belongs to this species.

Specimens Examined. MATALE DISTRICT: Laggala Estate, 1340 m, 19 Feb 1985, *Jayasuriya & Balasubramaniam s.n.* (PDA); Arratene, Madulkele, 27 July 1946, *Worthington 2019* (BM); Campbells Lane, 1220 m, 18 Jan 1975, *Waas 947* (K, PDA). KANDY DISTRICT: Dotulugala, 1460 m, 1 Feb 1975, *Waas 1036* (K, PDA); Madugoda, 640 m, 31 Aug 1944, *Worthington 1610* (PDA); Laxapana, 15 Mar 1975, *Waas & W.D.A. Gunatilake 1187* (K, PDA); Near Fishing Hut above Moray Estate, 1700 m, 14 Nov 1978, *Kostermans 27062, 27068* (PDA); Above Moray Estate, 1000–1600 m, 15 Oct 1972, *Nooteboom 3376* (PDA); Maskeliya, Moray Estate, near Fishing Hut, 27 Mar 1974, *Sumithraarachhi & Jayasuriya 185* (PDA); Between Corbet's Gap and Ferndale, 1280 m, 22 Nov 1977, *Huber 678* (PDA); Ginigathena, 11 Sept 1946, *Worthington 2073* (BM); Rangala to Corbet's Gap, 1000 m, 13 May 1969, *Kostermans 23495* (K); Knuckles, Madulkele area, 900 m, 14 June 1973, *Kostermans 25026* (PDA); Near Pussellawa, Nuwara Eliya road, 21 Nov 1974, *Sumithraarachchi & Davidse 544* (PDA). BADULLA DISTRICT: Passara, 25/11 mile, 5 June 1939, *Worthington 461* (PDA). KALUTARA DISTRICT: Morapitiya, ±300 m, 30 Aug 1976, *Waas 1897* (K, PDA). RATNAPURA DISTRICT: Gilimale forest reserve, along Carney road, 20 July 1970, *Meijer 419* (K, PDA); 69/1 mile marker, between Pelmadulla and Ratnapura, 18 Sept 1971, *Balakrishnan & Jayasuriya 917* (K, PDA); Mile 10, Ratnapura-Carney road, 300 m, 27 Apr 1965, *Kostermans 23314* (K); Marathagala, ±1200 m, 26 June 1976, *Waas 1757* (K, PDA); At Temple, Kalugala, Agalawatte, 150 m, 5 Dec 1975, *Bernardi 15754* (K); Sinharaja forest, Weddagala entrance, 200 m, 5 Oct 1979, *Kostermans 27835* (K, PDA); Below Gongala Cardamom Plantation, field no. 14, 6°23′ N, 80°40′ E, 1050 m, 14 Mar 1985, *Jayasuriya & Balasubramaniam 3384* (PDA); Dodangala near Ratnapura, 15 Aug 1932, *Simpson 9994* (BM, PDA); Gilimale, S. side Adam's Peak, 400–500 m, 17 Sept 1977, *Huber & Nooteboom 3166* (PDA); Ratnapura swimpool, 2 Mar 1948, *Worthington 3613* (PDA); Sinharaja Forest Reserve, March 1982, *de Zoysa s.n.* (PDA), 420 m, *Yapa 07* (PDA), *G. Ranasinghe 43* (PDA), *Kanakratne 09* (PDA), *Kulatunga 28* (PDA). NUWARA ELIYA DISTRICT: Kabaragala Estate, 7°04′ N, 80°45′ E, Cardamom Plantation, field no. 4, 1350 m, 30 Oct 1984, *Jayasuriya & Gunatilleke 3447* (PDA). GALLE DISTRICT: Kanneliya Forest Reserve, 26 July 1971, *Meijer 1034* (PDA), alt. sea level, 28 July 1970, *Cramer 3076* (PDA), 120 m, 6 Apr 1970, *Balakrishnan 270* (PDA); Hiniduma, sea level, 28 July 1970, *Cramer 3090* (PDA), 11 Sept 1974, *Kostermans 25603* (K, PDA), *25588A* (K);

Bambarawana forest, Liyanagamakanda, 17 Oct 1992, *Jayasuriya 6802* (PDA); Goda kande, 300–900 m, 25 Sept 1977, *Nooteboom 3200* (PDA); Pitigalle, alt. low, 10 Sept 1974, *Kostermans 25573* (K, PDA); Naunkitaela in Kaneliya, 9 March 1948, *Worthington 3682* (PDA, BM); Kottawa, 60 m, 24 Oct 1939, *Worthington 636* (BM). MATARA DISTRICT: Dediyagala Reserve, 14 Dec 1991, *Jayasuriya 6012* (PDA), alt. low, 1 July 1975, *Waas 1348* (K, PDA). HAMBANTOTA DISTRICT: Kaduyamrawa, Udawila, 545 m, 20 Apr 1943, *Worthington 1267* (BM). LOCALITY UNKNOWN: *Walker s.n.* (K); *s. coll. C.P. 1225* (PDA, type; CAL, K, isotypes); *Kelaart s.n.* (K); *s. coll. s.n.* (K); *s. coll. C.P. 409* (type of *Kurrimia ceylanica* var. *montana* Thw. PDA, type; BM, K, isotypes); *s. coll. C.P. 2477* (K).

2. Bhesa nitidissima Kostermans, Ceylon J. Sci., Biol. Sci. 12(2): 125, t. 1. 1977. Type: Sri Lanka: Galle district, Hiniduma, 100 m, 11 Sept 1974, *Kostermans 25601* (L).

Large tree, 20–30 m tall, up to 60 cm in diam.; bark deeply cracked, 3 mm thick, brown; live bark whitish; branchlets terete, glabrous. Leaves thinly coriaceous, glabrous, ovate, 7.1–12.0 (14.0) × 3.5–7.0(–8.5) cm, acute or shortly acuminate, base acute, lateral nerves up to c. 12 pairs; leaves in vivo narrowly oblong with acute base, glossy green beneath; petioles 1.5–2.5 cm long; stipules slender, ovate-lanceolate, c. 4 mm long. Flowers sessile in white paniculate racemes, 5-merous, bracteate, 3–4 mm across. Calyx lobes ovate, 1 mm wide, obtuse. Petals oblong-spathulate, 2.0–2.5 × 1.0 mm. Disk fleshy, thick, lobed. Stamens inserted in the sinuation of the disk; filaments subulate; anthers ovate, basifixed, extrorse. Ovary ± globose, c. 1 mm in diam., 1-celled, with a dense tuft of hairs on top; styles free. Capsules ovoid-oblong, up to 2.2 cm long, 1-celled, 1-seeded, glabrous, yellow. Seeds oblong, brownish-green; aril thick.

Distr. Endemic.

Ecol. In wet zone forests mostly at the lower altitudes, extending to 1600 m, fairly common; fls Sept–Oct, frts Nov–Dec.

Specimens Examined. KANDY DISTRICT: Rajamallay forest, Moray Estate, 25 Oct 1975, *Sohmer & Sumithraarachchi 9827* (K, PDA); Batula Oya stream near Fishing Hut, Moray Estate, 1370 m, 25 Oct 1978, *Fosberg 58011* (PDA); SW of Corbet's Gap junction, 1380 m, 17 Nov 1954, *Davidse 8501* (PDA); Nilambe, Hantane Ridge, 700 m, 25 Apr 1969, *Kostermans 23281* (K); Maskeliya Valley, 500 m, 2 May 1974, *Kostermans 24194* (K, PDA); Royal Botanic Gardens, Peradeniya, *Wadhwa 208* (PDA). BADULLA DISTRICT: Passara, 760 m, 20 Dec 1954, *Worthington 6707* (PDA); Thotulagale Estate above Haputale, 1600 m, 7 May 1969, *Kostermans 23376* (BM, PDA). KALUTARA DISTRICT:

Hallawakelle forest, Welipenna, 28 Jan 1972, *Balakrishnan 1173* (K, PDA). RATNAPURA DISTRICT: Between Opanake and Wewelwatta, alt. low, 8 Oct 1977, *Nooteboom 3300* (PDA). KEGALLE DISTRICT: Kitulgala, 2 June 1939, *Worthington 397* (PDA); Roadside near Kitulgala, 200 m, June 1971, *Kostermans s.n.* (K); Pig Ear, 4 May 1966, *Worthington 7134* (PDA). GALLE DISTRICT: Hiniduma, road to Kanneliya forest, 100 m, 11 Sept 1974, *Kostermans 25601* (BM, K, PDA, isotypes); Udugama Rest House compound, 3 May 1951, *Worthington 5268* (PDA); Hiniduma, close to Godegala, 300 m, 13 May 1972, *Cramer, Balasubramaniam & Jayasuriya 3779* (PDA); Gode Kande, E. of Hiniduma, 80°20′ E, 6°18′ N, 100 m, 11 May 1972, *Jayasuriya, Cramer & Balasubramaniam 805* (K, PDA); Hiniduma, near river, 100–150 m, Aug 1974, *Kostermans 25460* (BM, K, PDA); *Kostermans 25725* (BM, PDA), 150 m, 6 May 1979, *Kostermans 26762* (K, PDA), *Kostermans 27625* (K, PDA), *Kostermans 27686* (PDA). LOCALITY UNKNOWN: *Col. Walker s.n.* (K).

2. CELASTRUS

L., Sp. Pl. 196. 1753; L., Gen. Pl. ed. 5, 91. 1754; Thw., Enum. Pl. Zeyl. 72. 1858; Benth. & Hook. f., Gen. Pl. 1: 364. 1862; Lawson in Hook. f., Fl. Br. Ind. 1: 617. 1875; Trimen, Handb. Fl. Ceylon 1: 272. 1893; Loes. in Pflanzenfam. ed. 2, 20b: 131. 1942; Ding Hou, Ann. Missouri Bot. Gard. 42. 227. 1955; Ding Hou in Fl. Males. ser. 1, 6: 233. 1963. Lectotype species: *C. scandens* L.

Celastrus sect. *Eucelastrus* Wight & Arn., Prod. 158. 1834.

Scandent shrubs, sometimes very large, usually dioecious; branchlets terete, glabrous, almost always lenticillate. Leaves alternate, thin or subcoriaceous, elliptic to orbicular, serrate or subentire; stipules small, usually laciniate, caducous. Inflorescences terminal or axillary raceme-like thyrses with cymose-arranged flowers or paniculate cymes. Flowers 5-merous, small, usually light greenish, usually polygamous. Calyx campanulate, persistent; lobes imbricate, sometimes valvate. Petals 5, spreading. Disk usually membranous and cupular or fleshy and flat, entire or 5-lobed, lobes alternate with the stamens. Stamens in male flowers as long as the petals, inserted on the margin of the disk or slightly under the outer margin; filaments smooth or papillose; anthers ovoid or oblong-elliptic, versatile, dorsifixed, latrorse or introrse; in female flowers stamens rudimentary. Ovary in female flowers globose, usually free, 3-celled; style short, usually columnar; stigma usually 3-lobed; in male flowers ovary much smaller, stigma not lobed; ovules 2 in each cell, collateral, erect, with a cupular aril at its base. Capsules 3-celled, 3-valved, usually subglobose or ovoid, crowned by the remnants of persistent style. Seeds 1–2 in each cell, enclosed in a fleshy crimson aril; albumen copious; embryo erect.

About 30 species mostly in the tropical and subtropical areas of eastern Asia, N. America, New Caledonia, Australia and Madagascar; one in Sri Lanka.

Celastrus paniculatus Willd., Sp. Pl. 1: 1125. 1798; Roxb., Fl. Ind. ed. Wall. 2: 388. 1824; Roxb., Fl. Ind. ed. Carey 1: 621. 1832; Wight, Ic. Pl. Ind. Or. t. 158. 1839; Wight, Ill. Ind. Bot. 1: 179. t. 72. 1840; Thw., Enum. Pl. Zeyl. 72. 1858; Lawson in Hook. f., Fl. Br. Ind. 1: 617. 1875; Trimen, Handb. Fl. Ceylon 1: 272. 1893; Gamble, Fl. Pres. Madras 1: 208. 1918; Loes. in Pflanzenfam. ed. 2, 20b: 132. 1942; Ding Hou in Fl. Males. ser. 1, 6: 235. 1963; Matthew, Fl. Tamilnadu Carnatic 1: 255. 1983; Kulkarni, Fl. Sindhudurg 76. 1988. Type: from India (B-W).

Celastrus multiflorus Roxb., [Hort. Beng. 18. 1814, nom. nud.] Fl. Ind. ed. Wall. 2: 389. 1824; Roxb., Fl. Ind. ed. Carey 1: 622. 1832, non Lam., 1785. Type: Nepal, *Wallich n. 4302.*

Celastrus nutans Roxb., [Hort. Beng. 18. 1814, nom. nud.] Fl. Ind. ed. Wall., 2: 390. 1824; Roxb., Fl. Ind. ed. Carey 1: 623. 1832. Type: India, *Wallich n. 4301B* (K; BM, PDA, isotypes).

A large woody climber, up to 10 m; young shoots usually lenticellate, puberulous. Leaves very variable, elliptic to elliptic-oblong, broadly ovate, obovate or sub-orbicular, 4.5–12.0 × 3.0–6.5 cm, glabrous, thin, coriaceous, base cuneate, obtuse or rounded, apex acute to shortly acuminate, margin serrate; midrib elevated on both surfaces, lateral nerves 5–8 pairs; petioles 0.8–1.5 cm long. Panicles terminal, usually branching into compound cymes, (2)5–12 cm long, drooping, usually puberulous when young; peduncles 6–12 mm long. Flowers polygamous; pedicels 1.5–3 mm long (3–6 mm in fruits). Male flowers minute, pale-green. Calyx lobes sub-orbicular, 1.0 × 1.5 mm, toothed. Petals oblong or obovate-oblong, obtuse, entire, 2.5–3.0 × 1.0–1.5 mm, reflexed. Disk cupular, the lobes obscure or slightly triangular. Stamens c. 3 mm long, inserted on the margin of the disk; filaments short, subulate; anthers ovoid, obtuse. Sterile carpels conical, c. 1 mm long. Female flowers: sepals, petals and the disk similar to those of the male flowers; sterile stamens c. 1 mm long; carpels 2–3 mm long; ovary globose, glabrous, 3-celled; ovules 2 in each cell; style short, columnar; stigma 3-lobed. Capsules subglobose, 0.5–1.0 × 0.5–0.8 cm, 3-valved, 3–6-seeded, bright-yellow. Seeds ovoid or ellipsoid, brownish-red, each enclosed in a scarlet-red, fleshy aril.

Distr. India, Pakistan (foothills of the Himalayas), Nepal, Burma, Thailand, Indo-China, S. China, Malaysia, Australia, New Caledonia and Sri Lanka.

Ecol. In moist zone, up to 600 m, not common; fls May.

Uses. Trimen (l. c.) has mentioned that the bark and oil from seeds are used locally for medicinal purposes.

Vern. Duhundu (S).

Note. Plate No. 419 in PDA, based on specimen *C.P. 1232*, belongs to this species.

Specimens Examined. KANDY DISTRICT: Hantane, 610–915 m, *Gardner 173* (BM, K); Kandy, *Macrae 41* (K); Deltota, May 1851, *s. coll. C.P. 1232* (BM, CAL, PDA); Royal Botanic Gardens, Peradeniya, *Alston s.n.* (PDA). LOCALITY UNKNOWN: Aug 1896, *Willis s.n.* (CAL).

3. MAYTENUS

Molina, Saggio Chile 177, 349. 1781 ("1782"), emend Bosc, Nouv. Dict. Hist. Nat. 14: 211. 1803; Loes. in Pflanzenfam. ed. 2, 20b: 134. 1942; Ding Hou in Fl. Males. ser. 1, 6: 238. 1963. Type species: *M. boaria* Molina.

Celastrus sect. *Gymnosporia* Wight & Arn., Prod. 1: 159. 1834.
Gymnosporia (Wight & Arn.) Hook. f. in Benth. & Hook. f., Gen. Pl. 1: 365.
1862; Thw., Enum. Pl. Zeyl. 409. 1864; Lawson in Hook. f., Fl. Br. Ind. 1:
618. 1875; Trimen, Handb. Fl. Ceylon 1: 273. 1893; Loes. in Pflanzenfam.,
ed. 2, 20b: 147. 1942. Type species: *G. montana* (Roth ex Roem. & Schult.)
Benth. (= *Celastrus montana* Roth ex Roem. & Schult.).

Shrubs or small trees, usually armed, spiny, sometimes unarmed; spines ending in short shoots and/or axillary. Leaves alternate, or in fascicles especially on short shoots, variable in shape and texture, margin serrate or entire, generally exstipulate, if stipulate, stipules minute, lanceolate, caducous. Flowers usually bisexual, in axillary, solitary or fascicled dichotomous cymes, sometimes on the uppermost part of spiny branchlets; peduncles slender or stout; pedicels jointed. Sepals 4 or 5, imbricate in bud. Petals 4–5, usually spreading, imbricate in bud. Disk thick, fleshy, flat, sometimes cupular, sinuate or lobed. Stamens inserted on the margin of the disk or slightly below it; filaments slender, broad; anthers ± introrse. Ovary immersed in the disk, 2–3-celled; ovules 2 in each cell, basal; style short, stout; stigma capitate, 2–3-lobed. Capsule subglobose, turbinate or obovoid, 2–3-gonous, loculicidal, 2–6-seeded. Seeds ellipsoid, arillate.

A genus of over 200 species, mostly in the tropical and subtropical areas of Africa and Australia; in Sri Lanka 2 species.

KEY TO THE SPECIES

1 Plants armed, spiny; leaves obovate to oblanceolate, cuneate at base, apex rounded or obtuse,
emarginate, margin usually entire; flowers in condensed, axillary or terminal cymes on short
lateral shoots, often on the spines .... **1. M. emarginata**
1 Plants unarmed, without spines; leaves oblong-spathulate, tapering to short petiole at base,
apex obtuse or retuse, not emarginate, margin crenate-serrate; flowers in corymbose, axillary
cymes .... **2. M. fruticosa**

1. Maytenus emarginata (Willd.) Ding Hou in Fl. Males. ser. 1, 6(2): 241. 1963; Matthew, Fl. Tamilnadu Carnatic 1: 258. 1983; Ramamurthy in Nair & Henry, Fl. Tamilnadu 1: 74. 1983.

Celastrus emarginatus Willd., Sp. Pl. 1: 1128. 1798; Roxb., Fl. Ind. ed. Wall. 2: 387. 1824; Roxb., Fl. Ind. ed. Carey 1: 620. 1832; Wight & Arn., Prod. 1: 160. 1834. Type: from India (WB).

Celastrus montanus Roth ex Roem. & Schult., Syst. Veg. 5: 427. 1819; Roxb., Fl. Ind. ed. Wall. 2: 387. 1824 "montana"; Roxb., Fl. Ind. ed. Carey 1: 620. 1832; Wight & Arn., Prod. 1: 159. 1834; Wight, Ic. Pl. Ind. Or. t. 382. 1840. Type: from India.

Catha montana (Roth) G. Don, Syst. 2: 9. 1832.

Gymnosporia montana (Roth) Benth., Fl. Austr. 1: 400. 1863; Lawson in Hook. f., Fl. Br. Ind. 1: 621. 1875; Gamble, Fl. Pres. Madras 1: 209. 1918; Matthew, Mat. Fl. Tamilnadu Carnatic 171. 1981.

Gymnosporia emarginata (Willd.) Thw., Enum. Pl. Zeyl. 409. 1864; Lawson in Hook. f., Fl. Br. Ind. 1: 621. 1875; Trimen, Handb. Fl. Ceylon 1: 273. 1893; Gamble, Fl. Pres. Madras 1. 210. 1918; Loes. in Pflanzenfam. ed. 2, 20b: 150. 1942; Matthew, Mat. Fl. Tamilnadu Carnatic 171. 1981.

Armed shrubs, 2–5 m tall; young branches sometimes puberulous; spines either from the main branch or apically on short shoots 2–3.5 cm long. Leaves glabrous, chartaceous to coriaceous, obovate to oblanceolate, rarely elliptic-oblong, 2.5–8.0 × 1.5–4.0 cm, apex rounded or obtuse, usually emarginate, margin usually entire, nerves usually obscure; petioles 3.5–8 mm long. Cymes fascicled in the axils of the leaves, or terminal on short lateral shoots, often on the spines, much condensed; peduncle O or very short. Flowers 4 mm across, white, bracteate; bracts deltoid, short fimbriate; pedicels slender, 3.5–5 mm long. Calyx lobes deltoid, acute, sometimes obtuse, slightly erose or short laciniate, c. 1 mm long. Petals obovate-oblong or ovate, 2.0–2.5 × 1.0–1.5 mm, obtuse, entire. Disk fleshy, sinuate-lobed. Stamens usually 2–3 mm long, inserted slightly beneath the disk margin; anthers broadly ovoid, c. 1 mm long, sometimes very small and abortive, slightly apiculate or obtuse. Ovary semi-immersed in the disk, 3-celled, narrowed into a short cylindric but distinct style; ovules 2 in each cell; stigmas 3, distinct, reflexed or obscure. Capsules broadly obovoid or subglobose, 10–12 × 8–9 mm, 3-celled. Seeds ovoid or ellipsoid, 2.5–3.5 × 2–2.5 mm, red; aril fleshy.

Distr. S. India, Sri Lanka, S.E. Asia, Malesia and N. Queensland.

Ecol. In dry region in open scrub forests, common; fls Oct–Jan, frts Dec–Apr.

Note. Plate No. 421 in PDA, based on specimen *C.P. 1235*, belongs to this species.

Specimens Examined. JAFFNA DISTRICT: Keerimilai, Kankesan-thurai, 14 Nov 1970, *Kundu & Balakrishnan 659* (PDA); S.E. Point Pedro near Kaddaikadu, 16 March 1973, *Bernardi 14253* (PDA); Jaffna, in 1846, *Gardner C.P. 1235* (BM, CAL, K, PDA); Point Pedro, 12 March 1932, *Simpson 9275* (BM). MANNAR DISTRICT: Mannar Island, Feb 1890, *s. coll. s.n.* (PDA); Parayalankulam, alt. sea level, 22 March 1973, *Cramer s.n.* (K). VAYU-NIYA DISTRICT: 113th mile post Medawachchiya-Mannar, 80° 16′ E, 8° 41′ N, alt. low, 22 Jan 1972, *Jayasuriya, Balakrishnan, Dassanayake & Bala-subramaniam 584* (PDA). ANURADHAPURA DISTRICT: Ritigala, 11 Nov 1974, *Waas 327* (K, PDA); Wilpattu National Park, 1/4 mile S. E. of Mardan Maduwa, 28 Dec 1968, *Fosberg, Mueller-Dombois, Wirawan, Cooray & Balakrishnan 50751* (K); Near Thulankudah, 4 Jan 1969, *Fosberg, Mueller-Dombois, Wirawan, Cooray & Balakrishnan 51001* (K); Borupanwila, alt. low, 24 Jan 1974, *Jayasuriya & H.N. & A.L. Moldenke 1430* (PDA); 1 mile E. base of Ritigala, 6 March 1971, *Wheeler 12406* (PDA). TRINCOMALEE DISTRICT: Uppuveli, alt. sea level, 28 March 1977, *Cramer 4933* (PDA); Somapura, alt. sea level, 2 Feb 1978, *Cramer 5116* (K, PDA). MATALE DISTRICT: Laggala, beside Dikpatana-Illukkumbura road, +900 m, 25 Nov 1977, *Cramer 5005* (K); Milepost 37 between Laggala and Illukkumbura, 1000 m, 10 Feb 1971, *Jayasuriya 281* (PDA); Dikpatana, on road from Rattota to Illukkumbura, N.E. Knuckles Mts, 600 m, 5 Sept 1980, *Kostermans 28713* (K), Mile 38, Rattota to Illukkumbura road, 920 m, 13 Aug 1978, *Huber 721* (PDA). POLON-NARUWA DISTRICT: Polonnaruwa Sacred Reserve, 61 m, 4 Aug 1969, *Ripley 160* (PDA). COLOMBO DISTRICT: Henaratagoda (Gampaha) Garden, 1 Feb 1918, *Petch s.n.* (PDA). KANDY DISTRICT: Royal Botanic Gardens, 450 m, 3 Oct 1979, *Kostermans 27823* (K, PDA), *Wadhwa & Weerasooriya 209, 210* (PDA), *Kostermans 27803* (PDA). BADULLA DISTRICT: Patana, on way to Fort Macdonald, Oct 1896, *A.M. Silva s.n.* (PDA); Lunugala, Jan 1888, *s. coll. s.n.* (PDA). KALUTARA DISTRICT: Kalutara, *Gardner 174* (BM). RATNA-PURA DISTRICT: Sri Palabaddala, Mapalena ella, 19 Feb 1974, *Waas 432* (K). HAMBANTOTA DISTRICT: Ruhuna National Park, Andunoruwawewa, 3–5 m, 2 Apr 1968, *Fosberg & Mueller-Dombois 50141* (K, PDA), Block II, 19 June 1967, *Comanor 392* (PDA), Block I, N. of Buttuwa, 10 Dec 1967, *Mueller-Dombois & Cooray 67121038* (PDA), Patanagala Beach area, 4–6 m, 29 Jan 1968, *Comanor 895* (PDA); Block II, 30 Aug 1967, *Mueller-Dombois & Comanor 67083009* (PDA); Bundala Sanctuary, 10 m, 8 Mar 1973, *Bernardi 14162* (PDA); Near Hambantota salterns, 26 Nov 1974, *Sumithraarachchi & Davidse 577* (PDA). LOCALITY UNKNOWN: *Kelaart s.n.* (K).

2. Maytenus fruticosa (Thw.) Loes. in Pflanzenfam. ed. 2, 20b: 140. 1942.

Catha fruticosa Thw., Enum. Pl. Zeyl. 72. 1858. Type: Sri Lanka: Kandy District, Nawalapitiya, *Thwaites C.P. 3386* (PDA; BM, isotype).

Gymnosporia fruticosa (Thw.) Thw., Enum. Pl. Zeyl. 409. 1864; Lawson in Hook. f., Fl. Br. Ind. 1: 619. 1875; Trimen, Handb. Fl. Ceylon 1: 273. 1893.

Unarmed shrubs, up to 3 m tall; young branches terete, glabrous. Leaves spathulate-oblong, tapering to short petiole, 2.0–4.5 × 1.3–1.8 cm, thick, glabrous, apex obtuse or retuse, margin crenate-serrate except at base, prominently reticulately veined on the lower surface; petioles 3–5 mm. Cymes axillary, corymbose, many-flowered, c. 3.5 cm long. Flowers yellowish-white, 2.5–3.0 mm across; pedicels slender, filiform, 6–9 mm long. Calyx lobes minute, ovate, 1 mm broad, obtuse. Petals ovate, 1.5–2.0 mm wide, obtuse. Disk fleshy. Stamens arranged on the margin of the disk, 1.5 mm long; anthers ovoid, c. 1 mm long. Ovary semi-immersed in the disk, 3-celled, with 2 ovules in each cell; style small; stigma 3-lobed, recurved. Capsules obovoid, apiculate, 5–7 × 3–5 mm, 3-celled, faintly transversely striate, reddish-brown or crimson. Seeds globular, shining, scarlet, covered partly by white aril.

Distr. Endemic.

Ecol. Along river Mahaveli in wet montane region, rare; fls Oct, frts Mar.

Note. On the type sheets in PDA, Ambagamuwa has been given as locality, while Nawalapitiya as locality has been given both by Thwaites (l. c.) & Trimen (l. c.).

Plate No. 420 in PDA, based on specimen *C.P. 3386*, belongs to this species.

Specimens Examined. KANDY DISTRICT: Nawalapitiya, along Mahaveli river, Mar & Oct 1855, *Thwaites C.P. 3386* (PDA, holotype; BM, isotype).

4. CASSINE

L., Sp. Pl. 268. 1753; Loes. in Pflanzenfam. ed. 2, 20b: 176. 1942; Ding Hou in Fl. Males. ser. 1, 6: 284. 1963; Kostermans, Gard. Bull. Singapore 39(2): 177. 1987. Lectotype species: *C. peragua* L.

Elaeodendron Jacq. f. ex Jacq., Ic. Pl. Rar. 1: t. 48. 1782; Murray, Syst. ed. 14, 241. 1784 as *"Elaeodendrum"*; Jacq. f., Nov. Act. Helvet. 1: 36. 1787; Thw., Enum. Pl. Zeyl. 73. 1858; Benth. & Hook. f., Gen. Pl. 1: 367. 1862; Lawson in Hook. f., Fl. Br. Ind. 1: 623. 1875; Trimen, Handb. Fl. Ceylon 1: 271. 1893; Loes. in Pflanzenfam. ed. 2, 20b: 172. 1942. Type species: *E. orientalis* Jacq. f.

Shrubs or trees. Leaves opposite, subopposite or alternate, subcoriaceous, entire or crenulate; stipules small, caducous. Cymes axillary or extra-axillary, distinctly peduncled. Flowers bisexual, sometimes unisexual, 4–5-merous. Calyx 4–5-lobed; lobes imbricate, spreading. Petals 4–5, imbricate, spreading.

Disk fleshy, flat, entire to lobed. Stamens inserted at or under the margin of the disk; filaments subulate; anthers subglobose, versatile, introrse. Ovary usually partly immersed in the disk, sometimes slightly united at base with the disk, short, conical, 2-celled; ovules 1 or 2 in each cell, basal; style very short; stigma 2-lobed, sometimes obscure. Fruit a dry or fleshy, indehiscent drupe, 1-2-celled. Seeds 1-3, exarillate.

About 40 species in the tropical areas of Asia, America, Australia and S. Africa; 3 in Sri Lanka.

KEY TO THE SPECIES

1 Drupes narrow ellipsoid, 1-1.2 cm long, juicy, glossy white; putamen thin, crustaceous, spindle-like, conspicuously pointed at both ends; bark smooth, greyish or greyish-white, yellowish on the inner side .... 1. C. glauca
1 Drupes globose, subglobose or ellipsoid globose, 1.8-3.0 cm diam., dry, green; putamen thick, very hard, woody, sharply pointed; bark smooth or rough, grey, brown to yellowish-brown
2 Drupes globose or ellipsoid-globose, up to 2.4 cm diam., smooth; putamen with 2 longitudinal lines on each side, slightly compressed; bark smooth, grey to light brown, often lenticellate; trees c. 20 m tall .... 2. C. balae
2 Drupes subglobose, up to 3 cm diam; putamen longitudinally wavily grooved with appressed fibres; bark rough, brown to yellowish-brown, pustular; trees c. 30 m tall ....
.... 3. C. congylos

1. Cassine glauca (Rottb.) Kuntze, Rev. Gen. Pl. 1: 114. 1891; Ding Hou in Fl. Males. ser. 1, 6(2): 286. 1963, p.p.; Kostermans, Gard. Bull. Singapore 39(2): 181. 1986 (1987).

Mangifera glauca Rottb., Nye Saml. Kongel. Danske Vidensk. Selsk. Skr. (Nov. Act. Haffn.), Anden Deel 2: 533-555. t. 4, f.1. 1783 (excl. *Melmedya*); Roxb., Fl. Ind. ed. Carey 1: 639. 1832 (as syn. of *Elaeodendron glaucum*); Lawson in Hook. f., Fl. Br. Ind. 1: 623. 1875 (as syn. of *E. glaucum*); Ding Hou in Fl. Males. ser. 1, 6(2): 286. 1963 (as syn. of *Cassine glauca*). Type: Sri Lanka, *Koenig s.n.* (C).

Celastrus glaucus (Rottb.) Vahl, Symbol. Bot. 2: 42. 1971 (non R. Br.), excl. descript; Moon, Cat. 17. 1824; Thw., Enum. Pl. Zeyl. 73. 1858 (as syn. of *Elaeodendron glaucum*); Trimen, Handb. Fl. Ceylon 271. 1893, p.p. (as syn. of *E. glaucum*); Alston in Trimen, Handb. Fl. Ceylon 6: 42. 1931, p.p. (as syn. of *E. glaucum*).

Elaeodendron glaucum (Rottb.) Pers., Syn. 241. 1805; Thw., Enum. Pl. Zeyl. 73. 1858, excl. var. *montanum*; Roxb., fl. Ind. ed. Carey 1: 638. 1832, p.p.; Lawson in Hook. f., Fl. Br. Ind. 1: 623. 1875, pro minime parte; Trimen, Handb. Fl. Ceylon 1: 271. 1893, p.p.; Worthington, Ceylon trees 135. 1959; Weeratunga, Kumar, Sultanbawa & Balasubramaniam, Proc. Sri Lanka Assoc. Adv. Sci. Sect. E, 74. 1983.

Small to medium sized tree, glabrous in all parts; bark smooth, greyish-white, yellowish on the inner side; live bark red. Leaves opposite, thin, coriaceous, glabrous, shining above, glaucous beneath, oblong or ovate-oblong, 3.0–7.6(–9.5) × 2.2–4.5(–5.5) cm, apex acute or subacuminate, base acute; margin crenate-serrate; young leaves conspicuously acuminate; petioles 1–1.5 cm long. Panicles axillary, lax, dichotomously branched, 7–8 cm long, many-flowered, on very slender and thin peduncles; peduncles 5.5 cm long. Flowers 4–5 mm across, white to whitish-green; pedicels 5–6 mm long. Calyx lobes rounded, c. 1.5–2 mm broad, margin membranous. Petals ovate-oblong, 2.5–3.5 × 1.5 mm, spreading, margins incurved. Disk thick, cushion-like, sinuate. Stamens inserted below the disk; filaments 1.5 mm long, recurved. Ovary 2-celled, adnate to the disk; style very short, persistent. Fruit a drupe, narrow-ellipsoid, 1.0–1.2 cm long, 1-celled, 1-seeded, glossy porcelain white, with soft, juicy, sweetish mesocarp, and thin, crustaceous, spindle-like, pale yellowish putamen conspicuously pointed at both ends.

Distr. Endemic.

Ecol. In open forests of dry zone, fairly common; fls May–July, frts July–Oct.

Uses. The leaves have a sternutatory and fumigatory action. They are also used as a snuff to relieve headaches. The gum from the tree dissolves in water to give an adhesive (Kostermans, l. c.).

Note. The slender, elongate, glossy, porcelain white, juicy fruitlets of *Cassine glauca* with slender spindle-like, thin putamens, gradually strongly pointed at both ends, make it easily recognizable from other species.

Specimens Examined. MANNAR DISTRICT: Near Mannar, on Madhu road, alt. low, 28 May 1973, *Kostermans 24887* (BM, K). VAVUNIYA DISTRICT: Vicinity of Nanthikadal lagoon, alt. low, 8 July 1971, *Jayasuriya 233* (K). PUTTALAM DISTRICT: Manampuri, alt. sea level, 20 May 1976, *Cramer 4665* (CAL). ANURADHAPURA DISTRICT: Andiyagala, 10 July 1965, *Amaratunga 963* (PDA). TRINCOMALEE DISTRICT: Trincomalee, sandy beach forest of Nilaveli Hotel, Mar 1980, *Kostermans 28310* (K, PDA); Beach near Trincomalee, alt. sea level, Apr 1979, *Kostermans 27719a* (K). MATALE DISTRICT: Kandalama-Dambulla, 4 May 1987, *Jayasuriya 3875* (PDA). POLONNARUWA DISTRICT: Attanakadawala-Jamburawala road, alt. low, 6 Oct 1971, *Jayasuriya 323* (K). LOCALITY UNKNOWN: *Thwaites C.P. 1227* (BM, CAL. K).

2. Cassine balae Kostermans, Gard. Bull. Singapore 39(2): 185. 1986 (1987). Type: Sri Lanka: Hambantota, along coast, *Balasubramaniam 2213* (L).

Elaeodendron glaucum Auct.: Roxb., Pl. Corom. 2: 2. 1798; Roxb., Fl. Ind. ed. Carey 1: 638. 1832 (excl. *Schrebera albens* Retz., *Celastrus glaucus* Vahl & *Mangifera glauca* Rottb.); Voigt, Hort. Suburb. Calcut. 167. 1834; Lawson in Hook. f., Fl. Br. Ind. 1: 623. 1875, pro minime parte quo ad cit. Ceylon; Trimen, Handb. Fl. Ceylon 271. 1893, p.p. (excl. *Schrebera albens* Retz., *Celastrus glaucus* Vahl, *Elaeodendron roxburghii* Wight & Arn.); Ding Hou in Fl. Males. ser. 1, 6(2): 286. 1963, non Pers.

Elaeodendron balae Kostermans ex Weeratunga, Kumar, Sultanbuwa & Balasubramaniam, J. Chem. Soc. Perkin Trans. 1: 2457. 1982, nom. invalid.-without latin diagnosis.

Tree, up to 20 m tall, glabrous in all parts; branchlets slender, terete, smooth; bark grey to light brown, often lenticellate, live bark red. Leaves opposite, subcoriaceous, obovate-elliptic to obovate-oblong, 4.0–6.5(–9.0)× 2.2–3.5 (–5.0) cm, with an obtuse mucro, generally bent down, or obtuse, base gradually cuneate, margin shallowly remotely serrate; midrib above thin, subimpressed, prominent on the lower surface, with generally 5–6 pairs of erecto-patent veins, arcuately merging before the margin, sometimes the apical veins running out, connecting by a loose, irregular reticulation; petioles thin, glossy, 1.0–2.0 cm long, becoming flat above. Stipules scaly, very minute, caducous. Panicles áxillary, repeatedly dichotomous, up to 6 cm long, few to many-flowered, on slender peduncles and branchlets, thicker than those of *C. glauca*. Flowers green or whitish-green, 5–7 mm in diam., pedicels filiform, 1.0–1.2 cm long. Calyx lobes 5, roundish, 1–1.5 mm across, concave. Petals 5, oblong, 2–2.5 mm long, obtuse or subacute. Disk fleshy, 5-lobed. Stamens 5; filaments short, inserted on the disk, at first erect, later curving down to below the calyx. Ovary 2-celled, immersed in the disk, with 2 ovules in each cell attached to the bottom of the cell; styles very short, conical; stigma simple. Drupes globose or ellipsoid-globose, 1.8 × 2.4 cm, shortly sharp pointed, 1-celled, smooth, green; putamen with 5 mm thick, hard woody wall, with 2 longitudinal lines on each side, slightly compressed.

Distr. Endemic.

Ecol. In dry zone in coastal and inland savannah vegetation.

Vern. Nareloo, Neraloo (S), Perun, Piyaree (T).

Note. Kostermans (l. c.) states "In North-east Sri Lanka on dunes and sandy coastal areas, this occurs as a many short boled bushy shrub with leaves often large, sharply prickly serrate, and is called in Tamil "Kurativa or Kuruka vaichchi". It is this which Roxburgh records as Ceylon Tea, under which name it was sent from Ceylon to Calcutta Botanic Garden by General McDowall."

Specimens Examined. MANNAR DISTRICT: Mannar Island, Feb 1890, *s. coll. s.n.* (PDA). VAVUNIYA DISTRICT: Kokkavil, 28 June 1931,

Simpson 8280 (BM). ANURADHAPURA DISTRICT: Wilpattu National Park, Plot W23, Way to Kolankanatta beach, 3 July 1969, *Wirawan, Cooray & Balakrishnan 955* (K, PDA), *Ripley 222, 467* (PDA), *Hladik 788* (PDA). TRINCOMALEE DISTRICT: Uppuveli, alt. sea level, 29 March 1977, *Cramer 4938* (K, PDA); Trincomalee, 30 m, 24 May 1947, *Worthington 2793* (BM). POLONNARUWA DISTRICT: Alut Oya, 180 m, 21 Apr 1947, *Worthington 2703* (BM); Polonnaruwa Ruin area, 30 Apr 1971, *Kostermans 24309* (K, PDA); Polonnaruwa Sacred area, 61 m, 23 July 1968, *Ripley 67* (PDA). BATTICALOA DISTRICT: Kalkudah, Vandeloos point, 7 Dec 1942, *Worthington 1238* (BM); Mankeni, 29.5 miles N. of Batticaloa, 18 Aug 1950, *Worthington 4894* (PDA); Batticaloa near Unichchai tank, 14 Aug 1967, *Mueller-Dombois 67081406* (PDA); Kalkudah Rest House compound, c. 60 m from sea, 9 June 1974, *Waas 639* (K, PDA). KANDY DISTRICT: Royal Botanic Gardens, Peradeniya, *Wadhwa 211* (PDA), *Kostermans 27409, 27793* (K, PDA). BADULLA DISTRICT: Jamburagala, 4 Jan 1969, *Fosberg, Mueller-Dombois, Wirawan, Cooray & Balakrishnan 51012* (PDA). AMPARAI DISTRICT: Wellawaya, +100 m, 5 Aug 1947, *Worthington 2992* (BM, K, PDA); along road in Gal Oya National Park, Inginiyagala, 26 June 1970, *Meijer & Balakrishnan 162* (PDA); Nuwaragala forest reserve, Divulana, alt. low, *Jayasuriya 2085* (K); Kumana, Kumbukkan Oya river mouth, 2 May 1975, *Jayasuriya 2016* (K). MONERAGALA DISTRICT: Near Wellawaya, mile 187, 28 June 1970, *Meijer 199* (K); Yala, 13 June 1965, *Worthington 7107* (PDA). HAMBANTOTA DISTRICT: W. of Kirinda, Estuarine fringing forest, 12 Aug 1932, *Simpson 9943* (BM); Buttawa, Ruhuna National Park, 28 Feb 1952, *Worthington 5681* (PDA); Tissamaharama-Kataragama road, *Cooray s.n.* (PDA); Ruhuna National Park, beach E. of Buttawa-modera, 6 Apr 1968, *Fosberg 50333* (K, PDA), Behind Yala Bungalow, Plot R26, 30 Aug 1968, *Mueller-Dombois & Cooray 68083002* (PDA), Padikema, Block I, *Mueller-Dombois 69010808* (PDA), Behind Bungalow, Block I, 29 May 1968, *Cooray 68052909* (PDA), Near Buttawa, Block I, 10 Dec 1967, *Mueller-Dombois & Cooray 67121021* (PDA), Mayagala, 10 Km from Talgasmankada, 7 Sept 1992, *Jayasuriya 6696* (PDA), Yala, dune in plot R26, Block I, 18 Jan 1969, *Cooray 69011801* (K, PDA), Pilinnawa, 9 Sept 1992, *Jayasuriya 6715* (PDA), *Nooteboom 3332* (PDA), Block I, Yala Bungalow, 28 Oct 1968, *Wirawan 681* (K, PDA), Buttawa beach area, plot R29, 30 Jan 1968, *Comanor 899* (K, PDA). LOCALITY UNKNOWN: *Thwaites C.P. 1227* (BM).

3. Cassine congylos Kostermans, Gard. Bull. Singapore 39(2): 187. 1986 (1987). Type: Sri Lanka: Matale District, Knuckles Mts., at base of Dusingalle near Kalupahane peak, Aug 1973, *Kostermans 27791* (L).

Elaeodendron glaucum var. *montanum* Thw., Enum. Pl. Zeyl. 73. 1858; Lawson in Hook. f., Fl. Br. Ind. 1: 623. 1875; Trimen, Handb. Fl. Ceylon 1: 272. 1893. Type: Sri Lanka, *Thwaites C.P. 2520* (PDA; BM, CAL, K, isotypes).

Cassine glauca var. *montana* (Thw.) Pierre, Fl. Cochinch. 4. t. 196. 1893.

Large trees, up to 30 m tall, all parts glabrous; bark rough, brown to yellowish-brown, pustular, c. 1 mm thick, underneath orange, inflammable (burns like kerosene); live bark red, 10 mm thick. Leaves opposite, rigidly coriaceous, broadly elliptic, 3.5–6.5 × 2.5–4.0 cm, apex ± mucronate, base acute or obtuse, margins rather remotely serrate; midrib and lateral nerves 5–7 pairs, ± patent, laxly reticulated; petioles slender, 0.5–1.5 cm long, concave above. Panicles axillary, up to 7 cm long, few-flowered, with few, remote, slender branchlets up to 1 cm long. Flowers green, bracteate; pedicels slender, 3 mm long. Calyx lobes orbicular, 2.5 mm wide. Petals oblong, c. 5 mm long, 2–2.5 mm wide, green, margin lighter green. Disk large, thick, c. 3 mm diam., slightly lobed. Stamens inserted in the incision of the disk; filaments rather broad, 1.0–1.5 mm long, reflexed. Ovary immersed in the disk, 2-celled; style short, conical; stigma ± obscure. Fruit a drupe, subglobose, up to 3 cm diam., with short sharp point, green at maturity, 1-celled (by abortion); mesocarp none; putamen very hard, thick, sharply apiculate, longitudinally wavily grooved with appressed fibres.

Distr. Endemic.

Ecol. In wet zone forests, up to 1000 m, rare; fls June–Aug, frts Aug.

Note. Lawson (l. c.) erroneously gives the distribution as the drier zone of Sri Lanka; it is a wet zone plant.

Specimens Examined. KANDY DISTRICT: Knuckles Mts, Madulkele area, +1000 m, 12 June 1973, *Kostermans 25070* (BM, K, PDA); Knuckles Mts, Perera Cardamom Estate, at base of Dusingalle, near Kalupahane peak, Aug 1973, *Kostermans 27791* (K, isotype); Deltota, in 1853, *Thwaites C.P. 2520* (BM, CAL, K, PDA). NUWARA ELIYA DISTRICT: Dimbula, 1515 m, Apr 1852, *Thwaites C.P. 2520* (PDA).

5. PLEUROSTYLIA

Wight & Arn., Prod. 1: 157. 1834; Thw., Enum. Pl. Zeyl. 71. 1858; Benth. & Hook. f., Gen. Pl. 1: 363. 1862; Lawson in Hook. f., Fl. Br. Ind. 1: 617. 1875; Trimen, Handb. Fl. Ceylon 1: 270. 1893; Loes. in Pflanzenfam. ed. 2, 20b: 180. 1942. Lectotype species: *P. wightii* Wight & Arn.

Shrub or small tree. Leaves opposite, ± entire, subcoriaceous to coriaceous, usually exstipulate, if stipules present very small, caducous. Flowers in axillary, few-flowered cymes, bisexual, usually 5-merous. Calyx 5-lobed;

lobes imbricate. Petals imbricate. Disk thick, fleshy, cupular. Stamens 5, inserted below the margin of the disk; filaments shorter than the petals, flat; anthers ovoid or short apiculate, sub-basifixed, introrse, with thick and broad connectives on the dorsal side. Ovary conical or flask-shaped, nearly half-immersed in the disk or slightly united with the disk at the base, 2-celled or usually only 1-celled by abortion; style short, thick, slightly dilated at the top; stigma broadly peltate; ovules usually 2 in each cell, collateral. Fruit an indehiscent drupe or nut, 1-celled, with a prominent, hardened, persistent lateral style; exocarp thin, coriaceous; endocarp thin, crustaceous, irregularly ridged. Seeds 1 or 2, erect, covered by aril-like endocarp; albumen fleshy, copious.

About 6 species in the tropics and subtropics of Africa, Madagascar, Mascarenes, Sri Lanka, Indo-Malaysia, Queensland and New Caledonia; 1 in Sri Lanka.

Pleurostylia opposita (Wall.) Alston in Trimen, Handb. Fl. Ceylon 6: 48. 1931; Worthington, Ceylon Trees 134. 1959; Ding Hou in Fl. Males. ser. 1, 6: 288. 1963; Matthew, Fl. Tamilnadu Carnatic 1:261. 1983; Ramamurthy in Nair & Henry, Fl. Tamilnadu 1: 75. 1983; Nair & Nayar, Fl. Courtallum 1: 156. 1986.

Celastrus opposita Wall. in Roxb., Fl. Ind. ed. Wall. 2: 398. 1824. Type: India, *Wallich* n. *4314* (K; BM, isotype).
Celastrus wightiana Wall., Cat. n. 4322. 1831, nom. nud.
Pleurostylia wightii Wight & Arn., Prod. 1: 157. 1834; Wight, Ic. Pl. Ind. Or. t. 155. 1839; Thw., Enum. Pl. Zeyl. 71. 1858; Lawson in Hook. f., Fl. Br. Ind. 1: 617. 1875; Trimen, Handb. Fl. Ceylon 1: 271. 1893; Gamble, Fl. Pres. Madras 1: 211. 1918; Loes. in Pflanzenfam. ed. 2, 20b: 180. 1942. Type: India, Narthamala, *Wight* n. *481* (K; BM, PDA, isotypes).
Pleurostylia heynei Wight & Arn., Prod. 1157. 1834. Type: India, *Wight* n. *486.*

Small unarmed trees, 6–10 (–15) m tall; branchlets subterete, glabrous. Leaves obovate-oblong or lanceolate, 3.0–8.0 × 1.5–3.5 cm, thin, coriaceous, base cuneate or attenuate, apex obtuse or acute, margin entire; nerves about 6 pairs; petioles 2–3.5 mm long. Cymes axillary, sometimes terminal, 1–several flowered; peduncles short, 2–4 mm long. Flowers 3 mm across, green; pedicels 1–2 mm long. Calyx lobes 5, subreniform, rounded or triangular, 0.5–1.0 mm long. Petals 5, elliptic or broadly ovate, 1.5–2.0 × 1.0 mm, obtuse to slightly acute, reflexed at anthesis. Disk thick, fleshy, cupular, margin faintly crenate. Stamens 5, 1.0–1.5 mm long, attached beneath the margin of the disk; filaments subulate, fleshy, c. 1 mm long; anthers ± ovoid, white. Ovary flask-like, the base adnate to the disk. Usually 1-celled by abortion;

style very short; stigma capitate; ovules 2 in each cell, erect. Drupe ellipsoid or slightly obovoid, 5.0–6.5 × 3.5–4.5 mm, obtuse, 1(–2)-seeded, sustained by persistent floral parts, pale yellow.

Distr. Sri Lanka, India (S. India), Thailand, China, Malaysia, Madagascar, Mauritius, Australia and New Caledonia.

Ecol. In dry region in open scrub forest, common; fls June–Oct, frts Aug–onwards.

Vern. Piyari, Panakka (S), Chiru Piyari (T).

Note. On specimen *C.P. 329* in K & CAL, Balangoda (Ratnapura Dist.) has been mentioned as a locality.

Plate No. 417 in PDA, based on specimen *C.P. 329*, belongs to this species.

Specimens Examined. JAFFNA DISTRICT: Jaffna, *Dyke s.n.* (K). VAVUNIYA DISTRICT: 4 miles north of Mankulam, 10 Dec 1970, *Fosberg & Balakrishnan 53541 & 53545* (K); Madhu-Panditurichchan road, c. 2 miles off Mannar Dist., 21 June 1975, *D.B. & D. Sumithraarachchi 759* (K, PDA). PUTTALAM DISTRICT: Pallugaturai, 30 Dec 1968, *Fosberg, Mueller-Dombois, Wirawan, Cooray & Balakrishnan 50891, 50892* (K, PDA), 30 m, 10 July 1969, *Wheeler 12093* (PDA), 1 m, 1 Nov 1974, *Davidse & Sumithraarachchi 8292* (K, PDA); Kalpitiya Island, Aug 1883, *s. coll. s.n.* (PDA); Wilpattu National Park, 20 m, 13 March 1973, *Bernardi 14208* (PDA). ANURADHAPURA DISTRICT: Ritigala Strict Natural Reserve, SW slopes below Weweltenna plain in Plot 13/5, 28 May 1974, *Jayasuriya 1684* (PDA); Arippu Road, 90 m, 20 March 1940, *Worthington 847* (PDA); Muriyakadawala, 18 Oct 1971, *Balakrishnan & Jayasuriya 1116* (K, PDA); Ritigala, 13 Jan 1974, *Waas 365* (K, PDA), 455 m, 9 Aug 1973, *Jayasuriya 1309* (K, PDA); Anuradhapura-Puttalam road, 6 May 1940, *Worthington 939* (BM); Galapitagala Wewa, 13 Jan 1974, *Waas 361* (K, PDA); Wilpattu National Park, 20 m, 13 March 1973, *Bernardi 14208* (PDA), *15378* (K), close to Plot W39, N. of Maraivillu, 30 June 1969, *Wirawan, Cooray & Balakrishnan 886* (K, PDA), at Kelvette en route Kokmotai, 2 Dec 1969, *Cooray 69100205R* (K, PDA), 3 mile west of office, 14 July 1970, *Meijer 359* (K, PDA). TRINCOMALEE DISTRICT: Foul Point, 80°19′ E, 8°31′ N, alt. low, 5 Feb 1972, *Jayasuriya, Dassanayake & Balasubramaniam 667* (K, PDA); 20 May 1932, *Simpson 9672* (PDA, BM); 32 mile post, Trinco road, 26 June 1940, *Worthington 989* (BM); Illankatarai camp on east coast, 3 Aug 1929, *Worthington 528* (PDA); Uppuveli, alt. sea level, 28 March 1977, *Cramer 4939* (K, PDA); Alut Oya, Trincomalee road, 76.5 mile post, *Worthington 972* (BM); N. of Trincomalee, en route to Mullaittivu, 13 Oct 1974, *Davidse 7544* (K, PDA). KURUNEGALA DISTRICT:

Amaragalla, Kurunegala, 240 m, 29 June 1948, *Worthington 4019* (PDA, BM). MATALE DISTRICT: Nalanda, 425 m, 17 Jan 1948, *Worthington 3436* (BM); Arangalla, Nalanda, 365 m, 19 July 1952, *Worthington 5975* (PDA); Lenadora near Dambulla, alt. low, 3 Aug 1974, *Kostermans 25317* (K); Pitawala, along road to Illukkumbura, 900 m, 23 July 1974, *Jayasuriya & Bandaranayake 1770* (PDA); Illukkumbura, 500 m, 23 July 1974, *Jayasuriya & Bandaranayake 1778* (PDA); Sigiriya, 200 m, 2 Nov 1949, *Worthington 4353* (BM). POLONNARUWA DISTRICT: Mutugala, access road, 9.5 mile, East of Mahaweli bridge on Polonnaruwa to Valaichchenai road, 1 Feb 1969, *Ripley 261* (K, PDA); Giritāle, 90 m, 5 March 1970, *Worthington 7226* (PDA). BATTICALOA DISTRICT: Batticaloa, *Gardner C.P. 329* (BM, CAL, K, PDA); Kalkudah, alt. low, 3–4 May 1977, *Waas 2121 & 2143* (PDA). BADULLA DISTRICT: Mount Kokegala at Mahiyangana, 650 m, 11 Nov 1975, *Bernardi 15693* (K, PDA). AMPARAI DISTRICT: Panawamoderagala, c. 1 mile N.W. of Panama, 20 m, 2 Dec 1974, *Davidse & Sumithraarachchi 8972* (K, PDA); Kumana, Panama coast, 10 Jan 1971, *Balakrishnan 591* (K, PDA), 5–10 m, 1 Nov 1975, *Bernardi 15586* (PDA); Amparai Rest House yard, 10 m, 8 July 1971, *Ripley 436* (K, PDA). KALUTARA DISTRICT: Coast of Kalutara, *Gardner 172* (K, PDA). MONERAGALA DISTRICT: Kataragama, 10 Km off Buttala, 550 m, 28 Oct 1975, *Bernardi 15525* (K, PDA); Amaduwa near Yala National Park, alt. low, 16 Sept 1979, *Balasubramaniam 2200, 2201, 2202, 2203* (PDA); 4 miles N. of Bibile, 310 m, 8 Sept 1978, *Huber 924* (PDA); Yala near Jamburagala, 7 March 1969, *Ripley 115* (PDA). GALLE DISTRICT: Kanneliya, 10 May 1974, *Waas & Peeris 523A* (K, PDA); Pilimagala, 29 Km from Talgasmankade via Mayagala, 8 Sept 1992, *Jayasuriya 6709* (PDA); Galle, *Gardner 171* (BM); Redcliff on sea coast, Weligama, 20 m, 25 June 1980, *Kostermans 28563* (PDA). HAMBANTOTA DISTRICT: Ruhuna National Park, Block I, at Buttawa, Plot R27, 30 May 1968, *Cooray 68053018R* (PDA), N. of Buttawa, Plot R27, 10 Dec 1967, *Mueller-Dombois & Cooray 67121022* (PDA), S.E. Buttawa, S. of Uraniya Kalapuwa, Plot R29, 10 Dec ·1967, *Mueller-Dombois & Cooray 67121075* (PDA), Block I, Karugaswala, Plot R28, 30 May 1968, *Cooray 68053001* (K, PDA), Block I, 200 m, west of Karugala at Yala road, 19 Oct 1968, *Mueller-Dombois 68101910* (K, PDA), beach E. of Buttawa Modera, 2–3 m, 6 Apr 1968, *Fosberg 50317* (K, PDA), Pilinnawa, Block II, 17 July 1975, *Jayasuriya & Austin 2249* (K, PDA), Patanagala Beach area, 3 m, 29 Jan 1968, *Comanor 893A* (K), Block III, N. of Kataragama, junction to Pilmagala, 25 Oct 1968, *Mueller-Dombois 68102503* (PDA), Block I, Gonagala Plain, 10 Jan 1969, *Wirawan 797, 798* (K, PDA). LOCALITY UNKNOWN: in 1839, *Mackenzie s.n.* (K); *Col. Walker s.n.* (CAL); Aug 1836, *s. coll. 124* (CAL).

6. MICROTROPIS

Wall. [Num. List 152, n. 4337–40. 1830, nom. nud.] ex Meisn., Pl. Vasc. Gen. Tabl. Diagn. 68. 1837, nom. cons.; Endl., Gen. Pl. 1087. 1841; Thw., Enum. Pl. Zeyl. 71. 1858; Benth. & Hook. f., Gen. Pl. 1: 361. 1862; Lawson in Hook. f., Fl. Br. Ind. 1: 613. 1875; Trimen, Handb. Fl. Ceylon 1: 268. 1893; Merr. & Freem., Proc. Amer. Acad. Arts 73: 276. 1940; Loes. in Pflanzenfam. ed. 2, 20b: 126. 1942 (non E. Mey., 1836). Type species: *M. discolor* (Wall.) Meisn. (= *Cassine discolor* Wall.).

Paracelastrus Miq., Fl. Ind. Bat. 1(2): 590. 1859. Type species: *P. bivalvis* (Jack.) Miq. (= *Celastrus bivalvis* Jack.).

Ortherodendron Makino, Bot. Mag. Tokyo 23: 62. 1909. Type species: *O. japonicum* (Franch. & Sav.) Hall. f. (= *Elaeodendron japonicum* Franch. & Sav.).

Chingithamnus Hand.-Maz., Sinesnia 2: 128. 1932. Type species: *Chingithamnus osmanthoides* Hand.-Maz.

Shrubs or small trees. Leaves opposite, usually glabrous, entire, coriaceous, generally exstipulate. Flowers sessile or subsessile in axillary or extra-axillary dichotomous or paniculate cymes, sometimes condensed to sessile clusters, bisexual, sometimes unisexual, 5- or 4-merous. Calyx deeply lobed; lobes almost free, often unequal in size, persistent, imbricate. Petals slightly united at the base, sometimes free, imbricate, erect. Stamens dorsifixed; filaments subulate, united at the base into a short tube or a ring, the ring sometimes interpreted as a disk, generally free from the petals; anthers ± broadly ovoid, introrse, sometimes extrorse. Ovary free, ± 2-celled; Ovules 2 in each cell, erect, collateral; style very short; stigma discoid, slightly 2–4-lobed or obscure. Capsule ovoid or oblong, short-apiculate to beaked, 1-celled, longitudinally striate, splitting laterally along one side, 1-seeded, with persistent calyx. Seed usually erect, oblong, albuminous, fleshy, enveloped by the aril, usually wrinkled; testa ± fleshy; cotyledon foliaceous.

About 70 species, distributed in C. America, Sri Lanka, India (S. India, Assam), Bangladesh, eastwards to S. China, Formosa, Central Japan, southwards through Burma, Thailand, Indo-China to Malesia (Sumatra, Malay Peninsula, West Java, Borneo) and the Philippines. Two species in Sri Lanka.

KEY TO THE SPECIES

1 Leaves thin, coriaceous, elliptic or oblanceolate, base cuneate to attenuate, apex acute or short acuminate, margin flat; flowers in lateral, axillary, sessile fascicles; capsules oblong-ellipsoid
. . . . **1. M. wallichiana**
1 Leaves rigidly coriaceous, oblong or elliptic, base rounded to abruptly obtuse, apex obtuse, rounded, sometimes retuse; margin ± revolute; flowers in lateral, axillary, sessile, subsessile, or very often short-peduncled, capitate fascicles; capsules ovoid or cylindric-ellipsoid, acute
. . . . **2. M. zeylanica**

1. Microtropis wallichiana Wight ex Thw., Enum. Pl. Zeyl. 71. 1858; Lawson in Hook. f., Fl. Br. Ind. 1: 613. 1875; Trimen, Handb. Fl. Ceylon 1: 269. 1893; Gamble, Fl. Pres. Madras 1: 206. 1918; Merr. & Freem., Proc. Amer. Acad. Arts 73: 283. 1940; Ding Hou in Fl. Males. ser. 1, 6: 279. 1963, excl. syn. *M. zeylanica* Merr. & Freem.; Ramamurthy in Nair & Henry, Fl. Tamilnadu 1: 75. 1983. Type: Sri Lanka, in 1836, *Wight 528* (K).

Paracelastrus wallichianus (Wight ex Thw.) F.N. Williams, Bull. Herb. Boiss. ser. 2, 5: 224. 1905.

Shrub or small tree, up to 8 m tall; branchlets terete, sometimes 4-angular. Leaves thin, coriaceous, glabrous, elliptic or oblanceolate, 4.6–13.8 × 2.6–5.6 cm, base cuneate to attenuate, apex acute or shortly acuminate, margin entire, flat; nerves 8–12 pairs, distinct, spreading; petioles 0.8–1.2 cm long. Flowers very small in lateral, axillary or extra-axillary, dense clusters or fascicles 5–6 mm across, yellowish. Bracteoles small, ovate, 1–2 mm long. Calyx lobes subreniform, c. 1.0–1.5 × 2.0–2.5 mm, short-lacerate. Petals rather fleshy, elliptic-oblong or elliptic, 2.0–3.0 × 1.5–2.0 mm, obtuse. Stamens c. 2 mm long; filaments flat, 1.0–1.5 mm long, united in the lower half; anthers broadly ovoid, obtuse, slightly apiculate. Ovary short-conical, obtuse, 2-locular, furrowed; ovules 2 in each cell; style very short. Capsules oblong-ellipsoid, 1–1.5 cm long, 1-celled, 1-seeded, furfuraceous, obtuse, crowned by the persistent style.

Distr. Sri Lanka, India and Malaysia, Sumatra and Borneo.

Ecol. In forests of wet zone, up to 1600 m, fairly common; fls Nov–Mar, frts Jan–Apr.

Note. On the specimens *C.P. 43*, kept in PDA, Watagoda, Aug 1854; Adam's peak, Galagama, Gardner, Hantane, 1851, Deltota & Hunnasgiriya, 1851 are annotated, while on specimens in K & BM, Adam's Peak, March 1846, and in HAK Haputale, are annotated as localities.

Plate No. 414 based on specimen *C.P. 43* in PDA belongs to this species.

Specimens Examined. KEGALLE DISTRICT: Kitulgala, 5 Nov 1974, *Waas 883* (PDA); Kalugamuwa, *J.M. Silva s.n.* (PDA). KANDY DISTRICT: Kandy, 21 March 1819, *Moon 521* (BM); Madulkele, 8 Oct 1887, *s. coll. s.n.* (PDA); Laxapana, 27 Dec 1931, *Simpson 9001*(BM, PDA); Adam's Peak, March 1846, *Thwaites C.P. 374* (K); Hunnasgiriya, 1370 m, 19 Jan 1975, *Waas 985* (K, PDA); Loolecondera Estate, S. of Deltota, 1325 m, 25 Oct 1984, *Jayasuriya & N. Gunatilleke 3428* (PDA); 4 miles SW of Maskeliya, 1400 m, 21 Nov 1974, *Davidse & Sumithraarachchi 8611* (PDA); NW of Loolecondera, 1220 m, 80°42' E, 7°11' N, 11 March 1977, *B. & K. Bremer 1050* (PDA); Rajamallay, Moray Estate, 25 Oct 1975, *Sohmer & Sumithraarachchi 9871*(PDA). BADULLA DISTRICT: Haputale,

Sept 1890, *s. coll. s.n.* (PDA); C¬aig Estate, Bandarawela, 25 March 1906, *A.M. Silva s.n.* (PDA). RATNAPURA DISTRICT: Horagulkande-Sinharaja, +300 m, 25 Feb 1977, *Waas 2020* (PDA); Sinhagala, Sinharaja, +600 m, 23 Jan 1977, *Waas 1956* (K, PDA); Maratenna, 1500 m, 24 Aug 1976, *Waas 1785* (PDA); W. of Tummodara (Seethagangula), trail to Adam's Peak from Eratne, 1025 m, 2 March 1985, *Jayasuriya & N. Gunatilleke 3211* (PDA); Sudugala near Kuruwita, 510 m, 1 Nov 1977, *Huber 528* (PDA); Peak wilderness, Henagastenne to Gartmore, s. slope, 1400 m, 8 Jan 1981, *Werner 197* (PDA); Kiribathgala, 17 March 1975, *Waas 1222* (PDA); Sinharaja towards Dotalugala range from Ellawala Ganga, 26 Sept 1991, *Jayasuriya 5748* (PDA); Sri Palabaddala, IBP plot, 16 Feb 1974, *Waas 421* (K, PDA); Marathagala, +1200 m, 26 June 1976, *Waas 1746* (K, PDA); Mannikkawatta forest, +1300 m, *Waas 1774* (K, PDA); Ensalwatte, near Radio transmitting station, 975 m, 15 March 1985, *Jayasuriya & Balasubramaniam 3250*(CAL, PDA). NUWARA ELIYA DISTRICT: Nuwara Eliya, 1830 m, *Gardner 177* (BM, K, PDA). GALLE DISTRICT: Hinidum Pattu, Kodigaha kanda, 19 Oct 1992, *Jayasuriya 6845*(PDA); Kaneliya, 17 July 1992, *Jayasuriya 6562*(PDA); Darakulkande forest reserve off Athuwela, 7 Oct 1992, *Jayasuriya 6778*(PDA). MATARA DISTRICT: Sinharaja-Deniyaya range, 27 March 1992, *Jayasuriya 6271, 6276* (PDA); Kekunadura forest range, 19 Oct 1991, *Jayasuriya 5829*(PDA); North Enselwatta, 8 Oct 1975, *Sohmer & Waas 10438* (PDA); N. of Deniyaya, Sinharaja forest, 1060 m, 12 Nov 1977, *Huber 614* (PDA). LOCALITY UNKNOWN: *Col. Walker s.n.* (K), *Thwaites C.P. 402, C.P. 202* (K); *s. coll. s.n.* (CAL); *s. coll. C.P. 43*(PDA, BM, HAK, K, PDA); Herb. *Wight 528* (Type, K).

2. Microtropis zeylanica Merr. & Freem., Proc. Amer. Acad. Arts. 73: 282. 1940. Type: Sri Lanka, *Thwaites C.P. 1228* (P).

Microtropis ramiflora sensu Thw., Enum. Pl. Zeyl. 72. 1858; Trimen, Cat. 18. 1885; Trimen, Handb. Fl. Ceylon 1: 268. 1893, non Wight, 1845.

Mich-branched shrubs or small trees; branches terete, sometimes angular or sulcate. Leaves variable, rigidly coriaceous, oblong or elliptic, 2.0–6.4× 1.5–4.0 cm, apex obtuse, rounded, sometimes retuse, base rounded to abruptly obtuse, margin often very strongly revolute, brown-olivaceous or olive-coloured above, pale beneath; slightly rugose on upper surface when dry, laxly reticulate; petioles short, thick, c. 2 mm long. Inflorescence axillary, sessile, subsessile or very often short-peduncled capitate fascicles; peduncles 3(–6) mm long. Flowers sessile, 5-merous, white. Calyx lobes coriaceous, sub-orbicular, 1.5–2.0 mm diam., margin erose. Petals subovate, obtuse, c. 2 mm diam. Stamens short, inserted on the margin of the disk; filaments 1 mm long; anthers ovoid, introrse. Ovary slightly ovoid, 2-celled, ovules 2 in each cell; styles short. Capsules ovoid or cylindric-ellipsoid, tapering at apex, c. 1 cm long, 1-celled, 1-seeded.

Distr. Endemic.

Ecol. In forests of wet, montane zone, fairly common; fls Feb–Sept, frts Sept–Jan.

Note. On 2 specimens of *C.P. 148* in PDA, Horton Plains has been given as locality, while on another specimen of *C.P. 148* in PDA, and also on specimens in BM, CAL & K, Adam's Peak as locality has been annotated, hence separately listed.

Plate No. 415 based on specimen *C.P. 1228,* and annotated as *M. ramiflora* Wight, in PDA belongs to this species.

Specimens Examined. KANDY DISTRICT: Hunnasgiriya, 1370 m, 19 Jan 1975, *Waas 984* (K, PDA); Adam's Peak, Feb 1889, *s. coll. s.n.* (PDA), 2135 m, Mar 1846, *Thwaites C.P. 148* (BM, CAL, K, PDA). BADULLA DISTRICT: Southern slope of Namunukula Kande, 1930 m, 6 Sept 1978, *Huber 905* (PDA); Namunukula Peak, 2020 m, 28 July 1980, *Werner 94 & 97* (PDA), 24–28 Oct 1910, *J.M. Silva s.n.* (PDA). NUWARA ELIYA DISTRICT: Towards summit of Pidurutalagala, 2500 m, 24 Jan 1973, *Trivengadam 4020 & Cramer 273* (PDA), ±2035 m, 24 Jan 1973, *Cramer & Trivengadam 4020*(K), 2285 m, 8 Oct 1973, *Waas 181* (PDA), 2200 m, 19 Nov 1977, *Huber 669* (PDA); Horton Plains, Feb 1846, *Thwaites C.P. 297,* (K), *s. coll. C.P. 148* (PDA), 25 May 1911, *J.M. Silva s.n.*(PDA), 2 May 1906, *A.M. Silva s.n.*(PDA); Trail Farr Inn to Big Worlds End, 2200 m, 11 Oct 1977, *Nooteboom 3333* (PDA); "Totapella", 22–24 Feb 1882, *s. coll. s.n.*(PDA); Elephant Plains, 1830 m, *Gardner 179,* (BM, K, PDA); Nuwara Eliya, 1830 m, *Gardner 178* (BM, K). LOCALITY UNKNOWN: *s. coll. 121* (K); *Col. Walker s.n.*(K); 2100 m, *Thwaites C.P. 1228* (CAL, K, isotypes).

7. GLYPTOPETALUM

Thw., Hooker's J. Bot. Kew Gard. Misc. 8: 267. t. 7B. 1856; Thw., Enum. Pl. Zeyl. 73. 1858; Benth. & Hook. f., Gen. Pl. 1: 361. 1862; Lawson in Hook. f., Fl. Br. Ind. 1: 612. 1875; Trimen, Handb. Fl. Ceylon 1: 268. 1893; Loes. in Pflanzenfam. ed. 2, 20b: 125. 1942. Type species: *G. zeylanicum* Thw.

Erect shrubs or trees; branchlets 4-angled, sometimes terete, glabrous. Leaves opposite, sometimes sub-opposite in the upper part of young branchlets, entire or crenulate, usually exstipulate. Flowers in axillary or extra-axillary dichasial cymes, sometimes reduced to 3 flowers only, bisexual, 4-merous, bracteate. Calyx with 4 short lobes, spreading. Petals rather fleshy, usually smooth, with 2 pit-like depressions on the upper surface. Disk large, fleshy, flat, 4-angled. Stamens 4, inserted on the disk; filaments very short, persistent, connectives dilated; anthers divergent, dehiscent at the top or introrse. Ovary immersed in the disk, pyramidal or 4-angled, 4-celled; style very short or obscure; stigma minute, capitate or obscure; ovules one in each cell, pendulous from the top of each cell. Capsules subglobose, dehiscing loculicidally, valves leaving a persistent columella, 1–4-celled, 4- or 1-seeded by abortion. Seeds nearly covered by fleshy aril; albumen fleshy; cotyledons leafy, flat.

About 20 species, in India, Burma, China, Sri Lanka, Thailand, Indo-China and Malaysia.

Glyptopetalum zeylanicum Thw., Hooker's J. Bot. Kew Gard. Misc. 8: 268. t. 7B: 1856; Thw., Enum. Pl. Zeyl. 73. 1858; Lawson in Hook. f., Fl. Br. Ind. 1: 612. 1875; Trimen, Handb. Fl. Ceylon 1: 268. 1893; Gamble, Fl. Pres. Madras 1: 204. 1918; Manilal, Fl. Silent Valley 55. 1988. Type: Sri Lanka: *C.P. 589* (PDA; K, isotype).

var. zeylanicum

Small bushy tree, 8–10 m tall; branches terete. Leaves subcoriaceous, glabrous, elliptic-oblong or ovate-oblong, 5.5–13.6 × 3.2–6.4 cm, base cuneate, apex acute to short acuminate, shallowly crenate or serrate except at base, sometimes entire, distinctly reticulated on both surfaces; petioles 4–6 mm long. Flowers 4-merous, bracteate, 8–12 mm across, in axillary or extra-axillary, lax, dichasial cymes up to 10 cm long; peduncles 3–5 cm long; flower stalk articulated at the lower one-fifth, and pedicel proper 0.8–1.2 cm long. Calyx lobes suborbicular, c. 1 mm broad, slightly erose at the margin. Petals oblong, 5–6 mm long, obtuse, fleshy, reflexed, bifoveolate on the inner side near the apex. Disk large, conspicuous, 4-angled. Stamens inserted on the disk, c. 1 mm long; filaments with broad dilated connectives. Ovary immersed in the disk, 4-angled, 4-celled; style very short; stigma ± obscure; ovule one in each cell, pendulous; capsules globose, up to 1.4 cm diam., green, smooth. Seeds ovoid; aril crimson.

Distr. India (Annamalai Hills) and Sri Lanka.

Ecol. In submontane forests in wet zone, up to 1400 m, fairly common; fls Feb, frts Nov.

Note. Ding Hou (in Fl. Males. ser. 1, 6: 257. 1963) has described a new variety, viz. *G. zeylanicum* var. *brevipedicellatum* Ding Hou from Malaysia, which differs from var. *zeylanicum* by the more lax, spreading dichotomously branched inflorescences; flower stalks articulated at about the upper one-fifth, and the pedicel proper being only 2–3 mm long.

On one herbarium sheet of *C.P. 589* in PDA, Hantane *Gardner* Feb 1855 (in fls) and Nov 1854 (in Frts); Galagama May 1964 and Ambagamuwa Aug 1835 are annotated.

Plate No. 413, based on specimen *C.P. 589* in PDA, belongs to this species.

Specimens Examined. KANDY DISTRICT: Hantane, in 1847, *Gardner 170* (K, BM); Rangala, 29 Aug 1936, *Alston 465*(PDA). NUWARA ELIYA DISTRICT: Ohiyagama, close to culvert 56/8, Hakgala-Keppetipola road, 16 May 1975, *Cramer 4477* (PDA). BADULLA DISTRICT: Ekiriyankumbura, Jan 1888, *s. coll. s.n.* (PDA). MATARA DISTRICT: Deniyaya, Morawak Korale, Mar 1881, *s. coll. s.n.* (PDA); Climb to Deniyagala, Sinharaja Reserve forest, 13 May 1990, *Jayasuriya & Balasubramaniam 5050* (PDA). RATNAPURA DISTRICT: Mannikkawatta forest, ±1300 m, 27 June 1976, *Waas 1773* (PDA); Dotalugala forest, ±1000 m, 27 Aug 1976, *Waas 1841* (K).

8. EUONYMUS

L., Sp. Pl. 197. 1753 "Evonymus"; L., Gen. Pl., ed. 5, 91. 1754; Thw., Enum. Pl. Zeyl. 73. 1858 "Evonymus"; Benth. & Hook. f., Gen. Pl. 1: 360. 1862; Lawson in Hook. f., Fl. Br. Ind. 1: 607. 1875; Trimen, Handb. Fl. Ceylon 1: 266. 1893; Loes. in Pflanzenfam. ed. 2, 20b: 115. 1942. Type species: *E. europaeus* L.

Erect shrubs or small trees, sometimes scandent. Leaves opposite, rarely spiral or alternate; stipules lanceolate, caducous. Inflorescence axillary dichasical cymes, rarely fascicles; pedicels articulated; flowers bisexual, 4–5-merous. Calyx 4–5, deeply lobed, imbricate, entire or crenulate, spreading or recurved, persistent. Petals imbricate, spreading or reflexed, entire, erose, short-ciliate or long-fimbriate, smooth, areolate, often with coloured veins. Disk distinct, fleshy or thin, flat or 4–5-lobed or rounded, smooth or covered with subulate processes. Stamens as many as the petals, alternate with them, inserted on the disk or at the margin; filaments distinct, short or obscure; anthers oblong or broadly ovoid, 2-celled, dehiscence latrorse or introrse. Ovary partly or wholly immersed in the disk, the free part terete or slightly 4–5-angled, 3–5-locular; style short; stigma obscure; ovules 2–8 in each cell, in 2 rows. Fruit a loculicidal capsule, often coloured, usually 3–(4)–5-angled or lobed or winged, sometimes echinate, apex obtuse, acute, more or less truncate. Seeds 1-several per cell, arillate.

176 species, mostly in tropical and subtropical Asia, Pakistan and India (in the Himalayas), China and Japan; 4 species in Europe, 1 in Africa, 2 in Madagascar, 1 in N.E. Queensland and 9 in N. & C. America; 3 in Sri Lanka.

KEY TO THE SPECIES

1 Inflorescences *sessile* or *subsessile*; capsule clavate or pyriform-turbinate; leaves distinctly serrate-crenate in the upper two-thirds .... **1. E. walkeri**
1 Inflorescences clearly *pedunculate*; capsule obovoid, turbinate or clavate; leaves entire or obscurely serrate in the upper part
  2 Leaves ovate, rotundate to obovate, margin strongly revolute; flowers up to 8 mm diam. .... **2. E. revolutus**
  2 Leaves elliptic-lanceolate or oblong-lanceolate, margin flat, not revolute; flowers 8–10 mm diam. .... **3. E. thwaitesii**

1. Euonymus walkeri Wight, Ill. Ind. Bot. 1: 178. 1849; Thw., Enum. Pl. Zeyl. 73. 1858; Lawson in Hook. f., Fl. Br. Ind. 1: 608. 1875; Trimen, Handb. Fl. Ceylon 1: 267. 1893; Blakelock, Kew Bull. 6: 251, 254. 1951. Type: Sri Lanka, *Col. Walker s.n.* (K).

Euonymus zeylanicus Moon, Cat. 17. 1824, nom. nud.

Shrubs or small trees, 2.5–4 m tall; branches terete, extreme ones somewhat 4-angled, glabrous; outer bark dark. Leaves subcoriaceous, glabrous, very variable, oval-lanceolate, subovate to ovate or oblong-lanceolate, 3.1–10.8 × 2.4–4.6 (–5.2 cm), tapering at base, apex acute or acuminate, crenate-serrate towards upper two-thirds, entire below; petioles 2–3 mm long. Inflorescences axillary, sessile or subsessile, 1–3 flowered cymes; peduncles 2 mm long. Flowers small, 5-merous, sessile, bisexual. Calyx lobes ovate, 1–1.5 mm long, very obtuse, recurved, persistent. Petals 5, orbicular, 2.0–2.5 × 1.5–2 mm, margin undulate, white with dark-red tips. Disk fleshy, distinct, 5-angled, green. Stamens inserted on the disk, alternate with petals; filaments short; anthers ± ovoid, 2-celled. Ovary ± immersed in the disk, the free part 5-angled; style short. Capsules clavate or pyriform-turbinate, 1.2–1.8 cm long, 5-angled, short-apiculate with persistent style, young fruits green, mature red. Seeds with a very large aril, smooth, yellow.

Distr. Endemic.

Ecol. In low-country forests of moist zone on eroded soil near streams, common; fls Aug–Apr; frts Sept–June.

Note. Plate No. 412, based on specimen *C.P. 478* in PDA, belongs to this species.

Specimens Examined. KURUNEGALA DISTRICT: Galagama, Feb & May 1836, *Gardner C.P. 478* (BM, CAL, K, PDA). MATALE DISTRICT: Rattota-Illukkumbura road, 1110 m, 16 Aug 1978, *Huber 752* (PDA). KEGALLE DISTRICT: Salgala, 11 May 1963, *Ameratunga 611* (PDA); Alagalla, Dekanda, 610 m, 15 Oct 1939, *Worthington 597*(BM). KANDY DISTRICT: Deltota Estate, Galaha, 915 m, 24 Sept 1950, *Worthington 4972* (PDA); Madugoda-Nugatenna, 14 June 1926, *Alston 1201* (PDA); Near Madugoda, 8 Nov 1931, *Simpson 8781* (PDA); Hill Crest,

Kandy, ±650 m, 7 May 1969, *Worthington 7200* (PDA); Murutalawa, 11 May 1962, *Amaratunga 166* (PDA); Madugoda-Mahiyangana road, 600 m, 6 Aug 1978, *Kostermans 26781* (PDA); 4 miles NE of Madugoda, 80° 54' E, 7° 20' N, 6 Jan 1972, *Jayasuriya, Dassanayake & Balasubramaniam 502* (PDA), *493* (K); Trail to Nitre Cave from St. Martin's Estate, base of Dumbangala spur, 14 Nov 1975, *Sohmer & Jayasuriya 10689* (PDA); Nitre Cave, Sept 1888, *s. coll. s.n.* (PDA); Maskeliya, 1000 m, 19 Oct 1977, *Nooteboom 3411*(PDA); Summit of Hantane ridge, 750 m, 14 Apr 1980, *Kostermans 28288*(PDA); Kadugannawa near Kandy, 150 m, 5 June 1973, *Kostermans 24944* (K, PDA); Madugoda, 18 Apr 1932, *Simpson 9472* (BM); 3 miles N. of Corbet's Gap, 750 m, 17 Nov 1974, *Davidse 8330* (K, PDA); Boyagoda kanda near Galgadera, 80° 30' E, 7° 23' N, 300 m, 13 Feb 1975, *Sumithraarachchi 603* (K, PDA); Hantane, 915 m, *Gardner 175*(BM, K); Deltota Tea Estate, 1000 m, 30 Sept 1974, *Kostermans 25684* (K, PDA); Hunnasgiriya forest above Tea Estate, *Wirawan B-10* (K); Rangala, 29 Aug 1926, *Alston 1202* (PDA). KALUTARA DISTRICT: Yagirala forest range, 1 July 1980, *Jayasuriya 5179* (PDA), 1 June 1980, *Jayasuriya 5193* (PDA); Trail Kudawa-Pitakale, 22 Aug 1988, *Jayasuriya 4448* (PDA); Morapitiya, *Sumithraarachchi & Foster 1001* (K, PDA); Danihena, ±150 m, 29 Aug 1976, *Waas 1872* (K, PDA); Kalugala, 335 m, 28 June 1975, *Waas 1284* (K, PDA); Pelawatta, Mar 1887, *s. coll. s.n.* (PDA). RATNAPURA DISTRICT: Kurulugala forest, Bulutota Pass, 920 m, 5 Nov 1977, *Huber 563* (PDA); Gawarahena forest, Balangoda, +500 m, 25 Aug 1976, *Waas 1800, 1811* (K, PDA); Yakinidola-Sinharaja, ±250 m, 26 Feb 1977, *Waas 2044* (K, PDA); Mannikkawatta forest, ±1300 m, 27 June 1976, *Waas 1776* (K, PDA); Waturawa main trail, Sinharaja forest, 26 Oct 1987, *Jayasuriya 4121*(PDA), Trail to Moulewella, 26 Sept 1987, *Jayasuriya & Balasubramaniam 4057* (PDA); Delgoda-Kalawana range, 24 May 1990, *Jayasuriya 5160* (PDA). NUWARA ELIYA DISTRICT: Forest on way to Fort Mcdonald, Hakgala, 11 March 1906, *A.M. Silva s.n.* (Acc. No. 000101-HAK); High forest Estate, Maturata, 1830 m, 9 March 1906, *s. coll. s.n.* (Acc. No. 000102-HAK); Forest at s. boundary, Hakgala, 1665 m, 13 Apr 1906, *A.M. Silva s.n.* (Acc. No. 000103-HAK). MONERAGALA DISTRICT: Koslanda, base of Diyaluma Falls, 26 Apr 1974, *Sumithraarachchi & Waas 290* (PDA). GALLE DISTRICT: Ascent to Haycock, W. of Hiniduma, 305 m, 21 Oct 1974, *Davidse 7841* (K); Hinidum kande, 18 Oct 1992, *Jayasuriya 6818* (PDA), 200 m, 31 Aug 1974, *Kostermans 254888* (K), 3 July 1979, *Kostermans s.n.* (PDA); Kanneliya forest near Hiniduma, 200 m, 3 Sept 1974, *Kostermans 25527* (K, PDA), alt. low, 9 Aug 1974, *Kostermans 25366* (K); Hiniduma, 10 Aug 1974, *Cramer 4298* (PDA); Nelluwa, 120–150 m, 3–4 Apr 1970, *Balakrishnan 240, 253* (PDA), 400 m, 27 Jan 1979, *Gunatilleke 289X* (PDA); Sinharaja forest, 600 m, 13 July 1974, *Kostermans 26799* (PDA); Hills N.

of Nelluwa, 260 m, 26 Sept 1977, *Huber 355* (PDA). MATARA DISTRICT: Masmulla forest range, 18 Aug 1991, *Jayasuriya 5351* (PDA); Diyadawa forest range, 16 June 1990, *Jayasuriya 5292* (PDA); Viharakele forest reserve, 7 Nov 1991, *Jayasuriya 5912A* (PDA). LOCALITY UNKNOWN: *Col. Walker 79* (K, PDA); *s. coll. s.n.* Herb. Hooker (K); Aug 1836, *s. coll. 153* (CAL); *Col. Walker s.n.* (K, Type).

2. Euonymus revolutus Wight, Ill. Ind. Bot. 1: 178. 1840; Thw., Enum. Pl. Zeyl. 73. 1858; Lawson in Hook. f., Fl. Br. Ind. 1: 608. 1875; Trimen, Handb. Fl. Ceylon 1: 267. 1893; Loes. in Pflanzenfam. ed. 2, 20b: 121. 1942; Blakelock, Kew Bull. 6: 251, 254. 1951. Type: Sri Lanka, *Col. Walker 360* (K).

Small to moderate-sized trees, up to 6 m high; branches terete, young ones obtusely 4-angled, glabrous; bark soft, brown, fissured. Leaves coriaceous, glabrous, ovate, rotundate to obovate, 2.0–5.6(–7.8) × 1.8–3.8(–4.6) cm, apex obtuse or with a short, blunt acumen, base rounded or acute, margin entire or obscurely serrate at apex, strongly revolute; petioles 2–3 mm long. Inflorescences axillary 1–3-flowered cymes, when 1-flowered, then furnished with 2 minute bracts; peduncles 1.4–2.2 cm long. Flowers 5-merous, bisexual, 7–8 mm across, bright brick-red; pedicels 2–3 mm long. Calyx lobes 5, rotundate, 1 mm wide. Petals 5, orbicular, 3.5–4.0 × 1.5–2.0 mm, entire, somewhat reflexed. Disk fleshy, 5-lobed or angled. Stamens as many as the petals, alternate with them, inserted at the margin of the disk; anthers ovoid, 2-celled. Ovary immersed in the disk, 5-celled, 5-angled, with 2-ovules in each cell; style very short. Capsule turbinate or obovoid, 1.0–1.6 cm long, flat-topped, green when young, red when ripe. Seeds angular, smooth, brown.

Distr. Endemic.

Ecol. In montane moist forests, 1500–2000 m, fairly common; fls Sept–Mar, frts Oct–Nov.

Note. Plate No. 410 in PDA, based on specimen *C.P. 579* belongs to this species.

Specimens Examined. KANDY DISTRICT: Adam's peak, Mar 1846, *Thwaites C.P. 22* (K), Mar 1883 & Feb 1884, *s. coll. s.n.* (PDA); Trail to Adam's peak, above Moray Estate, 1900 m, 13 Nov 1978, *Kostermans 27025* (K, PDA), 15 Nov 1973, *Sohmer & Waas 8661* (PDA), ±2000 m, 23 June 1976, *Waas 1694* (K, PDA); Rajamallay Estate, Maskeliya, ±1500 m, 12 Dec 1974, *Cramer 4396* (PDA). BADULLA DISTRICT: Namunukula, 25 Apr 1974, *Sumithraarachchi & Waas 267* (PDA), ±2000 m, 6 May 1976, *Waas 1617* (K, PDA), near summit, 29 Apr 1907, *Willis s.n.* (PDA). NUWARA ELIYA DISTRICT: Nuwara Eliya, Feb 1846, *Gardner C.P. 579*

(BM, CAL, HAK, PDA), in 1845, *Thomson s.n.* (K); *Gardner 176* (K); Forest along ridge trail from Gonapitiya Tea Estate, 29 Oct 1975, *Sohmer & Waas 10177* (K); Forest in front of Hakgala Botanic Gardens, 2000 m, 27 Nov 1971, *Balakrishnan & Dassanayake 1133* (K, PDA); Hakgala, 9 Jan 1932, *Simpson 9070* (BM); Above Hakgala Botanic Gardens, 1800–2000 m, 30 Sept 1977, *Nooteboom 3245* (PDA); Hakgala Natural Reserve, 1810 m, 20 Oct 1977, *Huber 489* (PDA); Trail to Pidurutalagala, 31 Oct 1973, *Sohmer, Jayasuriya & Eliezer 8442* (PDA); Sita Eliya forest Reserve, near 52/5 mile marker, culvert no. 5, 31 Oct 1973, *Sohmer, Jayasuriya & Eliezer 8449* (PDA); Mt. Hakgala, 1850 m, 9 Dec 1975, *Bernardi 15838* (PDA); Northern slope of Pidurutalagala from Gonapitiya Tea Estate, 29 Oct 1975, *Sohmer & Sumithraarachchi 10161* (PDA); Kandapola forest reserve, 1657 m, 9 Feb 1952, *Worthington 5637* (PDA); Hakgala forest reserve, 1 Nov 1973, *Sohmer, Jayasuriya & Eliezer 8480* (PDA), 20 Oct 1977, *Huber 489* (PDA); Jungle near Hakgala Garden, ±1680 m, 12 March 1952, *Worthington 5740* (PDA), 23 May 1911, *J.M. Silva s.n.* (PDA); Hakgala Strict Natural Reserve, 1750 m, 18 Apr 1973, *Stone 11236* (PDA); Hakgala, at Sita Eliya, 6 Apr 1906, *A.M. Silva s.n.* (PDA). LOCALITY UNKNOWN: *Thomson s.n.* (K); *s. coll. s.n.* Herb. S. Kurz (CAL-Acc. No. 85049); *Col. Walker 360* (K, type).

3. Euonymus thwaitesii Lawson in Hook. f., Fl. Br. Ind. 1: 608. 1875; Trimen, Handb. Fl. Ceylon 1: 267. 1893; Loes. in Pflanzenfam. ed. 2, 20b: 118. 1942; Blakelock, Kew Bull. 6: 251, 254. 1951. Type: Sri Lanka, *s. coll. C.P. 171* (PDA).

Enonymus dichotomus sensu Thw., Enum. Pl. Zeyl. 73. 1858, non Heyne, 1824.

Small trees, 5–7 m tall; branchlets slender, glabrous; bark rough, dark. Leaves subcoriaceous, glabrous, elliptic-lanceolate or broadly oblong-lanceolate, 3.5–6.5 × 1.6–2.6 cm, tapering at base, obtusely acuminate, margin entire or undulate, flat, not revolute; petioles 3–4 mm long. Inflorescences axillary dichasial cymes, 3-flowered; peduncles 1.3–2.0 cm long. Flowers 8–10 mm across, bisexual; pedicels slender, 1.0–1.3 cm, articulated. Calyx lobes 5, ovate, 1 mm long, very obtuse. Petals 5, orbicular, 3.5–4.0 × 2.0 mm, margin undulate, slightly reflexed. Disk fleshy, 5-lobed. Stamens as many as the petals, alternate with them, inserted at the margin of the disk; filaments short; anthers oblong, 2-celled. Ovary nearly completely immersed in the disk, 5-angled, 5-locular; style short. Capsules clavate-turbinate, 1.0–1.3 cm long, glabrous, 5-angled, smooth. Seeds angular, smooth, brown.

Distr. Endemic.

Ecol. In low country forest of moist zone along the banks of streams, up to 1000 m, rare; fls Feb–May, frts May–Nov.

Note. On specimen of *C.P. 171* in PDA, the localities Ambagamuwa (*Gardner*) and Galagama, May 1856 are annotated. *Euonymus dichotomus* Heyne has been advertently given by Lawson (in Hook. f. in Fl. Br. Ind. 1: 609. 1875) for Ceylon; the locality quoted being Galagama, above given for *E. thwaitesii.*

Specimens Examined. KANDY DISTRICT: Kandy-Mahiyangana road, 18 Oct 1973, *Waas 215* (K, PDA); Kalutenne, off Madugoda, 915 m, 3 Feb 1975, *Waas 1093* (PDA). BADULLA DISTRICT: Namunukula, 25 Apr 1974, *Sumithraarachchi & Waas 265* (PDA). RATNAPURA DISTRICT: S.E. of Depedene, 80° 33' E, 6° 27' N, 610 m, 22 Feb 1977, *B. & K. Bremer 883* (PDA); Morapitiya logging area, E. of Matugama, 100 m, 12 Aug 1974, *Kostermans 25399* (PDA). MATARA DISTRICT: Viharakele forest reserve, 7 Nov 1991, *Jayasuriya 5912* (PDA); Gombadala, Mulatiyana forest reserve, 22 Nov 1991, *Jayasuriya 5957* (PDA); Enselwatte, ±1000 m, 14 Feb 1976, *Waas 1482* (K, PDA). LOCALITY UNKNOWN: 610 m, *s. coll. C.P. 171* (PDA, type; BM, CAL, K, isotypes), *Col. Walker s.n.* (K).

9. KOKOONA

Thw., Hooker's J. Bot. Kew Gard. Misc. 5: 379. 1853; Thw., Enum. Pl. Zeyl. 52. 1858; Benth. & Hook. f., Gen. Pl. 1: 362. 1862; Lawson in Hook. f., Fl. Br. Ind. 1: 616. 1875; Trimen, Handb. Fl. Ceylon 1: 269. 1893; King, J. Asiat. Soc. Bengal 65 (2): 346. 1896; Loes. in Pflanzenfam. ed. 2, 20b: 164. 1942. Type species: *K. zeylanica* Thw.

Large evergreen trees, sometimes buttressed; inner bark yellow or pale-coloured; branchlets flat at the nodes. Leaves opposite, obovate, rounded or retuse, coriaceous, the midrib usually convex; stipules small, ± persistent. Flowers in axillary panicles, bisexual, bractete; pedicels usually articulated. Calyx cup-shaped, shallowly 5-lobed, persistent. Petals 5, contorted in aestivation, slightly concave, ± coriaceous, glandular punctate. Disk large, fleshy, cupular, corrugated, obscurely 5-angled. Stamens 5, inserted inside the inner edge of the disk, alternate with the petals; filaments usually fusiform, narrowed towards the apex and transparent at the upper end; anthers oblong, introrse, usually with lengthened connectives, dehiscing longitudinally. Ovary partly immersed in the disk, 3-celled; style short; stigma thick, capitate or orbicular and flat, sometimes 3-lobed; ovules 4–16 in each cell in 2 series, adnate to the axis, ascending. Capsules oblong, 3-celled, woody, 3-angled, dehiscing loculicidally by 3 valves. Seeds 4–8(–16) in each cell, imbricate, exalbuminous, with a large wing at the apical end; the wing very broad, oblong, truncate or blunt.

About 8 species in the tropical forests of Malesia, Sri Lanka and Burma; 1 in Sri Lanka.

Kokoona zeylanica Thw. Hooker's J. Bot. Kew Gard. Misc. 5: 380. t. 6. 1853; Thw., Enum. Pl. Zeyl. 52. 1858; Bedd., Fl. Sylv. t. 146. 1871; Lawson in Hook. f., Fl. Br. Ind. 1: 616. 1875; Trimen, Handb. Fl. Ceylon 1: 270. 1893. Type: Sri Lanka: Ambagamuwa, Mar 1853, 1220 m, *Thwaites C.P. 2584* (K).

A large tree, 20–35 m high, much branched, especially towards the top; outer bark rough, corky, grey, when cut bright yellow or dark orange within, young parts glabrous. Leaves obovate, 3.4–8.4 × 2.4–4.6 cm, coriaceous, glabrous, paler underneath, with numerous dark-red glandular dots covered by a red scale, cuneate at base, rounded or retuse at apex, margin entire or faintly serrate; petioles 8–10 mm long; young leaves elliptic-lanceolate or oblong-lanceolate, 15.0–16.8 × 4.1–6.1 cm, acuminate, serrate. Flowers 6–8 mm diam. in axillary or extra-axillary panicles shorter than the leaves, bisexual, dull yellowish-brown, bracteate; pedicels usually articulated, 4–5 mm long. Bracts ovate, acute, minute. Calyx with 5 shallow lobes; lobes ovate, 1 mm long, obtuse, persistent. Petals semi-orbicular, 2 mm broad, concave, thick, glandular punctate within. Disk large, fleshy, cupular. Stamens 5, inserted on the inner edge of the disk, alternate with the petals; filaments fusiform; anthers oblong, introrse, dehiscing longitudinally. Ovary 3-celled, with 4 ascending ovules in each cell; style short; stigma globular, somewhat 3-lobed. Capsules oblong-ovoid, 10.2 × 2.9 cm, bluntly 3-gonous, 3-valved, 3-celled, black, glabrous. Seeds compressed, light brown, 6.5–8.0 cm long (incl. wing); wing broadly oblong, erect, truncate, veined, exalbuminous.

Distr. Endemic.

Ecol. In forests of moist region, 200–520 m, not common; fls Mar–May, frts May–June.

Vern. Kokoon (S).

Uses. The pounded inner bark is used by jewellers for polishing gold jewellery, and also used as a kind of cephalic snuff, being mixed with ghee (butter-oil) and introduced into the nostrils to relieve severe headache, by encouraging a copious secretion from the nose.

Note. On the type sheets *C.P. 2584* in PDA, the localities Ambagamuwa, Mar–Apr 1853; "Saffragam", 1820 (*Moon*), And Ambagamuwa, Dimbula, Mar–Apr 1852 are annotated. Though Lawson (l. c.) has given Anamallay Hills as the locality from India, Beddome (l. c.) did not mention any locality with his plate. I have not seen any specimen in BM, K or PDA herbaria from Anamallay Hills. Gamble (Fl. Pres. Madras 1, 1918) has not mentioned this genus and species from S. India. In my opinion, it is endemic to Sri Lanka.

Plate No. 416 in PDA, based on specimen *C.P. 2584*, belongs to this species.

Specimens Examined. RATNAPURA DISTRICT: Sinharaja Reserve forest, N. of Pelawatta, 300 m, 7 Oct 1979, *Kostermans 27844* (K, PDA), near Weddagala old camp site, 26 July 1970, *Meijer 526* (K, PDA), Atheha Dola, ±300 m, 27 Feb 1977, *Waas 2073* (PDA); Gilimale forest, Ratnapura to Carney Estate road, 200 m, 3 June 1978, *Kostermans 26601, 26602* (PDA); Peak Wilderness above Devonford and Maratenne Estate, 6°45' N, 80°39' E, 1650 m, 15 Aug 1984, *Jayasuriya, Balasubramaniam, Greller, S. & N. Gunatilleke 2818* (PDA); Peak Wilderness, Carney, 520 m, 2 Oct 1953, *Worthington 6469* (PDA); Udabage, Kelani valley, 270 m, 12 Sept 1946, *Worthington 2099* (PDA); Naunkite Elabode Forest Reserve, 120–135 m, Mar & Sept 1948, *Worthington 4140, 3698* (PDA); Climb to Deniyaya, Sinharaja forest reserve, 13 May 1990, *Jayasuriya & Balasubramaiam 5059* (PDA). GALLE DISTRICT: Hinidum kande (Haycock) near Hiniduma, 400 m, 5 May 1979, *Kostermans 27601* (K, PDA); Nakiyadeniya forest, 120 m, 1 May 1951, *Worthington 5231* (PDA). LOCALITY UNKNOWN: Herb. *Miers s.n.* (BM).

CERATOPHYLLACEAE
(by M.D. Dassanayake*)

S.F. Gray, Nat. Arr. Brit. Pl. 2: 395, 554. 1821 ("Ceratophyllae"). Type genus: *Ceratophyllum* L.

Submerged aquatic herbs floating in still fresh water with branching stems (one branch per node) and without roots. Leaves whorled, sessile, somewhat cartilaginous, flat, bifurcating a few times into narrow linear serrulate segments, without stomata, exstipulate, tipped by bristles. Flowers solitary or clustered, submerged, axillary, small and inconspicuous, regular, unisexual. Plants monoecious, staminate and pistillate flowers at different nodes on the same plant, the staminate flowers often at higher nodes. Perianth in staminate flower of (8–) 12 (–15), in pistillate flower of 9 or 10, persistent, valvate, sepaloid, tepals connate at the base. Staminate flower with (5–) 10–20 (–27) free stamens spirally arranged on a flat receptacle. Filaments very short and broad, not clearly differentiated from anthers. Anthers subsessile, 2-celled, erect, extrorse, longitudinally dehiscent, with connective prolonged into a point, tip of anther with a pair of bristles. Pollen grains smooth, thin-walled, uniaperturate. Pistillode absent in staminate flower. Pistillate flower with superior ovary consisting of a single carpel tapering above into a subulate style with a simple lateral stigma; staminodes absent. Ovule solitary, laminar, pendulous, orthotropous, unitegmic. Fruit an achene, tipped with the persistent style. Seed exalbuminous; embryo large, straight.

One genus, *Ceratophyllum*.

CERATOPHYLLUM

L., Sp. Pl. 992. 1753; L., Gen. Pl. ed. 5. 428. 1754. Type species: *C. demersum* L.

Characters of the family. Two species, cosmopolitan.

Ceratophyllum demersum L., Sp. Pl. 992. 1753; Alston in Trimen, Handb. Fl. Ceylon 6: 271. 1931. Type: from Europe.

Ceratophyllum verticillatum Roxb., Fl. Ind. 3: 624. 1832; Trimen, Handb. Fl. Ceylon 4: 120. 1898. Type: from India.

---

* Flora of Ceylon Project, Peradeniya.

Much-branched aquatic herb. Stems to c. 3 m long, c. 2 mm thick, green, with internodes to c. 1 cm long. Leaves in whorls of 6–12, bright green, 1–4 cm long, mostly twice forked: segments (2–) 4(–5), c. 1 mm wide at their widest. Male flowers solitary (rarely 2) at a node, c. 3 mm broad, pink. Pedicel to c. 0.5 mm long, pink. Tepals c. 8, translucent, with 2 bristles near the apex, lanceolate or narrowly ovate, c. 1 mm long. Stamens 5-c. 23, c. 1.5 mm long, with 2 bristles at the apex. Female flower: pedicel c. 0.25 mm, pink. Tepals 6–8, c. 1.5 mm long, lanceolate, with 2 bristles at the apex, translucent with red markings.·Ovary 0.5–1 mm long, without lateral spines; style c. 2 mm long, bent, tapering. Fruit black, 5–7 mm long, ovoid, compressed, sparsely tuberculate, with apical spine (style) 7–8 mm long and 2 lateral spines c. 2 mm above the base, slightly recurved, 4–5 mm long.

Distr. In most temperate and tropical countries of the world.

Ecol. Ponds, tanks, slow streams, in the lowlands, particularly in the dry region. Flowering March, April.

Specimens Examined. ANURADHAPURA DISTRICT: Ulukku-lama Tank, 25 March 1985, *Jayasuriya 3303, 3304* (PDA). POLON-NARUWA DISTRICT: Minneriya wewa, 23 July 1931, *Simpson 8382* (PDA). POLONNARUWA & BADULLA DISTRICTS: Habarane & Bintenne, *s. coll. C.P. 2311* (PDA).

DILLENIACEAE
(by B.M. Wadhwa*)

Salisbury, Parad. Lond. 2 (1): sub. t. 73. 1807 'Dilleneae'.

Trees, shrubs, lianas or perennial herbs. Leaves alternate or spirally arranged, rarely opposite, simple, sometimes pinnatipartite or pinnatisect (as in *Acrotrema*), exstipulate, but in *Acrotrema* and in a number of species of *Dillenia* petiole with stipule-like, often wholly or partly caducous, wings. Inflorescence cymose or racemose, sometimes reduced to a single flower, terminal or axillary. Flowers bisexual, usually regular, hypogynous, mostly yellow or white, often showy. Sepals (3–) 4–5, imbricate, persistent. Petals (2–) 3–5 (–7), caducous, imbricate in bud, equal, apex rounded or emarginate. Stamens numerous, often partly staminodial, free or partly coherent by their filaments; anthers basifixed, oblong to linear, opening by a terminal pore or a longitudinal slit. Carpels 1–many, free or connate along the central axis; styles free, distinct; ovules 1–many, anatropous, sometimes apotropous, on an axile placenta. Fruits dehiscent or indehiscent, in the latter case enclosed by the sepals. Seeds 1–many, usually arillate, with abundant endosperm; embryo minute, straight.

About ten genera, chiefly tropical, one circumtropical (*Tetracera*), three confined to tropical South America, one (*Dillenia*) in the Old World tropics from Madagascar to Fiji Islands, one (*Schumacheria*) endemic to Sri Lanka, one (*Acrotrema*) in S. India, Sri Lanka and Malay Peninsula, one (*Didesmandra*) endemic to Borneo, one (*Pachynema*) endemic to Australia, and one (*Hibbertia*) distributed in the Southern hemisphere from Madagascar to the Fiji Islands, mainly in Australia. In Sri Lanka four genera and 16 species.

KEY TO THE GENERA

1 Carpels distinctly coherent along their adaxial side; receptacle conical .... **1. Dillenia**
1 Carpels either completely free or loosely coherent along their adaxial side; receptacle flat
  2 Filaments with dilated or broad connectives; anther-cells diverging towards the base .....
  .... **2. Tetracera**
  2 Filaments with linear connectives; anther-cells parallel
    3 Small herbs, perennial; carpels 3, slightly coherent along the adaxial side ....
    .... **3. Acrotrema**
    3 Shrubs or small trees; carpels 3, free .... **4. Schumacheria**

* Royal Botanic Garden, Kew.

1. DILLENIA

L., Sp. Pl. 535. 1753; L., Gen. Pl. ed. 5, 239. 1754; Thunb., Trans. Linn.
Soc. London 1: 198. 1791; Hook. f. & Thoms., Fl. Ind. 1: 59. 1855; Thw.,
Enum. Pl. Zeyl. 5. 1858; Hook. f. & Thoms. in Hook. f., Fl. Br. Ind. 1: 36.
1872; Trimen, Handb. Fl. Ceylon 1: 12. 1893; Hoogl., Blumea 7: 1. 1952.
Type species: *Dillenia indica* L.
Wormia Rottb., Nye Saml. Kongel. Danske Vidensk. Selsk. Skrift. 2: 531.
1783; Hook. f. & Thoms., Fl. Ind. 1: 56. 1855; Thw., Enum. Pl. Zeyl. 4.
1858; Hook. f. & Thoms. in Hook. f., Fl. Br. Ind. 1: 35. 1872; Trimen,
Handb. Fl. Ceylon 1: 11. 1893. Type species: *Wormia triquetra* Rottb.
(= *Dillenia triquetra* (Rottb.) Gilg).
Lenidia Thou., Gen. Nov. Madag. 17. 1806; Poir., Enc. Suppl. 3: 330. 1813;
Poir., Dict. Sci. Nat. 25: 447. 1822. Type species: *Lenidia madagas-
cariensis* Poir. (= *Dillenia triquetra* (Rottb.) Gilg).

Usually trees, some species shrubs only, evergreen or deciduous; branches
in evergreen species sympodial, branching from axil of uppermost leaf below
inflorescence or solitary flower, monopodial in deciduous species. Leaf-scars
either completely amplexicaul or clasping up to 3/4 of branch. Leaves spirally
arranged, simple, entire, undulate or distinctly serrate at margin, pinnately
nerved; petioles channelled above, in some species with completely amplexicaul
caducous wings. Flowers solitary or in few—(up to 20)—flowered racemose
inflorescences, terminal (subsequently leaf-opposed lateral) or in few—(up to
7)—flowered fascicles on the branches below the leaves. Sepals usually 5,
persistent in fruit. Petals 5, much larger than sepals, early caducous. Stamens
numerous, all of the same length or the innermost ones distinctly longer, in
some species the outer ones staminodial; anthers linear, the cells opening with
a terminal pore. Carpels 4–20, attached to a conical receptacle, otherwise free,
each with 5–80 ovules attached to 2 marginal adaxial placentae; styles linear,
curved outward. Pseudocarps consisting of gynoecium and persistent calyx, the
enlarged fleshy sepals permanently enclosing the gynoecium or spreading on
maturity. Fruiting carpels indehiscent or spreading star-like on maturity. Seeds
1–few, dark brown to black, arillate or exarillate.

A tropical genus with about 65 species, from Madagascar through South-
East Asia and Malaysia to Northern Australia and Fiji Islands; represented in
Sri Lanka by 4 species, 3 undoubtedly native, 1 introduced but now widely
naturalized.

Dillenia bracteata. Two Wight collections, labelled from Ceylon, and
identified as *Dillenia pentagyna* Roxb. and *Dillenia bracteata* Wight respectively, are available in the herbarium of the Royal Botanic Garden, Edinburgh
(E). Most probably these specimens are wrongly labelled, as these are two
species not found in Sri Lanka, hence not included in this work.

KEY TO THE SPECIES

1 Petioles with broad fully amplexicaul wings, leaving an annular leaf scar around the branches; inflorescence a simple or compound raceme, usually 5–20-flowered
  2 Petiolar wings sharply separated from leaf-blade or ± constricted below the blade, caducous; flowers white; all stamens ± of same length, at most few staminodes on outer side of androecium; carpels indehiscent, off-white at maturity . . . . **1. D. triquetra**
  2 Petiolar wings gradually merging into leaf-blade, or not constricted below the blade, ultimately falling with the leaf; flowers yellow; stamens of different shape, staminodes on outer side of androecium, innermost stamens distinctly longer than outer ones; carpels dehiscent, purple-red at maturity . . . . **2. D. suffruticosa**
1 Petioles without broad wings, never completely amplexicaul; leaf scar embracing about 3/4 of the branches; flowers solitary, in 1–3-flowered racemes or in fascicles on the branches below the leaves
  3 Leaves obovate, 10–14-nerved; inflorescences 2–3-flowered; flowers 6–9 cm across; pseudocarps 2–3 cm in diameter . . . . **3. D. retusa**
  3 Leaves oblong, 20–70-nerved; flowers solitary, 15–20 cm across; pseudocarps 8–10 cm in diameter . . . . **4. D. indica**

1. Dillenia triquetra (Rottb.) Gilg in Pflanzenfam. 3(6): 123. 1893; Hoogl., Blumea 7: 30. 1952.

Wormia triquetra Rottb., Nye Saml. Kongel. Danske Vidensk. Selsk. Skrift. 2: 532. t. 3. 1783; Hook. f. & Thoms., Fl. Ind. 1: 67. 1855; Thw., Enum. Pl. Zeyl. 4. 1858; Hook. f. & Thoms. in Hook. f., Fl. Br. Ind. 1: 35. 1872; Trimen, Handb. Fl. Ceylon 1: 11. t. 3. 1893; Worthington, Ceylon Trees t. 1. 1959; Type: Ceylon, Herb. Burman (C, holotype; L, isotype).
Dillenia dentata Thunb., Trans. Linn. Soc. London 1: 201. t. 20. 1791. Type: Ceylon, Herb. Thunberg (UPS, holotype).
Lenidia madagascariensis Poir., Enc. Suppl. 3: 330. 1813. Type: Madagascar, Herb. du Petit-Thouars (P, lectotype; K, isolectotype).
Wormia dentata (Thunb.) DC., Syst. 1: 434. 1818.
Lenidia dentata (Thunb.) Poir., Dict. Sci. Nat. 25: 448. 1822.
Lenidia triquetra (Rottb.) Poir., Dict. Sci. Nat. 25: 448. 1822.
Dillenia aquatica Moon, Cat. 42. 1824, nom. nud.

Trees up to 20 m tall, but usually 5–8 m, 30 cm thick; young branches glabrous to densely sericeously villose. Leaves subcoriaceous, oval to oblong or obovate, 6–30 × 4–19 cm, rounded or often ± retuse to truncate at apex, obtuse to rounded at base, 5–14-nerved, margin entire to distinctly undulate, glabrous and feebly shining above, glabrescent on midrib and nerves beneath; petioles 3–13 cm long, with caducous, linear-lanceolate, amplexicaul wings, 3–8 mm broad near base, 0.5–1.5 mm at apex. Inflorescences terminal, a simple or once, rarely twice, branched raceme, 3–20-flowered, up to 25 cm long; axis 2–4 mm thick, glabrescent; bracts triangular to lanceolate, up to 15 × 4 mm, glabrous above, glabrous to densely sericeously villose beneath.

Flowers 4–7 cm across; pedicels 7–40 mm, distinctly thickened at apex. Sepals usually 5, thick, oval, 17 × 13 mm. Petals 5, obovate-spathulate, 30 × 20 mm, white. Stamens 125–140, of equal length, 8 mm long. Carpels 4–5 (–7), elliptic-oblong, 3–3.5 × 1.5–2 mm, glabrous, with 7–12 ovules in each; styles 6–8 mm long, slightly spreading. Pseudocarps oval, c. 25 × 20 mm, opening by spreading of sepals, but carpels apparently indehiscent. Seeds obovoid, 5 × 3 mm, dark-glossy brown, enclosed by membranous aril.

Distr. Sri Lanka and Madagascar.

Ecol. In rain forests throughout the wet zone, sea level to 1,200 m; fls throughout the year.

Vern. Diyapara (S); the name Godapara (S), the usual name for *D. retusa* Thunb., is also used for·this species, where *D. retusa* is absent, especially in Nawalapitiya (Kandy district).

Uses. The timber is used for panels in cabinet making and plywood.

Note. Plate No. 20 in PDA based on *C.P. 1013*, belongs to this species.

Specimens Examined. COLOMBO DISTRICT: Maturajawela, *Amaratunga 369* (PDA); Colombo, *s. coll. s.n.* (MH 60332). KANDY DISTRICT: St. Katherine Estate, c. 1 mile S. of Dolosbage, 1,000 m, 6 Feb 1969, *Hoogland 11416* (A, BR, E, K, CANB, L, PDA, US); Peradeniya, 450 m, Apr 1887, *Gamble 18745* (K), Peradeniya, in 1831, *Macrae 301* (K, CGE), Peradeniya, cultivated, 15 Feb 1907, *J.M. Silva s.n.* (PDA), *Worthington 6792* (PDA), in 1844, *Serell s.n.* (K), *Pearson 962* (CGE); Attabagie, *Worthington 2816* (PDA); Imboolpitiya, Nawalapitiya, *Worthington 898* (PDA). RATNAPURA DISTRICT: Monarakakila near Weddagala, 300 m, 20 Feb 1969, *Hoogland 11454* (B, BM, CAL, C, CANB, K, US), near Belihul-Oya, 100 m, 25 May 1969, *Kostermans 23648* (K); Mile 10, Ratnapura-Carney road, 200 m, 28 Apr 1969, *Kostermans 23325* (K). GALLE DISTRICT: Hiniduma, 7 May 1973, *Kostermans 24715* (K); Udamalwatte, ± 300 m, 13 May 1972, *Cramer, Balasubramaniam & Jayasuriya 3783* (K); Hiniduma, 250 m, 19 Jan 1973, *Trivengadum & Jayasuriya 263* (K); Elpitiya, *Pearson 935* (CGE); Neluwa, *Hoogland 11459* (A, BISH, C, CANB, G, L, NY, P, PDA, UC, US, W); Nakiyadeniya, *Worthington 5233* (K); Kottawa, *Worthington 6016* (BM); Sinharaja Forest, Aug 1936, *J.R. Baker 2002* (BM); Delalure, Nov 1931, *Simpson 8942* (IFI); Kanneliya forest reserve, along Dediyagala road, 1 Sept 1974, *Jayasuriya & Bandaranayake 1806* (K). LOCALITY UNKNOWN: *Col. Walker s.n.* (G, K, OX); *Col. Walker 120* (E, P); *Champion s.n.* (CGE, GL); in 1841, *Kelaart s.n.* (G); *Koenig s.n.* (BM, C, L); *Rostrup 120* (C); in 1866, *Talmy s.n.* (P); *Wight 7, 12* (E); March 1849, *Wight s.n.* (E); *s. coll. C.P. 1013* (BM, CAL, CGE, G, K, NY, P, PDA); *s. coll. s.n.* (K, NY).

2. Dillenia suffruticosa (Griff.) Martelli in Becc., Malesia 3: 163. 1886; Hoogl., Blumea 7: 70. 1952; Hoogl. in Fl. Thailand 2 (2): 101. 1972.

Wormia suffruticosa Griff. [Not. 4: 706. 1854, Griff., Icon. Pl. As. t. 649. 1854, nom. illegit.] ex Hook. f. & Thoms. in Hook. f., Fl. Br. Ind. 1: 35. 1872, nom. cons. prop. Lectotype: Griffith Kew Distr. 55, Malacca, 1845 (K) (the same as previously given as holotype for *W. suffruticosa* Griff., nom. invalid.: Hoogland, Blumia 7: 70. 1952).
Wormia subsessilis Miq., Fl. Ned. Ind. Eerste Bijv. 619. 1860, nom. rej. prop. Teysm. & Binn., J. Bot. Neerl. 1: 364. 1861; Ridley, J. Straits Branch Roy. Asiat. Soc. 54: 4. 1910. Type: Banka, *Teysmann 3203* HB (U, holotype; CAL, K, L, isotypes).
Wormia burbidgei Hook. f. in Curtis' Bot. Mag. 106: t. 6531. 1880. Type: Borneo, *Burbidge s.n.* (K, holotype).

Large shrubs or small spreading trees up to 8 m tall; branches sympodial, ± glabrescent, younger ones glabrous to densely villose. Leaves deep green above, pale green below, elliptic to ovate, 10–45 × 5–26 cm, 7–27-nerved, rounded to obtuse at apex, obtuse at base, margin entire to dentate; petioles 2–6 cm long with persistent, 4–15 mm broad, amplexicaul wings merging into the leaf-blade. Inflorescences terminal, 4–18-flowered, up to 30 cm long, simple or once (rarely twice) branched racemes; bracts triangular, 6–15 × 3–5 mm, caducous. Flowers 8–12 cm across; pedicels 8–30 mm long, slightly thickened at apex. Sepals 5, obovate, 1.5–2.2 × 0.8–1.2 cm, fleshy, glabrous to rather densely villose outside. Petals 5, obovate-spathulate, 4–5 × 2.5–3 cm, bright yellow. Androecium with c. 100 staminodes on outer side; staminodes linear, 4–6 mm long, yellow, obtuse at apex; stamens c. 175, outermost ones slightly curved, c. 8 mm long, innermost ones with apical part reflexed outward, 13 mm long, with stamens of intermediate length between. Carpels 5–8, elliptic, 5 × 2 mm, glabrous, each with 7–10 ovules; styles spreading, filamentous, c. 10 mm long. Pseudocarps dehiscent, opening up by spreading of sepals and unfolding of carpels. Seeds obovoid, 3 × 2 mm, enclosed by scarlet, membranous aril.

Distr. Native to Sumatra, the Malay Peninsula and Borneo, widely cultivated in the tropics and, apart from Sri Lanka, naturalized also in Java and Jamaica.

Ecol. Naturalized in the wet lowlands along ditches in rice fields, on creek banks in neglected plantations, on roadside and other similar locations, particularly common in the Ratnapura District. Earlier introduced to the Royal Botanic Garden, Peradeniya in the late nineteenth century under the name of *Wormia burbidgei*, and still grown as an ornamental there and in private gardens in the wet zone up to 900 m. Fls throughout the year.

Vern. Diyapara (S) and Godapara (S), the names for *D. triquetra* (Rottb.) Gilg. and *D. retusa* Thunb. respectively, are used locally for this species.

Specimens Examined. KANDY DISTRICT: Peradeniya, cultivated, Nov 1887 and June 1891, *s. coll. s.n.* (PDA); Galboda, *Worthington 6550*

(PDA). KALUTARA DISTRICT: c. 3 miles from Alutgama along road to Matugama, 31 Jan 1969, *Hoogland 11415* (A, B, BISH, BM, BR, CAL, CANB, E, G, K, L, PDA, US). RATNAPURA DISTRICT: Ratnapura-Avissawella road, 150 m, 20 March 1968, *Comanor 1133* (K, PDA, US); Kuruwita, 100 m, 28 Feb 1971, *A.G. Robyns 7254* (K).

3. Dillenia retusa Thunb., Trans. Linn. Soc. London 1: 200. t. 19. 1791; Wight & Arn., Prod. 6. 1834; Thw., Enum. Pl. Zeyl. 5. 1858; Hook. f. & Thoms. in Hook. f., Fl. Br. Ind. 1: 37. 1872; Trimen, Handb. Fl. Ceylon 1: 13. 1893; Gamble, Fl. Pres. Madras 1: 7. 1915; Hoogl., Blumea 7: 100. 1952; Worthington, Ceylon Trees t. 3. 1959. Type: Same as for *Dillenia integra* Thunb.

Dillenia integra Thunb., Trans. Linn. Soc. London 1: 199. t. 18. 1791; Thw., Enum. Pl. Zeyl. 5. 1858, 398. 1864. Type: Sri Lanka, Herb. Thunberg (UPS, holotype).
Wormia integra (Thunb.) Hook. f. & Thoms., Fl. Ind. 1: 68. 1855.
Wormia retusa (Thunb.) Hook. f. & Thoms., Fl. Ind. 1: 68. 1855.
Dillenia retusa Thunb. var. *integra* (Thunb.) Boerl., Cat. Hort. Bogor. 7. 1899.

Evergreen trees, up to 20 m tall, 70 cm thick; branches sympodial, glabrescent. Leaves subcoriaceous, obovate, 8–21 × 5–12 cm, 10–14-nerved, rounded to retuse at apex, acute and decurrent at base, margin entire or slightly dentate, shining glabrous above, sparsely strigose on nerves and midrib beneath; petioles 1.3–2.2 cm long, glabrous above, densely to sparsely strigose beneath. Inflorescences terminal, usually 2–3–(rarely 1) flowered simple raceme, with 2–4 cm long axis; bracts triangular, 4–8 × 1–2 mm. Flowers 6–8 cm across; pedicels 10–35 mm long, slightly thickened at apex. Sepals 5, fleshy, circular to oval, 2 outermost ones smaller, 1.6 × 1.4 cm, 3 innermost ones 2.0 × 2.0 cm, glabrous within, rather densely appressedly silky hairy outside, all densely ciliate at margin. Petals 5, white, obovate-spathulate, 3.5 × 2.5 cm. Stamens c. 180, linear, gradually increasing in length towards the centre of the flower, the outermost ones c. 6.5 mm long, slightly curved, the innermost ones c. 11 mm long, with apical part strongly reflexed; anthers white, linear, opening with pore near the apex on outer side. Carpels 5–6, lanceolate, 8 × 2.5 mm, glabrous, each with 22–34 ovules; styles spreading, cylindric, 7–8 mm long. Pseudocarps indehiscent, orange, 2–3 cm diam. (including sepals). Seeds ovoid, 2.5 × 2 mm, exarillate.

Distr. Sri Lanka and India (Coimbatore and Annamalai Hills).

Ecol. In primary rain forest and on creek banks, also occasionally in secondary forests in the wet zone, apparently absent N.E. of the Adam's Peak range; from near sea level to 450 m; fls almost throughout the year.

Uses. The timber is used for rafters, posts and plywood.

Vern. Godapara (S).

Specimens Examined. COLOMBO DISTRICT: Colombo, *Beckett 963* (MEL). KANDY DISTRICT: Royal Botanic Garden, Peradeniya, 450 m, Apr 1969, *Kostermans s.n.* (K); Peradeniya, cultivated, 22 May 1901, *s. coll. s.n.* (PDA). KALUTARA DISTRICT: Ingiriya-Kurnanamada Kada Reserve, *R.F.O. Ingiriya 590* (Forest Dept., Colombo); Reigam Korale, Sept 1864, *Thwaites C.P. 2860* (BM, BO, CAL, CGE, G, GH, K, MEL, P, PDA). RAT-NAPURA DISTRICT: Morapitiya logging area, 100 m, 29 Apr 1973, *Kostermans 24652* (K); Near Ratnapura, *Hoogland 11431* (BR, CANB, E, PDA, US); Indikada, *Liyanage 1491* (Forest Dept., Colombo). GALLE DISTRICT: Galle, 12 March 1896, *Herb. Rottler s.n.* (K); Elpitiya, *Pearson 935* (CGE); Kunalida, Hiniduma, *Worthington 2320* (PDA); Kanneliya Forest, Udugama, *Worthington 6031* (BM); 2 miles N of Udugama, 60–250 m, 27 Jan 1969, *Hoogland s.n.* (CANB, K, L, PDA,USA); Kottawa, *Worthington 5226* (Forest Dept, Colombo), *s. coll. s.n.* (Forest Dept, Colombo); Galle, *Champion s.n.* (CGE); Kanneliya forest, near Hiniduma, 7 May 1973, *Kostermans 24728* (K). MATARA DISTRICT: Deniyaya, *Worthington 2179* (BM), Nannikila Forest Reserve, *Worthington 4152* (BM). LOCALITY UNKNOWN: Dr. Kelaart *s.n.* (K); *Wight s.n.* (E); *Col. Walker 55* (K, G); *Walker s.n.* (G); *Rudolf s.n.* (MICH); *Thwaites C.P. 2960* (K). Cultivated and introduced from Ceylon in 1821, ex. Herb. Hort. Bot. Calcutta, in 1861, *T. Anderson s.n.* (K); ex. Herb. Hort. Bot. Bogor No. IV. G. 28—(native of Ceylon) Nov 1889, *s. coll. 15* (K); No. IV. G. 22, Aug 1890, *s. coll. 17* (K).

4. Dillenia indica L., Sp. Pl. 535. 1753; Burm. f., Fl. Ind. 124. 1768; Hook. f. & Thoms. in Hook. f., Fl. Br. Ind. 1: 36. 1872; Trimen, Handb. Fl. Ceylon 1: 12. 1893; Gamble, Fl. Pres. Madras 1: 7. 1915; Blatter & Millard, Some Beautif. Ind. Trees 45. 3tt. 1937; Hoogl., Blumea 7: 108. 1952; Worthington, Ceylon Trees t. 2. 1959. Type: India, Rheede, Hort. Malab. 3: 39. tt. 38, 39. 1683.

Dillenia speciosa Thunb., Trans. Linn. Soc. London 1: 200. 1791, nom. illegit.; Roxb., Fl. Ind. ed. Carey 2: 650. 1832; Wight & Arn., Prod. 5. 1834; Hook. f. & Thoms., Fl. Ind. 1: 69. 1855; Hook. f. in Curtis' Bot. Mag. 83: t. 5016. 1857; Thw., Enum. Pl. Zeyl. 5. 1858. Type: India, as for *Dillenia indica* L.

Evergreen trees, up to 30 m tall, 120 cm thick. Leaves oblong, 8–40 × 4–15 cm, 20–70-nerved, acute or obtuse, sometimes slightly acuminate at apex, rounded to acute and decurrent at base, margin serrate; petioles 2.5–10 cm long, glabrous above, ± densely silky strigose beneath. Flowers solitary, terminal, 15–20 cm across; pedicels 4–8 cm long, thickened up to 8 mm at apex, appressedly silky hairy. Sepals 5, yellowish-green, oval, 4–6

× 3–5 cm, up to 10 mm thick at base. Petals 5, white, obovate-spathulate, 7–9 × 5–6.5 cm. Stamens in 2 distinct groups, those of outer group c. 550, slightly curved, 13–15 mm long, those of inner group c. 25, with apical part strongly reflexed, 20–22 mm long. Carpels 14–20, lanceolate, 14 × 3 mm, glabrous, each with 40–80 ovules; styles 25 mm long. Pseudocarps ± globular, 8–10 cm diam. (including sepals), indehiscent, yellowish green. Seeds reniform, 4 × 6 mm, finely echinate, exarillate.

Distr. India, Burma, S. China, Thailand, Indo-China, Sumatra, Malay Peninsula, Borneo and Java.

Ecol. Widespread in low country, below 900 m, generally on creek banks or in smaller creeks; also sometimes planted in gardens.

Uses. The rather acidic fruits are sometimes used in curries or jellies. The timber is used for window frames, gun stocks and beams.

Vern. Hondapara, Wampara (S), Akku (T).

Note. Plate No. 21 in PDA, based on specimen *C.P. 2961*, belongs to this species.

Specimens Examined. ANURADHAPURA DISTRICT: Vavuniya-Horowupotana road, 23/2 mark, *Worthington 4515* (BM). KURUNEGALA DISTRICT: Omeragalla, *Worthington 1648* (BM). MATALE DISTRICT: Nalanda, *Worthington 6377* (PDA). KEGALLE DISTRICT: Kegalle-Bulatkohupitiya road, *Amaratunga 284* (PDA). KANDY DISTRICT: Haragama, *s. coll. C.P. 2961* (DD, K, PDA). RATNAPURA DISTRICT: Belihuloya, *Worthington s.n.* (PDA). LOCALITY UNKNOWN: *Herb. Burman s.n.* (L); *Herb. Houttuyn s.n.* (L); *Col. Walker s.n.* (G); *s. coll. s.n.* (MH 60333).

2. TETRACERA

L., Sp. Pl. 533. 1753; L., Gen. Pl. ed. 5, 237. 1754; Hook. f. & Thoms., Fl. Ind. 1: 62. 1855; Thw., Enum. Pl. Zeyl. 2. 1858; Hook. f. & Thoms. in Hook. f., Fl. Br. Ind. 1: 31. 1872; Trimen, Handb. Fl. Ceylon 1: 6. 1893; Hoogl., Reinwardtia 2: 185. 1953. Type species: *T. volubilis* L.

Delima L., Gen. Pl. ed. 5, 231, et pag. ult. 1754; Hook. f. & Thoms., Fl. Ind. 1: 61. 1855; Thw., Enum. Pl. Zeyl. 2. 1858; Hook. f. & Thoms. in Hook. f., Fl. Br. Ind. 1: 31. 1872; Trimen, Handb. Fl. Ceylon 1: 5. 1893. Type: *D. sarmentosa* L. (= *Tetracera sarmentosa* (L.) Vahl).

High climbing or creeping lianas with flexuous stems. Leaves spirally arranged, simple. Inflorescences terminal or axillary, few to many-flowered, panicles. Flowers actinomorphic, bisexual. Sepals 4 or 5, imbricate, generally reflexed in flower and fruit. Petals 3–4, early caducous, ovate-spathulate, with slightly emarginate apex, white, often with a red tinge. Stamens 8, with thin filaments and broad connectives. Carpels 1–4, free, each with a short style; stigma hardly differentiated, 10–12 ovules in each loculus, attached to 2 marginal adaxial placentae. Fruits coriaceous, opening by longitudinal slits along ventral and dorsal sutures, into 2 valves, 1–few-seeded. Seeds arillate; aril fleshy, fimbriate for half to quarter of its length.

A pantropical genus of 40 species; represented by 2 species in Sri Lanka.

KEY TO THE SPECIES

1 Inflorescence terminal, with numerous (up to 150) flowers; flowers small, 7–10 mm diam., with 1, rarely 2, carpels; sepals 5, glabrous inside; capsules ovoid, attenuate at apex, 6–10 × 4–5 mm .... **1. T. sarmentosa**
1 Inflorescence terminal or axillary, with few (5–10) flowers; flowers large, 25–30 mm diam., with 3 or 4 carpels; sepals 4, sericeous inside; capsules globular, rounded and apiculate at apex, 10 × 10 mm .... **2. T. akara**

1. Tetracera sarmentosa (L.) Vahl, Symb. Bot. 3: 70. 1794; Hoogl., Blumea 9: 588. 1959.

Delima sarmentosa L., Gen. Pl. ed. 5, pag. ult. 1754, sphalm. '*sparmentosa*'; Hook. f. & Thoms., Fl. Ind. 1: 61. 1855, p.p.; Thw., Enum. Pl. Zeyl. 2. 1858; Hook. f. & Thoms. in Hook. f., Fl. Br. Ind. 1: 31. 1872, p.p.; Trimen, Handb. Fl. Ceylon 1: 5. 1893. Type: Sri Lanka, Herb. *Hermann* 2: 19. n. *205* (BM, holotype; L, isotype, folio 74).
Seguieria asiatica Lour., Fl. Cochinch. 341. 1790. Type: Cochinchina, *Loureiro s.n.* (lost); Neotype: Tonkin, *d'Alleizette s.n.* (L, vide Hoogl., Reinwardia 2: 194. 1953).
Tetracera asiatica (Lour.) Hoogl. in Steenis, Fl. Mal. ser. 1, 4: 143. 1951; Hoogl., Reinwardtia 2: 193. f. 2. 1953.

subsp. **sarmentosa** Hoogl., Blumea 9: 588. 1959.

Tetracera asiatica (Lour.) Hoogl. subsp. *zeylanica* Hoogl., Reinwardtia 2: 196. 1953. Type: Sri Lanka, *s. coll. 69* (L).
Tetracera scandens Auct. (non (L.) Merr.): Alston in Trimen, Handb. Fl. Ceylon 6: 3. 1931.

Tall climbers or creeping lianas, reaching at least 15 m in tall trees; branches scabrid with 1 mm long hairs. Leaves oblong, 3–15 × 1.5–7.5 cm, 5–20-nerved, apex obtuse to acute, sometimes rounded, base acute, margin entire to distinctly dentate, deep lustrous green, ± shining and slightly scabrid above, pale green and strongly scabrid below; petioles 4–15 mm long, glabrous to hirsute above, strigose beneath. Inflorescences terminal, 10–25 × 5–15 cm, 30–150-flowered; branches of inflorescence scabrid or sparsely strigose; bracts lanceolate, 3· × 1 mm, with acuminate apex, usually caducous. Flowers 7–10 mm across; pedicels 1–5 mm long. Sepals 5, ovate, 2–4 mm long, reflexed, glabrous inside, slightly strigose outside. Petals 3,

3–4 × 2–3 mm. Stamens 100–125, 3–4 mm long. Carpels usually 1, rarely 2, ovoid, c. 1 × 0.5 mm, with very small style; ovules c. 10. Capsules ovoid, 6–10 × 4–6 mm, acute, with 2–5 mm long beak, 1 (–2)-seeded. Seeds ovoid, 4 × 3 mm, glossy black; aril up to 5 mm long.

Distr. Subspecies endemic.

Ecol. In primary rain forests throughout the wet zone, from near sea level to 610 m, common; fls sporadic from March–June.

Vern. Korasa, Korasawel (S).

Note. It appears that the species flowers rarely and possibly only in a few individuals, though it has been observed in many localities. Possibly the species may be flowering during the wet season, when little collecting is done. On *C.P. 1015*, two more localities, Ambagamuwa (Kandy District), and Kurunegala are also annotated. Hoogland (l. c.) has described 3 more subspecies, viz. subsp. *andamanica* (Hoogl.) Hoogl., subsp. *asiatica* (Lour.) Hoogl. and subsp. *sumatrana* (Hoogl.) Hoogl., based on geographical distribution.

Specimens Examined. PUTTALAM DISTRICT: Puttalam, Aug 1848, *Thwaites C.P. 1015* (BM, BO, CAL, CGE, G, K, MEL, NY, P, PDA). KURUNEGALA DISTRICT: Weudakanda, ± 400 m, 18 June 1976, *Waas 1936* (K). COLOMBO DISTRICT: Colombo, *Beckett 1287* (DD, MEL). KEGALLE DISTRICT: Kalugammane, *J.M. Silva 274* (NY); Kalugala, 8 June 1911, *J.M. Silva s.n.* (PDA). GALLE DISTRICT: Kalubowitiyana Kanda, *Hoogland 11456* (US). LOCALITY UNKNOWN: *Herb. Burman s.n.* (G, L, M); *Champion s.n.* (CGE); *Herb. Houttuyn s.n.* (G); *Koenig s.n.* (BM, C, L); *Macrae 492* (BM, CGE); *Herb. Meeb. s.n.* (L); *Col. Walker 141* (G), *1021* (E), *425* (K); *Col. Walker s.n.* (G, OX); *Wight 3–100* (E); *s. coll. 69* (L, type of *Tetracera asiatica* (Lour.) Hoogl. subsp. *zeylanica* Hoogl.).

2. Tetracera akara (Burm. f.) Merr., Philipp. J. Sci. 19: 366. 1921; Alston in Trimen, Handb. Fl. Ceylon 6: 3. 1931; Hoogl., Reinwardtia 2: 208. 1953.

Calophyllum akara Burm. f., Fl. Ind. 121. 1768. Type: India, Malabar, Akara-patsjoti Rheede, Hort. Malab. 5: 15. t. 8. 1685.
Tetracera rheedii DC., Syst. 1: 402. 1818; DC., Prod. 1: 68. 1824; Wight & Arn., Prod. 1: 5. 1834; Wight, Ic. Pl. Ind. Or. 1: t. 70. 1838. Type: As for *Calophyllum akara* Burm. f.
Tetracera laevis Auct.: Hook. f. & Thoms., Fl. Ind. 1: 62. 1855; Thw., Enum. Pl. Zeyl. 2. 1858; Hook. f. & Thoms. in Hook. f., Fl. Br. Ind. 1: 31. 1872; Trimen, Handb. Fl. Ceylon 1: 6. 1893; Gamble, Fl. Pres. Madras 1: 7. 1915, non Vahl.

Tall climbers or creeping lianas, reaching up to 25 m; young branches strigose, older ones glabrous. Leaves oblong to lanceolate, 5–22 × 1.5–11 cm,

5–10-nerved, apex distinctly acuminate, base acute, margin entire to slightly undulate or dentate, glossy bright green above, paler and dull below; petioles 3–7 (–10) mm long, glabrous to slightly pubescent above, sparsely strigose beneath. Inflorescences terminal or axillary, up to 8 × 6 cm, 2–12-flowered; peduncles 0.5–3 cm, sparsely hirsute to glabrous; bracts lanceolate, 2 × 1 mm, acute, glabrous to slightly hirsute above, sparsely to densely strigose hirsute beneath. Flowers 2.5–3 cm across; pedicels 1–2.5 cm long, hirsute to glabrous. Sepals 4, 8–10 × 6–8 mm, densely sericeous within, glabrous without, margin ciliate, often reflexed in fruit. Petals 3–4, 12–15 × 6–8 mm, white or whitish-green. Stamens c. 250, 7–8 mm long, yellowish white with grey tips. Carpels 3 (–4), 1.5 × 1 mm, with 5 mm long style. Capsules almost globular, c. 10 mm in diam., with 1–3 mm long beak, glabrous, glossy, chocolate-brown, 1–2-seeded. Seeds ovoid, 4 × 3 mm, glossy black; aril 6 mm long.

Distr. India (Malabar, Travancore), Sri Lanka, Cambodia, Malesia (Malay Peninsula, Sumatra, Borneo); nowhere common and generally restricted to lowland forests.

Ecol. In the lowlands of the wet zone, but now mainly under cover of planted trees in cleared areas; fls Aug–Sept; frts Apr.

Vern. Ethkorassawel (S).

Note. On *C.P. 1016*, two localities, viz. Puttalam 'Puttam' and Kalutara 'Caltura' are annotated.

Specimens Examined. PUTTALAM DISTRICT: Panirendawa Forest Reserve, alt. low, 27 July 1973, *Jayasuriya 1240* (K), 79° 53′E, 7° 33′N, alt. low, 5 June 1972, *Jayasuriya & Maxwell 813* (K). COLOMBO DISTRICT: Kirillawala, *Amaratunga 967* (PDA); *Miriswatte, Amaratunga 839* (PDA); Near Colombo, in 1859, *Thomson s.n.* (CAL, MEL). LOCALITY UNKNOWN: *s. coll. C.P. 1016* (BM, BR, CAL, CGE, G, GH, K, MEL, NY, P, PDA); *Col. Walker 1128* (E); *Col. Walker s.n.* (G).

3. ACROTREMA

Jack in Malay. Misc. 1 (Art. 5): 36. 1820; Hook. f. & Thoms., Fl. Ind. 1: 64. 1855; Thw., Enum. Pl. Zeyl. 2. 1858; Benth. & Hook. f., Gen. Pl. 1: 13. 1862; Hook. f. & Thoms. in Hook. f., Fl. Br. Ind. 1: 32. 1872; Trimen, Handb. Fl. Ceylon 1: 6. 1893. Type Species: *A. costatum* Jack.

Perennial herbs, stemless with woody rootstock or with small, horizontal, woody rhizome. Leaves all radical or on a short stem, simple or pinnatisect or pinnatipartite, the petiole with sheathing, membranous, caducous wings. Inflorescence a terminal raceme, sometimes reduced to a single flower, with membranous bracts. Flowers 5-merous, yellow. Stamens 15-many, usually in

3 bundles alternating with the carpels. Carpels usually 3, slightly coherent on the adaxial side or in the centre; styles linear or subulate, recurved; ovules 2–6 (–20). Fruit of 3 follicles, dehiscing irregularly. Seeds 1–15, with a white, membranous aril; testa crustaceous, pitted.

About 9 species, one in India (Deccan Peninsula), one in lower Burma, Peninsular Thailand and the Malay Peninsula, and 7 in Sri Lanka.

KEY TO THE SPECIES

1 Racemes with 2–12 cm long peduncles; leaves lyrate .... 1. A. lyratum
1 Racemes with peduncles not exceeding 3 cm; leaves not lyrate
  2 Leaves pinnate, pinnatifid or pinnatisect throughout
    3 Leaves pinnatisect; leaflets or leaf-segments very unequal; stamens 30 ....
    ....2. A. dissectum
    3 Leaves pinnatifid; leaflets or leaf-segments ± equal, linear-oblong; stamens 20 ....
    ....3. A. thwaitesii
  2 Leaves not pinnate, rarely pinnatisect at the very base
    4 Leaves usually variegated, white along the veins, ovate-oblong .... 4. A. walkeri
    4 Leaves not variegated; variable in shape, narrowly spathulate to lanceolate, oblong-lanceolate or obovate-oblong
      5 Leaves narrowly spathulate to lanceolate or oblong-lanceolate, auriculate at the base, not bullate; stamens 15–40
        6 Leaves oblong-lanceolate, acute-acuminate, crenate-toothed, ± hirsute above, appressed hairy on veins beneath; stamens 40 .... 5. A. intermedium
        6 Leaves narrowly spathulate or narrowly lanceolate, sinuate-toothed, glabrous above, sub-strigose on the costa and nerves beneath; stamens 15–40 .... 6. A. lanceolatum
      5 Leaves obovate-oblong, obtuse, rounded or retuse at the apex, not auriculate at the base, sometimes bullate; stamens 20–50 .... 7. A. uniflorum

1. Acrotrema lyratum Thw. ex Hook. f. in Hooker's J. Bot. Kew Gard. Misc. 8: 242. 1856; Thw., Enum. Pl. Zeyl. 3. 1858; Hook. f. & Thoms. in Hook. f., Fl. Br. Ind. 1: 34. 1872; Trimen, Handb. Fl. Ceylon 1: 9. t. 1. f. 1. 1893. Type: Sri Lanka, "Hinidoon Korle near Nellowe", May 1855, *Thwaites C.P. 3392* (K, holotype; BM, CAL, PDA, isotypes).

Perennial herbs, stemless; rootstocks stout, ascending. Leaves radical, erect, coriaceous, large, 20–25 cm long, lyrate-pinnatifid, glabrous, except the nerves beneath with appressed hairs, the terminal lobe very large, green above, purple beneath, oblong, 6–18 × 4.5–9.5 cm, obtuse at apex, cordate at base, margin sub-denticulate and long ciliate; nerves thick, horizontal, patent, branching near apices, the lower lobes in 3–5 pairs, small, rounded, 1.5 × 1.2 cm, adnate by a broad base, deflexed, with a few sharp prickly teeth and long cilia along margin; petioles c. 2.5 cm long, broadly winged. Inflorescence a terminal, erect raceme, with 8–12 cm long, glabrous peduncles; bracts imbricate, recurved, lanceolate, 5–6 mm long. Flowers large, 1.8–2.0 cm across, yellow, 5-merous; pedicels 2.5–5.0 cm long, glabrous, often recurved. Sepals 5, oblanceolate, 6 × 3 mm, glabrous or sparsely puberulous without.

Petals 5, obovate, 7–8 × 5–6 mm. Stamens c. 35, in 3 bundles alternating with the carpels. Carpels 3, slightly coherent on the adaxial side; ovules c. 20 in each; styles linear, recurved. Fruit of 3 follicles, enclosed by the sepals, dehiscing irregularly, many-seeded. Seeds small, finely echinate.

Distr. Endemic.

Ecol. In moist low country in rocky crevices, rare; fls & fts Dec–May.

Vern. Binberu (S).

Note. Plate No. 16 in PDA, based on *C.P. 3392*, belongs to this species.

Specimens Examined. RATNAPURA DISTRICT: Moulawella Nature Trail, Sinharaja forest, 760 m, 26–27 Nov 1992, *Wadhwa & Weerasooriya 343, 351* (K, PDA). GALLE DISTRICT: "Hinidoon Korle near Nellowe", May 1855, *Thwaites C.P. 3392* (K, holotype; BM, CAL, PDA, isotypes); Ellabode Kande, 25 March 1919, *Lewis & J.M. Silva s.n.* (PDA).

2. Acrotrema dissectum Thw. ex Hook. f. in Hooker's J. Bot. Kew Gard. Misc. 8: 242. t. 4B. 1856; Thw., Enum. Pl. Zeyl. 3. 1858; Hook. f. & Thoms. in Hook. f., Fl. Br. Ind. 1: 34. 1872; Trimen, Handb. Fl. Ceylon 1: 9. 1893. Type: Sri Lanka: Hewesse "Hellesse, Pasdun Korle", May 1855, *Thwaites C.P. 3393* (K, holotype; BM, CAL, PDA, isotypes).

Perennial herbs, stemless; rootstocks small, thick, horizontal. Leaves lanceolate in outline, deeply pinnatisect or pinnate, 5–12.5 × 2.5–3.0 cm; rachis stout, softly hairy; segments very unequal, the larger ones with acute lobes, with 2 or more pairs of very small segments between each pair, white sericeous or silky hairy beneath, pinnae small, lanceolate, acute, the small segments ovate-lanceolate, 2–5 mm long, sometimes shortly petiolate. Inflorescence terminal, erect racemes with 2.0 cm long peduncles, pilose. Flowers small, c. 5 mm across, yellow; pedicels slender, 2.5–3.0 cm long, with long, spreading hairs. Sepals 5, orbicular, 3 × 2 mm, with spreading hairs on outer side. Petals 5, ovate, 4 × 2 mm. Stamens c. 30 in 3 bundles. Carpels 3, slightly coherent adaxially; ovules many in each. Follicles dehiscing irregularly, many-seeded. Seeds large, pitted.

Distr. Endemic.

Ecol. Among mosses under shade in wet zone lowlands, rare; fls April.

Note. Plate No. 15 in PDA, based on specimen *C.P. 3393*, belongs to this species.

Specimens Examined. KALUTARA DISTRICT: Gulanawatte forest near Pelawatta, 200 m, 29 Jan 1969, *Hoogland 11412* (PDA). GALLE DISTRICT: Hewesse "Hellesse, Pasdun Korle", May 1855, *Thwaites C.P. 3393* (K, holotype; BM, CAL, PDA, isotypes).

3. Acrotrema thwaitesii Hook. f. & Thoms. ex Hook. f. in Hooker's. J. Bot. Kew Gard. Misc. 8: 241. t. 4A, f. 1–5. 1856; Thw., Enum. Pl. Zeyl. 3. 1858;

Hook. f. & Thoms. in Hook. f., Fl. Br. Ind. 1: 34. 1872; Trimen, Handb. Fl. Ceylon 1: 9. 1893. Type: Sri Lanka, Dolosbage, Feb 1855, *Thwaites C.P. 3664* (holotype K; isotypes BM, CAL, PDA).

Perennial herbs; rhizome small, thick, horizontal. Leaves lanceolate in outline, 7.5–12.5 × 2.5–3.5 cm, pinnatifid; the segments linear-oblong, 10–15 × 2–3 mm, acute, entire or subdentate, hairy above and on the nerves beneath; rhachis strong; petioles very short, 4–5 mm long, with patent hairs. Racemes terminal, small, 2 cm long, hairy. Bracts lanceolate, 5 × 3 mm, hirsute. Flowers 6–10 mm across, yellow; pedicels slender, 2.5–4 cm long, with spreading hairs. Sepals 5, broadly ovate, 2–2.5 mm long, hirsute without. Petals 5, obovate, 7 × 4 mm. Stamens 15–25 in 3 bundles. Carpels 3, united at the base; styles slender, recurved; ovules many.

KEY TO THE VARIETIES

1 Plants without stolons .... **1. var. thwaitesii**
1 Plants with stolons .... **2. var. stolonifera**

1. var. thwaitesii

Plant without stolons; leaves lanceolate in ouline, 7.5–12.5 × 2.5–3.5 cm, pinnatifid.

Specimens Examined. KANDY DISTRICT: Dolosbage, *Thwaites C.P. 3364* (K, holotype; BM, CAL, PDA, isotypes). GALLE DISTRICT: "Hinidoon Corle", *Thwaites 3393a* (K).

2. var. stolonifera Trimen, Handb. Fl. Ceylon 1: 9. 1893. Type: Sri Lanka, Dolosbage, 905 m, *Thwaites C.P. 3969* (K, holotype; PDA, isotype).

Plants with long, trailing runners or stolons, rooting at the nodes. Leaves lanceolate in outline, 6–7 × 1.2–1.8 cm, pinnatifid; segments more deeply cut in the lower portion.

Distr. Endemic (both varieties).

Ecol. Both varieties occur in moist country, up to 1,000 m, very rare; fls Feb–Mar.

Note. Plate No. 14 in PDA, based on specimen *C.P. 3364*, belongs to var. *thwaitesii.*

Specimens Examined. KANDY DISTRICT: Dolosbage, 915 m, *Thwaites C.P. 3969* (K, holotype; PDA, isotype); *Thwaites C.P. 3364b* (K, PDA). GALLE DISTRICT: "Pasdoon Corle", *Thwaites 3393* (K, PDA).

4. Acrotrema walkeri Wight ex Thw., Enum. Pl. Zeyl. 3. 1858; Hook. in Curtis' Bot. Mag. 88: t. 5353. 1862; Hook. f. & Thoms. in Hook. f., Fl. Br. Ind. 1: 33. 1872. Lectotype: Sri Lanka, Ambagamuwa, Feb 1846, *Thwaites C.P. 694* (K, holotype; BM, CAL, PDA, isolectotypes).

Acrotrema walkeri Wight ex Thw. var. *argyroneuron* Thw., Enum. Pl. Zeyl. 398. 1864; Hook. f. & Thoms. in Hook. f., Fl. Br. Ind. 1: 33. 1872. Type: Sri Lanka, Hewesse, "Pasdoon Corle", *Thwaites C.P. 3831* (K; BM, PDA, isotypes).

Acrotrema uniflorum Hook. var. *walkeri* (Wight ex Thw.) Trimen, Handb. Fl. Ceylon 1: 8. 1893.

Perennial herbs; rhizome very small, woody. Leaves narrowly obovate-oblong, 4.8–12.5 × 1.50–4.8 cm, base subauriculate, margin crenate-dentate to sharply toothed, ciliate, deeply plaited and laxly hairy above, dull red and hairy on the veins below, frequently variegated with white along the midrib and veins above; petioles short, 6–8 mm long, winged. Racemes axillary, short, laxly few-many-flowered. Flowers yellow, 1.8–2 cm across; pedicels 3–5 cm long, laxly hairy or villous. Sepals 5, ovate, c. 7 × 2 mm, patent, ciliate-hairy without. Petals 5, obovate, 8–10 × 4–5 mm. Stamens c. 15, filaments in 3 bundles, dilated at the apex; anthers erect, ± linear. Carpels 3, more or less united with 2-many ovules; on maturity irregularly dehiscent. Seeds with membranous arils.

Distr. Endemic.

Ecol. In the montane forests in the wet zone, 610–1,300 m, common; fls June–Oct.

Vern. Binberu (S).

Note. Plate No. t. 10 in PDA, based on specimen *C.P. 694*, belongs to this species.

The leaves of this species are frequently marked on the upper surface with a pale-white area occupying a more or less wide space on either side of the midrib and primary veins, and are thus easily recognizable in the field.

Specimens Examined. KEGALLE DISTRICT: Nagastenne Group, c. 6 miles W. of Nawalapitiya, 200 m, 1 Apr 1969, *Hoogland 11571, 11572, 11573* (PDA). KANDY DISTRICT: Dalhousie, c. 3 miles W. of Maske-liya, 1300 m, 2 Mar 1969, *Hoogland 11469* (PDA); St. Katherine Estate, c. 1 mile S. of Dolosbage, 1000 m, 6 Feb 1969, *Hoogland 11417, 11419* (PDA); Maskeliya, foot of Adam's Peak, 1200 m, 21 Sept 1969, *C.F. & R.J. van Beusekom 1538* (PDA); Rajamally, Moray Estate, 25 Oct 1975, *Sohmer & Sumithraarachchi 9817, 9851* (PDA); Near Maskeliya, 1200 m, 4 Mar 1969, *Hoogland 11479* (PDA); Surroundings of Maskeliya, c. 1, 200 m, 20 Sept 1969, *C.F. & R.J. van Beusekom 1530* (PDA); Adam's Peak trail, 1000 m, 28 March 1956, *Chatterjee 599* (BM, CAL); Gartmore, Rassmalayi, 29 July 1924, *J.M. Silva s.n.* (PDA); Near Gartmore Estate, Maskeliya, 29 Apr 1926, *J.M. Silva s.n.* (PDA); Castlereagh, 3 July 1927, *Alston 1897* (PDA); "Saffragam", *Wight 16* (K); Adam's Peak, Mar 1846, *Thwaites C.P. 345, C.P. 446* (K); Laxapana, Maskeliya, 24 Dec 1931, *Simpson 8990* (BM).

RATNAPURA DISTRICT: Sinharaja Forest, 425 m, 29 Jan 1960, *Rathnapala 50* (PDA); forest path, Sinharaja Gala rock, 28 Nov 1992, *Wadhwa & Weerasooriya 363* (K, PDA). NUWARA ELIYA DISTRICT: Ambagamuwa, Feb 1846, *Thwaites 694* (K, lectotype; BM, CAL, K, PDA, isolectotypes). GALLE DISTRICT: Hiniduma, 680 m, 25 Oct 1975, *Bernardi 15443* (PDA). KALUTARA DISTRICT: Hewesse, "Pasdoon Corle", Aug 1864 & Sept 1865, *Thwaites 3831* Type of *A. walkeri* Wight ex Thw. var. *argyroneuron* Thw. (K, holotype; BM, PDA, isotypes).

5. Acrotrema intermedium Thw., Enum. Pl. Zeyl. 3. 1858; Hook. f. & Thoms. in Hook. f., Fl. Br. Ind. 1: 33. 1872; Trimen, Handb. Fl. Ceylon 1: 8. 1893. Type: Sri Lanka, Ambagamuwa, Feb 1854, *Thwaites C.P. 3114* (K, holotype; BM, CAL, K, PDA, isotypes).

Acrotrema uniflorum Hook. var. *membranaceum* Trimen, Handb. Fl. Ceylon 1: 8. 1893. Type: Sri Lanka, Dolosbage, Jan 1866, *Thwaites C.P. 3897* (CAL, K, PDA, syntypes).

Perennial, stemless herbs; rootstocks elongated, up to 6.0 cm, woody. Leaves obovate-lanceolate, 10–30 × 2.5–5.6 cm, deep green above, pale green below, acute or acuminate, much attenuate below, base auriculate, margin crenate-toothed or strongly serrate, more or less hirsute above, adpressed hairy on the veins beneath; petioles 2.5–3.5 cm long, winged. Racemes short, subsessile, many-flowered. Flowers large, 7–8 mm across, yellow; pedicels c. 4 cm long, hirsute or with long spreading hairs. Sepals 5, ovate, patent, pale-green to purplish, hirsute without. Petals 5, obovate. Stamens c. 40 in 3 bundles, alternating with the carpels. Carpels 3, coherent on the adaxial side; styles linear, recurved; ovules many. Follicles ovoid, dehiscing irregularly. Seeds with membranous aril.

Distr. Endemic.

Ecol. In moist low country, 100–600 m, common; fls Dec–Apr.

Specimens Examined. KEGALLE DISTRICT: Near entrance to Ingoya Estate, 50 m, 2 Apr 1969, *Hoogland 11577* (PDA); Ingoya Estate near Kitulgala, 450 m, 2 Apr 1969, *Hoogland 11578* (PDA); Kitulgala Forest Reserve across Kelani Ganga, c. 150 m, 25 Feb 1971, *Robyns 7225* (K, PDA); Kitulgala, Makandawa Makulana, ± 100 m, 10 Dec 1977, *Cramer 5038* (K); Hills S. of Kitulgala across Kelani Ganga, 180 m, 20 Nov 1974, *Davidse & Sumithraarachchi 8525* (PDA); Kitulgala, 28 Aug 1927, *Alston 896* (K, PDA); Kalugammana, 15 Sep 1927, *J.M. Silva s.n.* (PDA); Dambemulatenna Makulana, Kalugammana, 1 Feb 1927, *J.M. Silva s.n.* (PDA). KANDY DISTRICT: Dolosbage, Jan 1866, *Thwaites C.P. 3897.* Syntype of *Acrotrema uniflorum* Hook. var. *membranaceum* Trimen (CAL, K, PDA). RATNAPURA DISTRICT: Adavi Kande, Eratane, 19 Feb 1927, *J.M. Silva 117* (PDA).

NUWARA ELIYA DISTRICT: Ambagamuwa, *Thwaites C.P. 3114* (K, holotype; BM, CAL, K, PDA, isotypes).

6. Acrotrema lanceolatum Hook., Ic. Pl. 2: sub t. 157. 1837; Hook. f. & Thoms., Fl. Ind. 1: 65. 1855; Thw., Enum. Pl. Zeyl. 3. 1858; Hook. f. & Thoms. in Hook. f., Fl. Br. Ind. 1: 33. 1872; Trimen, Handb. Fl. Ceylon 1: 8. 1893. Type: Sri Lanka, *Wight 15* in Herb. Hookerianum (K).

Acrotrema gardneri Thw., Enum. Pl. Zeyl. 3. 1858; Hook. f. & Thoms. in Hook. f., Fl. Br. Ind. 1: 33. 1872; Trimen, Handb. Fl. Ceylon 1: 8. 1893. Type: Sri Lanka, *Thwaites C.P. 253* (PDA, holotype; BM, CAL, K, isotypes).

Perennial, stemless herbs; rootstock horizontal to erect, short, woody. Leaves linear-spathulate to linear-lanceolate, 5–30 × 1–3 cm, sub-acute to acute, base auriculate, margin sinuate-toothed, glabrous above, adpressed hairy on veins beneath; petioles 1.5–3.0 cm long, winged. Racemes short, erect, few-flowered. Flowers 5-merous, 4–5 mm across, yellow; pedicels 2.5-5 cm long, reddish with appressed hairs. Sepals 5, ovate, 3–3.5 × 2 mm, patent, hairy outside. Petals 5, obovate, 3.5–4 × 2.5–3 mm. Stamens variable, 15–40, in 3 bundles. Carpels 3, coherent on the adaxial side; styles linear; ovules 2–many. Follicles dehiscing irregularly.

Distr. Endemic.

Ecol. In moist low country forests by streams, not common; fls Jan–Mar.

Note. Possibly, *Wight No. 15* in Herb. Hookerianum is the type of *A. lanceolatum* Hook., and not *Mrs. Col. Walker* (No. 15), as given in the protologue, as nowhere else in literature, is there reference to Mrs. Walker's collection. Both Thwaites (l. c.) and Benth. & Hook. f. (l. c.) have mentioned the Wight collection only.

Usually the leaves are linear-spathulate and in *C.P. 2660* only, the leaves are linear-lanceolate.

Hooker. fil. and Thomson (l. c.), while describing *A. gardneri* have commented 'very near to *A. lanceolatum* Hook., but differs in the number of stamens, perhaps not a constant character'.

Specimens Examined. KEGALLE DISTRICT: Banks and Forest along tributary to Kelani Ganga, ½ mile S. of Kitulgala, 80° 24'E, 6° 59'N, 60 m, 3 March 1977, *B. & K. Bremer 933* (PDA); Kitulgala 'Kittoolgulle', March 1853, *Thwaites C.P. 2660* (BM, CAL, K, PDA). KALUTARA DISTRICT: Tiniawelle-Telangalkota forest along Maguruganga river, N. of Nelluwa, 200 m, 28 Jan 1969, *Hoogland 11407* (PDA). RATNAPURA DISTRICT: Track in Gilimale forest from Carney Road to Kalu Ganga, 6° 49'N, 80° 25'E, 3 Dec 1976, *R.D. & A.J. Faden 76/460* (K, PDA). LOCALITY UNKNOWN: *Wight No. 15* (K, type of *A. lanceolatum* Hook.); *Thwaites C.P. 253* (BM, CAL, K, PDA, Type of *A. gardneri* Thw.).

7. Acrotrema uniflorum Hook., Ic. Pl. 2: t. 157. 1837; Thw., Enum. Pl. Zeyl. 2. 1858; Hook. f. & Thoms., Fl. Ind. 1: 64. 1855; Hook. f. & Thoms. in Hook. f., Fl. Br. Ind. 1: 33. 1872; Trimen, Handb. Fl. Ceylon 1: 7. 1893. Type: Sri Lanka, Adam's Peak, *Mrs. Col. Walker s.n.* (K, isotypes).

Perennial herbs with woody rootstocks; stem very short, erect, simple. Leaves subradical, membranous, obovate-oblong or narrowly to broadly obovate, 10–23 × 2.5–10.0 cm, usually obtuse at apex and narrowed or attenuate towards the base, the actual base generally cordate or auriculate, usually crenate-dentate, sometimes ± entire, sometimes deeply cut or with separate leaflets at base, glabrous or pilose or scabrid above and beneath, the veins always hairy, sometimes bullate; petioles 1.2–5.0 cm long, narrowly winged, hirsute. Racemes very short, usually 1–2-flowered, sometimes up to 8-flowered. Flowers small, 3–4 mm across; yellow; pedicels 2.5–7 cm long. Sepals 5, oval, 4–5 × 1.5–2 mm; hirsute outside. Petals 5, ovate, longer than the sepals. Stamens 30–50. Carpels 3, coherent on the adaxial side; styles 3, recurved. Capsules 3-locular. Seeds many, reniform, punctate.

Distr. Endemic.

Ecol. In forests of wet zone in shady moist places, 450–1,300 m.

Note. A very variable species, of which the following varieties seem the best marked.

KEY TO THE VARIETIES

1 Leaves with long petioles, up to 7 cm. .... **1. var. uniflorum**
1 Leaves with short petioles, 1–1.5 cm long
  2 Leaves broadly obovate, ± entire .... **2. var. rotundatum**
  2 Leaves narrow to broadly obovate, crenate-serrate to coarsely serrate
    3 Leaves bullate above
      4 Leaves bullate and with 2 pairs of deeply incised, often stalked leaflets at base; stamens
        50 .... **3. var. appendiculatum**
      4 Leaves very bullate, without deeply incised, stalked leaflets at base, stamens 20 ......
      .... **4. var. bullatum**
    3 Leaves not bullate
      5 Leaves large, broadly obovate, 10–15 × 5–8.5 cm, membranous, crenate-serrate, pilose-hirsute above; stamens 50 .... **5. var. sylvaticum**
      5 Leaves narrowly obovate, 5–16.5 × 2–5.0 cm, not membranous, coarsely crenate-dentate in the upper part, teeth becoming larger and deeper downwards, forming at the base 1 or 2 pairs of separate leaflets; pilose-hairy above; stamens c. 27 ..... **6. var. dentatum**

1. var. uniflorum

Acrotrema uniflorum Hook. var. *petiolare* Thw., Enum. Pl. Zeyl. 2. 1858 'petiolaris'; Benth. & Hook. f. in Hook. f., Fl. Br. Ind. 1: 33. 1872 'petiolaris'; Trimen, Handb. Fl. Ceylon 1: 7. 1893. Type: Sri Lanka, *Thwaites 265* (K, holotype; BM, CAL, K, MH, PDA, isotypes).

Leaves narrow, obovate-oblong, glabrous to scabrid above; petioles up to 7 cm long. Stamens 30.

Note. In the protologue of the species, *Mrs. Col. Walker* (No. *16*), has been mentioned as the type specimen, but on the specimen in K herbarium, no number is given, though the locality Adam's Peak and collector's name are annotated.

There are 3 specimens of *C.P. 265* in K, on one sheet, Horton Plains (Feb 1846) and on the other sheets "Saffragram" (Sept. 1865) and Central Province (Mar 1866) as localities are annotated. On the Herbarium specimen *C.P. 3896* in PDA, "Pasdoon Corle" (Dec 1848), *Gardner* has been annotated; while in K, Central Prov. (Mar 1866) has been annotated.

Plate No. 10 in PDA, based on specimen *C.P. 265*, belongs to this variety.

Specimens Examined. KEGALLE DISTRICT: Kitulgala, 11 Nov 1975, *Sohmer & Waas 10541, 10544* (PDA). KANDY DISTRICT: Maskeliya, above Moray Estate, 1,350–1,700 m, 14 Sept 1977, *Nooteboom & Huber 3121* (PDA); Maratenne, c. 2 miles W. of Norton Bridge, 900 m, 3 March 1969, *Hoogland 11474* (PDA). KALUTARA DISTRICT: Rainforest near Ihala Hewessa, 150 m, 80°27'E, 6°21'N,, 17 Feb 1977, *B. & K. Bremer 823* (PDA). RATNAPURA DISTRICT: Along Kalu Ganga above Carney, 450 m, 18 Feb 1969, *Hoogland 11423* (PDA); Along Karuwalagulla Creek, Bambarabotuwa Forest near Hettikanda, c. 10 miles E. of Ratnapura, c. 500 m, 8 Apr 1969, *Hoogland 11587* (PDA); Gilimale, Adam's Peak S. side, 400–500 m, 17 Sept 1977, *Nooteboom & Huber 3176* (PDA); Kitulgala, Makandawa Mukulana, ± 100 m, 10 Dec 1977, *Cramer 5038* (K, PDA); Along Kaluganga near Palabaddale, along track from Carney to Adam's Peak, 400 m, 7 Apr 1969, *Hoogland 11581* (PDA); Sinharaja Forest boundary near Martin's bungalow, 26 Nov 1992, *Wadhwa & Weerasooriya 349* (K, PDA); Between Pawanella and Linihela, along track from Carney to Adam's Peak, 750 m, 7 Apr 1969, *Hoogland 11583* (PDA); Katussagala Hill, 450–600 m, 6°43'N, 80°29'E, 5 Dec 1976, *R.B. & A.J. Faden 76/486* (PDA); Along Dorenella Creek near Kudawe, c. 2 miles S.W. of Weddagala, 400 m, 20 Feb 1969, *Hoogland 11436* (PDA); Tithaweraluwa Kotha, 27 March 1919, *s. coll. s.n.* (PDA); Warukandeniya-Sinharaja, ± 400 m, 24 Feb 1977, *Waas 1988* (PDA); Sinharaja Forest, S.E. of Weddagala, 20 July 1971, *Meijer 895* (K, PDA); Hettikanda, Bambarabotuwa Forest Reserve, 550 m, 19 Feb 1969, *Hoogland 11435* (PDA); Near Hettikanda, c. 10 miles E. of Ratnapura, c. 450 m, 8 Apr 1969, *Hoogland 11589* (PDA); Tumbagoda, above Balangoda, 200 m, 11 June 1971, *Kostermans 24499* (K, PDA); Sinharaja Forest, Sept 1865, *Thwaites C.P. 3882* (K, PDA); Gatanitulla Kanda, along track from Carney to Adam's Peak, c. 1,300 m, 7 Apr 1969, *Hoogland 11585, 11586* (PDA), *Meijer 921* (PDA). GALLE DISTRICT: Along Pandola

Ella on S. slope of Haycock, Hiniduma Kanda, 150 m, 1 March 1969, *Hoogland 11466* (PDA); Near Hiniduma, 750 m, 26 Oct 1975, *Bernardi 15489* (K, PDA); Hiniduma Kanda, 400 m, 6 Oct 1974, *Tirvengadum & Kostermans 638* (K). LOCALITY UNKNOWN: *Col. Walker s.n.* (K); *Thwaites C.P. 3896* (K, PDA); *Thwaites C.P. 265* (Type of *Acrotrema uniflorum* Hook. var. *petiolaris* Thw., K, holotype; BM, CAL, MH, PDA, isotypes).

2. var. **rotundatum** Thw., Enum. Pl. Zeyl. 2. 1858 'rotundata'; Hook. f. & Thoms. in Hook. f., Fl. Br. Ind. 1: 33. 1872; Trimen, Handb. Fl. Ceylon 1: 7. 1893. Type: Sri Lanka, "Pasdoon Corle", 150 m, *Thwaites C.P. 1014* (K, lectotypes; BM, CAL, K, MH, PDA, isolectotypes).

Leaves broadly obovate, 7–16.5 × 4.5–10 cm, obtuse or retuse at apex, more or less entire; hairs stiff; petioles very short, c. 6 mm long.

Note. The sheet bearing *C.P. 3486* in PDA is wrongly labelled. It should read as *C.P. 3484*, as given in the protologue.

Specimens Examined. KALUTARA DISTRICT: Thotaha-Kalutara road, 30 Oct 1970, *Kundu & Balakrishnan 471* (PDA); Gulanawatta Forest near Pelawatte, 200 m, 29 Jan 1969, *Hoogland 11413* (PDA); Pelawatte, alt. sea level, 15 Sep 1970, *Cramer 3178* (PDA). RATNAPURA DISTRICT: Adam's Peak descent, 1525 m, 8 Apr 1970, *Balakrishnan 286* (PDA); Kehelwatupola near Kudawe, Weddagala, 300 m, 20 Feb 1969, *Hoogland 11449* (PDA); Sinharaja Forest, way to Maullawella, 26 Nov 1992, *Wadhwa & Weerasooriya 342* (K, PDA). GALLE DISTRICT: Kaneliya Forest, close to burnt forest bungalow, 30 Nov 1992, *Wadhwa & Weerasooriya 377* (K, PDA); Near Mawita, c. 2 miles N.W. of Nelluwa, Hinidum Pattuwa, c. 250 m, 28 Feb 1969, *Hoogland 11460* (PDA); Hiniduma, 650 m, 25 Oct 1975, *Bernardi 15437* (PDA); Kaneliya Parwita, Dec 1917, *Lewis s.n.* (PDA); "Pasdoon Corle", 150 m, *Thwaites C.P. 1014* (K, lectotype; BM, CAL, MH, PDA, isolectotypes). LOCALITY UNKNOWN: *Thwaites C.P. 3484* (K, PDA); *Col. Walker s.n.* (marked B, K).

3. var. **appendiculatum** Hook. f. & Thoms. in Hook. f., Fl. Br. Ind.1: 33. 1872; Trimen, Handb. Fl. Ceylon 1: 8. 1893. Type: Sri Lanka: "Kittoolgalle", *Thwaites 3880* (K, lectotype; K, PDA, BM, isotypes).

Acrotrema uniflorum Hook. var. *membranaceum* Trimen, Handb. Fl. Ceylon 1: 8. 1893, p.p.

Leaves with 2 pairs of deeply incised, often stalked leaflets at base; petioles short, c. 1 cm long. Stamens 50.

Note. Plate No. 10 in PDA, based on specimen *C.P. 3880*, belongs to this variety.

Specimens Examined. KEGALLE DISTRICT: "Kittoolgalle", Sept 1865, *Thwaites C.P. 3880* (K, holotype; BM, CAL, K, PDA, isotypes). RATNAPURA DISTRICT: W. of Adam's Peak, Mar 1866, *Thwaites C.P. 3898* (Syntype of *A. uniflorum* var. *membranaceum* Trimen, K, PDA).

4. var. bullatum (Thw.) Hook. f. & Thoms. in Hook. f., Fl. Br. Ind. 1: 33. 1872; Trimen, Handb. Fl. Ceylon 1: 7. 1893.

Acrotrema bullatum Thw., Enum. Pl. Zeyl. 2. 1858. Type: Sri Lanka, Ratnapura, *Thwaites C.P. 239* (K, holotype; BM, CAL, MH, PDA, isotypes). *Acrotrema uniflorum* Hook. var. *rugatum* Trimen, Handb. Fl. Ceylon 1: 8. 1893. Syntypes: Sri Lanka, W. of Adam's Peak, Mar 1866, *s. coll. C.P. 3905* (BM, K, PDA); Locality unknown, *s. coll. C.P. 3899* (BM, CAL, K, PDA).

Leaves very bullate above and excavated between the veins beneath, softly pubescent or hispid hairy; petioles short, 5 mm long. Pedicels with adpressed hairs. Stamens 20.

Note. Plate No. 10 in PDA, based on specimen *C.P. 239*, belongs to this variety.

Specimens Examined. KANDY DISTRICT: Moray Estate near Fishing Hut, 27 March 1974, *Sumithraarachchi & Jayasuriya 176* (PDA); Along road to Nagastenne Group, c. 3.5 miles W. of Nawalapitiya, 850 m, 1 Apr 1969, *Hoogland 11575* (PDA); Forest remnants, 1 mile W. of Herondale, 670 m, 80°29'E, 7°23'N, 8 March 1977, *B. & K. Bremer 1016* (PDA); Delgoda, 23 March 1919, *Lewis & J.M. Silva s.n.* (PDA). KALUTARA DISTRICT: Mandagala, 100 m, 4 Dec 1975, *Bernardi 15722* (PDA). RATNAPURA DISTRICT: Near Lellopitiya, c. 5 miles E. of Ratnapura, c. 150 m, 8 Apr 1969, *Hoogland 11590* (PDA); Ratnapura Hill, 11 Apr 1930, *A. de Alwis s.n.* (PDA); Weddagala Entrance, Sinharaja forest, 21 Sept 1977, *Meijer & Gunatilleke 1408* (PDA); Getanagama, c. 1 mile S.W. of Ratnapura, 30 m, 15 Apr 1969, *Hoogland 11595* (PDA); Diganekanda, c. 5 miles S. of Ratnapura, 2 miles N.W. of Karawita, 150 m, 6 Apr 1969, *Hoogland 11580* (PDA); Ratnapura, Mar 1846, *Gardner C.P. 239* (K, holotype; BM, CAL, MH, PDA, isotypes); Gilimale Forest Reserve, 120–180 m, 27 Feb 1971, *Robyns 7244* (PDA); Ratnapura-Pelmadulla, 68 mile post, 27 Aug 1963, *Amaratunga 711* (PDA); Pagoda on Gilimale Forest road, 80°25'E, 6°45'N, c. 90 m, 3 Dec 1976, *R.B. & A.J. Faden 76/454* (PDA); Near Kahawatta along road, 15 Aug 1932, *Simpson 9995* (BM); Ratnapura to Gilimale, 27 June 1972, *Hepper, Maxwell & Fernando 4521* (PDA); W. of Adam's Peak, Mar 1866, *s. coll. C.P. 3905* (Syntype of *A. uniflorum* var. *rugatum* Trimen, BM, K, PDA). GALLE DISTRICT: Kanneliya Forest, 20 Nov 1973, *Sohmer, Waas & Eleizer 8933* (PDA); Kanneliya Forest, c. 2 miles N. of Udugama, Hinidum Pattuwa, c.

60 m, 27 Jan 1969, *Hoogland 11400* (PDA); Kanneliya Forest, near mile 2, 25 July 1971, *Meijer 1001* (K, PDA). LOCALITY UNKNOWN: *s. coll. C.P. 3899* (Syntype of *A. uniflorum* var. *rugatum* Trimen, BM, CAL, K, PDA).

5. var. sylvaticum (Thw.) Hook. f. & Thoms. in Hook. f., Fl. Br. Ind. 1: 33. 1872; Trimen, Handb. Fl. Ceylon 1: 8. 1893.

Acrotrema sylvaticum Thw., Enum. Pl. Zeyl. 2. 1858. Type: Sri Lanka, "Eknalagodde", Ratnapura, Mar 1853, *Thwaites C.P. 2659* (K, holotype; BM, CAL, MH, PDA, isotypes).

Leaves large, broadly obovate, 10–15 × 5–8.5 cm, membranous, pilose-hirsute above, with rigid hairs on the nerves beneath, base deeply cordate; petioles short, 8–10 mm long. Stamens 50.

Specimens Examined. KEGALLE DISTRICT: Lassanagama, between Dehiovita and Deraniyagala, 500 m, 16 Nov 1974, *Jayasuriya & Bandaranayake 1856* (PDA). KALUTARA DISTRICT: Hallawakellae Forest, Welipenna, 28 Jan 1972, *Balakrishnan 1183* (PDA). RATNAPURA DISTRICT: Adam's Peak to Carney, 1700 m, 21 Dec 1975, *Bernardi 16081* (PDA); Adam's Peak, N.E. slope, 750–1200 m, 26 Feb 1971, *Robyns 7240* (PDA); Tittaweraluwa Kotha, 27 March 1919, *s. coll. s.n.* (PDA); Ellaboda Kande, 24 March 1919, *Lewis & J.M. Silva s.n.* (PDA); Gikiyana Kande, *Willis s.n.* (PDA); "Eknalegodda", Ratnapura, Mar 1853, *Thwaites C.P. 2659* (K, holotype; BM, CAL, MH, PDA, isotypes).

6. var. dentatum Trimen, Handb. Fl. Ceylon 1: 8. 1893. Type: Sri Lanka, Morowak Korle, Sept 1865, *Thwaites C.P. 3881* (PDA, holotype; BM, K, isotypes).

Acrotrema uniflorum Hook. var. *membranaceum* Trimen, Handb. Fl. Ceylon 1: 8. 1893, p.p.

Leaves narrowly obovate, 5–15.5 × 2–5 cm, coarsely crenate-dentate in the upper part, teeth becoming larger and deeper downwards, forming at the base 1 or 2 pairs of separate leaflets. Racemes short, 6–7-flowered. Stamens c. 27.

Specimens Examined. KANDY DISTRICT: Dolosbage, *Thwaites C.P. 3897+* (K, PDA); Dolosbage, May 1866, *Thwaites C.P. 3897++* (K, PDA). RATNAPURA DISTRICT: ·Along Kudulgaldole Creek, Karawita Kanda Proposed Reserve, c. 7 miles S. of Ratnapura, 350 m, 9 Apr 1969, *Hoogland 11592* (PDA); Pawanella, track from Carney to Adam's Peak, 600 m, 7 Apr 1969, *Hoogland 11582* (PDA); Karawita Kande, 17 March 1881, *s. coll. s.n.* (PDA). MATARA DISTRICT: Morawak Korale, Sept 1885, *Thwaites C.P. 3881* (PDA, holotype; BM, K, isotypes).

4. SCHUMACHERIA

Vahl, Skr. Naturhist. Selsk. 6: 122. 1810; Wight, Ill. Ind. Bot. 1: 9. 1831; Arn., Edinburgh New Philos. J. 16: 315. 1834; Hook. f. & Thoms., Fl. Ind. 1: 65. 1855; Thw., Enum. Pl. Zeyl. 4. 1858; Hook. f. & Thoms. in Hook. f., Fl. Br. Ind. 1: 34. 1872; Trimen, Handb. Fl. Ceylon 1: 10. 1893. Type: *S. castaneifolia* Vahl.

Pleurodesmia Arn., Edinburgh New Philos. J. 16: 316. 1834, nom. illegit. pro syn.

Erect or scrambling evergreen shrubs or small trees; branches mostly zigzag, glabrous, with completely amplexicaul leaf-scars. Leaves alternate, coriaceous, glabrous, simple, exstipulate, repando-serrate, the serratures mucronate; petioles channelled above, with completely amplexicaul wings enclosing the terminal bud in young leaves, subsequently splitting in leaf-opposed centre and persistent with the leaf. Flowers in axillary or terminal, simple or panicled spikes, sessile or nearly so, 2-bracteate. Sepals 5, 2 exterior, 3 interior, imbricate, persistent in fruit. Petals 5, imbricate, yellow, little larger than the sepals, early caducous. Stamens numerous, monadelphous, the connate filaments forming a semicylindrical column at abaxial side of flower; anthers linear, elongate, 2-celled; connective with apical mucro, cells opening with apical pore. Carpels 3, free, densely hairy, with a single basal ovule; style terete, curved. Fruiting carpels dehiscent, spreading star-like. Seeds globular at base, with small, membranous aril.

Endemic, represented by 3 species.

KEY TO THE SPECIES

1 Panicles terminal and in the axils of the upper few leaves only, with numerous flowers, at least half as long as the nearest leaf; leaves broad-lanceolate to ovate or elliptic, acute at apex .... **1. S. castaneifolia**
1 Panicles mostly in the axils of the leaves, rarely terminal, with relatively few (up to 20, rarely 30 flowers), less than half as long as the nearest leaf
2 Leaves ovate-elliptic to nearly orbicular, rounded or retuse at apex .... **2. S. alnifolia**
2 Leaves oblong-lanceolate, acute to acuminate at apex .... **3. S. angustifolia**

1. Schumacheria castaneifolia Vahl, Skr. Naturhist. Selsk. 6: 122. 1810; Wight, Ill. Ind. Bot. 1: 9. t. 4. 1831; Arn., Edinburgh New Philos. J. 16: 316. 1834; Hook. f. & Thoms., Fl. Ind. 1; 66. 1855; Thw., Enum. Pl. Zeyl. 4. 1858; Hook. f. & Thoms. in Hook. f., Fl. Br. Ind. 1: 35. 1872; Trimen, Handb. Fl. Ceylon 1: 10. 1893. Type: Sri Lanka, *Koenig s.n.*

Schumacheria castaneifolia var. *vahlii* Arn., Edinburgh New Philos. J. 16: 316. 1834; Hook. f. & Thoms., Fl. Ind. 1: 66. 1855, pro var. Type: As for the species.

Schumacheria castaneifolia var. *grahamii* Arn., Edinburgh New Philos. J. 16: 316. 1834; Hook. f. & Thoms., Fl. Ind. 1: 66. 1855. Type: Sri Lanka, *Graham s.n.* (not seen).

Pleurodesmia grahamii Arn., Edinburgh New Philos. J. 16: 316. 1834. nom. illeg. pro. syn.

Large, often straggling shrubs or small trees up to 8 m tall; branches cylindric, hoary, at length glabrate. Leaves ovate or ovate-oblong or elliptic, 10–20 × 5–12 cm, 18–25-nerved, obtuse to acute at apex, obtuse, rounded or retuse at base, sometimes acute at either end, glabrous, hoary or puberulous on the veins beneath, margin sinuate-dentate; serratures mucronate; petioles 5–12 mm long, channelled above. Inflorescences usually terminal spreading panicles, sometimes axillary also in the upper few leaves, up to 40 × 30 cm, many-flowered (up to 500 or more). Flowers 10–14 mm across; bi-bracteolate; bracts sericeous on both sides. Sepals 5, coriaceous, sericeous on both sides, 6–8 × 4–5 mm, 3 outer longer than 2 inner and inner wider than the 3 outer ones. Petals usually 4, crisped, oblong, caducous. Stamens c. 25–30, the filament column c. 1 mm long; anthers c. 4 mm long. Carpels c. 2 mm long with 4 mm long styles, when dehisced c. 4 × 4 mm.

Distr. Endemic.

Ecol. In primary rain forest, but more commonly in disturbed or secondary forests in the wet zone, 100–760 m; fls almost throughout the year.

Vern. Kekiriwara (S).

Note. The leaves are variable, usually obtuse to acute at apex, obtuse or rounded at the base, but sometimes they are acute both at the base and apex. Sometimes leaves of both shapes are found on the same specimen, as in *Meijer 2196, Meijer 494, Waas 9, Townsend 73/48* and *Hoogland 11402*; hence maintaining the varieties on leaf shape alone is not considered correct. Plate No. 17 in PDA, based on specimen *C.P. 230*, belongs to this species.

Specimens Examined. COLOMBO DISTRICT: Labugama, in 1882, *Ferguson s.n.* (PDA); Indikada, *Anandam 1469, 1474* (Herb. Forest Dept, Colombo), *de Silva 1198* (Herb. Forest Dept, Colombo). KALUTARA DISTRICT: Kalawana-Kalutara road, at Morapitiya, 29 Oct 1973, *Sumithraarachchi & Ashton 2* (K); Gulanawatta Forest near Pelawatta, *Hoogland 11414* (A, BR, CANB, E, G, PDA, US). RATNAPURA DISTRICT: Sinharaja MAB Reserve near Kudawa Entrance, Aug 1978, *Meijer 2196* (K); Gilimale Forest Reserve, 25 July 1970, *Meijer 494* (K), *Worthington 4829* (PDA); Mile 10 mark, Ratnapura—Carney road, 200 m, 26 Apr 1969, *Kostermans 23294* (K), Morapitiya Logging area, 100 m, 29 Apr 1975, *Kostermans 24660* (K); Mallaga Ela, east effluent of Kura Ganga above Kuruwita, 8.5 miles N. of Ratnapura, 760 m, 8 Dec 1977, *Fosberg 27332* (K); Weeragama Estate, W. of Ratnapura, 29 Nov 1969, *Read & Jayasuriya*

2311 (K); Kudawe near Weddagala, 300 m, 20 Feb 1969, *Hoogland 11438* (A, CANB, K, L, PDA, US); I.B.P. Plot, Sri Palabaddala, 29 Sep 1973, *Waas 9* (K); Ratnapura, Mar 1846, *Thwaites C.P. 230, 248* (K), *C.P. 232* (K); Indurawa, *Hoogland 11426* (B, BH, CANB, P, PDA, US, PDA); 12 Apr 1930, *de Alwis s.n.* (PDA), *Amaratunga 403* (PDA); Karawita Kande, *J.M. Silva 132* (PDA); Ballahela, Kelani Valley, *Worthington 2101* (PDA); Madampe, *Alston 1115* (PDA); Tundole Ela, Sinharaja Forest, 25 Nov 1992, *Wadhwa & Weerasooriya 336* (K, PDA). GALLE DISTRICT: Kanneliya Forest, 2 miles N. of Udugama, Hinidum Pattuwa, 100 m, 27 June 1969, *Hoogland 11401* (CANB, K, L, PDA, US), *Hoogland 11402* (CAL, CANB, K, L, PDA, US); Kottawa, *Alston 1114* (PDA), *Hoogland 11398* (CANB, PDA, USA), *Hoogland 11399* (CANB, L, US, PDA), *Worthington 626* (Herb. Forest Dept, Colombo), *s. coll. 78, 264* (Herb. Forest Dept, Colombo); Galle, *Wight 4/3* (Leg. Champion, PDA); Hingalgoda, Hiniduma Forest, 30 Nov 1992, *Wadhwa & Weerasooriya 370* (K, PDA); Beraliya Proposed Reserve, 20 Nov 1974, *Tirvengadum 725* (K); Hiniduma, Haycock Hill, 14 Feb 1973, *Townsend 73/48* (K); Kanneliya Forest Station, near Udugama, 30 March 1973, *Bremer, Kers & Thorans 63* (K); Remale Forest, 5 May 1968, *Comanor 1202* (K); Hiniduma, ± 350 m, 31 March 1978, *Cramer 5212* (K). MATARA DISTRICT: Oliyagankele, *Mendis 1229* (PDA). LOCALITY UNKNOWN: *Col. Walker s.n.* (K); *Gardner 8* (K); *Mrs. Gen. Walker s.n.* (K); *s. coll. s.n.* (K, PDA).

2. Schumacheria alnifolia Hook. f. & Thoms., Fl. Ind. 1: 66. 1855; Thw., Enum. Pl. Zeyl. 4. 1858; Hook. f. & Thoms. in Hook. f., Fl. Br. Ind. 1: 34. 1872; Trimen, Handb. Fl. Ceylon 1: 10. 1893. Type: Sri Lanka, *Gardner s.n.* (not seen).

Schumacheria alnifolia Hook. f. & Thoms. var. *scabra* Thw., Enum. Pl. Zeyl. 4: 1858; Hook. f. & Thoms. in Hook. f., Fl. Br. Ind. 1: 34. 1872. Type: Sri Lanka, Adam's Peak, Mar 1846, *Thwaites C.P. 9* (K, PDA).
Schumacheria alnifolia Hook. f. & Thoms. var. *dentata* Thw., Enum. Pl. Zeyl. 4: 1858; Hook. f. & Thoms. in Hook. f., Fl. Br. Ind. 1: 35. 1872; Trimen, Handb. Fl. Ceylon 1: 11. 1893. Type: Sri Lanka, Adam's Peak, Mar 1846, *Thwaites C.P. 339* (K, PDA).
Schumacheria alnifolia Hook. f. & Thoms. var. *subglabra* Thw., Enum. Pl. Zeyl. 4. 1858; Hook. f. & Thoms. in Hook. f., Fl. Br. Ind. 1: 35. 1872; Trimen, Handb. Fl. Ceylon 1: 11. 1893. Type: Sri Lanka, Hunnasgiriya, *Thwaites C.P. 2416* (PDA, syntype), *Thwaites C.P. 397* (K, syntype).

Erect shrubs, with spreading branches; branches when young strigose, old glabrous. Leaves glabrous above, strigose on the nerves beneath and pubescent or glabrous between them, ovate-elliptic to nearly orbicular, 6–10

× 5–13 cm, rounded to retuse at apex, rounded to subcordate, rarely broadly obtuse at base, margin undulate to distinctly dentate; petioles 6–15 mm long, deeply channelled above, scabrid or hirsute beneath. Inflorescences axillary, relatively few-flowered (20–30-flowered), up to 5 × 3 cm, once or twice dichotomously branched; branches silky-hairy, much shorter than the upper nearest leaf. Flowers 8–10 mm across. Sepals subcoriaceous, ovate, 5–6 × 4–5 mm with obtuse apex, silky hairy without. Stamens c. 20–25, the filamental column c. 1 mm, the anthers c. 5 mm long. Carpels c. 1.5 mm long, with 3 mm long styles, when dehisced c. 4 mm long and wide.

Distr. Endemic.

Ecol. In montane forests in wet zone, 1200–2000 m; fairly common; fls & frts June–Apr.

Vern. Kekeriwara (S).

Note. As already suggested by Trimen (1. c.) the varieties are hardly worth maintaining. The indumentum is only moderately variable within the species and the acceptance of the scabrous-pubescent and subglabrous variants as distinct taxa at the varietal level is not justified. *C.P. 339* represents either a young shoot or a luxuriant individual rather than a distinct taxon. Plate No. 18 in PDA, based on *C.P. 2416*, belongs to this species.

Specimens Examined. KANDY DISTRICT: Upper part of Moray Estate near Fishing hut, 1600–1700 m, 16 May 1971, *Kostermans 24122, 24127* (K); Adam's Peak Wilderness bordering Moray Estate, 80°31'E, 6°48'N, 1500 m, 6 March 1974, *Jayasuriya & Sumithraarachchi 1570* (K); Moray Estate forest edge, 6 June 1970, *Balakrishnan 580* (K); Gartmore, Maskeliya, *J.M. Silva 28* (PDA); Maskeliya, 2 March 1883, *s. coll. s.n.* (MH Acc. No. 6033/1, PDA); Galboda, *Worthington 6545* (PDA); Above Blair Athol, Maskeliya, *Alston 1904* (PDA); Brownlow, Maskeliya, *Worthington 2753* (PDA); Rajamallay forest, Moray Estate, 25 Oct 1975, *Sohmer & Sumithraarachchi 9880* (K); Double Cut Junction, Maskeliya, 19 Oct 1971, *Dassanayake 550* (K); Adam's Peak Jungle, above Moray Estate, 2000 m, 22 June 1971, *Kostermans 24215* (K); Adam's Peak, Mar 1846, *Thwaites C.P. 9* (Type of *S. alnifolia* var. *scabra* Thw. PDA, K); Adam's Peak, Mar 1846, *Thwaites C.P. 339* (Type of *S. alnifolia* var. *dentata* Thw. K, PDA); Hunnasgiriya, *Thwaites C.P. 2416* (Syntype of *S. alnifolia* var. *subglabra* Thw. PDA); Peak Wilderness, trail up to Adam's Peak from Moray Estate, ± 2000 m, 23 June 1976, *Waas 1704* (K). BADULLA DISTRICT: Way to Namunukula, 28 Apr 1924, *J.M. de Silva s.n.* (PDA). LOCALITY UNKNOWN: *Thwaites C.P. 397* (Syntype of *S. alnifolia* var. *subglabra* Thw. K).

3. Schumacheria angustifolia Hook. f. & Thoms., Fl. Ind. 1: 66. 1855; Thw., Enum. Pl. Zeyl. 4. 1858; Hook. f. & Thoms. in Hook. f., Fl. Br. Ind. 1: 34.

1872; Trimen, Handb. Fl. Ceylon 1: 11. t. 2. 1893. Lectotype: Sri Lanka, *Col. Walker s.n.* (K).

Small shrubs, up to 2 m tall; branches slender, young silky, older glabrous. Leaves oblong lanceolate or narrow-lanceolate, 8–20 × 1.5–5.0 cm, 12–22-nerved, acute to acuminate at apex, obtuse at base, margin serrate-dentate; petioles 5–10 mm long. Inflorescences axillary, simple or panicled spikes, with only few branches, up to 5 × 4 cm, few-flowered (up to 20). Flowers 6–8 mm across. Sepals membranous, obovate, 4–5 × 3–4 mm, silky without. Stamens c. 20, the filament column c. 1 mm long, the anthers c. 2.5 mm long. Carpels c. 1 mm long, silky hairy, with 2 mm long styles; when mature c. 2 mm long and wide.

Distr. Endemic.

Ecol. In lowland hills in wet zone, 150–700 m; fls throughout the year.

Vern. Hinkekiriwara, Kekiriwara (S).

Note. Out of 2 syntypes, *Col. Walker s.n.*, and *Gardner s.n.* (K), the former representing all the characters, has been selected as the Lectotype in this work. Plate No. 19 in PDA, based on *C.P. 2992*, belongs to this species.

Specimens Examined. KALUTARA DISTRICT: East Kalugala forest, ± 200 m, 1 May 1976, *Waas 1544* (K); Denihena, ± 150 m, 29 Aug 1976, *Waas 1880* (K); Kalugala forest, Pahala Hewessa, ± 150 m, 6 Nov 1975, *Sohmer & Waas 10254, 10260* (K); Gulanawatta forest near Pelawatta, *Hoogland s.n.* (BR, CANB, E, L, PDA, USA). RATNAPURA DISTRICT: Kukul Korale, *s. coll. s.n.* (PDA). GALLE DISTRICT: Hiniduma, towards summit of Hiniduma Kande, 650–750 m, 25 Oct 1975, *Bernardi 15436* (K), 600 m, 2 Sep 1974, *Jayasuriya, Kostermans & Bandaranayake 1814* (K). *Worthington 2278* (Herb. Forest Dept, Colombo), 700 m, 6 Oct 1974, *Kostermans 25716* (K); Neluwa, 120 m, 4 Apr 1970, *Balakrishnan 245* (K); Top of Hinidumkande (Haycock), 700 m, 2 Sep 1974, *Kostermans 25514* (K), Hinidum Korle, *Thwaites C.P. 2992* (K, PDA). MATARA DISTRICT: Oliyagankele, *Mendis 1230* (Herb. Forest Dept, Colombo). LOCALITY UNKNOWN: *Col. Walker s.n.* (K, lectotype); *Gardner s.n.* (K); *Champion s.n.* (K).

ELAEAGNACEAE
(by M.D. Dassanayake[*])

Juss., Gen. Pl. 74. 1789 ("Elaeagni"). *Type genus: Elaeagnus* L.

 Much-branched erect or climbing shrubs, rarely trees, generally with symbiotic nitrogen-fixing bacteria in root nodules, frequently with thorns which are modified shoots. Vegetative parts and flowers with peltate scales or stellate hairs. Leaves simple, entire, spirally arranged or rarely opposite, often leathery, without stipules. Flowers solitary or in small umbels with small ca-ducous bracts, regular, bisexual or unisexual, mostly 4-merous. Hypanthium in bisexual and pistillate flowers tubular, constricted above the ovary, the lower part persistent, the upper part deciduous. Perianth lobes uniseriate, val-vate, arising from edge of hypanthium, often somewhat petaloid. Stamens as many, or twice as many, as perianth lobes, arising from the edge of the hypan-thium. Filaments free, generally short. Anthers dorsifixed, 2-lobed, introrse, opening by longitudinal slits. Ovary sessile at the base of the hypanthium, with a single carpel, unilocular, with one basal, anatropous, bitegmic ovule. Style single with capitate or simple stigma. Fruit a nut or achene, remain-ing enclosed within the persistent hypanthium tube which becomes partly or wholly fleshy, often with a hard inner layer, and the fruit thereby drupe-like. Seed without or with little endosperm; embryo straight, with thick, fleshy cotyledons.

 3 genera with about 45 species in the temperate and warm northern hemisphere, tropical Asia and northern Australia.

ELAEAGNUS

L., Sp. Pl. 1: 121. 1753; L., Gen. Pl. ed. 5. 57. 1754. Lectotype species: *E. angustifolia* L.

 Erect or climbing shrubs, rarely trees, often thorny, with stellate and peltate silvery or coppery brown scales. Leaves spirally arranged. Flowers bi-sexual, in small axillary inflorescences. Hypanthium tubular or funnel-shaped, more or less constricted above the ovary, basal part persistent, upper part with the 4 perianth lobes deciduous. Stamens 4, arising from the edge of the

---

[*] Flora of Ceylon Project, Peradeniya.

hypanthium and alternating with the perianth lobes. Style erect, with a lateral simple stigma. Fruit drupe-like.

About 40 species, in the northern hemisphere and Australia.

Elaeagnus latifolia L., Sp. Pl. 1: 121. 1753; Thw., Enum. Pl. Zeyl. 252. 1861; Trimen, Handb. Fl. Ceylon 3: 461. 1895. Type: Sri Lanka, *Hermann Herbarium 58* (BM).

Elaeagnus thwaitesii Schlecht. in DC., Prod. 14: 611. 1856–57. Type: Sri Lanka, *s. coll. C.P. 2724* (G-DC).

Shrub, about 3 m high, more or less scandent, with horizontal, often thorny branches. Thorns up to about 3 cm long, strong and woody. Bark greyish white with rusty patches or brown, covered with scales when young. Leaves 1–12 cm long, to 9 cm broad, very variable, elliptic to broadly elliptic, acute or obtuse at base, apex obtuse or rounded or suddenly acute, entire, margin sometimes slightly waved, sometimes undulate, midrib slightly impressed above, young leaves scaly above, older leaves more or less glabrous and dark green above; lower surface densely covered with silvery or rust brown scales. Upper surface sometimes with irregular purplish-brown patches. Petiole 5–15(–18) mm, scaly, channelled above. Flowers silvery white to greenish orange with scattered brown scales, in fascicles of 2–15 in leaf-axils. Bracts c. 2 mm long, filiform. Pedicel 1–3 mm, scaly, elongating in fruit to about 12 mm. Hypanthium about 7 mm long at anthesis; lower part 1.5–2 mm long, narrow, tubular, about 1 mm broad, upper part 3.5–5.5 mm long, 3–3.5 mm broad, campanulate, more or less 4-sided, densely covered with scales. Perianth lobes 2–4 mm long, very broadly ovate, acute, spreading, scaly. Filaments 0.3–1 mm long. Anthers grey-brown, or pale yellow, 0.5–1 mm long. Style 5–7 mm long, densely scaly, particularly at base, very pale green or pale yellow, straight, curved and hook-like at tip. Basal part of hypanthium surrounding achene to make up fruit, upper part and perianth crowning fruit, deciduous. Fruit ovoid-oblong, 1–2.5 cm long, reddish pink to bright red to reddish brown when ripe, scaly, with a pericarp of 3 layers—an outer fleshy layer, a middle hard layer with 8 longitudinal ribs and an inner felt-like layer of compacted hairs. Seeds narrowly ovoid, with a brown papery testa, about 12 mm long.

Distr. Also in peninsular India, Burma, Malesia, China.

Ecol. Moist and intermediate regions, common in the scrub-jungle around the patanas and at the edge of forests in the mid-country. Flowering periodically, throughout the year. Flowers fragrant.

Uses. Fruits edible, with a slightly sour taste.

Vern. Welembilla, Katuembilla (S).

Note. Trimen (1. c.) combines the species *Elaeagnus thwaitesii* Schlecht. (in DC., Prod. 14: 611. 1856–57) with *E. latifolia* as var. *β. thwaitesii*, the leaves being broader, rounded at base, more acute, thicker, the scales all ferruginous, and the plants restricted to the montane zone. However, as Thwaites (1. c.) observes, *E. latifolia* is an extremely variable species, the colour of the leaf-underside varying all the way from silvery in the lowlands to brown in the hills, with a mixture of the two in between. The shape and the thickness of the leaves also vary a great deal.

Specimens Examined. ANURADHAPURA DISTRICT: Ritigala, Weweltenna, "large climber, flowers green-white", 28 Sep 1972, *Jayasuriya 573* (PDA). MATALE DISTRICT: Illukkumbura, 30 Oct 1926, *J.M. Silva 148* (PDA). KANDY DISTRICT: Ambagamuwa, Jan 1853, *s. coll. C.P. 2264*, p.p. (PDA); Hantane and Deltota, Jan 1852, *Gardner in C.P. 2264*, p.p. (PDA); Peradeniya and Haragama, Jan 1853, *s. coll. C.P. 2263*, p.p. (PDA); Hunasgiriya-Mahiyangane, culvert 29/21, 14 Nov 1974, *Davidse & Jayasuriya 8418* (PDA); Hunasgiriya, 9 Sep 1974, *Tirvengadum & Waas 440* (PDA); Nawangala, near Hunasgiriya, *Dassanayake 301* (PDA); Peradeniya Campus, 7 Jan 1970, *Dassanayake 124* (PDA); Moray Estate, near Fishing Hut, 14 Nov 1978, *Kostermans 27045* (PDA); Kandy-Madugoda-Mahiyangane road, 6 Aug 1978, *Kostermans 26761* (PDA); Kandy-Madugoda-Hasalaka, 16 Oct 1977, *Huber 462* (PDA); Ramboda and Hewaheta, Oct 1853, *s. coll. C.P. 2264*, p.p. (PDA). NUWARA ELIYA DISTRICT: Totupola, *Gardner in C.P. 83* (PDA); Hakgala, 26 May 1926, *Alston 1346* (PDA); Sita Eliya, 25 Aug 1926, *J.M. de Silva s.n.* (PDA). BADULLA DISTRICT; Palugama-Boralande road, culvert 1/12, 5 Oct 1972, *Cramer 3881* (PDA), 5 Nov 1971, *Balakrishnan 1055* (PDA); Horton Plains, 10 Aug 1970, *Meijer et al. 627* (PDA). RATNAPURA DISTRICT: "Gangoola", Adams Peak, Feb 1853, *s. coll. C.P. 2724*, p.p. (PDA). COLOMBO DISTRICT: Colombo, *s. coll. C.P. 2263*, p.p. (PDA).

ERYTHROXYLACEAE

(by B.M. Wadhwa* and A. Weerasooriya**)

Kunth in H.B.K., Nov. Gen. Sp. ed. fol. 5: 135, ed. qu. 175. 1822 ("Erythroxyleae"). Type genus: *Erythroxylum* P. Browne.

Trees or shrubs. Leaves simple, alternate, often distichous, rarely opposite, entire. Stipules intrapetiolar, small, mostly connate, often early caducous, sometimes persistent. Flowers small, axillary, solitary or in fascicles, hypogynous, regular, bisexual, rarely unisexual, pentamerous, commonly heterostylous. Calyx campanulate, free or connate below, persistent, 5-lobed; lobes usually imbricate, sometimes valvate. Petals 5, free, imbricate, caducous, usually with a ligular appendage towards the base on the inner side. Stamens 10, usually in 2 whorls, persistent; filaments united at base into a staminal tube; anthers ellipsoid, basifixed, 2-celled, opening by longitudinal slits, latrorse. Gynoecium of 3 or less (often only 2) carpels united to form a compound ovary, with as many locules as carpels, usually one locule of the ovary ovuliferous, the other empty; ovules 1–2 in the fertile locule, pendulous, anatropous; styles 3, free or ± connate, with obliquely clavate stigmas. Fruit a drupe, usually 1-seeded. Seed with or without endosperm; embryo oblong, straight; cotyledons flat or planoconvex.

4 genera and 250 species distributed throughout the tropics and subtropics; 1 genus and 5 species in Sri Lanka.

ERYTHROXYLUM

P. Browne, Hist. Jamaica 1: 278. 1756; L., Syst. Nat. ed. 10, 2: 1035. 1759 *"Erythroxylon"*; Benth. et Hook. f., Gen. Pl. 1: 244. 1867 *"Erythroxylon"*; Hook. f., Fl. Br. Ind. 1: 414. 1874; Trimen, Handb. Fl. Ceylon 1: 190. 1893; O. Schulz in Pflanzenr. Heft. 29 (IV. 134): 1–176. 1907; O. Schulz in Pflanzenfam. ed. 2, 19a: 135–142. 1931. Type species: *E. areolatum* L.

Sethia Kunth in H.B.K., Nov. Gen. Sp. ed. fol. 5: 135 in adnot, ed. qu. 175. 1822; Thw., Enum. Pl. Zeyl. 53. 1858. Type species: *S. indica* DC.

* Royal Botanic Gardens, Kew.
** Flora of Ceylon Project, Peradeniya.

Shrubs or trees, usually glabrous; base of the lateral twigs often with small distichous bracts (ramenta), sometimes also occurring between the leaves. Leaves simple, alternate, often distichous, entire, involute in bud, margin leaving a more or less permanent trace as two longitudinal lines on the upper surface. Stipules intrapetiolar, mostly entirely connate, rarely bifid, often bicarinate, persistent or early caducous. Flowers axillary, solitary or in fascicles, often dimorphous or even 3–4-morphous, 5-merous, regular, bisexual. Pedicels with 2 bracteoles at the base, ± thickened, often only under the calyx. Calyx persistent, campanulate, ± halfway divided into 5 lobes, imbricate in bud. Petals 5, free, caducous, alternating with calyx lobes, always provided with an emarginate or 3-lobed ligule on the inner side at the apex of the claw of the petal. Stamens 10, in two whorls, persistent; filaments connate towards base into a staminal tube, often with toothed margin; anthers ellipsoid, basifixed, cordate at the base, 2-celled, opening longitudinally, latrorse. Ovary (1–) 3-celled, normally only 1 cell fertile; each cell with 1 pendulous, anatropous ovule; styles 3, erect, free or partly connate; stigmas flattened or often obliquely clavate. Fruit a drupe. Seed with or without endosperm; embryo oblong, erect.

About 200 species distributed throughout the tropical and subtropical regions of the world, chiefly in S. America and Madagascar; represented by 5 species in Sri Lanka.

Erythroxylum novogranatense (Morris) Hieron. [= *E. coca* Lam. var. *novogranatense* Morris and also *E. coca* Lam. var. *spruceanum* Burck.] has been introduced in the Royal Botanic Garden, Peradeniya and cultivated in home gardens in Sri Lanka.

KEY TO THE SPECIES

1 Stipules persistent; bracts (ramenta) present at the base of lateral twigs ....
....1. **E. monogynum**
1 Stipules caducous; bracts (ramenta) at the base of lateral twigs few, or rarely absent
  2 Staminal tube as long as or shorter than the calyx
    3 Leaves elliptic-lanceolate, apex usually obtusely caudate-acuminate, base acute; staminal tube as long as the calyx .... 2. **E. moonii**
    3 Leaves lanceolate or elliptic-lanceolate, apex acute or with short acumen, base truncate; staminal tube shorter than the calyx .... 3. **E. zeylanicum**
  2 Staminal tube longer than the calyx
    4 Staminal tube at the margin distinctly toothed; leaves oblong-lanceolate, obovate or broadly oblong, coriaceous, base obtuse, apex rotundate or slightly obtusely pointed, somewhat emarginate .... 4. **E. obtusifolium**
    4 Staminal tube at the margin obscurely crenate; leaves lanceolate or linear-elliptic, chartaceous, base acute, apex acute or short acuminate ....5. **E. lanceolatum**

1. Erythroxylum monogynum Roxb., Pl. Corom. 1: 62. t. 88. 1798; Hook. f., Fl. Br. Ind. 1: 414. 1874; Trimen, Handb. Fl. Ceylon 1: 190. 1893; O. Schulz in Pflanzenr. 29(IV. 134): 129. 1907; Dunn in Gamble, Fl. Pres. Madras 1: 127. 1915; Matthew & Britto in Matthew, Fl. Tamilnadu Carnatic 1: 182. 1983; Chithra in Nair & Henry, Fl. Tamilnadu 1: 47. 1983; K.K.N. Nair & M.P. Nayar, Fl. Courtallum 2: 210. 1987. Type: India, s. loc., *Roxburgh s.n.* (BM).

Sethia indica DC., Prod. 1: 576. 1824, nom. superfl.; Wight, Ill. Ind. Bot. 1: t. 48. 1840; Thw., Enum. Pl. Zeyl. 53. 1858. Type: same as for *Erythroxylum monogynum* Roxb.

Erythroxylum indicum (DC.) Bedd., Fl. Sylv. t. 81. 1871.

Shrubs or small trees, 2–4.5 m tall, much branched; bark dark brown, rough, lenticellate; branches with deciduous bracts at the base. Leaves alternate, obovate or cuneate-obovate, 1.4–4.2 (–5.2) × 0.7–1.9 (–2.2) cm, simple, apex rounded or very obtuse, emarginate, entire, slightly recurved at base, hardly shining above, glaucous brown beneath; veins reticulate, anastomosing, midrib prominent beneath; petioles 3–7 mm long; stipules triangular, 1.5–2 mm long, finely serrate, persistent. Flowers white or greenish-white, scented, 1–4 in axillary fascicles, dimorphous, c. 1 cm across; peduncles c. 2 mm long; pedicels 5–8 mm long, thickened below the calyx. Calyx campanulate, persistent, glabrous; lobes 5, ovate or triangular, 2 × 2 mm, acute. Petals 5, oblong, spreading, ligulate; ligules to 2 mm long. Stamens 10; staminal tube c. 2 mm long; filaments 1.5–1.8 mm long, subequal; anthers basifixed, dehiscing longitudinally. Ovary ellipsoid or oblong, 2 mm long, 10-ribbed, 1–3-celled; ovule 1, pendulous; styles 3, connate for nearly whole length; stigmas clavate. Drupes cylindric or oblong, straight, 9–10 × 3–4 mm, furrowed, green, becoming blood red or scarlet when ripe, tipped above with persistent styles and subtended below by persistent calyx and filaments, 1-seeded.

Distr. India (in drier tracts of Peninsular India) and Sri Lanka.

Ecol. Dry region, to 600 m, very common; fls & frts Aug–Feb.

Vern. Devadaram, chemmanatti (T).

Notes. "Heart wood very hard and heavy and has a pleasant resinous scent and yields, by distillation, a kind of tar, used by Moormen as a preservative for the wood of their boats"—Ondaatje in Obs. Veg. Prod. 13. 1853, quoted by Trimen, l. c. 1: 191. 1893. The leaves contain an alkaloid, which is bitter and astringent.

Specimens Examined. JAFFNA DISTRICT: Near Ponneryn, 5 m. 17 March 1973, *Bernardi 14284* (K, PDA); Jaffna, Feb 1890, *s. coll. s.n.* (PDA); Jaffna, *Gardner C.P. 1168* (BM, K, PDA); Elephant Pass, 11 March 1932, *Simpson 9260* (BM, PDA), *Simpson 9250* (BM). MANNAR DISTRICT: Madhu Road, 80°09′ E, 8°46′ N, 23 Jan 1972, *Jayasuriya,*

Balakrishnan, Dassanayake & Balasubramaniam 601 (PDA). VAVUNIYA DISTRICT: Alampil near lagoon, 7 July 1971, *Meijer 763* (K, PDA); 4 miles N. of Mankulam, 10 Dec 1970, *Fosberg & Balakrishnan 53546* (PDA). PUTTALAM DISTRICT: Chilaw, along coast near mangrove, 12 Apr 1995, *Wadhwa, Weerasooriya & Samarasinghe 584* (K, PDA); Kalpitiya, Aug 1883, *s. coll. s.n.* (PDA). TRINCOMALEE DISTRICT: Foul Point, 4 March 1995, *Weerasooriya & Samarasinghe 177* (K, PDA); Shell Bay, Foul Point, 8 June 1940, *Worthington 967, 1360* (K), *877* (BM, K); Foul Point, 80°19′ E, 8°31′ N, 4 Feb 1974, *Jayasuriya, Dassanayake & Balasubramanium 645* (PDA). ANURADHAPURA DISTRICT: Mihintale, Aug 1885, *s. coll. s.n.* (PDA); Borupangoda wewa, 12 Jan 1974, *Waas 341* (K, PDA); Ritigala, 11 June 1974, *Waas 332* (K, PDA); Wilpattu National Park, 15 Oct 1975, *Bernardi 15387* (K, PDA); Mihintale Sanctuary, 21 Nov 1975, *Sohmer & Sumithraarachchi 10741* (PDA); Road to Ritigala Reserve, 3 Dec 1994, *Wadhwa, Weerasooriya & Samarasinghe 507* (K, PDA); POLONNARUWA DISTRICT: Mutugala, 15 Apr 1953, *Worthington 6411* (K, BM). BATTICALOA DISTRICT: Kathiraveli at Manikakulam, 30 m, 4 Nov 1975, *Bernardi 15658* (PDA, K). AMPARAI DISTRICT: Kumana, 10 Jan 1971, *Balakrishnan 587* (PDA). HAMBANTOTA DISTRICT: Near Kirinda, Dec 1882, *s. coll. s.n.* (PDA); Ruhuna National Park, Block IV, 8 Oct 1994, *Wadhwa, Weerasooriya & Samarasinghe 414* (K, PDA), Block I, 7 Oct 1994, *Wadhwa, Weerasooriya & Samarasinghe 411* (K, PDA), 2 Oct 91 *Jayasuriya 5791* (PDA); Bundala, 20 July 1973, *Hepper & G. de Silva 4753* (PDA); Mahasi villu, South of Ruhuna National Park, 2 m, 20 Oct 1974, *Davidse 7785* (PDA, K); 1 mile W. of Bundala, 1–4 m, 25 Nov 1974, *Davidse & Sumithraarachchi 8871* (PDA, K); W. Boundary of Ruhuna National Park, 29 June 1970, *Meijer 205* (PDA, K); mile marker 7, 10 Dec 1967, *Mueller-Dombois & Cooray 67121045* (PDA), near Kotabande Wewa, 20 m, 20 June 1967, *Mueller-Dombois & Comanor 67062207* (PDA), Patanagala, 6 Apr 1968, *Fosberg 50336* (PDA), Buttawa Bungalow area, 8 m, 18 June 1967, *Comanor 367* (PDA), Patanagala beach area, 3 m, 29 Jan 1968, *Comanor 893* (K, PDA), Plot 24, 12 m, 25 Feb 1968, *Comanor 1042* (PDA, K), Andunoruwa area, 3–5 m, 3 Apr 1968, *Fosberg & Mueller-Dombois 50137* (PDA), 16 Nov 1965, *Cooray 69111604* (K, PDA), Block I, at Buttawa Plot R-27, 30 May 1968, *Cooray 68053013* (PDA), Buttawa Bungalow, 24 Nov 1970, *Fosberg & Sachet 52884* (PDA), E. of Ulpasse-wewa, 2 miles N. of Panama, 26 Nov 1970, *Fosberg & Sachet 52951* (PDA), Rakinawala, Block I, 22 Oct 1968, *Cooray 68102202* (PDA, K), next to Uraniya Lagoon, 18 July 1973, *Norwicke & Jayasuriya 393* (PDA); Yala Park Road, 10 Dec 1969, *Comanor 661* (PDA, K). LOCALITY UNKNOWN: *s. coll. s.n.* (K); *Macrae 794* (BM).

2. Erythroxylum moonii Hochr., Bull. Inst. Bot. Buitenzorg 22: 54. 1905; Henry & Roy, Bull. Bot. Surv. India 10: 274. 1976; Chithra in Nair & Henry, Fl. Tamilnadu 1: 47. 1983; Vajr., Fl. Palgh. 97. 1990; Sharma et al., Fl. India 3: 593, 1993. Type: same as for *E. lucidum* Moon ex Hook. f.

Sethia acuminata Arn., Nov. Actorum Acad. Caes. Leop.-Carol. Nat. Cur. 18(1): 324. 1836; Wight, Ill. Ind. Bot. 1: 135. 1840; Thw., Enum. Pl. Zeyl. 54. 1858 (incl. var. *B*). Type: from Sri Lanka, *s. coll. s.n.*

Erythroxylum lucidum Moon [Cat. 36. 1824, nom. nud.] ex Hook. f., Fl. Br. Ind. 1: 415. 1874 (non H.B.K., 1822); Trimen, Handb. Fl. Ceylon 1: 191. 1893, p.p. Type: from Sri Lanka, *Walker s.n.* (PDA, holotype; K, isotype).

Erythroxylum acuminatum (Arn.) Walp., Repert. 1: 407. 1842 (non Ruiz & Pavon, 1802); O. Schulz in Pflanzenr. Heft 29 (IV. 134): 145. 1907; Dunn in Gamble, Fl. Pres. Madras 1: 127. 1915.

Shrubs or small trees, up to 5 m tall, much branched, branches glabrous; bark pale brown, lenticellate; twigs often with a few deciduous bracts at the base. Leaves elliptic-lanceolate, 3.5–9.0 × 1.2–2.8 cm, obtusely caudate-acuminate at apex, acute at base, entire, opaque or dull, shining above, shining beneath; veins reticulate, horizontal, anastomosing, midrib prominent beneath; petioles 2.5–3.5 mm long, slender; stipules finely serrate, caducous. Flowers white, axillary, 1 or 2 together, c. 2 mm long, dimorphous; pedicels 8–10 mm long, stout; bracteoles 2, triangular. Calyx campanulate, 1.5 × 1.5 mm, persistent; lobes 5, triangular, acute. Petals elliptic-oblong, c. 2 mm long, obtuse, slightly exserted, ligulate; ligule 3-fid, near the base on inner side, persistent. Stamens 10, unequal; staminal tube as long as the calyx, margin dentate; filaments slender; anthers dehiscing longitudinally. Ovary ovoid, 3-locular; styles 3, connate almost throughout; stigma capitate. Drupe ellipsoid-oblong, slightly falcate, 13 × 3 mm, furrowed, glabrous, green, shining, 1-seeded.

Distr. India (Tamil Nadu—evergreen forests) and Sri Lanka.

Ecol. In wet region in evergreen forests, along streams, to 1,000 m, fairly common; fls & frts Aug–Dec.

Vern. Batakirilla (S).

Notes. The juice of the fresh leaves is reported to possess anthelmintic properties, and is used especially for diarrhoea.

Plate No. 309 in PDA based on *C.P. 3488* belongs to this species.

Specimens Examined. COLOMBO DISTRICT: Labugama Reservoir, 22 July 1994, *P.M. Arunareka s.n.* (PDA). KEGALLE DISTRICT: Salgala, 25 Aug 1965, *Amaratunga 973* (PDA), 19 Apr 1963, *Amaratunga 581* (PDA). BADULLA DISTRICT: Namunukula Hill, *Capt. Champion s.n.* (K). RATNAPURA DISTRICT: Sinharaja Forest, road to MAB Laboratory, 4

Nov 1994, *Wadhwa, Weerasooriya & Samarasinghe 464* (K, PDA); Palabaddala, Adam's Peak side, 20 Apr 1995, *Wadhwa, Weerasooriya & Samarasinghe 544* (K, PDA); Sinharaja Forest at Leopard Rock, 200 m, 3 Sept 1994, *Weersooriya, Wickramasinghe & Gunatilleke 187* (PDA); Mukkuwatta Forest, ± 1,000 m, 26 Aug 1976, *Waas 1820* (PDA); Ratnapura, Mar 1846, *Thwaites C.P. 222* (BM, K, PDA); Sinharaja Forest, Apr & Sept 1855, *Thwaites C.P. 3488* (BM, K, PDA); Trail Kudawa-Pitakele, Sinharaja, 335 m, 22 Aug 1988, *Jayasuriya & Balasubramaniam 4442* (PDA). GALLE DISTRICT: Hiniduma Kanda, 175 m, 31 Aug 1974, *Jayasuriya, Kostermans & Banadaranayake 1788* (K, PDA); Hiniduma, 240 m, 7 Nov 1975, *Sohmer & Waas 10318* (PDA K), *10350* (PDA); Hills N. of Nelluwa, 180 m, 20 Sept 1977, *Huber 348* (PDA); Lower slopes of Haycock towards Hiniduma, 160 m, 28 Sept 1977, *Huber 338* (PDA); Hiniduma Kanda (Haycock), near Hiniduma, 500 m, 6 Oct 1974, *Kostermans 25703* (K); Liyanagama Kanda, Bambarawana Other State Forest, 17 Oct 1992, *Jayasuriya 6804* (PDA); Habarakande Proposed Reserve, 23 Aug 1992, *Jayasuriya 6659* (PDA); Hiniduma Kanda, alt. low, 2 July 1975, *Waas 1355* (K, PDA,), c. 1,350 m, 6 Nov 1994, *Wadhwa, Weerasooriya & Samarasinghe 473* (K, PDA). MATARA DISTRICT: Diyadawa Forest Reserve, 12 March 1992, *Jayasuriya 6245* (PDA); Beraliya Forest Reserve, 9 Nov 1992, *Jayasuriya 6030* (PDA); Derangala Other State Forest, 29 June 1992, *Jayasuriya 6492* (PDA); Mulatiyana Forest Reserve, Dalukhinna atura Ridge top forest above Pambavila, 23 Nov 1993, *Jayasuriya 5954* (PDA); Kandewattegoda Forest Reserve, 29 July 1990, *Jayasuriya & Balasubramaniam 5347* (PDA); Panilkanda Forest Reserve, 16 June 1990, *Jayasuriya & Balasubramaniam 5308* (PDA). LOCALITY UNKNOWN : *Walker s.n.* (PDA, Holotype of *Erythroxylum lucidum* Moon; K, isotype).

3. Erythroxylum zeylanicum O. Schulz in Pflanzenr. Heft 29 (IV.134): 145. 1907; Alston in Trimen, Handb. Fl. Ceylon 6: 34. 1931; Abeywickrama, Ceylon J. Sci., Biol. Sci. 2(2): 178. 1959. Type: Sri Lanka, Ratnapura, *Thwaites C.P. 222*, p.p. (PDA, holotype: K, isotype).

Shrubs, 3 m tall, much branched; branchlets glabrous; bark pale-brown, lenticellate, twigs often with few deciduous scales at the base. Leaves simple, lanceolate or elliptic-lanceolate, 3–8.2 × 1.2–2.5 cm, base truncate, apex acute or slightly acuminate, entire, shining beneath; veins reticulate, anastomosing, midrib prominent beneath; petioles 3–4 mm long; stipules 3–4 mm long, lanceolate, subfalcately curved, apex 2-fid, caducous. Flowers axillary, 1–2 together, c. 2 mm long, yellowish-green; pedicels 5–8 mm long, thickened below the calyx. Calyx cupular, 2 × 2 mm, persistent; lobes 5, lanceolate, acute. Petals c. 2 mm long, elliptic, ligulate; ligule near the base of the petals. Stamens 10, unequal, episepalous stamens c. 1 mm long, epipetalous 1.8 mm long; staminal tube shorter than the calyx, c. 1.2 mm long; anthers 2-celled, dehiscing longitudinally. Ovary ovoid, much longer than the staminal tube, 3-celled; styles c. 2 mm long, connate for two-thirds of length. Drupes ovoid-oblong, 10×4 mm, furrowed, glabrous, subtended at base by persistent calyx and filaments and at the top with persistent styles, deep red when ripe, 1-seeded.

Distrib. Endemic.

Ecol. In the forests of the dry region, to 500 m, fairly common; fls & frts Sept–Dec.

Specimens Examined. ANURADHAPURA DISTRICT: Ritigala, 15 March 1943, *Worthington 281* (K); Forest W. of Ritigala Hill, 18 Nov 1971, *Balakrishnan & Jayasuriya 1104, 1113* (PDA); Ritigala Strict Natural Reserve, 3 Dec 1994, *Wadhwa, Weerasooriya & Samarasinghe 501* (K, PDA). TRINCOMALEE DISTRICT: Nîlaveli-Kuchchaveli Road, 11 July 1973, *Norwicke & Jayasuriya 288* (K, PDA), Foul Point, 20 May 1932, *Simpson 9684* (BM, PDA). KURUNEGALA DISTRICT: Ganewatta Medicinal Plant Garden, 100 m, 25 May 1993, *Dhanasekara 36* (PDA). MATALE DISTRICT: Dambulla Road, Nalanda, 32 Mile post, 305 m, 9 Nov 1946, *Worthington 2409* (K, BM), 26 June 1940, *Worthington 990* (K), 11 Nov 1940, *Worthington 1093* (K); Nalanda, 28/5 Mile post, 30 Oct 1939, *Worthington 652* (BM); Dambulla-Habarana, forest at 98 Mile post, 150 m 1 July 1971, *Balakrishnan 505* (PDA); Dambulla, *Thwaites C.P. 4011* (PDA), 12 May 1928, *Alston 2400* (K, PDA). BATTICALOA DISTRICT: Near Thulankudah, 4 Jan 1969, *Fosberg, Mueller-Dombois, Wirawan, Cooray & Balakrishnan 51002* (K, PDA). KANDY DISTRICT: Weragamtota, 14 Feb 1928, *Alston 1958* (PDA); Royal Botanic Gardens, Peradeniya, Medicinal Plant Beds, 4 Oct 1994, *Wadhwa & Weerasooriya 396* (PDA), July 1986, *Tennakoon, Tilakaratne, Rizvi & Madawala 24* (PDA); Pilimatalawa, in Home Garden, Jan 1993, *Hochegger & Ekanayake s.n.* (PDA). BADULLA DISTRICT: Uma Oya, Talpitigala, 425 m, 15 Sept 1971, *Balakrishnan & Jayasuriya 851* (PDA); Uma Oya, June 1881, *Trimen s.n.* (PDA). KALUTARA DISTRICT: Ithagoda, Matugama, 21 Oct 1971, *Balakrishnan 1014* (PDA). RATNAPURA DISTRICT: Ratnapura, *Thwaites C.P. 222* (PDA, Holotype; K, isotype); Gallehetota Other State Forest, 28 Nov 1993, *Jayasuriya 7742* (PDA); Kapugalla Other State Forest, 26 Nov 1993, *Jayasuriya 7706* (PDA). MONERAGALA DISTRICT: Malwathuhella, W. of Wellawaya, bank of Bibilihena ela, 13 Nov 1968, *Wirawan 732* (K, PDA); Sellaka Oya Sanctuary, Gal Oya National Park, 5 May 1973, *Jayasuriya 2073* (K, PDA); Ormbuwa Forest Reserve, beyond mile post 157/2 on MoneragalaWellawaya Road, c. 6°51′ N, 81°19′ E, 275 m, 25 Dec 1976, *R.B. & A.J. Faden 76/581* (K, PDA); Moneragala, 31 mile post, 150 m, 19

Aug 1951, *Worthington 5441* (K, PDA); Bibile, 23 Oct 1925, *J.M. Silva s.n.* (PDA). HAMBANTOTA DISTRICT: Ruhuna National Park, Block III, Tammitigala, 24 Sept 1992, *Jayasuriya 6755* (PDA), Block II, at Kumbukkan oya, 2 m, 1 Oct 1967, *Mueller-Dombois, Comanor & Cooray 67100108* (PDA), Block II, 100 m W. of Kumbukkan oya, 30 Aug 1967, *Mueller-Dombois & Comanor 67083016* (PDA); Kumbukkan oya sand flat, 31 July 1969, *Cooray 69073017* (PDA). LOCALITY UNKNOWN: *s. coll. 1418* (K).

4. Erythroxylum obtusifolium (Wight) Hook. f., Fl. Br. Ind. 1: 415. 1874; Trimen, Handb. Fl. Ceylon 1: 192. 1893; O. Schulz in Pflanzenr. Heft. 29 (IV. 134): 146. 1907; Dunn in Gamble, Fl. Pres. Madras 1: 127. 1915; Chithra in Nair & Henry, Fl. Tamilnadu 1: 48. 1983; Manilal, Fl. Silent Valley 37. 1988; Sharma et al., Fl. India 3: 593. 1993.

Sethia lanceolata Wight var. *obtusifolia* Wight, Ill. Ind. Bot. 1:136. 1840.
Holotype: India, Courtallum, July 1835, *Wight 290* (K).
Erythroxylum lanceolatum (Wight) Walp. var. *obtusifolium* (Wight) Walp., Repert. 1: 407. 1842.
Sethia obtusifolia (Wight) Thw., Enum. Pl. Zeyl. 54. 1858.
Sethia obtusifolia (Wight) Thw. var. *stylosa* Thw., Enum. Pl. Zeyl. 54. 1858.
Type: Sri Lanka, Dambulla, Aug 1852, *Thwaites C.P. 2613* (PDA, holotype; BM, K, isotype).
Sethia obtusifolia (Wight) Thw. var. *staminea* Thw., Enum. Pl. Zeyl. 54. 1858. Type: Sri Lanka, Adam's Peak, Mar 1846, *Thwaites C.P. 25* (K, holotype; BM, isotype).

Shrubs or small trees, to 6 m tall, much branched; branches angular, warty. Leaves oblong-lanceolate, obovate or broadly oblong, 2.5–8.5 × 1.5–4 cm, rounded or obtusely pointed at apex, emarginate, obtuse at base, entire, dark brown above, pale brownish beneath, shining on both surfaces; nerves rather obliquely elevated, midrib prominent beneath; petioles 3–7 mm long, thick; stipules lanceolate, 2.5–4 mm long, apex acute, falcately recurved, entire, caducous, leaving conspicuous scars. Flowers axillary, solitary, c. 5 mm long, white, greenish-white or pale-yellow, dolichostylous. Calyx divided up to three-fourths length; lobes 1.5 mm long, subovate, with small acumen, persistent. Petals 5, free, oblong, 4 mm long, unguiculate, obtuse, imbricate with long ligule on inner side at base. Stamens 10, unequal, episepalous stamens 1 mm long, epipetalous stamens c. 2 mm long; staminal tube longer than the calyx, margin 10-denticulate or toothed; anthers c. 1 mm long, longitudinally dehiscent. Ovary ovoid, slightly longer than the staminal tube; styles 3 mm long, united at the base, recurved at apex; stigmas capitate. Drupe oblong or linear-oblong, 1.2–1.6 cm long, 3–4 mm broad, straight, subtrigonous, immature green, ripe scarlet, subtended by persistent calyx and filaments below and with persistent styles above, 1-seeded.

Distr. India (in the montane region of Tamilnadu, up to 1,200 m) and Sri Lanka.

Ecol. In the forests of lower montane zone, to 1,200 m, rather common; fls & frts Sept–Mar.

Notes. Plate No. 311 in PDA based on *C.P. 2613* belongs to this species.

Specimens Examined. ANURADHAPURA DISTRICT: Top of Ritigala, July 1887, *s. coll. s.n.* (PDA), 22 & 24 Aug 1905, *Willis s.n.* (PDA); Ritigala Hill, July 1887, *s. coll. s.n.* (PDA); Ritigala strict Natural Reserve, 760 m, 22 Oct 1973, *Jayasuriya & Ashton 1346* (PDA), 700 m, 17 Nov 1971, *Balakrishnan & Jayasuriya 1097* (PDA). KANDY DISTRICT: Above Wariagala, Hantane, Dec 1889, *s. coll. s.n.* (PDA); Looloowatte, 1,100 m, 23 Oct 1973, *Jayasuriya 947* (K); Hunnasgiriya, June 1848 & 1851, *Gardner C.P. 25* (PDA), Aug 1852, *s. coll. C.P. 2613* (Type of *E. obtusifolium* var. *stylosa* Thw. PDA, Holotype; BM, K, isotypes); 3 miles NE of Madugoda, 80°54′ E, 7°20′ N, 825 m, 6 Jan 1972, *Jayasuriya, Dassanayake & Balakrishnan 485* (PDA); Loolkandura Estate, Cardamom Plantation, S. of Deltota, 7°09′ N, 80°42′ E, 1325 m, 25 Oct 1984, *Jayasuriya & Gunatilake 2968* (PDA); Adam's Peak, Mar 1846, *Thwaites C.P. 25* (Type of *E. obtusifolium* var. *staminea* Thw. K, Holotype; BM, isotype). BADULLA DISTRICT: Namunukula, 1,200 m, 23 Oct 1994, *Wadhwa, Weerasooriya & Samarasinghe 453* (K, PDA); Knuckles, 14 June 1926, *J.M. Silva s.n.* (PDA). KALUTARA DISTRICT: Pasdun Korale, May 1883, *s. coll. s.n.* (PDA); Vellihalure Other State Forest, 23 May 1993, *Jayasuriya 7331* (PDA). RATNAPURA DISTRICT: Selvakande, 28 June 1992, *Jayasuriya 6483* (PDA); Kudumariya Proposed Reserve, Kalawane Range, 22 Jan 1994, *Jayasuriya 7920* (PDA); Deniyaya-Ensalwatte, Sinharaja Trail (near ITN Tower), 800 m, 21 Apr 1995, *Wadhwa, Weerasooriya & Samarasinghe 547* (K, PDA); Handapanella, on Rakwana-Pothupitiya Road, 13 Feb 1994, *Jayasuriya 8022* (PDA); Handapanella, SW of Sinharaja Aranya, Suriyakanda, 16 Jan 1993, *Jayasuriya 7137* (PDA); Sinharaja Forest, Weddegale Entrance, 200 m, 30 March 1979, *Kostermans 27488* (PDA, K); Hillcrest between Balangoda & Paname, 800 m, 30 Aug 1978, *Huber 853* (PDA); Handapanella, 6 Oct 1994, *Wadhwa, Weerasooriya & Samarsinghe 402* (K, PDA); Suriyakanda above Aberfoyle Estate, 6°27′ N. 80°36′ E, Yakinidole-Sinharaja, ± 250 m, 26 Feb 1977, *Waas 2050* (PDA); Enselwatte, near Radio Transmitting Line, 975 m, 15 March 1985, *Jayasuriya & Balasubramaniam 3251* (PDA); Sinharaja Forest, road to MAB Laboratory, 4 Nov 1994, *Wadhwa, Weerasooriya & Samarasinghe 465* (K, PDA); Sinharaja, near Enselwatte, *Sohmer & Waas 10465* (PDA). GALLE DISTRICT: Hiniduma Kande, 18 Oct 1992, *Jayasuriya 6836* (PDA). LOCALITY UNKNOWN: *Thwaites C.P. 1169* (K).

5. Erythroxylum lanceolatum (Wight) Walp., Repert. 1: 407. 1842; Hook. f., Fl. Br. Ind. 1: 415. 1874; Trimen, Handb. Fl. Ceylon 1: 191. 1893; O. Schulz in Pflanzenr. Heft 29 (IV. 134): 146. 1907; Dunn in Gamble, Fl. Pres. Madras 1: 127. 1915; Chithra in Nair & Henry, Fl. Tamilnadu 1: 47. 1983; Sharma et al., Fl. India 3: 590. 1993.

Sethia lanceolata Wight, Ill. Ind. Bot. 1: 136. 1840; Thw., Enum. Pl. Zeyl. 54. 1858. Type: India, Courtallum, Aug 1835, *Wight 291* (K).
Sethia erythroxyloides Wight, Ill. Ind. Bot. 1: 136. 1840. Type: India, Courtallum, *Wight 292* (K).

Shrubs, 2–2.5 m tall, much branched; branches more or less smooth, striate. Leaves lanceolate, or linear-elliptic, 5.5–10.5 × 1.5–2.2 cm, acute at base, acute or short acuminate at apex, shining on both surfaces, chartaceous; veins reticulate, anastomosing, lateral veins somewhat oblique, recurved near margin, midrib prominent beneath; petioles 3–5 mm long; stipules subulate, shorter than the petioles, entire, caducous, scars prominent. Flowers axillary, solitary, 1.5 cm long, heterostylous; pedicels to 8 mm long, thickened above. Calyx 5-lobed, lanceolate, 1.5 × 1.5 mm, persistent. Petals 5, obtuse, free, imbricate. Stamens 10, unequal; episepalous stamens 1.2 mm long, epipetalous ones c. 2 mm long; staminal tube longer than the calyx, margin obscurely crenate or ± entire; anthers dehiscing longitudinally. Ovary ovoid; styles 3, varying in length, united two-thirds of the length; stigma capitate. Drupes oblong, 12 × 3 mm, straight, subtrigonous, scarlet red when ripe, 1-seeded.

Distr. India (Courtallum) and Sri Lanka.

Ecol. In the lower montane forests of the wet zone, to 1,500 m, rare; fls & frts Feb–Apr.

Specimens Examined. BADULLA DISTRICT: Haputale, 1,500 m, 17 Oct 1977, *Nooteboom 340* (PDA); Galagama, May 1846, *Gardner C.P. 493* (PDA, BM, K). RATNAPURA DISTRICT: Warathalgoda Proposed Reserve, Kalawana Range, 3 June 1990, *Jayasuriya & Balasubramaniam 5214* (PDA); Athwelthota, 16 May 1995, *Wadhwa, Weerasooriya & Samarasinghe 631* (PDA); Massena Proposed Reserve, Ratnapura Range, 29 July 1990, *Jayasuriya 5478* (PDA). GALLE DISTRICT: Tawalana Proposed Reserve, approach from Hiniduma, 8 Aug 1992, *Jayasuriya 6624* (PDA); Tibbotuwakota, Deniyaya Range, 10 Oct 1992, *Jayasuriya 6789* (PDA). LOCALITY UNKNOWN: *Ferguson s.n.* (PDA).

FABACEAE (LEGUMINOSAE)

Subfamily FABOIDEAE (PAPILIONOIDEAE) (continued)

A key to the tribes of Faboideae was published in the Revised Flora of Ceylon 7: 108. 1991.

Tribe DESMODIEAE

(by Les Pedley[*]; *Pseudarthria* by Velva E. Rudd[**])

(Benth.) Hutch., Gen. Fl. Pl. 1: 477. 1964; Ohashi, Polhill & Schubert in Polhill & Raven, Adv. Leg. Syst. Part 1: 292–300. 1981. Type: *Desmodium* Desv.

Hedysareae subtribe Desmodiineae Benth. in Benth. & Hook. f., Gen. Pl. 1: 449. 1865, as 'Desmodieae'.

Coronilleae subtribe Desmodiineae (Benth.) Schulze-Menz in Engler's Syllabus 2: 237. 1964.

Herbs, shrubs or rarely trees; leaves pinnately 1- or 3-foliolate, rarely to 9-foliolate, generally with stipels; stipules mostly striate; inflorescences terminal or axillary, often pseudoracemes with flowers in fascicles subtended by a bract with secondary bracts at the base of pedicels present or absent, or flowers in true racemes; bracteoles present or absent; calyx 5-lobed or 4-lobed with the upper lobe entire, bifid or deeply divided, occasionally distinctly 2-lipped; petals commonly white, blue or purplish; stamens 10, diadelphous with vexillary stamen free or monadelphous, sometimes forming closed tube; disk sometimes present within the stamens; filaments equal or alternating in length; anthers uniform; fruit transversely jointed breaking into dehiscent or indehiscent 1-seeded articles, rarely of only 1 article, or whole fruit though jointed dehiscing its full length by one suture or fruit without joints; seeds mostly reniform with well-developed radicular lobe, mostly with rim aril or occasionally aril well developed; hairs, particularly of inflorescence, calyx and fruit uncinate.

Of the three subtribes only the Desmodiineae is present in Ceylon.

* Queensland Herbarium, Indooroopilly, Queensland 4068, Australia.
** Smithsonian Institution, Washington, D.C., U.S.A.

KEY TO THE GENERA

1 Calyx glumaceous, striate with strong nerves from base of lobes; pod with relatively wide
   articles, more or less symmetrical along upper and lower margins .... **1. Alysicarpus**
1 Calyx not glumaceous, not strongly nerved; articles or whole pod usually more convex on
   lower margin (except in *Aphyllodium* where articles only 2) or pod not jointed or pod not
   breaking up into articles
  2 Pod persistently folded up within the calyx; pedicels apically abruptly turned upwards towards
     apex, particularly in fruit; leaflets 1–7 .... **2. Uraria**
  2 Pod not folded in calyx; pedicels not curved upwards in fruit; leaflets 1 or 3
      3 Pod boat-shaped, membranous with 1 oblong seed, borne on a long filiform pedicel; leaflets
         1, rarely 3, rounded or reniform .... **3. Eleiotis**
      3 Pod 2-many seeded; leaflets not rounded nor reniform
          4 Pod inflated, dehiscent, transversely veined, many-seeded .... **4. Pycnospora**
          4 Pod not inflated nor many-seeded
              5 Flowers 1 per bract, crowded into fasciculate, short-racemose or subumbellate inflo-
                 rescence, each subtended by a trifoliolate leaf or a foliar bract (usually 2-foliolate);
                 bracteoles present
                  6 Inflorescences fasciculate subtended by foliar bract (two leaflets with terminal seta);
                     articles not thick and corky .... **5. Dendrolobium**
                  6 Inflorescences fasciculate subtended by foliar bract (two leaflets with terminal seta);
                     articles not thick and corky .... **6. Phyllodium**
              5 Inflorescence a pseudoraceme; flowers 2-several per primary bract sometimes each sub-
                 tended by a secondary bract; bracteoles present or absent
                  7 Stipules united, amplexicaul, 3-fid; primary and secondary bracts similar in texture to
                     stipules; leaflets 3, subdigitate; pod with 2 rounded articles ..... **7. Aphyllodium**
                  7 Stipules free, entire; leaflets 1, or if 3, distinctly pinnate, pod with more than 2 articles,
                     or articles not rounded
                      8 Petioles winged; leaves 1-foliolate ....**8. Tadehagi**
                      8 Petioles not winged; leaves 1- or 3-foliolate
                          9 Fruit distinctly jointed, breaking into 1-seeded articles .... **9. Desmodium**
                          9 Fruit jointed or not, not breaking into 1-seeded articles
                              10 Fruit jointed but not breaking up into articles, dehiscing along lower suture;
                                 lateral leaflets reduced or 0, lateral veins looped within margin of leaflet; seeds
                                 conspicuously arillate .... **10. Codariocalyx**
                              10 Fruit not jointed, apparently indehiscent; leaves distinctly 3- foliolate, lateral
                                 veins extending to margin; seeds with rim aril only .... **11. Pseudarthria**

Circumscription of some genea within the Desmodieae is a matter of
some debate. Genera recognised by Ohashi et al. (Advances in Legume Sys-
tematics ed. Polhill & Raven, 1981) are adopted here. The familiar generic
name *Dicerma* is illegitimate and is replaced by *Aphyllodium*.

1. ALYSICARPUS

Desv., J. Bot. Agric. 1: 120. 1813; Wight & Arn., Prod. 232. 1834; Benth. in
Benth. & Hook. f., Gen. Pl. 1: 522. 1865; Hutch., Gen. Fl. Pl. 1: 482. 1964.
Type species: *A. bupleurifolius* (L.) DC.

Herbs; leaves 1-foliolate or rarely 3-foliolate; petiole sulcate; stipules scarious, long-pointed, persistent; inflorescence terminal, axillary or leaf-opposed pseudoracemes, flowers in pairs subtended by scarious primary bract; secondary bracts and bracteoles absent; calyx scarious, 4-lobed, the lobes subequal, the upper usually bifid; corolla small; stamens diadelphous, the upper one free at anthesis; free parts of filaments alternately long and short; ovary more or less sessile, with few to many ovules; pods linear in outline, several-jointed, the margins straight or constricted between the articles which are indehiscent, compressed, subcylindrical or rounded, smooth or variously reticulate; seeds subglobose or oblong; hilum minute, without a rim aril.

A genus of about 25 species in the tropics and subtropics of the Old World, one or two species introduced into the New World.

KEY TO THE SPECIES

1 Calyx scarcely longer than the lowest joint of pod, divided to the middle, lobes not overlapping at the base when in fruit

  2 Articles of pod orbicular, not reticulate, pod constricted between them .... **1. A. monilifer**

  2 Articles subcylindric, smooth (sometimes obscurely reticulate when dry), pod not constricted between them .... **2. A. vaginalis**

1 Calyx distinctly longer than the lowest joint of the pod, divided to near the base, lobes overlapping at the base when in fruit

  3 Articles glabrous, smooth or obscurely reticulately veined; pod not constricted between the articles .... **3. A. bupleurifolius**

  3 Articles puberulous, reticulate; pod constricted or not between articles

    4 Flowers large, standard 7–8 mm long, about 5 mm wide; pod not constricted between articles; articles not strongly transversely reticulate .... **4. A. longifolius**

    4 Flowers smaller, standard to 6 mm long, 3 mm wide; pods constricted between articles; articles strongly transversely reticulate

      5 Bracts and calyx lobes densely ciliate, the hairs about 1 mm or more long; bracts persistent; plants more or less prostrate (drying greyish); stems becoming glabrous ...
.... **5. A. scariosus**

      5 Bracts and calyx lobes with a few marginal hairs (less than 1 mm long) towards the tips; bracts deciduous after anthesis; plants erect (drying greenish); stems pilose or glabrous except for long hairs persisting in leaf axils .... **6. A. heyneanus**

1. Alysicarpus monilifer (L.) DC., Prod. 2: 353. 1825; Wight & Arn., Prod. 232. 1834; Thw., Enum. Pl. Zeyl. 412. 1864; Baker in Hook. f., Fl. Br. Ind. 2: 157. 1876; Trimen, Handb. Fl. Ceylon 2: 43. 1894; Gamble, Fl. Pres. Madras 338. 1918.

Hedysarum moniliferum L., Mant. 102. 1767. Type: from India.
Hedysarum moniliforme Burm. f., Fl. Ind. 168. t. 52. f. 3. 1768.

Prostrate perennial; stems hirsute, hairs 1 mm long; leaves unifoliolate; stipules 6–9 mm long; petioles 4–8 mm long; leaflets chartaceous, ovate, oblong or orbicular, truncate or cordate at the base, obtuse at the apex, glabrous on upper surface, a few appressed long hairs on veins and minute hooked hairs on lower surface, 7–25 mm long, 5–15 mm wide, usually 1.3–2 times longer than wide, rarely 1.1–2.4, smaller and only 5 mm long and 4 mm wide on depauperate plants; inflorescences terminal or in the upper axils, up to 30 mm long; bracts scarious, as long as the calyx, deciduous; pedicels 0.3–1 mm long; calyx ciliate, 3–4 mm long, lobed to about the middle, the lobes all about equal in length, the upper lobe divided to about the middle; corolla pink, standard ovate, about 5.5 mm long, wings and keel petals about equal in length, 4–4.5 mm long; pods moniliform, with 3–8 articles; articles more or less spherical, 4.5–5 mm diam., straw-coloured, smooth with indumentum of spreading hooked hairs 0.15 mm long; seeds subglobular, about 2 mm long and 1.5 mm wide.

Distr. Ceylon, India.

Ecol. In open places among grass.

Specimens Examined. HAMBANTOTA DISTRICT: Ruhuna National Park, *Mueller-Dombois 68050315* (K), *Cooray & Balakrishnan 69011028R* (K), *Mueller-Dombois et al. 69010537* (K).

2. Alysicarpus vaginalis (L.) DC., Prod. 2: 353. 1825; Wight & Arn., Prod. 233. 1834; Thw., Enum. Pl. Zeyl. 87. 1859; Baker in Hook. f., Fl. Br. Ind. 2: 158. 1876; Trimen, Handb. Fl. Ceylon 2: 44. 1894; Gamble, Fl. Pres. Madras 338. 1918; Steenis, Reinwardtia 6: 87. 1961; Verdcourt in Fl. Trop. East Africa, Leg.: Papilion. 491. 1971; Verdcourt, Man. New Guinea Leg. 420. 1979.

Hedysarum vaginale L., Sp. Pl. 746. 1753. Type: *Ceylon, Hermann* Herb. 1: 27, 29 (BM, syntypes).
Alysicarpus nummularifolius DC., Prod. 2: 353. 1825; Wight & Arn., Prod. 233. 1834, non *Hedysarum nummularifolium* L.
Alysicarpus vaginalis var. *nummularifolius* Baker in Hook. f., Fl. Br. Ind. 2: 158. 1876.

Perennial herb, more or less prostrate; branches glabrous, puberulous or occasionally with a few long hyaline curved hairs; leaves unifoliolate; stipules to 12 mm long; petioles 3–14 mm long; leaflets variable in size and shape on a single plant, obovate, broadly to narrowly oblong, subcordate or truncate at the base, acute to emarginate at the apex, 7–40 mm, rarely 50 mm long, 7–14 mm wide, only 5 mm wide on depauperate plants, 1.3–4 times longer than wide. Inflorescences terminal, dense, to about 50 mm long, the rachis puberulous with uncinate hairs; flowers in pairs subtended by stipule-like bract deciduous at anthesis; pedicels 0.5–1.5 mm long; calyx 5–5.5 mm, all the lobes about equal in length about 3 mm long, the upper shortly bifid, a few long hairs on the margins and at the base of the lobes; corolla usually reddish;

standard 4–6 mm long, wings and keel petals shorter, 3.5–4.5 mm long; pods to about 20 mm long, exserted from the calyx, not constricted between the articles; articles 4–7, subcylindric, 2.5–3 mm long, 2–2.5 mm wide, smooth, glabrous or sparsely puberulous; seeds 1.3–1.6 mm long, 1.1–1.3 mm wide.

Distr. Old world tropics; Africa, Ceylon, India, Indo-China, southern China, Malesia, northern Australia, introduced into South America.

Ecol. Open places in grass, often on roadsides, common in places.

Specimens Examined. JAFFNA DISTRICT: between Jaffna and Elephant Pass, *Rudd 3287* (K). BATTICALOA DISTRICT: north of Kalkudah on road to Elephant Point, *Rudd & Balakrishnan 3144* (K). POLONNARUWA DISTRICT: Polonnaruwa, sacred area, *Dittus 70011803* (K); between Habarane and Kantalai, *Rudd & Balakrishnan 3116* (K). KANDY DISTRICT: Kobbekaduwa, near Kandy, *Rudd 3337* (K). AMPARAI DISTRICT: between Amparai and Maha Oya, markers 34–35, *Rudd & Balakrishnan 3231* (K); Kalmunai, *Kostermans 24326* (K). GALLE DISTRICT: Koggala (SE of Galle), *Rudd 3086* (K). HAMBANTOTA DISTRICT: Tissamaharama to Wellawaya, mile 172, *Hepper & de Silva 4768* (K). LOCALITY UNKNOWN: *Thwaites C.P. 1427, p.p.* (K); *C.P. 1428* (K).

Note. *Alysicarpus vaginalis* is sympatric over most of its range in the Old World with *A. ovalifolius* (Schumach.) J. Leonard which is an erect annual with open inflorescences. *A. ovalifolius* has not yet been recorded from Ceylon.

3. Alysicarpus bupleurifolius (L.) DC., Prod. 2: 352. 1825; Wight & Arn., Prod. 233. 1834; Thw., Enum. Pl. Zeyl. 87. 1859; Baker in Hook. f., Fl. Br. Ind. 2: 158. 1876; Trimen, Handb. Fl. Ceylon 2: 44. 1896; Gamble, Fl. Pres. Madras 338. 1918; Steenis, Reinwardtia 6: 88. 1961; Verdcourt, Man. New Guinea Leg. 420. 1979.

Hedysarum bupleurifolium L., Sp. Pl. 745. 1753. Type: from India.

Erect or ascending annual herb; branches with covering of appressed hairs, becoming glabrous; leaves unifoliolate; stipules acute, 3–10 mm long; petioles 2–4 mm long; leaflets narrowly ovate on young plants, lanceolate to linear on older ones, rounded at the base, acute at the apex, usually 30–70 mm long, 3–4.5 mm wide, 10–20 times longer than wide, occasionally down to 5 mm long, up to 9 mm wide and only twice as long as wide, glabrous on upper surface, sparse appressed hairs on lower; prominently reticulately veined on both surfaces; inflorescences terminal, dense, to 20 mm long in flower, open and to 50 mm or more in fruit; rachis glabrous; pairs of flowers subtended by acute or acuminate bract, deciduous at anthesis; pedicels 0.8–1.5 mm long, with ascending brown hairs, becoming glabrous; calyx rather stiff, angled towards the base, 6–7 mm long, the lobes equalling each other in length, strongly overlapping at the base, 4–5 mm long, the upper entire or shortly bifid, glabrous; corolla orange and purple; standard oblong, 4–5 mm long, 1.5–2.5 mm wide, the wings and keel petals about equal in length, shorter than standard, about 4 mm long; pods to 15 mm long, slightly to markedly exserted from the calyx, not constricted between the articles; articles 3–8, cylindrical or slightly flattened, 1.6–2 mm long, 1.3–2 mm wide, smooth or obscurely reticulately veined, glabrous; seeds almost cubic, about 1 mm long and 0.8 mm wide.

Distr. Ceylon, India, Indo-China, southern China, Malesia, northern Australia.

Ecol. In open places, often a weed of cultivation.

Vern. Kutiraivale (T).

Specimens Examined. *Thwaites C.P. 1427* reported to have been collected, but not seen.

4. Alysicarpus longifolius (Rottler ex Spreng.) Wight & Arn., Prod. 233. 1834; Thw., Enum. Pl. Zeyl. 412. 1864; Baker in Hook. f., Fl. Br. Ind. 2: 159. 1876; Trimen, Handb. Fl. Ceylon 2: 45. 1894; Gamble, Fl. Pres. Madras 338. 1918.

Hedysarum longifolium Rottler ex Spreng., Syst. Veg. 3: 319. 1826. Type: Peninsular India, Rottler.

Erect, little-branched annual to about 2 mm tall; young stems puberulous with short uncinate hairs, becoming glabrous; leaves unifoliolate; stipules 10–30 mm long; petioles 4–10 mm long; leaflets lanceolate, subcordate at the base, acute at the apex, 80–125 mm long, 9–15 mm wide, 7–12 times longer than wide, or oblong, subcordate at the base, obtuse at the apex, 40–60 mm long, 5–20 mm wide, 2–8 times longer than wide towards the top of old plants or on young plants, glabrous on the upper surface, sparsely minutely puberulous, with uncinate hairs and with scattered longer appressed hairs; inflorescences terminal, dense, 15–30 mm long, rhachis puberulous with uncinate hairs, pairs of flowers subtended by coarsely striate bracts hirsute on the outside, deciduous at anthesis; calyx 6–7 mm long, the tube about 2 mm long, obscurely 4-angled at the base, lobes all of about equal length, the upper shortly bifid at the apex, all ciliolate with a few long hairs at the base; corolla with standard obovate, 6–7 mm long, about 5 mm wide, wings 6–6.5 mm long, slightly longer and wider than the keel petals; pods to about 8 mm long, as long as the calyx or shortly exserted with 3 or 4 articles, not constricted between the articles; articles cylindrical or slightly flattened, 2–2.4 mm long, 2–2.4 mm wide, puberulous, obscurely reticulate; seeds depressed globular, 1.5–1.8 mm long, 1.3–1.6 mm wide.

Distr. Southern India, Ceylon (?).

Specimens Examined. No specimens seen from Ceylon despite its being recorded by Thwaites and Trimen.

5. Alysicarpus scariosus (Rottler ex Spreng.) Graham ex Thw., Enum. Pl. Zeyl. 88. 1859.

Alysicarpus styracifolius DC., Prod. 2: 353. 1825; Wight & Arn., Prod. 234. 1834.

Hedysarum scariosum Rottler ex Spreng., Syst. Veg. 3: 319. 1826. Type: from India.

Alysicarpus rugosus var. *styracifolius* Baker in Hook. f., Fl. Br. Ind. 2: 159. 1876; Gamble, Fl. Pres. Madras 339. 1918, non *Hedysarum styracifolium* L.

Alysicarpus rugosus auct.: Trimen, Handb. Fl. Ceylon 2: 45. 1894, non (Willd.) DC.

Prostrate or perennial ascending herb; stems pilose or hairs appressed, becoming glabrous. Leaves unifoliolate; stipules 3–9 mm long; petioles 2–6 mm long; leaflets broadly or narrowly oblong or occasionally obovate, rounded at the base, rounded or somewhat retuse at the apex, with sparse scattered appressed hairs on both surfaces or glabrous above, 12–50 mm long, 6–15 mm wide, 1.2–6 or occasionally 8 times longer than wide. Inflorescences terminal or in the upper axils, to 50 mm long, rhachis pilose; pairs of flowers subtended by oblong acuminate bracts 5–7 mm long, minutely puberulent towards the apex, long ciliate (with hairs 1 mm long) on margins, persistent; pedicels 0.5–1 mm long; calyx stiff, 7–8.5 mm long, the lobes strongly imbricate at the base, all about equal in length, 6–7.5 mm long, the upper shortly bifid, densely ciliate on margins, corolla pink or purplish; standard obovate, 5–6 mm long, 2–2.5 mm wide, wings and keel petals shorter than the standard, equal in length, 4–5.3 mm long; pods of 3–6 articles, constricted between them, about equalling the calyx or shortly exserted; articles puberulous, 1.5–2 mm long, 2.5–3 mm wide, strongly reticulate, the nerves thick and anastomosing; seeds oblong in outline, 2–2.5 mm long, 1.3–1.6 mm wide.

Distr. Southern India and Ceylon.

Ecol. In open places in sandy soil, often near the sea.

Specimens Examined. JAFFNA DISTRICT: between Jaffna Causeway and Kayts, *Rudd 3267* (K). HAMBANTOTA DISTRICT: Ruhuna National Park, Patanagala Beach, *Comanor 912* (K), *Cooray 69113003R* (K).

6. Alysicarpus heyneanus Wight & Arn., Prod. 234. 1834; Thw., Enum. Pl. Zeyl. 88. 1859. Type: Peninsular India, *Wallich 5770a* (K, holotype).

Alsicarpus rugosus var. *heyneanus* Baker in Hook. f., Fl. Br. Ind. 2: 159. 1876; Trimen, Handb. Fl. Ceylon 2: 45. 1894; Gamble, Fl. Pres. Madras 339. 1918.

Erect annual herb, branched at the base; stems pilose, hairs usually persisting at the base, sometimes becoming glabrous except in axils of leaves; leaves unifoliolate; stipules 5–10 mm long; petioles 3–8 mm long; leaflets oblong, elliptic or obovate, becoming elongate in the upper part of the plant, attenuate at the base, rounded or retuse at the apex, 25–55 mm, rarely to 65 mm long, 9–33 mm wide, 1.5–3.5 times, rarely to 6 times, longer than wide, glabrous on upper surface, rather loose ascending hairs on lower; inflorescences terminal, axillary or leaf-opposed, to 350 mm long (in fruit), rather open; rhachis puberulous with short uncinate and scattered longer appressed hairs; pairs of flowers subtended by ovate acuminate bracts 4–5 mm long, 2.5–3 mm wide, ciliate towards the apex, deciduous at about anthesis; pedicels 1–3 mm long; calyx stiff, about 6 mm long, deeply lobed, the tube about 1 mm long, the upper lobe shortly bifid at the apex, the lobes strongly imbricate at the base, ciliate with sometimes a few long hairs along the midline; corolla with standard 4–4.5 mm long, 3–3.5 mm wide; wings 3.5–4 mm long, about as long as or a little shorter than the keel; pods of 3–6 articles, strongly constricted between them, equally or shortly exserted from the calyx; articles puberulous, strongly reticulate, the nerves thick and anastomosing, 1.5–2 mm long, 2.5–3 mm wide; seeds oblong in outline, 1.9–2.3 mm long, 1.7–1.8 mm wide.

Distr. Ceylon, southern India.

Ecol. Open situations among grass.

Specimens Examined. LOCALITY UNKNOWN: *Walker s.n.* (K).

2. URARIA

Desv., J. Bot. Agric. 1: 122. 1813; Wight & Arn., Prod. 221. 1834; Benth. in Benth. & Hook. f., Gen. Pl. 1: 521. 1865; Hutch., Gen. Fl. Pl. 1: 481. 1964. Type species: *U. picta* (Jacq.) DC.
Doodia Roxb., Fl. Ind. ed. 2. 3: 365. 1832, non R. Br. 1810. Type species: none designated.

Herbs; leaves pinnately 3–9-foliolate, sometimes on young plants 1-foliolate; petiole sulcate; stipules acuminate, stipels present; inflorescences mostly terminal spike-like racemes or panicles, pedicels in pairs; inflexed hooked at the apex; bracts subtending pedicels ovate, acuminate; calyx 5-lobed, the upper pair united to some extent, the lower 3 longer than the others, often twisted; stamens diadelphous, the upper one free at anthesis, anthers uniform; ovary sessile, 2-many-ovulate; style filiform, bent about the middle; pods in concertina-like folds, more or less enclosed in the persistent calyx, constricted between the segments; articles quadrate, inflated, 1-seeded, indehiscent; seeds subglobose or reniform; hilum lateral without a rim aril.

A genus of some 20 species in Old World tropics, best developed in south-east Asia.

KEY TO THE SPECIES

1 Inflorescence open (rhachis quite apparent when plant in fruit); pedicels pubescent but without
   hooked hairs; fruits obscurely reticulate, puberulent .... 3. U. rufescens
1 Inflorescence dense (rhachis obscured when plant in fruit); pedicels with dense hooked hairs
   towards the apex; articles of pod smooth, shining, glabrous
   2 Leaflets elongate, more than 4 times as long as wide, usually with pale line along centre
      above; pedicels 4–8 mm long .... 1. U. picta
   2 Leaflets less than 3.5 times longer than wide, without a pale central area; pedicels 10–14 mm
      long .... 2. U. crinita

1. Uraria picta (Jacq.) DC., Prod. 2: 324. 1824; Wight & Arn., Prod. 221. 1834;
Thw., Enum. Pl. Zeyl. 85. 1859; Baker in Hook. f., Fl. Br. Ind. 2: 155. 1876;
Trimen, Handb. Fl. Ceylon 2: 42. 1894; Gamble, Fl. Pres. Madras 336. 1918;
van Meeuwen, Reinwardtia 5: 452. 1961; Verdcourt in Fl. Trop. East Africa,
Leg.: Papilion. 479. 1971; Verdcourt, Man. New Guinea Leg. 416. 1979.

Hedysarum pictum Jacq., Collect. 2: 267. 1789. Type: from West Africa
   (Ghana).
Doodia picta Roxb., Fl. Ind. ed. 2. 3: 368. 1832.

   Much branched perennial herb; branches densely pubescent with uncinate hairs; leaves 3–9-foliolate, 1-foliolate on young plants; stipules triangular, sometimes long-pointed, 10–15 mm long, about 5 mm wide at the
base, pubescent, deciduous; rhachis 15–65 mm long; leaflets coriaceous,
linear-oblong or narrowly oblong (ovate or broadly oblong when 1-foliolate),
rounded or subcordate at the base, obtuse at the apex, glabrous with a pale
central line above, puberulous beneath; leaflets of upper leaves 5.5–15 cm,
rarely 21.5 cm, long, 7–15 mm, rarely to 20 mm, wide, 4–13 times longer
than wide, rarely to 15; inflorescences terminal, dense, 13–25 cm long,
rhachis covered with dense short straight and scattered uncinate hairs; bracts
10–20 mm long, 2–4.5 mm wide, long-pointed, early deciduous; pedicels
4–8 mm long, distal part curving upwards in fruit, with dense covering of
short straight and longer uncinate hairs, the latter dense towards the apex;
calyx tube 1–1.3 mm long, densely puberulous, lobes subulate, with long
(1–1.5 mm) spreading hyaline hairs, 1.3–2.2 mm long, the upper pair united
for a short distance; corolla reddish violet; standard obovate, 5.5–6.5 mm
long, 4–4.5 mm wide; wings 5–6.5 mm long, 1.5–2.5 mm wide, about as
long as, and wider than, keel petals; pods of 3–5 articles, brown becoming
pale grey, smooth, shining, each 3.2–3.5 mm long, about 3 mm wide; seeds
reniform, about 2.5 mm long, 2 mm wide.

   Distr. Old World tropics: Africa, Ceylon, India, Indo-China, southern
China, Malesia and northern Australia, introduced into the West Indies.

   Ecol. Open places, on roadsides, occasionally in forest, locally common.

Specimens Examined. TRINCOMALEE DISTRICT: 18 miles E. of Kantalai, *Davidse 7586* (K); about 2 miles W of Mahaweli Ganga in flood plain, about 20 km from Koddiyar Bay, *Fosberg 56400* (K); crossing Mahaweli Ganga and road E Potankadu, *Bremer et al. 103* (K).

2. Uraria crinita (L.) Desv. ex DC., Prod. 2: 324. 1825; Wight, Ic. Pl. Ind. Or. t. 411 (as *U. picta*). 1840; Baker in Hook. f., Fl. Br. Ind. 2: 155. 1876; Trimen, Handb. Fl. Ceylon 2: 42. 1894; van Meeuwen, Reinwardtia 5: 452. 1961.

Hedysarum crinitum L., Mant. Pl. 102. 1767; Burm. f., Fl. Ind. 169, t. 56. 1768. Type: Burmann's plate (?)
Doodia crinita Roxb., Fl. Ind. ed. 2. 3: 369. 1832.

Shrub to 1 m tall; branches densely pubescent with uncinate hairs; leaves 3–5-foliate (unifoliolate on young plants); stipules long-pointed, 10–15 mm long, 3–5 mm wide at the base, ciliate, deciduous; petioles 45–100 mm long, rhachis 12–18 mm long when 3-foliolate, to 30–75 mm when 5-foliolate; leaflets coriaceous, oblong, elliptic or ovate, rounded or truncate at the base, obtuse at the apex (cordate at the base when 1-foliolate), glabrous on upper surface, with short sparse hairs on veins on lower surface, 4.5–13.5 cm long, 2–5 cm wide, 1.5–3.5 times longer than wide; inflorescences terminal, dense, to 35 cm long, axis puberulous with scattered longer uncinate hairs; bracts lanceolate, about 12 mm long, 2.5 mm wide, ciliate, early deciduous; petioles 10–14 mm long, with indumentum of short sparse uncinate hairs, long (2 mm) stiff white hairs and towards the apex hyaline uncinate hairs 1 mm long; calyx tube 1–1.5 mm long, glabrous, lobes subulate, the upper pair united for a short distance, 2–3 mm long, the others about 4 mm long, all with spreading hyaline hairs about 2 mm long; corolla with standard orbicular, 7–9 mm long, 6–7 mm wide, wings 6–6.5 mm long, 2.2–2.5 mm wide, slightly shorter than the keel petals; pods of 5 or 6 articles, brown becoming pale grey, smooth, each 2.5–3.5 mm long, 2.3–2.7 mm wide; seeds reniform, 2.2–2.4 mm long, 1.7–1.9 mm wide.

Distr. India, Indo-China, southern China and Malesia except the Philippines and New Guinea. Trimen stated the plant in Ceylon was cultivated only and its absence from Peninsular India would support such a claim.

Specimen Examined. LOCALITY UNKNOWN: *Walker s.n.* (K).

3. Uraria rufescens (DC.) Schindl., Repert. 21: 14. 1925.

Desmodium rufescens DC., Ann. Sci. Nat. Bot. 4: 101. 1825. Type: from India.
Doodia hamosa Roxb., Fl. Ind. ed. 2. 3: 367. 1832. Type: India, *Wallich 5681 B* (?) (K).

Uraria hamosa (Roxb.) Wight & Arn., Prod. 222. 1834; Wight, Ic. Pl. Ind. Or. t. 284. 1840; Thw., Enum. Pl. Zeyl. 85. 1859; Baker in Hook. f., Fl. Br. Ind. 2: 156. 1876; Trimen, Handb. Fl. Ceylon 2: 43. 1894; Gamble, Fl. Pres. Madras 336. 1918.

Prostrate or erect shrub to 1 m tall; branches puberulous with long straight spreading hairs or pubescent with spreading brown uncinate hairs; leaves 1-or, usually, 3-foliolate; stipules subulate, 6–12 mm long; petioles 10–32 mm long, rhachis 5–18 mm long; leaflets rather coriaceous, terminal elliptic or less commonly ovate or obovate, laterals oblong, all cuneate at the base, obtuse at the apex, terminal 3.5–9 cm long, 25–52 mm wide, the laterals smaller, 2–6 cm long, 13–30 mm wide, all 1.4–2 times longer than wide, subglabrous on upper surface, pubescent with weak ascending hairs on lower, rather close lateral nerves prominent beneath; inflorescences terminal or in the upper axils, sometimes forming a terminal panicle, rather open; bracts 6–12 mm long, 2.5–4 mm wide, orbicular, abruptly acuminate; pedicels 2–4 mm with spreading brown hairs not uncinate; calyx tube about 1 mm long, glabrous, lobes narrow, the upper part united to about the middle, about 1.5 mm long, the others about 2 mm long, all with long straight white hairs on the margins; corolla purplish; standard broadly obovate, about 6 mm long, 4 mm wide; wings about 6 mm long, 1.5 mm wide, about as long as the keel petals; pods of 5–7 articles, brownish, not smooth, each about 2.5 mm long, 2.2–2.6 mm wide, puberulous; seeds quadrate in outline, about 1.6 mm long, 1.5 mm wide.

Distr. Ceylon, India, Indo-China, southern China and sporadically in western parts of Malesia.

Specimens Examined. LOCALITY UNKNOWN: *Thwaites C.P. 3590* (K).

3. ELEIOTIS

Wight & Arn., Prod. 231. 1834; Benth. in Hook. f., Gen. Pl. 1: 523. 1865; Hutch., Gen. Fl. Pl. 1: 487. 1964. Type species: *E. monophyllos* (Burm. f.)DC.

Prostrate perennial herbs; leaves 1-or 3-foliolate, leaflets reniform; petiole sulcate; stipules free, deltoid; stipels present; inflorescence a pseudo-raceme, flowers in pairs subtended by a scarious bract, secondary bracts and bracteoles absent; calyx truncate with short teeth; corolla small; stamens diadelphous with vexillary stamen free; anthers uniform; ovary sessile, 1-ovulate; style rather stout; pod compressed, boat-shaped, apiculate; seeds oblongo-cylindrical; aril not developed.

A genus of 2 species, southern India and Ceylon.

Eleiotis monophyllos (Burm. f.) DC., Mem. Legum. 7: 348. 1825; DC., Prod. 2: 348. 1825.

Glycine monophyllos Burm. f., Fl. Ind. 161. t. 50. f. 2. 1768. Type: Burmann's plate (?).

Hedysarum sororium L., Mant. 270. 1771; Roxb., Fl. Ind. ed. 2. 3: 352. 1832. *Eleiotis sororia* (L.) DC., Mém. Légum. 7: 348. 1825; DC., Prod. 2: 348. 1825; Wight & Arn., Prod. 231. 1834; Thw., Enum. Pl. Zeyl. 412. 1864; Baker in Hook. f., Fl. Br. Ind. 2: 153. 1876; Trimen, Handb. Fl. Ceylon 2: 40. 1894; Gamble, Fl. Pres. Madras 332. 1918.

Prostrate perennial herb; branchlets angular, glabrous or with appressed hairs on the angles; leaves 1-foliolate or rarely 3-foliolate, the lateral leaflets rudimentary on long pulvini; petioles 2.5 mm long or about 1 mm when leaves 3-foliolate, with long hairs; leaflets reniform or laterals elongate, glabrous on the upper surface, appressed pubescent on lower surface, 6–14 mm long, 10–30 mm wide, 1.5–2.2 times wider than long, the laterals about 5 mm long and 3 mm wide with pulvinus about 3 mm long; inflorescence axillary, very open, to 10 cm long; rhachis with short spreading uncinate hairs; bracts 1.5–2.5 mm long, about 1 mm wide; pedicels 1.5–2.5, rarely to 4 mm long; calyx about 1 mm long, more or less truncate, the lobes undulate, apiculate; corolla purplish with standard obovate, about 1.5 mm long, wings about 1.2 mm long, rather wide, about as long as the keel petals; pods 1-seeded, boat-shaped, the upper suture more or less straight, thickened, the lower convex, 5–6 mm long, 2–3 mm wide, glabrous, reticulately veined; seeds oblongoid, 3–3.5 mm long, 1.5–2.5 mm wide.

Distr. Ceylon and southern India.

Ecol. In shaded situations such as forest floors.

Specimen Examined. COLOMBO DISTRICT: Colombo, in 1862, *Ferguson* in *C.P. 3765* (K, PDA).

4. PYCNOSPORA

R.Br. ex Wight & Arn., Prod. 197. 1834; Benth. in Benth. & Hook. f., Gen. Pl. 1: 521. 1865; Hutch., Gen. Fl. Pl. 1: 399. 1964. Type species: *P. lutescens* (Poir.) Schindl.

Diffuse subshrub; leaves 3-foliolate or 1-foliolate at base of plant; petiole sulcate; stipules free, lanceolate; stipels present; inflorescence a terminal pseudoraceme, flowers in 2's or 3's subtended by membranous bracts; secondary bracts and bracteoles absent; calyx 4-lobed, the lobes subequal, the upper shortly bifid; corolla small, wings adhering closely to the keel; stamens diadelphous, upper free, intrastaminal disk present; anthers uniform; ovary sessile with to about 10 ovules, style short; pod (like that of *Crotalaria*) oblong, inflated, unilocular, not articulated, the valves thin; seeds small, with a thin rim-aril.

A genus of one species in the tropics of the Old World from Africa to Australia.

Pycnospora is rather anomalous. Its inflated pod without articulations technically removes it from the Desmodieae, but it cannot be placed in any other tribe of the Fabaceae.

Pycnospora lutescens (Poir.) Schindl., J. Bot. 64: 145. 1926; van Meeuwen, Reinwardtia 5: 436. 1961; Verdcourt in Fl. Trop. East Africa, Leg.: Papilion. 481. 1971; Verdcourt, Man. New Guinea Leg. 413. 1979.

Hedysarum lutescens Poir. in Lam., Enc. 6: 417. 1805. Type: from China; near Canton.
Desmodium lutescens (Poir.) DC., Prod. 2: 326. 1825.
Pycnospora nervosa Wight & Arn., Prod. 197. 1834; Thw., Enum. Pl. Zeyl. 92. 1859. Type: from Peninsular India.
Pycnospora hedysaroides R. Br. ex Benth., Fl. Austr. 2: 236. 1864; Baker in Hook. f., Fl. Br. Ind. 2: 153. 1876; Trimen, Handb. Fl. Ceylon 2: 41. 1894; Gamble, Fl. Pres. Madras 333. 1918. Type: from Australia.

Diffuse, prostrate or ascending herb; branches with rather dense covering of appressed or ascending hairs. Leaves 3-foliolate, sometimes 1-foliolate at base of plants; stipules long-pointed, 4–7 mm long; petioles 4–13 mm long, rhachis 2–5 mm long; leaflets oblong or obovate, cuneate at the base, rounded at the apex, sparse, with more or less appressed hairs on the upper surface, sparse to moderately dense hairs on lower surface; terminal leaflet 10–30 mm long, 5–20 mm wide. 1.1–1.8 times longer than wide, lateral leaflets smaller, 7–22 mm long, 5–14 mm wide, 1.2–2 times longer than wide; inflorescences terminal, to about 80 mm long, rather dense; bracts abruptly acuminate, 5–6 mm long, about 2 mm wide; pedicels 2–4 mm long; calyx 2.5–3.5 mm long, with rather long hairs on lobes; tube about 1 mm long; corolla blue or purplish; standard 4–5 mm long, 3–4 mm wide, obovate, obtuse or slightly retuse; wing and keel petals about equal in length, 4–5 mm long; ovary pubescent, style short; pod 8–14 mm long, 3–4 mm wide, drying black; seeds 2–2.3 mm long, 1.4–1.5 mm wide.

D i s t r. Central Africa, Ceylon, eastern India, Indo-China, southern China, Malesia and northern Australia.

E c o l. Open areas, particularly on roadsides.

S p e c i m e n s E x a m i n e d. COLOMBO DISTRICT: Kesbewa-Banda-ragama road, *Cooray 70012308R* (K). KANDY DISTRICT: 4 miles E. Hunnasgiriya, marker 24/14, *Jayasuriya et al. 1414* (K); between Madugoda and Hunnasgiriya, *Rudd & Balakrishnan 3243* (K). RATNAPURA DISTRICT: Ratnapura, Dec 1869, *Thwaites s.n.* (K). LOCALITY UNKNOWN: *Thwaites C.P. 279* (K); Walker 20 (K).

5. DENDROLOBIUM

(Wight & Arn.) Benth. in Miq., Pl. Jungh. 215 & 216. 1852; Schindl., Repert. 20: 278. 1924; Hutch., Gen. Fl. Pl. 1: 483. 1964; Ohashi, Ginkgoana 1: 50. 1973. Type species: *D. umbellatum* (L.) Benth.

Desmodium subg. *Dendrolobium* Wight & Arn., Prod. 223. 1834; Baker in Hook. f., Fl. Br. Ind. 2: 161. 1876.
Desmodium sect. *Dendrolobium* (Wight & Arn.) Benth. in Benth. & Hook. f., Gen. Pl. 1: 519. 1865.

Shrubs or rarely small trees, leaves 3-foliolate or rarely 1-foliolate; petioles sulcate; stipules free or connate on the side opposite the petioles, stipels present; inflorescences axillary, subsessile or shortly pedunculate, subumbellate, shortly racemose or (in one Australian species) distinctly racemose; flowers uniparous, pedicellate, a single deciduous bract at the base of the pedicel; bracteoles conspicuous at the base of the calyx; calyx deeply 4-lobed, the upper lobe entire or minutely bifid, wider than the others; corolla rather large, white; stamens monadelphous, the upper stamen free only in its upper half; ovary sessile with a slender style; pods more or less moniliform with up to about 8 seeds, the articles thick-walled, indehiscent; seeds with a distinct rim-aril.

A genus of 17 species mainly in the Asian-Australian region, one species extending to the Indian Ocean shores of east Africa and Madagascar, centres of diversity in Indo-China and northern Australia. Two species in Ceylon.

KEY TO THE SPECIES

1 Inflorescences distinctly pedunculate; terminal leaflets 55–85 mm long, less than twice as long as wide; 6–9 lateral veins prominent on each side of midrib; wings markedly narrower and shorter than keel petals; articles of pods 7–8 mm long . . . . **1. D. umbellatum**
1 Inflorescences subsessile; terminal leaflets 2–3 times longer than wide with 14–18 lateral veins prominent on each side of midrib; wings not markedly narrower and shorter than keel petals; articles of pods 4–5 mm long . . . . **2. D. triangulare**

1. Dendrolobium umbellatum (L.) Benth. in Miq., Pl. Jungh. 218. 1852; Thw., Enum. Pl. Zeyl. 86. 1859; Trimen, Handb. Fl. Ceylon 2: 47. 1894; Schindl., Repert. 20: 278. 1924; Ohashi, Ginkgoana 1: 82. 1973.

Hedysarum umbellatum L., Sp. Pl. 747. 1753. Type: Herb. *Hermann* 2: 26 & 4: 73 (BM).
Desmodium umbellatum (L.) DC., Prod. 2: 325. 1825; Wight & Arn., Prod. 224. 1834; Baker in Hook. f., Fl. Br. Ind. 2: 161. 1876; van Meeuwen, Reinwardtia 6: 263. 1962; Schubert in Fl. East. Trop. Africa, Leg.: Papilion. 455. 1971; Verdcourt, Man. New Guinea Leg. 410. 1979.
Meibomia umbellata (L.) Kuntze, Rev. Gen. Pl. 1: 197. 1891.

Shrub to about 3 m tall; branches with covering of silky appressed hairs becoming sparse; leaves 3-foliolate; stipules early deciduous, 6–9 mm long; petioles 18–35 mm long, rachis 7–12 mm long; leaflets ovate or elliptic, rounded or attenuate at the base, sometimes bluntly acuminate, obtuse at the apex, discolorous, glabrous above, sparsely appressed pubescent beneath, the hairs minute, 6–9 prominent lateral veins on each side of midrib; terminal leaflets 55–85 mm long, 37–55 mm wide, the laterals smaller, 40–55 mm long, 25–38 mm wide, all 1.3–1.8 times longer than wide; stipels setaceous, 1–4 mm long; pulvini 3–5 mm long; inflorescences axillary, usually densely 10–20-flowered; peduncles 5–10 mm long, bracts appressed sericeous, 2–3 mm long, with an acute apex; pedicels about 2 mm long at anthesis; calyx densely appressed sericeous, 4–5 mm long, the tube 2–2.5 mm long, lobes 2–3 mm long, the upper wider than the others; corolla with standard obovate or elliptic, 9–12 mm long, 6–9 mm wide, rounded at the apex, clawed; wings 10–12 mm long, 1–2 mm wide, distinctly shorter and narrower than the keel petals; ovary densely appressed sericeous; pods somewhat incurved, the lower suture deeply indented, the isthmus 2/3–4/5 width of articles, articles 2–5, each oblong, 7–8 mm long, 5–6 mm wide, sparsely appressed pubescent; seeds elliptic in outline, about 4 mm long and 3 mm wide.

Distr. Madagascar, East Africa, Ceylon, Indo-China, southern China, Malesia, north-eastern Australia and the Pacific Islands.

Ecol. On seacoasts, usually on sand just above high-water mark.

Specimens Examined. TRINCOMALEE DISTRICT: *Nilaveli, Cramer 4377* (K, PDA); Dead Man's Cove, Trincomalee, *Worthington 175* (K); Chapel Hill, *Worthington 1131* (K). BATICALOA DISTRICT: Pasikudah, *Sumithraarachchi & Sumithraarachchi 874* (PDA); Panichchankeni, on Baticaloa-Trincomalee road, *Balakrishnan 365* (K); Kalkudah, *Mueller-Dombois 68042003* (K, PDA), *Rudd & Balakrishnan 3152* (K). AMPARAI DISTRICT: Ruhuna National Park East, Kumana, Kumbukkan oya river mouth, *Jayasuriya 2015* (K); Arugam Bay, *Fosberg & Sachet 52932* (K). GALLE DISTRICT: Bentota, *Worthington 5275* (K), *s. coll s.n.* (PDA). LOCALITY UNKNOWN: *Thwaites C.P. 1436* (K, PDA).

2. Dendrolobium triangulare (Retz.) Schindl., Repert. 20: 279. 1924; Ohashi, Ginkgoana 1: 77. 1973.

Hedysarum triangulare Retz., Obs. 3: 40. 1783. Type: from Java, Thunberg.
Hedysarum cephalotes Roxb., Fl. Ind. ed. 2. 3: 360. 1832. Type: from India, cult. Calcutta.
Desmodium cephalotes Wight & Arn., Prod. 224. 1834; Wight, Ic. Pl. Ind. Or. 373. 1840; Baker in Hook. f., Fl. Br. Ind. 2: 161. 1876; Trimen, Handb. Fl. Ceylon 2: 47. 1894; Gamble, Fl. Pres. Madras 344. 1918.

Desmodium congestum Wight & Arn., Prod. 224. 1834; Wight, Ic. Pl. Ind. Or. t. 209. 1840. Type: from India.

Dendrolobium cephalotes (Roxb.) Benth. in Miq., Pl. Jungh. 216 & 218. 1852; Thw., Enum. Pl. Zeyl. 86. 1859.

Meibomia cephalotes (Roxb.) Kuntze, Rev. Gen. Pl. 1: 195. 1891.

Desmodium triangulare (Retz.) Merr., J. Arnold Arbor. 23: 170. 1942; van Meeuwen, Reinwardtia 6: 261. 1962.

Shrub to 1 m tall; branches covered with dense silky appressed hairs; leaves 3-foliolate; stipules long-pointed, 3–12 mm long; petiole 10–30 mm long, rhachis 8–16 mm long; leaflets ovate or elliptic, rounded at the base, acuminate at the apex, glabrous on upper surface, glabrous or with appressed hairs on veins on lower surface, 14–16 lateral veins prominent on each side of midrib; terminal leaflet 50–75 mm long, 18–35 mm wide, laterals smaller, 30–55 mm long, 14–23 mm wide, all 2–3 times longer than wide; pulvini about 2 mm long; stipels setaceous, 3–6 mm long; inflorescences subsessile, densely 20–30-flowered; bracts long-pointed, 2–4 mm long, appressed silky outside; pedicels 2–3 mm long, appressed silky; calyx 5–9 mm long, sparsely to densely sericeous; tube 2–3 mm long, upper and lateral lobes triangular, 1.5–3.5 mm long, lower lobe with acuminate apex, 3–6.5 mm long; corolla with standard obovate, about 9 mm long; wings about as long as the standard, slightly longer than the keel petals; ovary densely pubescent; pods straight, the lower suture more deeply indented than the upper, the isthmus 1/2–3/4 width of pod, articles 3–6, each quadrate, 4–5 mm long, 3–4 mm wide with sparse appressed silky hairs; seeds reniform, 2.5–3.5 mm long, 1.8–2.5 mm wide.

Distr. Ceylon, India, Indo-China, southern India, Malesia but absent from New Guinea.

Ecol. Openings in forest.

Specimens Examined. KANDY DISTRICT: Kandy, Oct. 1845, *Thomson* (K). BADULLA DISTRICT: Dambarawa Wewa, *Jayasuriya 400* (PDA). LOCALITY UNKNOWN: *Thwaites C.P. 2780* (K).

Note. The Ceylon plant is *D. triangulare* subsp. *triangulare;* the other subspecies recognised by Ohashi is restricted to eastern India, Burma and Indo-China.

6. PHYLLODIUM

Desv., J. Bot. Agric. 1: 123. t. 5. f. 24. 1813; Benth. in Miq., Pl. Jungh. 217 & 218. 1852; Schindl., Repert. 20: 269. 1924; Hutch., Gen. Fl. Pl. 1: 479. 1964; Ohashi, Ginkgoana 1: 260. 1973. Type species: *P. pulchellum* (L.) Desv.

Dicerma sect. *Phyllodium* (Desv.) DC., Prod. 2: 339. 1825; Wight & Arn., Prod. 230. 1834.

Desmodium sect. *Phyllodium* (Desv.) Benth. in Benth. & Hook. f., Gen. Pl.
1: 519. 1865; Merr., Philipp. J. Sci. 5: 79. 1910.
Desmodium subg. *Phyllodium* (Desv.) Baker in Hook. f., Fl. Br. Ind. 2: 162.
1876.

Shrubs; leaves 3-foliolate, distinctly pulvinate; stipules long-acuminate at
the apex; petioles sulcate; stipels present; inflorescence racemose with fascicles of flowers in the axils of conspicuous foliar bracts, each stipulate and with
stipels with 2 leaflets and a terminal bristle or with a terminal oblong leaflet
or, in species not in Ceylon, foliar bracts absent; flowers uniparous, pedicellate, with a single bract at the base of each pedicel; bracteoles present at the
base of the calyx; calyx 4-lobed, the upper lobe entire or minutely notched,
wider than the others; corolla white or yellowish; stamens monadelphous, the
upper stamen free only in its upper half; ovary sessile, surrounded at the base
by a minute disk, style slender; pods jointed with up to 8 articles, separating,
the articles indehiscent, rather thin-walled, reticulate; seeds with a distinct
rim-aril.

A genus of 8 species in the Asian-Australian region with the centre of
diversity in Indo-China; one species in Ceylon.

Phyllodium pulchellum (L.) Desv., J. Bot. Agric. 1: 124. t. 5. f. 24. 1813;
Thw., Enum. Pl. Zeyl. 86. 1859; Schindl., Repert. 20: 270. 1924; Ohashi,
Ginkgoana 1: 276. 1973.

Hedysarum pulchellum L., Sp. Pl. 747. 1753. Type: Herb. *Hermann 3: 46*
(BM).
Dicerma pulchellum (L.) DC., Prod. 2: 339. 1825; Wight & Arn., Prod. 230.
1834; Wight, Ic. Pl. Ind. Or. 418. 1841.
Desmodium pulchellum (L.) Benth., Fl. Hongk. 83. 1861; Baker in Hook. f.,
Fl. Br. Ind. 2: 162. 1876; Trimen, Handb. Fl. Ceylon 2: 48. 1894;
Gamble, Fl. Pres. Madras 344. 1918; van Meeuwen, Reinwardtia 6: 256.
1962; Verdcourt, Man. New Guinea Leg. 404. 1979.
Meibomia pulchella (L.) Kuntze, Rev. Gen. Pl. 1: 197. 1891.

Shrub to 2 m high, branches with covering of moderately dense appressed
or ascending hairs; leaves trifoliolate; stipules long-pointed, 5–8 mm long; petioles 4–10 mm, rarely to 14 mm, long; leaflets ovate or oblong, rounded at the
base, obtuse at the apex, margins somewhat undulate, discolorous, sparsely
pubescent with ascending hairs, usually becoming glabrous on upper surface,
moderately densely pubescent with ascending hairs on lower surface; terminal
leaflet 55–120 mm long, 25–60 mm wide, 2–2.5 times longer than wide, lateral
leaflets often orbicular and about half as long as terminal leaflet, 20–70 mm long,
12–35 mm, rarely to 45 mm, wide, 1.4–3 times longer than wide; pulvini 2–3 mm

long, stipels setaceous, 1.5–3 mm long; foliar bracts with stipules narrowly triangular, 2–4 mm long, petioles 2–3 mm long, lateral leaflets orbicular, oblique, 10–24 mm long, 10–18 mm wide, ascending hairs on both surfaces, terminal seta 5–10 mm long or terminal leaflet about 16 mm long, 8 mm wide; fascicles of 3–5 flowers; bracts narrowly ovate, about 1 mm long, pedicels 2–3 mm long, bracteoles 0.5–1 mm long; calyx pilose, the tube about 1 mm long, the lobes 1.3–1.5 mm long, the lower slightly longer than the others; corolla with standard ovate, 5–6 mm long, 2.5–3 mm wide, shortly clawed; wings about 6 mm long and 1 mm wide; keel petals about the same length as the wings, wider; pods of 2–4 articles, the articles glabrous on the faces, appressed pubescent on the upper and lower sutures, each 3–4.5 mm long, 3–3.5 mm wide; seeds elliptic in outline, about 2.5 mm long and 1.8 mm wide.

Distr. Ceylon, eastern India, Indo-China, southern China, Malesia to northern Australia.

Ecol. In open forest and on the margins of jungle.

Specimens Examined. PUTTALAM DISTRICT: Bingiriya, *Jayasuriya 632* (PDA). AMPARAI DISTRICT: Bibile-Batticaloa, *Worthington 5137* (K, PDA), *Worthington 4908* (K). MONERAGALA DISTRICT: Galoya National Park, Nilgala, *Jayasuriya 1960* (K, PDA); about 7 miles E of Bibile, *Fosberg & Sachet 53157* (K, PDA). GALLE DISTRICT: Point de Galle, March 1818, *s. coll. s.n.* (K). LOCALITY UNKNOWN: *Thwaites C.P. 1729* (K, PDA).

7. APHYLLODIUM

(DC.) Gagnep., Notul. Syst. (Paris) 3: 254. 1916. Type species: *A. biarticulatum* (L.) Gagnep.

Dicerma sect. *Aphyllodium* DC., Prod. 2: 339. 1825, non sect. *Phyllodium* (Desv.) DC.

Dicerma DC., Prod. 2: 339. 1825; Wight & Arn., Prod. 238. 1834; Benth. in Miq., Pl. Jungh. 217 & 219. 1852; Schindl., Repert. 20: 267. 1924; Hutch., Gen. Fl. Pl. 1: 483. 1964; Ohashi, Ginkgoana 1: 251. 1973.

Desmodium sect. *Dicerma* (DC.) Benth. in Benth. & Hook. f., Gen. Pl. 1: 519. 1865.

Desmodium subg. *Dicerma* (DC.) Baker in Hook. f., Fl. Br. Ind. 2: 163. 1876.

Subshrub; leaves subdigitately 3-foliolate; stipules amplexicaul, divided to about the middle into three; petioles sulcate; stipels present, minute; inflorescences terminal and axillary, racemose, primary bracts subtending secondary bracts, similar in texture and shape to the primary bracts, bearing 2–5 pedicellate flowers in their axils; bracteoles at base of calyx, persistent; calyx scarious, 4-lobed, the upper lobe entire or minutely bifid at the apex, wider than the others; corolla mauve to purple, rarely white; stamens monadelphous, the upper stamen free in its upper half; ovary sessile, surrounded at the base by a minute disk; pod of 2 articles, rarely 1, both sutures deeply indented, the articles broadly elliptic to orbicular, conspicuously reticulately veined; seeds with a rim aril.

A genus of probably 7 species in the Asian-Australian region with the centre of diversity in northern Australia; one species in Ceylon.

Aphyllodium biarticulatum (L.) Gagnep., Notul. Syst. (Paris) 3: 254. 1916.

Heydysarum biarticulatum L., Sp. Pl. 1054. 1753. Type: Herb. *Hermann 1: 25 & 3: 15* (BM).
Dicerma biarticulatum (L.) DC., Prod. 2: 339. 1825; Wight & Arn., Prod. 230. 1834; Wight, Ic. Pl. Ind. Or. t. 419. 1841; Thw., Enum. Pl. Zeyl. 86. 1859; Ohashi, Ginkgoana 1: 252. 1973.
Desmodium biarticulatum (L.) F. Muell., Fragm. Phyt. Austr. 2: 121. 1861; Baker in Hook. f., Fl. Br. Ind. 2: 163. 1876; Trimen, Handb. Fl. Ceylon 2: 48. 1894; Gamble, Fl. Pres. Madras 344. 1918; van Meeuwen, Reinwardtia 6: 246. 1962.
Meibomia biarticulata (L.) Kuntze, Rev. Gen. Pl. 1: 197. 1891.

Prostrate or weakly ascending; branches sparsely appressed pubescent, becoming glabrous, usually reddish; leaves with petioles 2.5–5 mm, rarely to 8 mm, long; rhachis only very slightly longer than pulvini of the leaflets; leaflets obovate or oblong, attenuate at the base, rounded or rarely emarginate at the apex, discolorous, glabrous on upper surface, appressed pubescent on lower; terminal leaflet 6.5–15.5 mm long, 3–6 mm, rarely to 7.5 mm, wide, 1.6–3 times longer than wide, the lateral leaflets usually a little smaller than the terminal ones; pulvini 0.5–1 mm long; stipels 0.3–0.5 mm long, often hidden in the hairs of the pulvinus; primary bracts long-pointed, 2–4 mm, rarely to 7 mm, long, secondary bracts 2–3 mm, rarely to 7 mm, long; pedicels 1–2 mm long; bracteoles 1.5–3.5, rarely to 5 mm, long; calyx with indumentum of minute uncinate hairs, tube 1.5–2 mm long, the lobes 1–2 mm long, the lower sometimes slightly longer than the others; corolla mauve to purple; standard obovate, 5.5–7.5 mm long, 2.5–4 mm wide; wings 4–6 mm long; keel petals about as long as the wings, wider; pod of 2 articles, (rarely 1), densely appressed pubescent when young, the hairs confined to the raised central area when mature, each 3.2–5 mm long (to 6 mm when only one formed), 3–4 mm wide; seed elliptic in outline, about 2.5 mm long and 1.8 mm wide.

Distr. Ceylon, southern India, Indo-China, sporadically in Malesia (absent from New Guinea) and northern Australia.

Ecol. Open situations usually in sandy soils on roadsides and near the sea.

Specimens Examined. JAFFNA DISTRICT: Chunavil, Feb 1890, *s. coll. s.n.* (PDA). VAVUNIYA DISTRICT: Mullaittivu, *Townsend 73/96* (K). PUTTALAM DISTRICT: *Wilpattu National Park, Cooray 70020235R* (K), Kali Villu, *Fosberg et al. 50960* (K), 2 miles N of Marai Villu, *Cooray 70020107R* (K). BATTICALOA DISTRICT: near Vakaneri, Apr 1907, *de Alwis s.n.* (PDA); Panichchankerni, *Jayasuriya et al. 702* (PDA). COLOMBO DISTRICT: Katunayake near Negombo, *Alston 2382* (K, PDA); Colombo, *Macrae 377* (K). AMPARAI DISTRICT: On road to Damana and Akkaraipattu, S of Batticaloa, Aug 1928, *Stockdale s.n.* (PDA). KALUTARA DISTRICT: Kalutara, *Thwaites C.P. 1450* (K, PDA).

Note. The unfamiliar name *Aphyllodium biarticulatum* has been used instead of the more familiar *Dicerma biarticulatum* because the generic name *Dicerma* is illegitimate.

8. TADEHAGI

Ohashi, Ginkgoana 1: 280. 1973. Type species: *T. triquetrum* (L.) Ohashi.
Desmodium sect. *Pleurolobium* DC., Prod. 2: 326. 1825.
Pteroloma Desv. ex Benth. in Miq., Pl. Jungh. 217, 219. 1852; Schindl., Repert. 20: 271. 1924.
Desmodium sect. *Pteroloma* (Desv. ex Benth.) Benth. in Benth. & Hook. f., Gen. Pl. 1: 519. 1865.
Desmodium subg. *Pteroloma* (Desv. ex Benth.) Baker in Hook. f., Fl. Br. Ind. 2: 163. 1865.

Shrubs; leaves 1-foliolate; stipules free; petioles winged; inflorescence racemose, primary bracts subtending secondary bracts; secondary bracts subtending 2 or 3 pedicellate flowers; bracteoles present or absent; calyx 4-lobed, the upper minutely notched at the apex, the lower often longer than the others; petals with well-developed veins; stamens diadelphous, the upper stamen free almost to the base; ovary sessile, surrounded at the base by a distinct disk; pod straight, exserted, the upper suture straight, the lower shallowly or deeply constricted, dividing into indehiscent articles; seeds with a rim aril.

A genus of 3 species ranging from India to northern Australia and the western Pacific; one polymorphous species throughout the range of the genus, including Ceylon.

Tadehagi triquetrum (L.) Ohashi, Ginkgoana 1: 290. 1973.

Hedysarum triquetrum L., Sp. Pl. 746. 1753. Type: Herb. *Hermann 3: 39 & 4: 18* (BM).
Desmodium triquetrum (L.) DC., Prod. 2: 326. 1825; Wight & Arn., Prod. 224. 1834; Baker in Hook. f., Fl. Br. Ind. 2: 163. 1876; Trimen, Handb.

Fl. Ceylon 2: 49. 1894; Gamble, Fl. Pres. Madras 345. 1918; van Meeuwen, Reinwardtia 6: 262. 1962; Verdcourt, Man. New Guinea Leg. 409. 1979.

Pteroloma triquetrum (L.) Desv. ex Benth. in Miq., Pl. Jungh. 220. 1852; Thw., Enum. Pl. Zeyl. 86. 1859.

Meibomia triquetra (L.) Kuntze, Rev. Gen. Pl. 1: 197. 1891.

Erect or ascending shrubs to 2 m; branches sharply angular, glabrous or with a few coarse appressed hairs on the angles; stipules persistent, glabrous, 12–18 mm long, 3–5 mm wide at base; petiole 10–30 mm long, distinctly winged, the wing, 4–10 mm wide, widest in distal third; leaflet discolorous, glabrous except for some appressed strigose hairs on veins beneath; narrowly ovate, rounded or cordate at the base, acute at the apex, 65–125 mm, rarely to 185 mm, long, 10–40 mm, rarely 55 mm, wide, 2.5–7 times longer than wide; inflorescences terminal or in the upper axils, rather dense, to 250 mm long; primary bracts scarious, long-pointed, 3–7 mm long; secondary bracts similar in shape and texture, 1.5–2.5 mm long; pedicels with minute uncinate hairs, 2–3.5 mm long; bracteoles 1.5–2 mm long; calyx with long white hairs, the tube 1.5–2 mm long, the upper lobe 2.5–4 mm long, the lower up to 5 mm long; corolla purplish; standard transversely elliptic, emarginate, 4.5–5 mm long, 6–6.5 mm wide; wings 4–4.5 mm long; keel petals slightly wider and shorter than wings; ovary densely pubescent, style short; pod to about 30 mm long, both sutures almost straight, with up to 8 articles, each 3–4 mm long, 3–4 mm wide, softly pubescent with hairs 1–2 mm long; seeds elliptic, 2–3 mm long, 1.5–2 mm wide.

D i s t r. Ceylon, western India, Indo-China, southern China, Malesia, northern Australia and western Pacific.

E c o l. Reported to be common in patana.

V e r n. Baloliya (S).

N o t e. The description covers subsp. *triquetrum*; other subspecies occur outside the Flora area.

S p e c i m e n s   E x a m i n e d. COLOMBO DISTRICT: Muthurajawela, *Amaratunga 617* (PDA). KANDY DISTRICT: Kandy, *Thomson s.n.* (K), *Worthington 5009* (K, PDA); Hantane, *Rudd & Balakrishnan 3055* (K, PDA); Gampola-Nawalapitiya, road marker 17/12, *Comanor 528* (K). NUWARA ELIYA DISTRICT: Nawalapitiya-Udahentenne road, *Amaratunga 53* (PDA). LOCALITY UNKNOWN: *Gardner 236* (K).

9. DESMODIUM

Desv., J. Bot. Agric. 1: 122. t. 5. 1813, nom. cons.; DC., Prod. 2: 325. 1825; DC., Mém. Lég. 315. 1826; Benth. in Miq., Pl. Jungh. 220. 1852; Hutch., Gen. Fl. Pl. 1: 481. 1964; Ohashi, Ginkgoana 1: 87. 1973. Type species: *D. scopiurus* (Sw.) Desv.

Nicolsonia DC., Prod. 2: 325. 1825. Type species: *N. barbata* (L.)DC. (= *D. barbatum* (L.) Benth.).

Dollinera Endl., Gen. 1285. 1840. Type species: *D. sambuensis* (D. Don) Endl. ex Walpers (= *Desmodium multiflorum* DC.).

Sagotia Duchass. & Walpers, Linnaea 23: 737. 1850. Type species: *S. triflora* (L.) Duchass. & Walpers (*D. triflorum* (L.) DC.).

Catenaria Benth. in Miq., Pl. Jungh. 217 & 220. 1852. Type species: *C. laburnifolia* (Poir.) Benth. (= *D. caudatum* (Thunb.) DC.).

Monarthrocarpus Merr., Philipp. J. Sci. 5: 88. 1910. Type species: *M. securiformis* (Benth.) Merr. (= *D. securiforme* Benth.).

Desmofischera Holthuis, Blumea 5: 188. 1942. Type species: *D.monosperma* Holthuis (*Desmodium securiforme* Benth.).

Numerous other descriptions of *Desmodium* have been published but for the most part they refer to the genus in a broad sense and include the genera *Aphyllodium, Dendrolobium, Phyllodium* and *Tadehagi* treated as distinct here.

Herbs, sometimes annuals, or shrubs; leaves 1- , 3- or occasionally 5–9-foliolate, distinctly pulvinate; stipules free or rarely connate, striate, stipels present; petioles sulcate; lateral leaflets usually smaller than the terminal leaflet; inflorescences racemose or paniculate, terminal and/or axillary; flowers pedicellate in fascicles of up to 8 flowers (sometimes reduced to a single flower), each subtended by a secondary bract, the whole fascicle subtended by a primary bract; secondary bracts sometimes absent; bracteoles present at base of calyx, or absent; calyx 4- or 5-lobed, if 4-lobed then upper lobe to some degree bifid; corolla commonly pink to pale purple; stamens diadelphous with upper stamen free or less commonly monadelphous; ovary sessile or stipitate, many ovuled; disk surrounding base of ovary present or absent; pod exserted from calyx, sessile or stipitate, compressed, jointed; articles often elliptic or quadrangular, indehiscent or occasionally dehiscent along lower suture; seeds compressed, more or less reniform, usually with only a rim-aril, occasionally distinctly arillate, or rarely with no aril of any kind.

A genus of some 300 species throughout the tropics and subtropics with centres of diversity in Mexico-Central America and in south-east Asia; 16 species recorded from Ceylon, one recently naturalised.

KEY TO THE SPECIES

1 Pod stipitate on stipe 2 mm or more long, deeply constricted, the isthmus 1/10 width of pod; seeds without aril of any sort; stamens monadelphous
  2 Stipe of pod 6–15 mm long, articles 8–15 mm long, asymmetric, widest near distal end; pedicels 5–8 mm long; flowers with standard 4.5–7.5 mm long .... **1. D. laxum**
  2 Stipe of pod 2–4.5 mm long, articles 6–8 mm long, symmetric, tapered equally to each end; pedicels filiform, 16–24 mm long; flowers with standard 7.5–9 mm long ....
   .... **2. D. repandum**

1 Pod sessile or on stipe less than 2 mm long, less deeply constricted, the isthmus more than 1/3
   width of pod; or if deeply constricted then pod articles twisted when young and inflorescences
   with spreading glandular hairs on rhachis and pedicels when young
   3 Bracteoles present; pod articles 12–13 mm long, 4–5 mm wide; seeds 8.5–9.5 mm long,
      scarcely rim-arillate .... **3. D. caudatum**
   3 Bracteoles absent; pod articles usually less than 6 mm long, but if 12 mm or more, then
      less than 3 mm wide and seeds less than 5 mm long; seeds rim-arillate or distinctly arillate
      4 Pod articles 13–15 mm long and more than 3 times longer than wide; leaves 1-foliolate,
         the leaflet 80–120 mm long, the margins often undulate .... **4. D. zonatum**
      4 Pod articles elliptic or quandrangular, less than 6 mm long and less than 3 times longer
         than wide; leaves 1- or 3-foliolate, terminal leaflet often less than 80 mm long, the margins
         not undulate
         5 Pod deeply incised on upper and lower margins, the isthmus about 1/8 width of pod;
            articles twisted when young, the margins alternately involute and revolute when mature;
            rhachis of inflorescence pilose with spreading glandular hairs when young ....
            .... **16. D. tortuosum**
         5 Pod with upper suture more or less straight or shallowly constricted, occasionally sharply
            but shallowly notched between articles; isthmus 1/3 or more width of pod; articles not
            twisted and margins neither involute nor revolute; rhachis of inflorescence never with
            glandular hairs
            6 Calyx 5-lobed, upper lobes united only at base (less than half length); pedicels slender,
               5–20 mm long; leaves 3-foliolate, rarely mixed with 1-foliolate leaves; pod articles
               2.5–3.7 mm long; plants herbaceous
               7 Seeds conspicuously arillate around the hilum; pod constricted along both margins,
                  isthmus about 1/2 width of pod; lateral leaflets usually elongate, 2–3 times longer
                  than wide .... **5. D. microphyllum**
               7 Seeds with rim-aril only; upper suture of pod straight, isthmus at least 2/3 width of
                  pod; lateral leaflets less than 1.5 times longer than wide
                  8 Terminal leaflet cuneiform, 4–7.5 mm long; pedicels 5–8 mm long, sparsely pilose
                     .... **6. D. triflorum**
                  8 Terminal leaflet elliptic or obovate, 10–27 mm long; pedicels 10–20 mm long,
                     glabrous or with short hairs near the tip .... **7. D. heterophyllum**
            6 Calyx 4-lobed, upper lobe entire or bifid, not divided to the middle; pedicels usually
               less than 5 mm long, but up to 18 mm and then leaves large with terminal leaflet
               15-40 mm long
               9 Pod articles more than twice as long as wide; pedicels 5–18 mm long ....
                  .... **8. D. adscendens**
               9 Pod articles less than twice as long as wide; pedicels usually less than 5 mm
                  long, rarely to 8 mm
                  10 Leaves 1-foliolate, rarely with occasional 3-foliolate leaves, but leaves not
                     predominantly 3-foliolate
                     11 Leaflets with sparse to moderately dense long stiff hairs on upper surface
                        and dense velvety pubescence on lower .... **11. D. velutinum**
                     11 Leaves glabrous or almost so, on upper surface
                        12 Leaflets orbicular, about as long as wide, dense white appressed pubescent
                           on lower surfaces; rhachis of inflorescence with small brown resinous dots
                           as well as uncinate and straight hairs .... **9. D. styracifolium**
                        12 Leaflets broadly elliptic or ovate, 1.1–2 times longer than wide; pubescence
                           of lower surface usually not dense and white; rhachis of inflorescence with
                           moderately dense uncinate hairs .... **10. D. gangeticum**

10 Leaves 3-foliolate, with occasional 1-foliolate leaves particularly on young
   plants, but leaves not predominantly 1-foliolate
   13 Leaves densely appressed pubescent and with about 11 prominently raised
      lateral veins on each side of midrib beneath ....**12. D. jucundum**
   13 Leaves not densely appressed pubescent and with prominently raised veins
      beneath
      14 Pod articles 5–5.5 mm long; rhachis of inflorescence covered with spread-
         ing ferrugineous hairs and shorter weak uncinate hairs ....
         .... **13. D. ferrugineum**
      14 Pod articles to 4 mm long; rhachis of inflorescence with uncinate hairs,
         long straight spreading or appressed hairs, but not ferrugineous appressed
         hairs
         15 Mature pods erect or ascending, upper suture straight; pedicels
            2–2.5 mm long .... **14. D. heterocarpon**
         15 Mature pods spreading, upper suture notched between articles; pedicels
            5–8 mm long .... **15. D. pryonii**

1. Desmodium laxum DC., Ann. Sci. Nat. Bot. 4: 102. 1825; DC., Prod.
2: 336. 1825; Prain, J. Asiat. Soc. Bengal 66: 138. 1897; Gamble, Fl. Pres.
Madras 345. 1918; Isely, Brittonia 7: 203. t. 29. 1951; van Meeuwen, Reinwardtia 6: 253. 1962; Ohashi, Ginkgoana 1: 134. 1973. Type: from Nepal,
Wallich.

Desmodium leptopus A. Gray ex Benth. in Miq., Pl. Jungh. 226. 1852; Isely,
   Brittonia 7: 199. t. 26. 1951. Type: from Philippines, *U.S. Explor. Exped.*
   (K, holotype).
Desmodium gardneri Benth. in Miq., Pl. Jungh. 226. 1852; Baker in Hook. f.,
   Fl. Br. Ind. 2: 165. 1876; Trimen, Handb. Fl. Ceylon 2: 50. 1894. Types:
   Ceylon, *Walker s.n.* (K); Ceylon, *Gardner 220* (K).
Desmodium podocarpum auct.: Thw., Enum. Pl. Zeyl. 87. 1859, non DC.
Desmodium podocarpum var. *laxum* (DC.) Baker in Hook. f., Fl. Br. Ind. 2:
   165. 1876.
Meibomia leptopus (A. Gray ex Benth.) Kuntze, Rev. Gen. Pl. 1: 198. 1891.
Meibomia gardneri (Benth.) Kuntze, Rev. Gen. Pl. 1: 198. 1891.

Annual stems from perennial rootstock to about 1 m high; branches
ribbed with more or less spreading hairs, becoming glabrous; leaves 3-
foliolate; stipules narrowly triangular to ovate, with a long point, 7–13 mm
long, 2–4 mm wide at base; petioles 25–75 mm long, rachis 10–30 mm
long; leaflets subcoriaceous, ovate, rounded at the base, acuminate at the
apex, glabrous on both surfaces or with some short stiff hairs on veins, 4–7
veins on each side of midrib; terminal leaflets 70–125 mm long, 35–75 mm
wide, 1.5–2.5 times longer than wide, the laterals smaller, 35–110 mm long,
25–50 mm wide, 2–2.5 times longer than wide; pulvini 2–6 mm long, stipels
subulate, 2–7 mm long; inflorescences usually terminal, sometimes on a separate shoot arising from base of plant, up to 80 cm long; rhachis with short

uncinate hairs; fascicles of 2 or 3 flowers; primary bracts narrowly ovate, acuminate, 2.5–4.5 mm long; secondary bracts 0.5–1 mm long, early deciduous; pedicels 5–8 mm long at anthesis; bracteoles absent; calyx 4-lobed, 1.5–2.5 mm long, the tube 1–2 mm long, glabrous, the lobes sparsely uncinate pubescent with some long soft hairs; corolla pinkish; standard orbicular to transversely elliptic, obtuse or emarginate at the apex, 4.5–7.5 mm long, 4–6.5 mm wide; wings 5–7.5 mm long, 1–2 mm wide; keel petals about as long as the wings, wider; stamens monadelphous; ovary stipitate, more or less glabrous; pods flat with stipe 6–15 mm long, upper suture thickened, the lower deeply incised, with 2–4 triangular articles, asymmetric, widest near distal end, each 8–15 mm long, 4–5 mm wide, with rather dense spreading uncinate hairs; seeds not seen.

Distr. Ceylon, India, Indo-China, China, Japan and Philippines.

Ecol. In shade, in rainforest at higher altitudes.

Specimens Examined. MATALE DISTRICT: Kalupahana forest in Knuckles Mountain range, *Jayasuriya & Balasubramaniam 1205* (K, PDA); Knuckles, Madulkelle area, *Kostermans 25043* (K); Gammaduwa, *Alston 674* (PDA). KANDY DISTRICT: "Rangale District", Sept 1888, *s. coll. s.n.* (PDA); Gallekelle, Medamahanuwara, *Alston 809* (PDA). RATNAPURA DISTRICT: Pinnawela, *Balakrishnan 553* (K, PDA); Tumbagoda road, Tamanawatte-Massenna, above Balangoda, *Kostermans 24459A* (K); Rassagala, *Amaratunga 707* (PDA). LOCALITY UNKNOWN: *Thwaites s.n.* (K); *Gardner 220* (K, syntype of *D. gardneri*); *Walker s.n.* (K, syntype of *D. gardneri*).

Note. *Desmodium laxum* is one of a complex of taxa that constitute subsect. *Podocarpium* which extends from Japan to Ceylon. Isely distinguished a number of species in the complex while Ohashi recognised only two, *D. laxum* and *D. podocarpum* DC., with numerous infraspecific taxa, subspecies and varieties. In Ceylon he accepted subsp. *laxum* (pods with articles 9–10 mm long) and subsp. *leptopus* (Benth.) Ohashi (pods with articles 12–18 mm long). The two appear to be easily distinguished in Ceylon but not elsewhere and the subspecies have not been maintained here. The two syntypes of *D. gardneri* represent the two subspecies; *Gardner 220* is subsp. *laxum* and *Walker s.n.* subsp. *leptopus*.

2. Desmodium repandum (Vahl) DC., Prod. 2: 334. 1825; Schubert, Fl. Congo Belge 5: 193. t. 14. 1954; van Meeuwen, Reinwardtia 6: 258. 1962; Schubert in Fl. Trop. East Africa, Leg: Papilion. 477. 1971; Ohashi, Ginkgoana 1: 160. 1973; Verdcourt, Man. New Guinea Leg. 404. 1979.

Hedysarum repandum Vahl, Symb. Bot. 2: 82. 1791. Type: from Yemen.

Desmodium scalpe DC., Prod. 2: 334. 1825; Baker in Hook. f., Fl. Br. Ind.
2: 165. 1876; Trimen, Handb. Fl. Ceylon 2: 50. 1894; Gamble, Fl. Pres.
Madras 345. 1918. Type: from Mauritius.
Desmodium strangulatum Wight & Arn., Prod. 228. 1834; Wight, Ic. Pl. Ind.
Or. 985. 1847; Thw., Enum. Pl. Zeyl. 87. 1859, excluding var. *minor*
Thw. which is *D. adscendens.* Type: from India (Wight Cat. 772, 773,
774).
Meibomia repanda (Vahl) Kuntze, Rev. Gen. Pl. 1: 197. 1891.
Meibomia scalpe (DC.) Kuntze, Rev. Gen. Pl. 3: 66. 1898.

Erect or ascending little-branched herb to about 1 m tall, somewhat
woody at the base; branchlets striate, tomentose with short erect uncinate
hairs and long straight spreading hairs; leaves 3-foliolate; stipules persistent,
long-pointed, 10–16 mm long, hirsute or fringed with long hairs; petioles
35–95 mm long, rhachis 8–15 mm long; leaflets chartaceous, rhomboidal or
broadly ovate, rounded or cuneate at the base, obtuse or acuminate at the apex,
discolorous, with sparse loose appressed hairs on the upper surface and weak
sparse appressed or ascending hairs on lower surface, sometimes becoming
glabrous; terminal leaflet 35–100 mm long, 25–70 mm wide, the laterals
smaller, 30–70 mm long, 15–45 mm wide, all 1.3–2.3 times longer than wide;
pulvini 2–3.5 mm long, stipels setaceous, 0.5–1.5 mm long; inflorescences
terminal and axillary, open, to 25 cm long, rhachis densely pubescent with
uncinate and long spreading hairs; fascicles of 2–4 flowers; primary bracts
ovate, 4–8 mm long, fringed with long hairs; secondary bracts 1–2 mm long,
both early deciduous; pedicels 16–24 mm long at anthesis; bracteoles absent;
calyx broadly campanulate, 4-lobed, rather scarious, densely uncinate-hairy,
2.5–3.5 mm long, tube about 1.5 mm long, the upper lobe slightly shorter
than the rest; corolla orange to pinkish red; standard orbicular, emarginate
at the apex, 7.5–9 mm long, 6–7 mm wide; wings narrowly elliptic, about
7 mm long and 2 mm wide; keel petals conspicuously longer and wider than
the wings, to about 10 mm long; stamens monadelphous; ovary stipitate,
appressed pubescent; pods flat, with stipe 2–4.5 mm long, the upper suture
thickened, straight, the lower deeply indented, with 3–5 articles, half-moon-
shaped, 6–8 mm long, 3–3.5 mm wide, covered with dense uncinate hairs;
seeds 5–6 mm long, 2.5–3.5 mm wide.

Distr. Old World tropics: Africa, Arabian Peninsula, Ceylon, India,
Indo-China, southern China, through Malesia to New Guinea where possibly
introduced.

Ecol. In shade, forest and margins, at high altitudes, up to 2000 m.

Specimens Examined. BADULLA DISTRICT: Namunukula moun-
tain, *Jayasuriya 128* (K, PDA); road from Haputale to Boralanda, just below
Haputale tea estate, *Koyama et al. 16045* (PDA); Haputale, *Alston s.n.* (PDA).

NUWARA ELIYA DISTRICT: Hakgala, *A.M.S. s.n.* (PDA); along Ohiya-Pattipola road, *Balakrishnan 478* (K). LOCALITY UNKNOWN: *Thwaites 1433* (K); *Walker s.n.* (K).

3. Desmodium caudatum (Thunb.) DC., Prod. 2: 337. 1825; van Meeuwen, Reinwardtia 6: 248. 1962; Ohashi, Ginkgoana 1: 91. 1973.

Hedysarum caudatum Thunb., Fl. Jap. 286. 1784. Type: from Japan.
Hedysarum laburnifolium Poir. in Lam., Enc. 6: 422. 1804. Type: from Java, Commerson.
Desmodium laburnifolium (Poir.) DC., Prod. 2: 337. 1825; Baker in Hook. f., Fl. Br. Ind. 2: 163. 1876; Trimen, Handb. Fl. Ceylon 2: 48. 1894; Gamble. Fl. Pres. Madras 344. 1918.
Catenaria laburnifolia (Poir.) Benth. in Miq., Pl. Jungh. 220. 1852; Thw., Enum. Pl. Zeyl. 86. 1859.
Meibomia laburnifolia (Poir.) Kuntze, Rev. Gen. Pl. 1: 196. 1891.
Meibomia caudata (Thunb.) Kuntze, Rev. Gen. Pl. 1: 197. 1891.
Catenaria caudata (Thunb.) Schindl., Repert. 20: 275. 1924.

Shrub to 1 m tall; branches ribbed, sparsely appressed pubescent; leaves 3-foliolate; stipules persistent, narrowly triangular, cuspidate, 6–9 mm long; petioles 20–35 mm long, rhachis 6–16 mm long; leaflets coriaceous, elliptic, attenuate at the base, acute (rarely obtuse) or acuminate at the apex, shining and glabrous or with a few hairs on the veins above, sparsely appressed pubescent, particularly on veins beneath; terminal leaflet 70–110 mm long, 30–40 mm wide, laterals smaller, 50–75 mm long, 17–30 mm wide, all 2.4–3.3 times longer than wide; pulvini 1.5–3 mm long, stipels setaceous, 2–4 mm long; inflorescences terminal and axillary, up to 25 cm long, rhachis pubescent with long straight and shorter uncinate hairs; fascicles of 2 or 3 flowers; primary bracts 3–4 mm long; secondary bracts 1.5–2 mm long; bracteoles on top of pedicel 1–1.8 mm long; calyx densely uncinate- and appressed-pubescent, 3.5–4.5 mm long, the tube 1.5–1.7 mm long, the lobes 1.5–2 mm long, the upper two united for 3/4 or more of their length, the lower slightly longer than the rest; corolla dull orange-red, distinctly veined; standard elliptic, about 6 mm long, 3.5 mm wide; wings about 6 mm long; keel petals distinctly longer and wider than the wings; stamens diadelphous; disk present; ovary stipitate, densely pilose; pods flat, linear, very shortly stipitate, the upper suture undulate, the lower suture more deeply incised, with 4–8 articles, each 12–13 mm long, 4–5 mm wide, densely covered with short uncinate hairs; seeds 8.5–9.5 mm long, about 3 mm wide, rim aril scarcely developed.

Distr. From Japan to eastern India with marked disjunctions in Ceylon and western Malesia.

Ecol. Rainforest at higher elevations.

Specimens Examined. BADULLA DISTRICT: Uma Oya off Ettampitiya, *Jayasuriya & Townsend 1178* (K). LOCALITY UNKNOWN: *Gardner 479* (K); *Macrae s.n.* (K); *Thwaites 1435* (K, PDA); *Walker s.n.* (K).

4. Desmodium zonatum Miq., Fl. Ind. Bat. 1(1): 250. 1855; van Meeuwen, Reinwardtia 6: 265. 1962; Ohashi, Ginkgoana 1: 107. 1973; Verdcourt, Man. New Guinea Leg. 412. 1979. Type: from Java.

Desmodium ormocarpoides auct.: Thw., Enum. Pl. Zeyl. 87. 1859; Baker in Hook. f., Fl. Br. Ind. 2: 164. 1876; Trimen, Handb. Fl. Ceylon 2: 49. 1894; Gamble, Fl. Pres. Madras 345. 1918, non DC.
Desmodium ormocarpoides var. *velutina* Prain, J. Asiat. Soc. Bengal 66: 147. 1897.

Erect or ascending subshrub to 60 cm high; young branches with short spreading brown uncinate hairs; leaves 1-foliolate; stipules long-pointed, 7–10 mm long; petioles 10–25 mm long; leaflets chartaceous, ovate, elliptic or oblong, rounded at the base, acute at the apex, the margin sometimes somewhat undulate, glabrous above, pubescent with sparse to moderately dense, weak ascending hairs beneath; 80–120 mm long, 25–50 mm wide, 1.8–2.6, rarely to 3.3, times longer than wide; pulvini hirsute, 2–3 mm long, stipels setaceous, 4–8 mm long; inflorescences terminal and axillary, up to 20 cm long, the rhachis with rather dense brown uncinate hairs and longer straight ones; fascicles of 2–5 flowers; primary bracts narrowly triangular, about 3 mm long; secondary bracts about 2 mm long; pedicels with spreading uncinate hairs, about 6 mm long; bracteoles absent; calyx 4-lobed, pubescent; the tube 1–1.5 mm long, the lobes as long as the tube, the upper bifid at the apex; corolla bluish white; standard obovate, rounded or slightly retuse at the apex; 6–7 mm long, 4.5–5.5 mm wide; the wings 5.5–6 mm long; keel petals distinctly longer and wider than the wings; pod shortly stipitate, deeply sinuate on each margin, with up to 8 articles, each 13–15 mm long, about 2.5 mm wide, striate, pubescent with short uncinate hairs; seeds narrowly oblong in outline, about 4.5 mm long, 1.5–2 mm wide.

Distr. Ceylon, eastern India, Indo-China, Taiwan and Malesia.

Ecol. Widely distributed in rainforest and secondary forest.

Specimens Examined. KANDY DISTRICT: Kandy, *Alston 1096* (K, PDA). LOCALITY UNKNOWN: *Gardner 219* (K); *Thwaites 3812* (K).

5. Desmodium microphyllum (Thunb.) DC., Prod. 2: 337. 1825; van Meeuwen, Reinwardtia 6: 254. 1962; Ohashi, Ginkgoana 1: 241. 1973; Verdcourt, Man. New Guinea Leg. 402. 1979.

Hedysarum microphyllum Thunb., Fl. Jap. 284. 1784; Poir. in Lam., Enc. 6: 417. 1804. Type: from Japan.

Hedysarum tenellum Hamilton ex D. Don, Prod. Fl. Nepal 243. 1825. Type: from Nepal.

Desmodium parvifolium DC., Ann. Sci. Nat. Bot. 4: 100. 1825; DC., Prod. 2: 234. 1825; Wight & Arn., Prod. 229. 1834; Thw, Enum. Pl. Zeyl. 86. 1859; Baker in Hook. f., Fl. Br. Ind. 2: 174. 1876; Trimen, Handb. Fl. Ceylon 2: 55. 1894; Gamble, Fl. Pres. Madras 348. 1918. Type: from Nepal.

Meibomia microphylla (Thunb.) Kuntze, Rev. Gen. Pl. 1: 198. 1891.

Weak, prostrate, sprawling or erect undershrub to 1 m high; young branches angular with spreading hairs about 1 mm long; leaves 1- or 3-foliolate, usually on the same plant; stipules persistent, narrowly triangular, 3–5 mm long; petiole 1–3 mm long, 2–9 mm when leaves 1-foliolate; rhachis, often longer than the petiole, 2–3 mm long; leaflets chartaceous, oblong, elliptic or obovate when narrow or orbicular when small, obtuse, sometimes mucronate, glabrous on upper surface, lower surface with covering of appressed or somewhat spreading sparse weak hairs; terminal leaflet 3.5–11 mm, rarely 16 mm long, 1.5–6 mm wide, 1.1–3 times, occasionally to 4 times, longer than wide, lateral leaflets 2.5–6 mm long, 1.2–2.2 mm wide, usually 2–3 times longer than wide; pulvini 0.3–0.5 mm long, stipels glabrous, 0.2–0.3 mm long; inflorescences terminal, often on short lateral branches up to 60 mm long, very open, with up to 8 nodes; rhachis with indumentum similar to the branchlets; flowers single; primary bracts early deciduous, ovate-acuminate, 3–4 mm long; secondary bracts and bracteoles absent; pedicels 8–12 mm long at anthesis; calyx deeply 5-lobed with indumentum of long dense appressed hairs, 3.5–4 mm long, the tube 0.7–1 mm long, the lobes all 2.5–3 mm long, the upper two united in the lower 1/4–1/3; corolla pink, white or purple; standard obovate, obtuse at the apex, 4–4.5 mm long, 2.5–3.5 mm wide; wings about 3 mm long; keel petals distinctly longer than the wings, incurved at right angles near the tip; stamens diadelphous; ovary pubescent; pods flat, linear, the upper suture thickened, somewhat incised, the lower more deeply so, isthmus about 1/2 width of pod, with 2–4 articles, each 2.5–3 mm long, 2–3 mm wide, with sparse, small erect uncinate hairs; seeds oblong in outline, about 2 mm long, 1 mm wide; aril well developed, not merely a rim around the hilum.

Distr. Ceylon, eastern India, Indo-China, China, Japan, Malesia, north-eastern Australia and islands of the Western Pacific.

Ecol. In open situations among grass, in patanas, sometimes a minor weed of cultivation.

Specimens Examined. BADULLA DISTRICT: Ambawela, *Alston 1031* (PDA); near Haputale, *de Alwis 14* (PDA). NUWARA ELIYA DISTRICT: Hakgala, *A.D.A. s.n.* (PDA). LOCALITY UNKNOWN: *Gardner 227* (K); *Macrae s.n.* (K); *Thwaites C.P. 1429* (K); *Walker s.n.* (K).

6. Desmodium triflorum (L.) DC., Prod. 2: 334. 1825; Wight & Arn., Prod. 229. 1834; Wight, Ic. Pl. Ind. Or. 292. 1839; Thw., Enum. Pl. Zeyl. 86. 1859; Baker in Hook. f., Fl. Br. Ind. 2: 173. 1876; Trimen, Handb. Fl. Ceylon 2: 54. 1894; Gamble, Fl. Pres. Madras 347. 1918; van Meeuwen, Reinwardtia 6: 261. 1962; Schubert in Fl. Trop. East Africa, Leg: Papilion. 459. 1971; Ohashi, Ginkgoana 1: 245. 1973; Verdcourt, Man. New Guinea Leg. 409. 1979.

Hedysarum triflorum L., Sp. Pl. 749. 1753. Type: Hedysarum no. 45, *Burmann* (LINN).

Hedysarum stipulaceum Burm. f., Fl. Ind. 168. t. 54. f. 2. 1768. Type: not seen.

Desmodium triflorum var. *minus* Wight & Arn., Prod. 228. 1834. Type: from India, Wight Cat. 779, in part, 777.

Desmodium triflorum var. *villosum* Wight & Arn., Prod. 228. 1834. Type: from India, Wight Cat. 778.

Meibomia triflora (L.) Kuntze, Rev. Gen. Pl. 1: 197. 1891.

Prostrate perennial often rooting at the nodes, stems to about 0.5 m long; stems sparsely pilose; leaves 3-foliolate; stipules ovate, 2.5–3.5 mm long; petioles 1.5–8 mm long; rhachis 1–2 mm long; leaflets chartaceous, obcordate, emarginate at the apex, upper surface glabrous, lower with sparse appressed hairs; terminal leaflet 4–7.5 mm long, rarely to 9 mm, 3–10 mm wide, usually about as long as wide, but a range of 0.8–1.3 times longer than wide; lateral leaflets smaller, 3–6 mm, rarely to 7.5 mm, long, 3–8 mm wide, 0.9–1.1 times, rarely to 1.6 times, longer than wide; pulvini 0.3–0.8 mm long, stipels 0.2–0.6 mm long; inflorescence sessile, leaf-opposed fascicles of 2–4 flowers; primary bracts narrowly ovate, acuminate, about 4 mm long; secondary bracts and bracteoles absent; pedicels 5–8 mm long; calyx 5-lobed, the upper two lobes united at the base, 2.5–3 mm long, appressed pubescent; tube about 1 mm long, lobes 1.5–2 mm long; corolla purplish; standard obovate, rounded at the apex, about 3 mm long and 1.8 mm wide; wings about 2.5 mm long; keel petals distinctly longer than the wings; ovary pubescent; stamens diadelphous; pods flat, linear, somewhat curved upwards, the upper suture not indented, the lower incised, the isthmus about 2/3 width of pods, with 3 or 4 articles, each about 3 mm long, 2.5–2.8 mm wide, with sparse uncinate hairs or glabrous; seeds rectangular in outline, about 2 mm long and 1.6 mm wide, with an obvious rim-aril.

Distr. Pantropical.

Ecol. On roadsides, in open places, sometimes in the shade, often a weed in lawns.

Specimens Examined. JAFFNA DISTRICT: between Jaffna and Elephant Pass, *Rudd 3290* (K). ANURADHAPURA DISTRICT: Mihintale,

Clayton 5316 (K). PUTTALAM DISTRICT: Jaela, Dec 1913, *Rajapakse s.n.* (PDA); Wilpattu National Park, Kali Villu, *Wirawan et al. 1002,* (K), *Fosberg et al. 50971* (K), 2.5 miles S of Madura Odai, 5 miles S of Marai Villu, *Fosberg et al. 50840* (K). POLONNARUWA DISTRICT: Polonnaruwa, *Townsend 73/230* (K). KANDY DISTRICT: Peradeniya, Nov 1927, *Singho s.n.* (PDA). GALLE DISTRICT: Elpitiya, Nov 1920, *s. coll. s.n.* (PDA).

7. Desmodium heterophyllum (Willd.) DC., Prod. 2: 334. 1825; Thw., Enum. Pl. Zeyl. 86. 1859; Baker in Hook. f., Fl. Br. Ind. 2: 173. 1876; Trimen, Handb. Fl. Ceylon 2: 55. 1894; Gamble, Fl. Pres. Madras 347. 1918.

Heydsarum triflorum var. β & γ L., Sp. Pl. 749. 1753. Types: β from Ceylon, γ from India.
Hedysarum heterophyllum Willd., Sp. Pl. 3: 1201. 1802. Based on *Hedysarum triflorum* var. β L.
Desmodium triflorum var. *majus* Wight & Arn., Prod. 229. 1834; Wight, Ic. Pl. Ind. Or. t. 291. 1839. Types: from India, Wight Cat. 779, in part, 781.
Meibomia heterophylla (Willd.) Kuntze, Rev. Gen. Pl. 1: 196. 1891.

Prostrate or ascending stoloniferous herb; young branches with covering of rather dense straight yellowish hairs about 1.5 mm long, rarely almost glabrous; leaves 3-foliolate; stipules ovate-acuminate, 4–6 mm long, fringed with long hairs; petiole 8–20 mm long, rhachis 2–5 mm long; leaflets chartaceous, elliptic or somewhat obovate, orbicular when small, rounded or slightly cordate at the base, rounded or somewhat emarginate at the apex, glabrous or with scattered long hairs on the upper surface, sparsely to moderately pubescent on the lower surface with longish hairs; terminal leaflet 10–27 mm long, 7–15 mm wide, laterals 8–20 mm long, 5–11 mm wide, all 1.3–1.9 times longer than wide, the terminal occasionally to 2.6 times longer than wide and the laterals as long as wide; pulvini 1–2 mm long, stipels 0.5–1 mm long, usually shorter than the pulvini; inflorescence usually a more or less sessile, leaf-opposed fascicle of 1–3 flowers or occasionally fascicles arranged in short racemes, the rhachis with sparse long spreading hairs; primary bracts ovate, acuminate, 3–4 mm long, fringed with long hairs; secondary bracts and bracteoles absent; pedicels 12–20 mm long, glabrous or with some short hairs towards the apex; calyx 4-lobed, the upper lobe divided to about the middle, with short uncinate pubescence and long hairs on lobes, 2.5–2.8 mm long; tube about 1 mm long, the lobes 1.5–1.8 mm long, the lower sometimes longer than the others; corolla purple; standard obovate, rounded or retuse at the apex, about 5 mm long and 3 mm wide; wings about 4 mm long; keel petals about as long as the wings, strongly incurved towards the tip; stamens diadelphous; ovary pubescent; pods flat, linear, the upper suture straight, the lower indented, both thickened, the isthmus 3/4–4/5 width of pod, articles 4 or 5, each 3–3.7 mm long, 3.2–4 mm wide, reticulately veined, with sparse short weak uncinate hairs; seeds reniform, 2.5–2.7 mm long, 1.5–2 mm wide, with obvious rim-aril.

Distr. Ceylon, India, Indo-China, southern China, Malesia and introduced into northern Australia and Micronesia.

Ecol. Roadsides and open places, moist grasslands.

Specimens Examined. KURUNEGALA DISTRICT: Mawatagama, *Amaratunga 1195* (PDA). COLOMBO DISTRICT: Muthurajawela, *Amaratunga 386* (PDA). KANDY DISTRICT: Hantane, Kandy, *Worthington 6134* (K). KALUTARA DISTRICT: Pamunugama, *Simpson 7955* (PDA). RATNAPURA DISTRICT: N of Ratnapura to Gilimale, *Maxwell et al. 945* (K); N of Ratnapura, 1 mile S of Carney, *Maxwell et al. 949* (K). NUWARA ELIYA DISTRICT: Dimbulla to Kotmale, *Maxwell et al. 923* (K); road from Kandy to Maturata, culvert 19/15, *Maxwell 996* (PDA). BADULLA DISTRICT: Fort Macdonald Valley, May 1906, *A.M.S. s.n.* (PDA).

8. Desmodium adscendens (Sw.) DC., Prod. 2: 332. 1825; Schindl., Repert. 21: 8. 1925; van Meeuwen, Reinwardtia 6: 245. 1962; Schubert in Fl. Trop. East Africa, Leg.: Papilion. 461. 1971; Ohashi, Ginkgoana 1: 199. 1973.

Hedysarum adscendens Sw., Prod. 1: 106. 1788. Type: from West Indies, *Swartz.*

Desmodium trifoliastrum Miq., Fl. Ind. Bat. 1(1): 248. 1855. Type: from Java, *Zollinger 904.*

Desmodium thwaitesii Baker in Hook. f., Fl. Br. Ind. 2: 169. 1876; Trimen, Handb. Fl. Ceylon 2: 51. 1894. Based on *Desmodium strangulatum* var. β *minor* Thw., Enum. Pl. Zeyl. 87. 1859. Type: Ceylon, *Thwaites 3327* (K, holotype).

Meibomia adscendens (Sw.) Kuntze, Rev. Gen. Pl. 1: 95. 1891.

Meibomia trifoliastrum (Miq.) Kuntze, Rev. Gen. Pl. 1: 198. 1891.

Meibomia thwaitesii (Baker) Kuntze, Rev. Gen. Pl. 1: 198. 1891.

Desmodium adscendens var. *trifoliastrum* (Miq.) Schindl., Repert. 21: 8. 1925.

Creeping or ascending herb often rooting near base of stems; branches with covering of rather sparse erect hairs; leaves 1- or 3-foliolate; stipules oblique, long-pointed, 5–12 mm long; petiole 6–22 mm long, rhachis 3–17 mm long; leaflets chartaceous, oblong, obovate, elliptic or occasionally orbicular, rounded or occasionally subcordate at the base, obtuse or slightly retuse at the apex, discolorous, glabrous or with scattered long hairs on upper surface, with covering of sparse to moderately dense appressed long silky hairs on lower surface; terminal leaflet 15–40 mm long, 12–26 mm wide, the laterals 14–22 mm, rarely to 30 mm, long, 7–16 mm wide, all 1.1–2 times longer than wide; pulvini 1–2 mm long, stipels 1–3 mm long; inflorescence terminal, open, rhachis with indumentum of short weak uncinate hairs, becoming sparse; flowers in pairs; primary bract concave, acuminate, fringed with long hairs, 3.5–6 mm long, about 1.5 mm wide; secondary bracts and bracteoles absent; pedicels 5–18 mm long; calyx 4-lobed with minute uncinate hairs and long hairs on lobes particularly, about 2.5–3.5 mm long, the tube about 1 mm long, the lobes all about equal in length or the lower slightly longer than the others; corolla mauve or purple; standard obovate or orbicular, rounded or retuse at apex, 4.5–5.5 mm long, 4–4.5 mm wide; wings about 4 mm long; keel petals about as long as the wings; stamens diadelphous; ovary densely minutely pubescent; pod flat, linear, upper margin thickened, slightly undulate or notched between the articles, lower deeply incised, the isthmus about 2/3 width of pod; with 2–4 articles (rarely 1), each 5–6 mm long, 2.5–3 mm wide with dense spreading uncinate hairs; seeds about 4–4.5 mm long, 2.5 mm wide with a distinct rim-aril.

Distr. Native of the West Indies and Central America, introduced into the Old World where it is widespread in tropical Africa and occurs sporadically in Ceylon, Indo-China and Malesia.

Ecol. Roadsides and clearings in rainforest, evidently rare.

Specimen Examined. LOCALITY UNKNOWN: *Thwaites C.P. 3327* (K, holotype of *Desmodium strangulatum* var. *minor, D. thwaitesii*).

9. Desmodium styracifolium (Osbeck) Merr., Amer. J. Bot. 3: 580. 1916; van Meeuwen, Reinwardtia 6: 259. 1962; Ohashi, Ginkgoana 1: 224. 1973.

Hedysarum styracifolium Osbeck, Dagbok Ostind. Resa 247. 1757; L., Syst. Nat. ed. 10. 2: 1169. 1759. Type: from China, *Osbeck.*
Hedysarum retroflexum L., Mant. 1: 103. 1767. Type: from India.
Hedysarum capitatum Burm. f., Fl. Ind. 167. t. 54. f. 1. 1768. Type: from India, *N.L. Burmann.*
Desmodium retroflexum (L.) DC., Prod. 2: 336. 1825; Baker in Hook. f., Fl. Br. Ind. 2: 170. 1876; Gamble, Fl. Pres. Madras 347. 1918.
Desmodium capitatum (Burm. f.) DC., Prod. 2: 336. 1825; Baker in Hook. f., Fl. Br. Ind. 2: 170. 1876; Schindl., Repert. Beih. 49. 143. 1928.
Uraria styracifolia Wight & Arn., Prod. 222. 1834.
Meibomia retroflexa (L.) Kuntze, Rev. Gen. Pl. 1: 197. 1891.
Meibomia capitata (Burm. f.) Kuntze, Rev. Gen. Pl. 1: 195. 1891.

Prostrate or sprawling herb to 1 m high; branches with covering of short uncinate hairs and long straight hairs; leaves 1- or rarely 3-foliolate; stipules long-pointed, oblique at the base, 8–13 mm long; petiole 15–25 mm long, rhachis 4–6 mm long; leaflets coriaceous, orbicular, the laterals oblong, rounded or cordate at the base, rounded at the apex, discolorous, glabrous above, with dense covering of white appressed hairs below; terminal leaflet 30–60 mm long, 30–55 mm wide, laterals 15–20 mm long, about 12 mm wide; pulvini 2–3 mm long, stipels setaceous, 3–9 mm long; inflorescence densely flowered, 10–30 mm long, terminal, often on short lateral branches; rhachis with small brown resinous dots, uncinate hairs and often dense long hairs; flowers in pairs; primary bracts acuminate, fringed with long hairs and often pubescent in the upper half, 3–5 mm long, 1.5–3 mm wide, secondary bracts and bracteoles absent; pedicels about 3 mm long; calyx 4-lobed, 3–3.5 mm long, with indumentum of short uncinate hairs and long straight hairs on lobes; tube about 1 mm long; upper lobe divided almost to the middle, 2–2.5 mm long, about as long as lower lobe; lateral lobes shorter, 1.5–2 mm long; corolla white or rose pink, fragrant; standard obovate, obtuse at the apex, about 4 mm long, 2.5–3 mm wide; wings 3–4 mm long; keel petals distinctly longer than the wings, 3.5–4.5 mm long, curved at right angles above the middle; stamens diadelphous; ovary puberulent; pods flat, upper suture thickened, flat; lower suture indented, isthmus about 3/4 width of pod; articles 4–6, each about 2.5 mm long, 2.2–2.5 mm wide, glabrous on faces with a fringe of erect uncinate hairs on lower margin; seeds about 2 mm long and 1.5 mm wide, with an obvious rim-aril.

Distr. Ceylon, India, southern China, western Malesia and Micronesia.

Specimen Examined. LOCALITY UNKNOWN: *s. coll. s.n.* (K).

Note. The species is recorded from Ceylon by Ohashi though he did not cite any specimen. Schindler considered that Baker in recording the species from Ceylon (as *D. capitatum*) confused N.L. Burmann with his father, J. Burmann, though the latter did not cite the species in his Thesaurus zeylanicus (1737). At K there is one poorly labelled specimen, allegedly from Ceylon.

10. Desmodium gangeticum (L.) DC., Prod. 2: 327. 1825; Wight & Arn., Prod. 225. 1834; Wight, Ic. Pl. Ind. Or. 271. 1840; Thw., Enum. Pl. Zeyl. 411. 1864; Baker in Hook. f., Fl. Br. Ind. 2: 168. 1876; Trimen, Handb. Fl. Ceylon 2: 51. 1894; Gamble, Fl. Pres. Madras 345. 1918; van Meeuwen, Reinwardtia 6: 249. 1962; Schubert in Fl. East Trop. Africa, Leg.: Papilion. 467. 1971; Ohashi, Ginkgoana 1: 184. 1973; Verdcourt, Man. New Guinea Leg. 397. 1979.

Hedysarum gangeticum L., Sp. Pl. 746. 1753. Type: Hedysarum no. 13 (LINN).

Hedysarum maculatum L., Sp. Pl. 746. 1753. Type: Hedysarum no. 14 (LINN).

Desmodium maculatum (L.) DC., Prod. 2: 327. 1825.

Desmodium gangeticum var. *maculatum* (L.) Baker in Hook. f., Fl. Br. Ind.
2: 168. 1876; Gamble, Fl. Pres. Madras 345. 1918.
Meibomia gangetica (L.) Kuntze, Rev. Gen. Pl. 1: 196. 1891.

Erect to ascending shrub to 2 m high, but often much smaller; branches
terete, with moderately dense hairs about 1 mm long, becoming glabrous;
leaves 1-foliolate; stipules subulate, 7–15 mm long; petiole 10–25 mm, rarely
to 40 mm, long; leaflets chartaceous, broadly elliptic to ovate, rounded or
cordate at the base, obtuse, acute or acuminate at the apex, more or less
glabrous on upper surface, with sparse to rather dense, weak, appressed or
subappressed hairs on lower surface, 17–85 mm, rarely to 160 mm, long,
24–60 mm wide (rarely 12–85 mm), 1.1–2 times longer than wide; pulvinus
1–4 mm long, stipels setaceous, 1–6 mm long; inflorescence terminal, rather
open, to 25 cm long; rhachis with moderately dense weak uncinate hairs;
fascicles of 2–5, rarely 7, flowers; primary bract 3–6 mm long, secondary
bracts 1.5–3 mm long, both deciduous; bracteoles absent; pedicels 1–4 mm
long; calyx broadly campanulate, 4-lobed, with dense indumentum of minute
uncinate hairs and longer straight hairs, particularly on the lobes, about 2 mm
long, lobed to about the middle; upper lobe wider than the others; corolla
white to pale pink; standard broadly obovate, about 3 mm long and wide,
emarginate at the apex; wings 3.5–4 mm long; keel petals distinctly longer and
wider than the wing petals; stamens diadelphous; ovary appressed pubescent;
pods flat, linear, occasionally very shortly stipitate, upper suture more or
less straight, lower deeply incised, isthmus about 2/3 width of pod; with 6–8
articles, each 2.5–3 mm long, 2–2.5 mm wide, reticulately veined with sparse
short erect uncinate hairs; seeds reniform, about 1.8 mm long, 1.5 mm wide,
with a distinct rim aril.

Distr. Old world tropics; Africa, Ceylon, India, Indo-China, southern
China, Malesia, northern Australia, naturalized in the West Indies (Jamaica,
St. Vincents).

Ecol. In open situations, grassland, roadsides.

Specimens Examined. JAFFNA DISTRICT: between Jaffna and
Elephant Pass, *Rudd 3285* (K, PDA). RATNAPURA DISTRICT: Wetagama
Estate, some miles W of Ratnapura, *Read & Jayaweera 2317* (PDA).
LOCALITY UNKNOWN: *Thwaites C.P. 3813* (K, PDA).

11. Desmodium velutinum (Willd.) DC., Prod. 2: 328. 1825; Schindl.,
Repert. 21: 6. 1925; van Meeuwen, Reinwardtia 6: 264. 1962; Schubert in
Fl. Trop. East Africa, Leg.: Papilion. 486. 1971; Ohashi, Ginkgoana 1: 192.
1973; Verdcourt, Man. New Guinea Leg. 411. 1979.

Hedysarum velutinum Willd., Sp. Pl. 3: 1174. 1802. Type: Tropical America,
Willdenow herb. *13763*.

Hedysarum latifolium Roxb. ex Ker-Gawl., Bot. Reg. 5. t. 355. 1819; Roxb., Fl. Ind. ed. 2. 3: 350. 1832. Type: from India (?).

Desmodium latifolium (Roxb. ex Ker-Gawl.) DC., Prod. 2: 328. 1825; Wight & Arn., Prod. 225. 1834; Thw., Enum. Pl. Zeyl. 87. 1859; Baker in Hook. f., Fl. Br. Ind. 2: 168. 1876; Trimen, Handb. Fl. Ceylon 2: 51.1894; Gamble, Fl. Pres. Madras 346. 1918.

Desmodium latifolium var. *roxburghii* Wight & Arn., Prod. 225. 1834. Type: from India, Wight Cat. 792.

Desmodium velutinum var. *roxburghii* (Wight & Arn.) Schindl., Repert. 21: 6. 1925.

Shrub to 3 m high; branches terete, densely clothed with long spreading brown rigid hairs and shorter weak uncinate hairs; leaves 1-foliolate; stipules striate, abruptly narrowed from broad base to long acuminate apex, 3–7 mm long; petiole 8–20 mm long; leaflets coriaceous, broadly ovate or occasionally oblong, truncate or shallowly cordate at the base, obtuse or less commonly short acuminate or acute at the apex, sparse to moderately dense, soft, long, ascending hairs on the upper surface, velutinous with dense, long, more or less spreading hairs on the lower surface, 60–125 mm long, 40–85 mm, rarely to 105 mm, wide, 1–1.7 times longer than wide; inflorescences terminal and axillary, up to 20 cm long, often paniculate; rhachis with dense brown spreading and uncinate hairs; fascicles of 2–4 flowers; primary bracts about 2.5 mm long; secondary bracts 1–1.5 mm long; bracteoles absent; pedicels 1.5–2 mm long, sparsely pubescent; calyx broadly campanulate, 4-lobed, densely pubescent with minute appressed and longer straight spreading hairs; the tube 0.8–1 mm long; the lobes all about equal in length to the tube except the lower one longer, to 1.5 mm long; corolla blue, pink or purplish; standard obovate, obtuse at apex, about 5 mm long, 3.3–4.2 mm wide; wings about 5 mm long; keel petals slightly longer and wider than the wings; stamens diadelphous; ovary densely strigose; pod flat, linear, more or less straight, the upper suture more or less straight, the lower indented, the isthmus about 3/4 width of pod, with 5–7 joints, each about 3 mm long, 2.5–3 mm wide, broadly oblong, reticulately veined and covered with dense straight hairs and shorter uncinate ones; seeds 1.8–2.4 mm long, 1.5–2 mm wide, with a distinct rim-aril.

Distr. Widely distributed in Old World tropics: Africa, Ceylon, India, Indo-China, southern China, Malesia and northern Australia, introduced into tropical America.

Ecol. In forest clearings, on tracks, etc.

Specimens Examined. POLONNARUWA DISTRICT: Minneriya, Aug 1885, *s. coll. s.n.* (PDA). LOCALITY UNKNOWN: *Mackenzie s.n.* (K); *Thwaites C.P. 3588* (K, PDA).

Note. *Desmodium plukenetii* (Wight & Arn.) Merr. (*D. virgatum* Prain nom. illeg., *D. latifolium* var. *plukenetii* Wight & Arn., *D. velutinum* var. *plukenetii* (Wight & Arn.) Schindl., *D. latifolium* var. *virgatum* Miq.), which is usually treated as being conspecific with *D. velutinum*, is at least varietally distinct. It is a native of southern China. It, *D. velutinum* subsp. *longibracteatum* (Schindl.) Ohashi and var. *sikkimense* (Schindl.) Ohashi are not included in the description above.

12. Desmodium jucundum Thw., Enum. Pl. Zeyl. 411. 1864; Baker in Hook. f., Fl. Br. Ind. 2: 172. 1876; Trimen, Handb. Fl. Ceylon 2: 54. 1894; Ohashi, Ginkgoana 1: 216. 1973. Type: *Thwaites 3778* (K, holotype; PDA, isotype).

Meibomia jucunda (Thw.) Kuntze, Rev. Gen. Pl. 1: 198. 1891.

Shrubs to about 1 m tall; branches with dense long appressed to spreading hairs; leaves 3-foliolate occasionally mixed with 1-foliolate; stipules early deciduous, about 7 mm long; petiole 10–30 mm long, rhachis 5–10 mm long; leaflets coriaceous, oblong or somewhat obovate, narrowed to the base, obtuse or slightly emarginate at the apex, discolorous, glabrous on upper surface, densely appressed pubescent on lower surface, about 11 raised lateral veins prominent on each side of midrib; terminal leaflet 35–55 mm long, 20–35 mm wide, 1.7–1.9 times longer than wide; lateral leaflets 20–50 mm long, 15–25 mm wide, 1.5–2 times longer than wide; pulvini 3–4 mm long, stipels 2.5–3.5 mm long; inflorescence terminal, rather dense when young, 8–12 cm long; rhachis with dense covering of reflexed ferruginous hairs with some straight hairs (to 1 mm long); fascicles of 2 or 3 flowers; primary bracts early deciduous, about 7 mm long and 4 mm wide; secondary bracts and bracteoles absent; pedicels about 5 mm long; calyx campanulate, 4-lobed, puberulent with short uncinate hairs and long hairs on lobes; 4.5–5 mm long, tube about 2 mm long, upper lobe divided to the middle, 2.5–3.2 mm long, lateral lobes about 2.5 mm long; lower lobes longer than the others, somewhat acuminate, 3–4 mm long. Corolla mauve to purple, not seen mature but reported to be rather large with standard distinctly larger than other petals, 7–10 mm long and wide; stamens diadelphous; ovary densely appressed sericeous; pods flat, linear, upper and lower sutures thickened, the upper straight, the lower undulate, isthmus 7/8 width of pod, articles 4–6, each 4.8–5.8 mm long, 4 mm wide, reticulately veined with sparse indumentum of weak uncinate hairs; seeds not seen.

Distr. Endemic.

Specimens Examined. MATALE DISTRICT: Lakkaigala-Matale East, *Thwaites C.P. 3778* (K, holotype; PDA, isotype).

13. Desmodium ferrugineum Wall. ex Thw., Enum. Pl. Zeyl. 87. 1859; Ohashi, Ginkgoana 1: 204. 1973. Type: southern India, *Wallich* Cat. 5732.

Desmodium rufescens Wight & Arn., Prod. 228. 1834; Wight, Ic. Pl. Ind. Or. 984. 1845; Benth in Miq., Pl. Jungh. 223. 1852; Baker in Hook. f., Fl. Br. Ind. 2. 171. 1876; Trimen, Handb. Fl. Ceylon 2: 52. 1894; Gamble, Fl. Pres. Madras 346. 1918, non *Desmodium rufescens* DC.

Erect shrub to 1.5 m tall; branches terete, covered with dense spreading ferruginous hairs; leaves 3-foliolate; stipules deciduous, narrowly triangular, 5–7 mm long, densely ferruginous; petiole 8–25 mm long, rhachis 6–13 mm, rarely only 2 mm long; leaflets chartaceous, obovate, attenuate at the base, obtuse or occasionally retuse at the apex, glabrous on upper surface, appressed pubescent, sparsely or occasionally densely on lower surface, 5–7 lateral veins prominent on each side of midrib; terminal leaflet 20–50 mm long, 12–30 mm wide, the laterals 15–40 mm long, 10–24 mm wide, all 1.3–2 times longer than wide; stipels subulate, 1.5–3.5 mm long, about the same length as the pulvini; inflorescences terminal and axillary, covered with large overlapping bracts when young; terminal inflorescences to 15 cm long; rhachis covered with spreading ferruginous hairs and shorter weak uncinate hairs; fascicles of 2–4 flowers; bracts ovate, acuminate, 8–10 mm long, 2.5–4 mm wide, covered with dense ferruginous hairs, early deciduous; secondary bracts and bracteoles absent; pedicels 3–7 mm long; calyx 4-lobed, densely ferruginous pubescent, the tube 1.7–2.2 mm long, upper lobe shortly bifid, 2.5–3 mm long, laterals 2–3 mm long, the lower longer than the others, 3.5–4 mm long; corolla purple; standard broadly obovate, emarginate at the apex, 7.5–8 mm long and wide; wings about 8 mm long; keel petals about 8 mm long, wider than the wings, bent at about right angles; stamens diadelphous; ovary densely covered with ferruginous hairs; pods flat, linear, usually reflexed; upper suture thickened, undulate, the lower incised; isthmus 2/3 width of pod, articles 6–8, each 5–5.5 mm long, 3–3.5 mm wide, oblong, reticulately veined, covered with minute straight hairs; seeds reniform, about 3 mm long and 2 mm wide.

Distr. Southern India and Ceylon.

Ecol. At higher altitudes, sometimes cultivated for its showy flowers.

Specimens Examined. BADULLA DISTRICT: Welimada, Apr 1921, *A.D.A. s.n.* (PDA). NUWARA ELIYA DISTRICT: Ohiya-Boralanda road near 9/5 marker, *Sohmer & Sumithraarachchi 10064* (K). HAMBANTOTA DISTRICT: Tissamaharama, *Alston 1642* (PDA). LOCALITY UNKNOWN: *Walker s.n.* (K); *Gardner 228* (K); *Thwaites C.P. 1426* (K, PDA), *C.P. 1425*, p.p. (PDA).

Note. *Desmodium wyaadense* Bedd. ex Gamble, from southern India, which Ohashi referred to *D. ferrugineum* as a subspecies, seems to be a distinct species.

14. Desmodium heterocarpon (L.) DC., Prod. 2: 337. 1825; Trimen, Handb. Fl. Ceylon 2: 53. 1894; van Meeuwen, Reinwardtia 6: 93. 1961 & 6: 251. 1962; Schubert in Fl. East Trop. Africa, Leg.: Papilion. 462. 1971; Ohashi, Ginkgoana 1: 210. 1973; Verdcourt, Man. New Guinea Leg. 399. 1879; Ohashi, J. Jap. Bot. 86: 14. 1991.

Hedysarum heterocarpon L., Sp. Pl. 74. 1753. Type: Ceylon, herb. *Hermann* (BM).

Hedysarum siliquosum Burm. f., Fl. Ind. 169. t. 55. f. 2. 1768. Type: from India, *N.L. Burmann*.

Hedysarum polycarpum Poir. in Lam., Enc. 6: 413. 1805. Type: from East Indies, herb. *Lamarck*.

Desmodium polycarpum (Poir.) DC., Prod. 2: 334. 1825; Wight & Arn., Prod. 227. 1834; Thw., Enum. Pl. Zeyl. 86. 1859; Baker in Hook. f., Fl. Br. Ind. 2: 171. 1876; Prain, J. Asiatic Soc. Bengal 66: 141. 1897; Gamble, Fl. Pres. Madras 346. 1918.

Desmodium tricholcaulon DC., Ann. Sci. Nat. Bot. 4: 101. 1825; DC., Prod. 2: 335. 1825. Type: from Nepal, *Wallich*.

Desmodium siliquosum (Burm. f.) DC., Prod. 2: 334. 1825.

Desmodium polycarpum var. *trichocaulon* (DC.) Baker in Hook. f., Fl. Br. Ind. 2: 172. 1876.

Meibomia heterocarpa (L.) Kuntze, Rev. Gen. Pl. 1: 198. 1891.

Prostrate or sprawling perennial herb; stems with covering of appressed or, rarely, spreading white hairs, often dense; leaves 3-foliolate or, on young plants, mixed with 1-foliolate leaves; stipules narrowly triangular, acuminate, 9–14 mm long, fringed with hairs; petioles 10–40 mm long, rhachis 3–9 mm long; leaflets chartaceous, obovate, oblong or elliptic, sometimes ovate when unifoliolate, attenuate or rounded at the base, obtuse, truncate or rarely emarginate at the apex, glabrous or rarely with scattered appressed hairs on the upper surface, sparse to moderately dense long appressed hairs on the lower surface; terminal leaflet 16–65 mm long, 12–32 mm wide, 1.3–2.5 times longer than wide, laterals smaller, sometimes markedly so on young plants, 15–55 mm long, 9–25 mm wide, 1.4–2.8 times longer than wide; pulvini 1.5–3 mm long, stipels setaceous, 3.5–7 mm long; inflorescences terminal and axillary, covered by primary bracts when young, 5–10 cm long, usually dense; rhachis with either sparse to moderately dense erect uncinate hairs or dense straight appressed white or yellowish hairs; flowers usually in pairs; primary bracts scarious, concave, ovate, long-acuminate, 5–7 mm long, fringed with hairs; secondary bracts and bracteoles absent; pedicels 2–2.5 mm long, glabrous or with short uncinate hairs; calyx 4-lobed, campanulate, 2–3 mm long, sparsely pubescent; tube 0.8–1.4 mm long; upper lobe minutely bifid at apex, 1.2–1.4 mm long, lateral lobes about 1 mm long; lower lobe 1.3–1.5 mm long; corolla white to purplish; standard broadly obovate, 4.5–6.5 mm long and wide, rounded or slightly emarginate at the apex; wings 4–5 mm long, distinctly curved; keel petals longer than the wings; stamens diadelphous; ovary pubescent or less commonly glabrous; pods erect or ascending when mature, flat, linear, the upper suture straight, the lower shallowly incised, isthmus 3/4–7/8 width of pod, with 5–8 articles, each 2.5–4 mm long, 2.2–3 mm wide, reticulately veined, glabrous or with long yellowish uncinate hairs on margins and faces; seeds reniform or rectangular in outline, 1.8–2.2 mm long, 1.4–1.6 mm wide, rim-aril prominent.

Distr. Widely spread in the Old World, Africa, Ceylon, India, Indo-China, China, Korea, Japan, Malesia, Australia and Pacific Islands.

Ecol. In grassland, on roadsides, occasionally in abandoned cultivation and on the margins of thickets.

KEY TO THE VARIETIES

1 Articles of pod pubescent, 2.5–3 mm long; inflorescence dense
    2 Rhachis of inflorescence with rather sparse short erect uncinate hairs . . . .
    . . . . 1. var. **heterocarpon**
    2 Rhachis of inflorescence with dense appressed long straight white or yellowish hairs . . . . .
    . . . . 2. var. **strigosum**
1 Articles of pod glabrous, 3–4 mm long; inflorescence tending to be elongate, open . . . .
    . . . . 3. var. **gymnocarpum**

1. var. **heterocarpon**

Dense inflorescences with short uncinate hairs on the rhachis, and rather small pubescent pod-articles characterise the variety.

Distr. Throughout the range of the species except possibly Africa (fide Schubert).

Specimens Examined. KANDY DISTRICT: Madugoda, *Worthington 4995* (K); Aladeniya, *Amaratunga 1490* (K). BADULLA DISTRICT: above Monamaya, Haputale, *Rudd & Balakrishnan 3206* (PDA); Haputale, *Mueller-Dombois & Cooray 68011506* (PDA). RATNAPURA DISTRICT: Dela town, *Cooray 70020801R* (K); N of Ratnapura to Gilimale, *Maxwell 944* (K). NUWARA ELIYA DISTRICT: Hakgala, Sept 1906, *A.M.S. s.n.* (PDA). COLOMBO DISTRICT: Angoda, Rajagiriya, *Cooray 69090102R* (PDA); Weboda, *Amaratunga 1453* (PDA). KEGALLE DISTRICT: Warakapola, *Cooray 70012401R* (PDA); Kitulgala, *Amaratunga 1189* (PDA). GALLE

DISTRICT: Kanneliya, *Balakrishnan 613* (K). LOCALITY UNKNOWN: *Gardner 227* (K); *Thwaites 3512* (K); *Walker 105* (K); *Mackenzie s.n.* (K).

2. var. **strigosum** van Meeuwen, Reinwardtia 6: 95. 1961 & 6: 251. 1962; Ohashi, Ginkgoana 1: 215. 1973. Type: from New Guinea, *Kalkman B.W. 3596.*

Dense inflorescences with long dense straight appressed hairs on the rhachis and rather small pubescent pod-articles distinguish this from the other two recognised.

Distr. Throughout the range of the species.

Specimens Examined. GALLE DISTRICT: Godakande, *Jayasuriya et al. 804* (PDA). RATNAPURA DISTRICT: N of Gilimale to Carney, *Maxwell et al. 948* (K). MATALE DISTRICT: Wiltshire Forest, Kandegedara, *Sumithraarachchi 396* (PDA). LOCALITY UNKNOWN: *Thwaites C.P. 1425* (K), *C.P. 1425*, p.p. (PDA).

3. var. **gymnocarpum** Schindl., Repert. 21: 5. 1925; Ohashi, Ginkgoana 1: 213. 1973. Type: Ceylon, *Thwaites.*

Characterized by the rather large glabrous pod articles and the inflorescence tending to elongate with short uncinate hairs on the rhachis.

Distr. Ceylon and southern India.

Specimens Examined. PUTTALAM DISTRICT: Galawatte, *de Silva s.n.* (PDA). KANDY DISTRICT: Kandy, *Worthington 6137* (K, PDA); near Peradeniya-Godawela road, marker 4/16, *Comanor 501* (K); Peradeniya, lower Hantana road, *Comanor 688* (K). LOCALITY UNKNOWN: *Gardner 266* (K); *Thwaites C.P. 1425* p.p. (PDA).

Note. *Desmodium heterocarpon* is a wide-ranging polymorphic species that has been treated in various ways by different authors. Ohashi, probably the only one to examine critically a large number of specimens from throughout its geographic range, has recognized eight infraspecific taxa, including one forma based principally on flower colour (Ohashi, 1991). His treatment, as far as it applies to plants from Ceylon, has been followed here. With var. *birmanicum* (Prain) Ohashi, the three varieties constitute subsp. *heterocarpon.* Detailed consideration of the application of the names to the different varieties used in the literature cited in the synonymy of the species name (above) would be both confusing and unproductive and has therefore been omitted. The papers of Ohashi and Knaap-van Meeuwen have detailed nomenclatural treatments.

15. Desmodium pryonii DC., Prod. 2: 334. 1825; Ohashi, Ginkgoana 1: 190. 1973. Type: from India, *Pryon* in herb. Burmann.

Desmodium wightii R. Grah. ex Wight & Arn., Prod. 266. 1834; Benth. in Miq., Pl. Jungh. 224. 1852; Thw., Enum. Pl. Zeyl. 87. 1859; Baker in Hook. f., Fl. Br. Ind. 2: 169. 1876; Trimen, Handb. Fl. Ceylon 2: 52. 1894; Gamble, Fl. Pres. Madras 346. 1918. Type: from peninsular India, *Wallich* cat. 5718.

Meibomia wightii (R. Grah. ex Wight & Arn.) Kuntze, Rev. Gen. Pl. 1: 198. 1891.

Sprawling shrub to 1 m tall; stems with indumentum of minute uncinate hairs and long spreading or appressed straight hairs; leaves 3-foliolate; stipules subulate, 7–13 mm long; petiole 10–45 mm long, rhachis 4–12 mm long; leaflets chartaceous, ovate or rhomboidal, cuneate or rounded at the base, acute or acuminate at the apex, glabrous or with scattered long hairs on the upper surface, the margin sometimes densely fringed; sparse to moderately dense, weak, appressed or somewhat spreading hairs on lower surface, 5–8 lateral veins on each side of midrib prominent; terminal leaflet 25–90 mm long, 13–40 mm, rarely to 50 mm, wide, the laterals usually smaller, 20–70 mm long, 10–33 mm wide, all 1.5–2.6 times longer than wide; pulvini 1.5–3 mm long, stipels 2.5–6 mm long; inflorescences terminal, to 30 cm long, sometimes rather dense; rhachis with short uncinate hairs and long stiff hairs; fascicles of 2 or 3 flowers; primary bracts 2.5–8 mm long; secondary bracts 1.5–3.5 mm long; pedicels 5–8 mm long, densely uncinate pubescent; bracteoles absent; calyx 4-lobed, 2.2–2.8 mm long, densely hairy with short uncinate hairs and long straight hairs, the tube about 1 mm long, the lobes all about the same length, 1.2–1.8 mm long; corolla pink to purplish red; standard obovate, slightly emarginate at the apex, about 4 mm long and 3 mm wide; wings about 4 mm long and 2 mm wide; keel petals slightly longer than the wings; stamens diadelphous; ovary minutely stiped, with dense appressed pubescence; pods linear, usually straight, sessile or minutely stiped, the upper suture notched between the articles, the lower deeply constricted, up to 7 articles, each 2.7–3.5 mm long, 2.4–2.7 mm wide, reticulately veined, glabrous or with some short weak uncinate hairs; seeds reniform, about 2.4 mm long, 1.6 mm wide, the rim aril prominent.

Distr. Ceylon and southern India.

Ecol. In disturbed situations in shallow sandy or rocky soils.

Specimens Examined. PUTTALAM DISTRICT: Wilpattu National Park, Mannar-Puttalam road, mile post 36, *Cooray 70020109R* (PDA). ANURADHAPURA DISTRICT: Ritigala Strict Nature Reserve, *Jayasuriya & Burtt 1707* (K), *1158* (PDA). MATALE DISTRICT: Dambulla, *Alston*

1206 (PDA). KANDY DISTRICT: between Madugoda and Weragantota, *Alston 1668* (K, PDA); Kandy, *Jayasuriya & Balasubramaniam 453* (PDA). BADULLA DISTRICT: below the Nuwara Eliya/Badulla road, *Townsend 73/138* (K). HAMBANTOTA DISTRICT: Tissamaharama, *Alston 1642* (PDA); Ruhuna National Park, Block 1 at mile marker 11 on Yala road; *Cooray 69011604R* (PDA), Kohombagaswala, *Cooray 69120323R* (PDA), near Rakinawala, *Cooray 69120708R* (PDA). LOCALITY UNKNOWN: *Walker 205* (K).

16. Desmodium tortuosum (Sw.) DC., Prod. 2: 332. 1825; van Meeuwen, Reinwardtia 6: 101. 1961 & 6: 260. 1962; Sundaraj & Nagarajan, J. Bombay Nat. Hist. Soc. 66: 658. 1969; Schubert in Fl. Trop. East Africa, Leg.: Papilion. 474. 1971; Verdcourt, Man. New Guinea Leg. 408. 1979; Smith, Fl. Vit. Nova 3: 191. 1985; McVaugh, Fl. Novo-Galiciana 5: 493. 1987; Howard, Fl. Less. Antilles 4(1): 482. 1988.

Hedysarum purpureum Mill., Gard. Dict. ed. 8. 1768. Type: from Mexico, *Houston*.
Hedysarum tortuosum Sw., Prod. Veg. Ind. Occ. 107. 1788. Type: Jamaica, *Swartz*.
Desmodium stipulaceum DC., Prod. 2: 330. 1825; Benth. in Miq., Pl. Jungh. 229. 1851. Type: from Mexico.
Meibomia tortuosa (Sw.) Kuntze, Rev. Gen. Pl. 198. 1891.
Meibomia purpurea (Mill.) Vail in Small, Fl. South-east United States 639. 1903.
Desmodium purpureum (Mill.) Fawcett & Rendle, Fl. Jamaica 4: 36. 1920, nom. illeg., non Hook. & Arn. 1832.

Erect herb to 1 m or more tall; young stems subangulate, finely striate, moderately uncinate pubescent; leaves 3-foliolate or towards the base of young plants 1-foliolate; stipules 10–20 mm, occasionally 30 mm, long; petioles 10–25 mm long, rhachis 5–12 mm long; leaflets chartaceous, elliptic to rhombic ovate, broadly cuneate at the base, obtuse mucronate at the apex, both surfaces puberulent to soft-appressed pilose, margins ciliate, occasionally subglabrous above; terminal leaflet 35–60 mm long, 15–33 mm wide, the laterals smaller, 15–35 mm long, 10–25 mm wide, all 1.7–2.3 times longer than wide; inflorescence racemose-paniculate, racemes terminal or axillary, open in fruit; rhachis pilose with spreading yellowish glandular hairs when young; flowers in pairs; primary bracts about 5 mm long; secondary bracts much smaller, both deciduous; bracteoles absent; pedicels moderately glandular-pilose, 10–17 mm long; calyx 4-lobed, puberulent and with long stiff hairs on the lobes; tube much shorter than the lobes; slightly less than 1 mm long; upper and lower lobe 1.5–2 mm, the upper bifid for about 1/3 length, lateral lobes shorter than the others; flowers mauve-pink with standard obovate, about 4 mm long and 3 mm wide; wings about 4 mm long and up to 1.5 mm wide; keel petals slightly longer than wings; pods sessile or with stipe less than 1 mm long, both upper and lower sutures deeply indented, the isthmus less than 1/8 width of pod, with 4–6 articles, twisted when young, rhomboidal, margins alternately involute and revolute, reticulately veined, rather densely uncinate pubescent, 3.5–4.5 mm long, 2.5–3.5 mm wide; seed reniform, about 3 mm long and 2 mm wide, with a small rim aril.

Distr. A native of tropical America naturalised in Africa, India, Malesia, Australia and Pacific Islands.

Ecol. It is reported to have been a 'weed around Kandy but only occasional' in 1917.

Specimens Examined. KANDY DISTRICT: Peradeniya, June 1917, *s. coll. s.n.* (K); Peradeniya, *Fosberg 50668* (K), *Fosberg 50672* (K), weed near herbarium building, *Rudd & Fernando 3318* (PDA).

As well as *D. tortuosum*, other plants have been introduced, probably as potential forage or cover crops but have not become naturalised. Specimens of three of these have been examined.

Desmodium cajanifolium (Kunth) DC., Prod. 2: 331. 1825. *Hedysarum cajanifolium* Kunth in H.B.K., Nov. Gen. Sp. 6: 525. t. 598. 1824. *Desmodium walkeri* Arn., Nov. Actorum Acad. Caes. Leop. Carol. Nat. Cur. 18: 331. 1836; Thw., Enum. Pl. Zeyl. 87. 1859. This is a native of South America similar in general appearance to *D. tortuosum*. It was considered to be endemic to Ceylon by Arnott. It is represented by several collections: *Thwaites C.P. 2973 & 3522* (K) and *Walker s.n.* (K), the last possibly the holotype or an isotype of *D. walkeri*.

Desmodium discolor Vogel, Linnaea 12: 103. 1838. This is a shrub with leaflets velvety on the undersurface and pods somewhat similar to those of *D. tortuosum*. It is a native of Brazil where it is widespread and common. It is represented by one specimen from Ceylon where it is reported to have been cultivated in the students' garden, Peradeniya, *Alston 2014* (K).

Desmodium piliusculum DC., Ann. Sci. Nat. Bot. 4: 101. 1825; DC., Prod. 2: 335. 1825. *Meibomia limensis* var. *piliuscula* (DC.) Schindl., Repert. 22: 275. 1926. *Desmodium intortum* var. *piliusculum* (DC.) Fosberg, Micronesica 4: 257. 1968. *Desmodium sandwichense* E. Meyer, Linnaea 24: 230. 1851; Ohashi in Wagner et al., Man. Fl. Pl. Hawai'i 1: 669. 1990. This is represented by one poorly labelled specimen in herb. K which may not be from Ceylon. A study of a rather small number of specimens from South America suggests that Schindler's alignment of the taxon is probably correct, but the necessary combination has not been made under *Desmodium limense*.

10. CODARIOCALYX

Hassk., Flora 25, Beibl. 2: 48. 1842; Schindl., Repert. 20: 280. 1924; Hutch., Gen. fl. Pl. 1: 479. 1964; Ohashi, Ginkgoana 1: 40. 1973. Type species: *C. motorius* (Houtt.) Ohashi.

Desmodium sect. *Pleurolobium* DC., Prod. 2: 326. 1825; DC., Mém. Lég. 322. 1825; Benth. in Miq., Pl. Jungh. 223. 1852.

Desmodium subg. *Pleurolobium* (DC.) Baker in Hook. f., Fl. Br. Ind. 2: 174. 1876.

Shrubs; 3-foliolate but lateral leaflets often rudimentary, or 1-foliolate, pulvinate; stipules free, early deciduous; petioles sulcate; inflorescences terminal and axillary, paniculate or racemose, young parts cone-like with large overlapping bracts; flowers pedicellate in 2–4-flowered fascicles; primary bracts entirely covering the flower-buds, deciduous; secondary bracts and bracteoles absent; calyx membranous, broadly campanulate, 4-lobed with the upper distinctly notched at the apex; corolla with broad wings widest at distal end and keel petals distinctly longer than the wings, clawed, with a membranous spur on the outside of the lamina at the base; stamens diadelphous, the upper shorter than the others, the free parts of the united stamens alternately a little longer and shorter; ovary with up to about 12 ovules; pod flat, linear, not dividing into distinct articles, dehiscing along the lower margin, the upper suture not indented, the lower undulate; seeds conspicuously arillate and broadly strophiolate.

A genus of two species occurring in southeastern Asia and Malesia with one species in Ceylon. It has affinities with both *Desmodium* and *Pseudarthria*.

Codariocalyx motorius (Houtt.) Ohashi, J. Jap. Bot. 40: 367. 1965; Ohashi, Ginkgoana 1: 46. 1973.

Hedysarum motorium Houtt., Nat. Hist. 11. 10: 246. 1779. Type: from Malesia.

Hedysarum gyrans L.f., Suppl. Pl. 332. 1781. Type: from Bengal, India.

Desmodium gyrans (L.f.) DC., Prod. 2: 326. 1825; Wight & Arn., Prod. 227. 1834; Thw. Enum. Pl. Zeyl. 87. 1859; Baker in Hook. f., Fl. Br. Ind. 2: 174. 1876; Trimen, Handb. Fl. Ceylon 2: 56. 1894; Gamble, Fl. Pres. Madras 348. 1918.

Codariocalyx gyrans (L.f.) Hassk., Flora 25, Beibl. 2: 49. 1842; Schindl., Repert. 20: 284. 1924.

Meibomia gyrans (L.f.) Kuntze, Rev. Gen. Pl. 1: 196. 1891.

Desmodium motorium (Houtt.) Merr., J. Arnold Arbor. 19: 345. 1938; van Meeuwen, Reinwardtia 6: 254. 1962; Deb, Sengupta & Malick, Bull. Bot. Soc. Bengal 22: 184. 1968.

Shrub to about 1.5 m tall; young stems ribbed, glabrous; leaves 1-foliolate or 3-foliolate with rudimentary lateral leaflets or lateral leaflets absent but stipels remaining; stipules 10–14 mm long; petioles 9–14 mm long; rhachis with long (to 1.5 mm) spreading hairs, 3–5.5 mm long; terminal leaflet chartaceous, oblong or narrowly obovate, attenuate at the base, obtuse at the apex, upper surface glabrous, lower with sparse short appressed hairs, 45–100 mm long, 15–30 mm wide, 2–3.3 times longer than wide, laterals linear, up to 10 mm long, 3–5.5 mm wide; pulvini about 2 mm long (laterals longer); stipels 2–4 mm long; inflorescences to 10 cm long, becoming open in fruit; rhachis with dense covering of short yellowish uncinate hairs; flowers paired; bracts ovate with short acuminate tip, 5–8 mm long, 3–6 mm wide, glabrous; pedicels with indumentum of inflorescence rhachis, 1–4 mm long at anthesis, strongly curved at base, 3–7 mm long in fruit; calyx minutely puberulous, 2–2.5 mm long; tube 1–1.3 mm long; upper lobe about 1 mm long and 2 mm wide, deeply divided at apex, lateral lobes 0.8–1 mm long and lower lobe 1–1.2 mm long; corolla mauve or lilac; standard obovate, rounded at the apex, about 8 mm long and 6 mm wide; wings about 6 mm long; keel petals slightly longer than wings; ovary puberulous; pods 35–45 mm long, 3.5–5 mm wide, with covering of rather sparse uncinate hairs, with 8–10 seeds, each segment 4–4.5 mm long; seeds 3.5–4 mm long, 2–2.5 mm wide, conspicuously arillate.

Distr. Ceylon, India, Indo-China, southern China, Malesia, not in Australia though reported to occur by Merrill and Ohashi.

Specimens Examined. KANDY DISTRICT: Peradeniya, *de Silva s.n.* (PDA). BADULLA DISTRICT: Pahala Pussellawa, *Alston s.n.* (PDA). NUWARA ELIYA DISTRICT: Mulhalkelle, *Worthington 5659* (K). LOCALITY UNKNOWN: *Thwaites 1430* (K, PDA).

Note. *Codariocalyx gyroides* (Link) Hassk. (*Desmodium gyroides* (Link) DC.) has been recorded from Ceylon, but Trimen noted that he knew it only from a cultivated plant. There is one specimen from the Botanic Gardens, Peradeniya (coll. 1887, PDA) which is probably the basis for Trimen's remark.

11. PSEUDARTHRIA
(by Velva E. Rudd*)

Wight & Arn., Prod. 1: 209. 1834. Type species: *P. viscida* (L.) Wight & Arn.

Hedysarum L., Gen. Pl. ed. 5. no. 793, p. 332. 1754, p.p.

Glycine sensu Willd., Ges. Naturf. Freunde Berlin, Neue Schriften 4: 209. 1803 [fide IK]; Willd., Nov. Actorum Acad. Caes. Leop.—Carol. Nat. Cur. 4: 208. 1803 (fide Wight & Arn., Prod. 1: 209. 1834). Note: In 1914, Schindler in Repert. 2. 1914, Das genus *Pseudarthria* Wight & Arn., stated that he could not find evidence that Willdenow had actually published the name *Glycine viscida* in 1803. In 1928, Repert. 49(2): 312, his reference to this name goes back to Wight & Arnott. He did note, however, in the 1914 reference that he found a specimen of *Pseudarthria viscida* in the Willdenow herbarium under the name *Glycine viscida*, q.v. See also Pers., Syn. Pl. 2: 300. 1907; Spreng., Syst. Veg. 3: 196. 1826.

Desmodium Desv., J. Bot. (Desvaux) 3: 122. t. 5, figs. 15, 22. 1813, p.p.; DC., Ann. Sci. Nat. Bot. 4: 19. 1825, p.p.; DC., Mém. Légum. 7: 320. 1826, p.p.; DC., Prod. 2: 325, 327. 1825, p.p.

Rhynchosia (Lour.) DC., Prod. 2: 384. 1825, p.p.

Anarthrosyne E. Mey., Comm. Pl. Afr. Austr. 124. 1835, p.p.; Klotzsch in Peters, Reise Mossamb. Bot. 1: 40, 41. 1862; Harvey in Harvey & Sonder, Fl. Capensis 2: 229. 1862. Note: The three authors cited in this paragraph treated only the species occurring in Africa, and therefore did not list *Pseudarthria viscida*.

Spreading or erect perennial herbs or subshrubs with pinnately compound 3-foliolate (or, according to Verdcourt, abnormally subpalmately 5-foliolate) leaves; stipules free, striate, lance-attenuate, stipels smaller, persistent; inflorescences axillary and/or terminal, pseudoracemose or paniculate; flowers mostly borne in fascicles of 3 or more, but often with only 2 maturing; one primary ovate-attenuate bract subtending each fascicle, a smaller secondary bract subtending each pedicel; bracteoles occasional or if occurring regularly, very early deciduous; calyx 2-lobed, upper lobe bifid, lower lobe 3-toothed, central tooth longest, laterals slightly shorter; corolla (white to) light pink to purple, glabrous, standard obovate, attenuate toward base; wings ± oblong, obtuse at apex, auriculate, abruptly narrowed to slender claw; keel petals longer, and longer clawed; stamens diadelphous (9+1), the vexillary stamen essentially free at maturity; ovary long-pilose and minutely uncinulate pubescent; style glabrous; stigma capitate, terminal; fruit oblong to linear (in African species) with sutures indented between seeds and surfaces, somewhat impressed but not articulated, continuous within; seeds subreniform with elongate funicles, 2-many per legume.

An Old World genus of 4–6 species, of which only one occurs in Ceylon, the remainder in Africa.

Although Hutchinson, in 1964 (The Genera of Flowering Plants 1: 398, 399) described a new tribe, the Pseudarthrieae, to which he assigned *Pseudarthria* and *Pycnospora*, this has not been generally adopted. We think that the true relationships of these two genera are more accurately expressed by maintaining them in subtribe Desmodiinae of the Desmodieae.

Pseudarthria viscida (L.) Wight & Arn., Prod. 1: 209. 1834; Wight, Ic. Pl. Ind. Or. 1: 286/651. 1840; Span., Prod. Fl. Timor in Linnaea 15: 190. 1841; Thw., Enum. Pl. Zeyl. 87. 1859; Baker in Hook. f., Fl. Br. Ind. 2: 153, 154.

1876; Trimen, Cat. 24. 1885; Trimen, Handb. Fl. Ceylon 2: 41. 1894; Merr., Philipp. J. Sci. 5. C. Bot. 90. 1910; Abeywickrama, Ceylon J. Sci., Biol. Sci. 2(2): 171. 1959.

Hedysarum viscidum L., Sp. Pl. 747. 1753; Poiret in Lam., Enc. 6: 409. 1804 (based on the description of Linnaeus, and on a specimen in hb. P-Lam, phot. GH); Roxb., Fl. Ind. 3: 350. 1832 (as *H. vescidum*). Type: Ceylon, *Hermann* herb. no. 1: 22 (BM, lectotype; *Hermann* herb. no. 3: 47 [fr.], 4: 21 BM, syntypes, fide Trimen, J. Linn. Soc. Bot. 24: 129–155. 1887, see especially p. 147, no. 295).

Glycine viscida sensu Willd. See note and references under *Glycine* in the synonymy of the genus *Pseudarthria*. The specimen found by Schindler in B-W may be seen on the microfiche IDC 7440. 959: III. 3, 4—Specimen no. 13456; Moon, Cat. 53. 1824, as *G. viscidum*.

Desmodium viscidum (L.) DC., Prod. 2: 336. 1825 (fide Schindler, who noted that the descriptions of bracts and stipules as glabrous did not agree with the original description. DeCandolle based his description on Burmann's Thes. Zeyl. t. 84, f. 1, where essentially no pubescence is indicated. There is no material under this name in the Prodromus Herbarium at G.—BGS); G. Don, Gen. Hist. 2: 276. 1832.

Hedysarum viscosum DC., Prod. 2: 336. 1825 in synon., non Burm. (fide Schindler); G. Don, Gen. Hist. 2: 296. 1832, in synon.

Rhynchosia viscida DC., Prod. 2: 387. 1825.

Desmodium leschenaultii DC., Ann. Sci. Nat. Bot. 4: 102. 1825; DC., Prod. 2: 336. 1825. Type: cult. in Hort. Calcutta and sent to DeCandolle by M. Leschenault; 2 sheets exist in the Prod. Herb. at G; GH, phot. of sheet no. 2.

Desmodium timoriense DC., Prod. 2: 327. 1825; DC., Mém. Légum. pt. 7: 323. 1826. Type: coll. on Timor on an expedition led by Capt. Baudin; see also the Merrill reference under *Pseudarthria viscida*.

Hedysarum timoriense (DC.) Spreng., Syst. Veg. 4. Cur. Post. 290. 1827.

Meibomia timoriensis (DC.) Kuntze, Rev. Gen. Pl. 1: 198. 1891.

Some other synonymy exists in the literature (chiefly in Wight & Arn., Prod. 1: 197. 1834, and Schindler in Repert. Sp. Nov. Beiheft 2: 1–20. 1914, Das Genus *Pseudarthria* Wight et Arn.), consisting largely of "sensu" names which would not be especially helpful in the context of this Flora.

Erect, trailing, or more or less twining perennial herbs to about 3.5 m long or high; stems slender, terete to angulate or angulate-sulcate, somewhat glandular and with abundant to dense pilosity of long white tapering trichomes; leaves trifoliolate, stipulate, stipules slenderly ovate- to linear-lanceolate, striate, pilose on dorsal surface with long tapering trichomes, more or less persistent, 3–7 mm long, 0.8–2 mm wide at base; stipels similar, linear, persistent, 1.5–4.5 mm long; petiole angular-sulcate, pilose with tapering trichomes and shorter uncinulate ones intermixed, 3–4 times as long as similar leaf rhachis; terminal leaflet ovate to ± rhombic in outline, base and apex acute, sides slightly repand and obtuse, lateral leaflets similar to terminal, or almost orbicular to more or less ovate-acute, rounded at base, only slightly oblique; inflorescence terminal, pseudoracemose or paniculate, flowers most often borne in pairs, (if borne in 3's the third one may or may not develop); primary bracts each subtending 1 fascicle of flowers, linear- to slenderly ovate-attenuate, 1.5–3 mm long, c. 1 mm wide; secondary bracts narrowly lanceolate, rather long-ciliate, 1–2 mm long, 0.5–1 mm wide; pedicels abundantly uncinulate-puberulent with recurved hairs, 4–6.5 mm long; calyx somewhat long-pilose with long tapering trichomes abundantly scattered and minute more or less glandular pubescence over whole outer surface, teeth ciliate, upper bifid lobe 2–2.8 mm long, central tooth of lower lobe 2.5–3.5 mm long, lateral teeth 2.3–3.2 mm long; corolla exceeding calyx, standard 3–4.5 mm long, 2–4 mm wide, wings 3.5–4 mm long, 1–1.5 mm wide, keel petals 3.5–4.5 mm long, 1–2.5 mm wide; fruit elliptic to oblong, short-stipitate, stipe 0.5–2 mm long, the persistent style appearing apiculate, sutures uncinulate-puberulent with stout hairs, surfaces minutely puberulent with stouter hairs as on sutures intermixed, glandular throughout with tiny glands, each legume 0.9–2 cm long, 0.4–0.6 cm wide; seeds 2–5 (or more), funiculate, funicle 1–1.8 mm long, seeds 1–4 mm long, c. 2 mm wide. Chromosome number 2n = 22.

Pseudarthria viscida differs from the African species in the characters of the leaves, the inflorescence, and the fruit. In general the leaflets of *P. viscida* are as long as broad, the terminal being essentially rhombic in outline, the laterals elliptic or almost orbicular, margins may be somewhat sinuate. In the African species the terminal leaflets are longer than broad. The inflorescence in all species is a pseudoraceme or panicle, in *P. viscida* it is more lax than in the other species. The fruit of *P. viscida* is wider in proportion to its length, being more nearly oblong than in the African species in which it is longer in proportion to the width and more nearly linear as sometimes described. In *P. viscida* also, both sutures of the legume are obviously somewhat indented between the seeds, but the fruit is not articulated and the seed cavity is continuous.

Some students of the genus would consider combining *Pseudarthria* with *Desmodium*. Its general aspect is similar and *P. hookeri* is frequently confused with *Desmodium salicifolium* in Africa. New studies of more material may help in reaching satisfactory conclusions about relationships within the subtribe.

We have seen one specimen of this species collected in Jamaica: Parish St. Andrew: gorge of the Hog Hole River, 1.25 miles due east of Gordon

Town, alt. c. 1900 ft, shaded earth bank, trailing to suberect herb, flowers rose; 17 Nov 1958, *G.R. Proctor 18370* (A). According to Dr. R.A. Howard this locality is close to the garden of Hinton East who was in charge of an early botanic garden on the island, from which seeds from cultivated plants may have escaped and plants from them have become naturalized. C.D. Adam in The Flowering Plants of Jamaica (1972) states that the species is very rare. He cited the collection noted above and adds the island of St. Vincent as a locality.

Specimens Examined. JAFFNA DISTRICT: Between Jaffna & Elephant Pass, "at edge of woods, fls light pink," *Rudd 3284* (GH, PDA, SFV, US). PUTTALAM DISTRICT: Wilpattu National Park, near Satpuda Kallu (Occapu junction), 2 miles E of Kattankandal Kulam, "shaded in scrub in irregularly semi-open scrub-forest", *Fosberg et al. 50815* (PDA, SFV, US), Mannar-Puttalam Road, 36th mile post, 1 Feb 1970, "creeper, flowers purple, common," *Cooray 70020109R* (PDA, US); Panirendawa Forest Reserve, SE of Chilaw, *Maxwell & Jayasuriya 837* (PDA, SFV, US). ANURADHAPURA/VAVUNIYA DISTRICTS: Issanbassawa, between Medawachchiya and Vavuniya, mile marker 97/2, *Rudd 3257* (GH, PDA, SFV, US). POLONNARUWA DISTRICT: between Habarane and Kantalai, "fls bright pink", *Rudd & Balakrishnan 3115* (PDA, SFV, US). KURUNEGALA DISTRICT: Uhumiya, "herb, fls bright pink", *Amaratunga 1445* (PDA). MATALE DISTRICT: Wahakotte, "vern. gas-gonika (S)", *Amaratunga 1520* (PDA); on top of Sigiriya rock, *Simpson 9195* (BM); below Dambulla temple, *Simpson 8126* (BM). KANDY DISTRICT: Peradeniya, *Thwaites* (?) *s. coll. s.n.* (PDA); Peradeniya, "herb about 4 ft, fls bright pink, common weed", 28 Dec 63, *Amaratunga 756* (PDA); between Mahiyangana and Madugoda, "on shady bank, fls bright pink," *Rudd & Balakrishnan 3237* (GH, PDA, SFV, US); Kobbekaduwa, near Kandy, *Rudd 3338* (PDA, SFV, US); Haragama, 500 m, "fls scarlet-pink", *Jayasuriya & Balasubramaniam 446* (PDA, US); BADULLA DISTRICT: Between Welimada & Badulla, mile marker 69, *Rudd & Balakrishnan 3198* (PDA, SFV, US); Passara, Jan. 1888, *s. coll.* [Trimen?] *s.n.* (PDA). MONARAGALA DISTRICT: 1 mile w. of Wellawaya, culvert 137/5, dry roadside 270 m alt., "woody prostrate, 2 m long, fls reddish", *Maxwell & Jayasuriya 758* (PDA, SFV, US). LOCALITY UNKNOWN: *Worthington 7230* (K-Worth. Herb.) *5011* (K-Worth. Herb.); *Thwaites* (?) *s. coll. C.P. 1432* (BM, P); "Ceylon", ex herb. Wight, Co. *Walker s.n.* (GH, P, PDA); *Col. Walker 311* (K); *Leschenault* (?) *s.n.*(P).

FLACOURTIACEAE
(by B. Verdcourt*)

DC., Prod. 1: 255. 1824; Hutch., Gen. Fl. Pl. 2: 201–232. 1967. Type genus: *Flacourtia* L.**

Trees or shrubs sometimes with spine-tipped branches or axillary spines. Leaves simple, alternate or less often opposite or whorled, entire or toothed, sometimes gland-dotted; stipules small to conspicuous, often deciduous. Flowers regular, hermaphrodite or sometimes unisexual, solitary or in racemes, cymes or fascicles, occasionally epiphyllous. Calyx-lobes 3–8(–15), imbricate or contorted. Corolla lobes free, (0–) 3–8(–15), sometimes merging with the sepals, sometimes with basal scale; disk sometimes present. Stamens 4–100, sometimes in groups opposite the petals, some often reduced to staminodes; filaments free or rarely united into a tube; anthers opening by slits save in *Kiggelaria* which has terminal pores. Ovary superior, half inferior or rarely completely inferior, 1-locular with parietal or almost basal placentas or placentas intrusive so that placentation is ± axile with ovary plurilocular; styles free or united; each placenta with 1-many anatropous to orthotropous bitegmic ovules. Fruit a berry, drupe or loculicidal capsule, sometimes dry and indehiscent, occasionally winged, horned or prickly; seeds often arillate, usually with copious endosperm.

A rather large family of 88 genera and 875 species of the tropics and subtropics now usually referred to Dillenidae-Violales; it shows affinities with the Passifloraceae and is often confused with various members of the Euphorbiaceae to which family Airy Shaw suggested they were related. Sleumer, the world authority on the family, recognized 85 genera and 1100 species (Fl. Trop. E. Africa: Flacourtiaceae, 1975).

Oncoba spinosa Forssk. a showy African shrub has been grown at Peradeniya (*Alston 820* (PDA); S. Garden, *Silva s.n.* PDA; Order area, *Trimen s.n.* (PDA)). It has been included in the generic key. *Gynocardia odorata* Roxb., the seeds of which yield an oil similar to chaulmoogra oil has also been cultivated

---

* Royal Botanic Gardens, Kew.
** Thwaites used Flacourtiaceae with Bixeae as a tribe but kept up Pangiaceae and Samydaceae as families; Trimen used Bixaceae for all; the present practice is to keep Bixaceae separate from an otherwise extended Flacourtiaceae.

at Peradeniya and is likewise included in the key. A single sheet of *Pangium edule* Reinw. has been seen (South Garden, *s. coll. s.n.*). It is a tree to 40 m with large ovate leaves up to 37 × 27 cm acuminate at apex and cordate at base, very long petioles and greenish flowers about 4 cm wide. The seeds are edible after the removal of hydrocyanic acid by washing. Burkill, Dict. Econ. Prod. Malay Pen. gives an account of its uses. It is a native of Malaysia and Papuasia.

KEY TO THE GENERA

1 Petals present but sometimes scarcely distinguishable from the sepals
  2 Flowers showy, about 4–6.5 cm wide (cult.)
    3 A ± large tree to 40 m, buttressed, without spines; leaves very long-petiolate; each pale green petal with ovate or ± round densely adpressed hairy scale .... **Pangium**
    3 Shrub or tree to 10 m with long spines; leaves very shortly petiolate; petals white, without scales .... **Oncoba**
  2 Flowers not showy, much smaller
    4 Petals with a scale at the base (*Pangieae*)
      5 Stamens numerous (about 100 fide Roxburgh); placentas with numerous ovules (cult.)
      .... **Gynocardia**
      5 Stamens 5–8 (but more numerous in cultivated *Hydnocarpus*); placentas with 1-many ovules
        6 Sepals ± united; stamens 5–6; placentas 1-ovulate .... **3. Trichadenia**
        6 Sepals free; stamens 5–9 (or 15–32 in cultivated *Hydnocarpus*); placentas 2–3-many-ovulate
          7 Fruit said to be ultimately dehiscent, certainly angled; flowers in short axillary 1-sided inflorescences, the short rachis marked with circular scars ....
          .... **4. Chlorocarpa**
          7 Fruit a berry, indehiscent; inflorescences not as above .... **2. Hydnocarpus**
    4 Petals without a scale at base
      8 Sepals, petals and stamens ± perigynous; stamens 1 or 3 opposite each petal (*Homalieae*)
      .... **6. Homalium**
      8 Sepals etc. hypogynous
        9 Stamens numerous (*Scolopieae*) .... **5. Scolopia**
        9 Stamens 5 (*Berberidopsideae*) .... **1. Erythrospermum**
1 Petals absent
  10 Plants usually dioecious, usually spiny; fruit a succulent berry (*Flacourtieae*)
    11 Ovary and fruit glabrous, the latter up to 2.5 cm diameter; leaves mostly distinctly crenate or toothed; inflorescence ± pubescent; stamens surrounded by a ring of disk-glands ..
    .... **7. Flacourtia**
    11 Ovary and fruit velvety pubescent (save in cult. *D. caffra*), the fruit 2.5–4 cm diameter; leaves mostly ± entire; inflorescence velvety tomentose (save in cult. *D. caffra*); stamens intermingled with disk-glands and alternating with them .... **8. Dovyalis**
  10 Plants with hermaphrodite flowers; not spiny; fruits eventually capsular (*Casearieae*)
    12 Flowers numerous in drooping spike-like inflorescences .... **9. Osmelia**
    12 Flowers fasciculate in leaf-axils .... **10. Casearia**

1. ERYTHROSPERMUM

Lam., Tabl. Enc. 2, t. 274. 1792; Poir. in Lam., Enc. Suppl. 2: 584. 1812; Lam. in Poir., Tabl. Enc. 2(5): 407. 1819; Gilg in Pflanzenfam ed. 2, 21: 396, fig. 167A–E. 1925; Sleumer in Fl. Males. 5: 6, map. 1954, nom. cons. Type species: *E. pyrifolium* Lam.

Trees or shrubs, occasionally scandent. Leaves alternate, subopposite or somewhat whorled, sessile or petiolate, ± coriaceous, persistent, entire, pinnately nerved; stipules small or absent. Flowers hermaphrodite or occasionally polygamous with rudimentary ovary in axillary or terminal fascicles, racemes or panicles, sometimes supraaxillary or rarely cauliflorous; pedicels articulated at base with small scale-like bract and 2 minute bracteoles. Perianth-parts 7–13, free, imbricate, disposed in a spiral, glabrous, becoming smaller and petaloid towards the centre. Stamens 5–15, uniseriate, free, hypogynous; filaments short; anthers elongate-sagittate; connective broadly dilated. Ovary with 3–4 placentas with numerous ovules; style short with entire or 2–5-lobed stigma. Capsule subglobose, apiculate with persistent style base; pericarp coriaceous, finely shagreened in dry state with 3–4 valves or only semi-dehiscent. Seeds 1–10, arillate, with copious endosperm.

Four species; one in Mauritius the others in Ceylon, China, New Guinea, Samoa and Fiji.

Erythrospermum zeylanicum (Gaertn.) Alston in Trimen, Handb. Fl. Ceylon 6: 14. 1931.

Pectinea zeylanica Gaertn., Fruct. 2: 136. 1791. Type: from seed collection in Herb. Leiden (L, holotype).

Erythrospermum phytolaccoides Gardn., Calcutta J. Nat. Hist. 7: 451. 1847; Thw., Enum. Pl. Zeyl. 18. 1858; Hook. f. & Thoms. in Hook. f., Fl. Br. Ind. 1: 191. 1872; Trimen, Handb. Fl. Ceylon 1: 72, t. 6. 1893. Type: Ceylon, woods between Balangoda and Pelmadulla in "Saffragam district", *Gardner s.n.* (PDA, holotype; K, isotype).

Tree (2.5–)6–18(–25) m tall, glabrous; bole 4–12 m long; girth 0.5–1.25 m; outer bark smooth or superficially fissured, greenish or light brown, or (fide Worthington) with green, black and verdigris patches (snake-like) peeling in thin strips 5 mm wide; live bark straw-coloured or light orange with reddish sap. Leaves elliptic, ovate-oblong or oblong-lanceolate, 6–20 cm long, 2.5–7.5 cm wide, acuminate to a very pointed apex, cuneate at the base, entire or very obscurely subserrate; petioles 0.45–2.5 cm long, ± channelled above; stipules reddish, small, deciduous. Racemes axillary and terminal, simple or ± paniculate, many-flowered, at first compact and ± oblong, lengthening in fruit, 3.5–13 cm long including peduncle; pedicels 4–8 mm long. Flowers white, fragrant; sepals 5, broadly elliptic, 3 mm long, 2 mm wide, obtuse; petals 5, similar, ciliate. Stamens 5, 3 mm long. Ovary 1-locular with 3 placentas; style reddish, 1 mm long; stigma 3-lobed. Capsule red or purplish, also reported to be red and green or white, globose,

0.8–1.2 cm diameter, 3–4-valved, 1–6-seeded. Seeds ovoid or globose with brilliant scarlet pulpy aril; testa hard, muriculate.

Distr. Endemic.

Ecol. Secondary wet evergreen forest, often by stream- and river-banks; 30–510 m.

Uses. None recorded.

Vern. Dodanwenna (S).

Specimens Examined. KURUNEGALA DISTRICT: Weudakanda, *Waas 1638* (K, PDA, US). MATALE DISTRICT: Rattota, Illukkumbura, along Rattota Oya, *Jayasuriya et al. 146* (PDA, US). COLOMBO DISTRICT: Indikada Mukulana, Waga, *Worthington 3545* (K). KEGALLE DISTRICT: opposite Kitulgala, near Kelaniya R., *Kostermans 28332* (K, L); Dickhena, *Waas 1661* (K, PDA, US); Kegalle to Ruanwella, 11/6 mi, Pindeniyoya, *Worthington 6483* (K, PDA). KANDY DISTRICT: cultivated at Peradeniya, Royal Botanic Garden, *Balakrishnan 504, 1025* (K, PDA, US), *Kostermans 23346* (BO, K), *27429* (K, L, PDA); Maskeliya Valley, E. of Adam's Peak, *Kostermans 24195* (PDA, US), *27636* (PDA); Morahenagama, *Waas 283* (PDA, US); Madulkelle, Hatale Estate, *Worthington 1987* (BM); W. of Knuckles, Madulkelle, Arratenne, *Worthington 2011* (K). KALUTARA DISTRICT: Matugama, Elladolla ganga, Kalawakalae, Lihiniyawa, *Balakrishnan 1009* (K, PDA, US); Hallawakellae Forest, Welipenna, *Balakrishnan 1174* (K, PDA, US); Delgoda, *Lewis & J.M. Silva s.n.* (PDA); Lihinigala, Hewesse, *s. coll. s.n.* (PDA); Kalugala, Agalawatte, *Bernardi 15757* (G, PDA, US); Badureliya, near culvert 32/11, Morapitiya-Badureliya road 75 mi, *Cramer 4170* (PDA, US); Atweltota, *Jayasuriya & Kostermans 2322* (K, PDA, US); Matugama-Morapitiya road, *Kostermans 24964* (PDA, US); Morapitiya, *Waas 916* (PDA), Deganda Forest, *Waas 1916* (K, US); Kalugala, *Waas 1280* (K, PDA, US); East Kalugala Forest, *Waas 1542* (K, PDA, US). RATNAPURA DISTRICT: "Saffragam District", Pelmadulla, *C.P. 468*, p.p. (PDA); Balangoda to Pelmadulla, *Gardner in C.P. 468*, p.p. (K, PDA); Gilimale Forest Reserve, *Jayasuriya & Gunatilleke 3071* (PDA), Carney Road, *Kostermans 28434* (K, L); Atweltota W. of Morapitiya, *Meijer 1078* (PDA); Ratnapura, *Trimen in C.P. 468* p.p. (PDA); Gilimale, *Worthington 6488* (K). NUWARA ELIYA DISTRICT: Watagoda, *Trimen in C.P. 468*, p.p. (PDA). GALLE DISTRICT: Pitigalle, *Kostermans 25574* (PDA); Kanneliya Forest, near Hiniduma, *Kostermans 25688* (PDA); Naunkita Ela, Udugama, *Worthington 2329* (K); Naunkita Ela in Kanneliya, *Worthington 3685* (K); Hiniduma Kanda, *s. coll. s.n.* (PDA). MATARA DISTRICT: Morawaka Korale, fide *Trimen*. LOCALITY UNKNOWN: *s. coll. s.n.* Herb. Beddome, (BM); *s. coll. C.P. 468*, p.p. (BM, PDA); *Col. Walker 1380* (K).

2. HYDNOCARPUS

Gaertn., Fruct. 1: 288, t. 60/3. 1788; Gilg in Pflanzenfam. ed. 2, 21: 407. 1925; Burkill, Dict. Econ. Prod. Malay Penins. 1: 1204–1209. 1935; Sleumer, Bot. Jahrb. Syst. 69: 1–94. 1938. Type species: *H. venenata* Gaertn.

Taraktogenos Hassk., Retzia 127. 1855; Hutch., Gen. Fl. Pl. 2: 219. 1967. Type species: *T. blumei* Hassk. (= *T. heterophylla* (Blume) van Slooten).

Mostly dioecious trees or shrubs. Leaves alternate, shortly petiolate, ± coriaceous, entire or slightly serrate, pinnately nerved; stipules deciduous. Male flowers in reduced dichasia or axillary pseudoracemes, rarely on the old wood or sessile and subfasciculate; female flowers usually solitary or paired; bracts minute. Sepals (3–) 4–5(–11), free or joined at base, imbricate. Petals 4–5(–14), free (or rarely joined in a ring) or joined at base, with a flat or thickened subfleshy often hairy scale at base inside, soon deciduous. Male flowers: stamens 5-numerous; rudimentary ovary present or almost or rarely quite absent. Female flowers (or hermaphrodite): staminodia 5-numerous; ovary unilocular with 3–6 parietal placentas bearing 2–3 (-numerous) anatropous ovules; style very short; stigmas 3–5, peltate and ± joined or free, dilated, radiate and bifid, often reflexed apically. Fruit usually ± large, globose or ovoid, indehiscent; wall woody or rarely crustaceous or fragile, mostly crowned with the persistent stigmas. Seeds few to numerous, arillate; endosperm copious.

Sleumer lists 43 species in his revision; they range from India and Ceylon, throughout Burma, Indo-China, Malesia to the Philippines, Celebes also S. China (Kwangsi and Hainan). The genus was formerly of great importance, several species supplying various chaulmoogra oils then the only treatment for leprosy. Hutchinson maintains *Taraktogenos* as a separate genus. Apart from the two native species, four others have been cultivated, the first mentioned being the source of the true chaulmoogra oils: *Hydnocarpus kurzii* (King) Warb. (Botanic Gardens, Peradeniya, *Jayasuriya & Bandara 3315* (PDA)), *H. laurifolia* (Dennst.) Sleumer (= *H. wightiana* Blume) (Botanic Gardens, Heneratgoda, *Simpson 8884* (BM)), *H. heterophylla* Blume (Botanic Gardens, Peradeniya, *Alston s.n.* from trees 103 & 104 (PDA); *Kostermans 24558* (BM, K, L, PDA), *27633* (K, L, PDA); *Silva 32* (PDA), *94* (PDA); *s. coll. s.n.* (PDA)) and *H. anthelmintica* Pierre (Botanic Gardens, Peradeniya, *Alston 968* (K, PDA); *Silva 7* (PDA); Botanic Gardens, Heneratgoda, *Simpson 8862* (BM)). Alston notes for the latter: "small tree with male and female flowers on the same tree; young leaves red; petals pale green rolled round staminodes; anthers cream, versatile; stamens clasping the ovary in female flowers. A report on their establishment as possible plantation crops at Peradeniya and Heneratgoda will be found in The Tropical Agriculturist 71: 204–210. 1928".

KEY TO THE SPECIES

1 Sepals 4 (sect. *Taraktogenos*, cult.)
  2 Pedicels slender, 1.5–3 cm long .... **H. heterophylla**
  2 Pedicels thicker, 0.5–1(–1.5) cm long .... **H. kurzii**
1 Sepals 5 (sect. *Hydnocarpus*)
  3 Stamens 5 with glabrous filaments; indumentum on young parts simple (subsect. *Oliganthera*)
    4 Petals about the same size as sepals or smaller; testa of seed rugose or sulcate-striate
      5 Fruit 2.5 cm in diameter, 2–3(–6)-seeded; scales about half the length of the petals
      .... **1. H. venenata**
      5 Fruit 0.5–1.2 cm in diameter, many-seeded; scales ± equalling petals (cult.) ....
      .... **H. laurifolia**
    4 Petals longer than sepals; testa ± smooth, not sulcate; leaves drying brownish, elongate-oblong, 10–30 × 3–7 cm (cult.) .... **H. anthelmintica**
  3 Stamens 7–9 with densely white-villous filaments; indumentum on young parts stellate (subsect. *Pleianthera*) .... **2. H. octandra**

1. Hydnocarpus venenata Gaertn., Fruct. 1: 288, t. 60/3. 1788; Hook. f. & Thoms. in Hook. f., Fl. Br. ind. 1: 196. 1872; Trimen, Handb. Fl. Ceylon 1: 75. 1893; Lewis, Veg. Prod. Ceylon 10. 1934; Worthington, Ceylon Trees 20. 1959; Sleumer, Bot. Jahrb. Syst. 69: 32. 1938. Type: Ceylon, seeds (L, syntype).

Hydnocarpus inebrians Vahl, Symb. 3: 100. 1794; DC., Prod. 1: 257. 1824; Thw., Enum. Pl. Zeyl. 18. 1858. Type: Ceylon, *Koenig s.n.* (C, holotype).

Tree 3–25(–30) m tall with girth 0.45–2.2 m; crown low and spreading; trunk usually fluted; bark greyish white or green and grey patched, smooth or sometimes ± rough and fissured, corky, peeling in strips about 5 mm wide; live bark pale pink, inner red-brown, smelling of bitter almonds; wood straw-coloured; very young shoots densely ferruginous velvety. Leaves oblong-elliptic to narrowly oblong-lanceolate, 10–19(–22.5) cm long, 3.5–7 cm wide, ± long acuminate at the apex, attenuate to the base, serrate, finely silky pubescent when very young, soon glabrous above and beneath or with minute pubescence on venation beneath; petioles 6–7 cm long. Male flowers in several-flowered axillary cymes; peduncles ± 1 cm long, pubescent; pedicels 5 mm long; sepals rounded, 2–3 mm long, ferruginous pubescent outside, glabrous inside; petals white, slightly shorter than the sepals, white-fimbriate at the margins, the scale broadly ovate, 2 mm long, obtuse at the apex, densely golden-villous; stamens 5; rudimentary ovary absent. Female flowers axillary, solitary or paired; pedicels up to 1.5 cm long; staminodes 5 with narrow sterile anthers; ovary globose, densely covered with golden hairs; stigmas 3, sessile, bilobed. Fruit white or green then brown (black fide Kostermans), globose, 2.5–3 cm in diameter, brownish velvety tomentose, Seeds 2–3(–6).

Distr. Endemic to Ceylon but cultivated at Calcutta, Bogor and elsewhere.

Ecol. Wet evergreen gallery forest along river and stream banks, dry deciduous forest, open forest and scrub on white dune sand; 300–800 m.

Uses. The timber is yellow and hard but of small dimensions; it has been used for building but is not looked on with much favour; fruits have been employed as a fish poison; oil from the seeds has been used for treating leprosy.

Vern. Makulu (S); Makul (T).

Specimens Examined. ANURADHAPURA DISTRICT: Ritigala Hill, *Balakrishnan & Jayasuriya 1075* (PDA); Mihintale, Hill of Temples, *Bernardi 14234* (G, PDA); Ritigala Strict Natural Reserve, Bandapokuna ancient pool, *Jayasuriya & Sumithraarachchi 1605* (K, PDA, US); Wilpattu National Park, NE. of Kuruttu Pandi Villu, *Koyama & Herat 13403* (K, PDA); Kuruttu Pandi Villu, *Mueller-Dombois & Cooray 68091004* (K), *Wirawan et al., 1054* (K, PDA, US). TRINCOMALEE DISTRICT: n. of Trincomalee, w. of Kuchchaveli, *Kostermans 24817* (K, L, PDA). KURUNEGALA DISTRICT: Kurunegala, *Gardner in C.P. 1630*, p.p. (PDA); Kurunegala Rock, *Huber 376* (PDA); Barigoda, *Waas 1150* (K, US). MATALE DISTRICT: Rattota to Illukkumbura road, mi. 26, *Huber 709* (PDA); hill SE of Dambulla, *Huber 761* (PDA); Sigiriya Tank, *Nowicke et al. 362* (K, PDA); Dambulla, on way to Wewala Tank, *Sumithraarachchi 478* (K, PDA, US); Naula-Elahera road, 5/10 mile marker, Kalugaloya stream edge, *Waas 569* (K, US). POLONNARUWA DISTRICT: about 1 mi. NE. of Elahera along the Amban Ganga, *Davidse 7362* (K); 4 mi. from Somawathiya Dagaba, *Sumithraarachchi 359* (PDA); Thunmodera, edge of Ambangaga, *Waas 592* (PDA). BATTICALOA DISTRICT: Batticaloa, Oopah, *s. coll. s.n.* (BM), *Worthington 479* (BM, K). COLOMBO DISTRICT: Kotugoda, *Amaratunga 2266* (PDA); Atulgala, *Amaratunga 966, 1323* (PDA); Botanic Gardens, Heneratgoda, *Simpson 8883* (BM, PDA); Mirigama, Kandalama, *Kostermans 25582* (K, L); Colombo, 28 Mar 1796, *Rottler s.n.* (BM, ex de Joinville, K); fide *Thwaites*. KEGALLE DISTRICT: Bank of Sitawaka Ganga, midway between Ruwanwella and Eheliyagoda, *B & K Bremer 784* (PDA, S, US); Kitulgala, *Kostermans & Wirawan 820* (K, L, PDA); Kitulgala, Kelani R., *Waas & Gunatilake 1189* (K, PDA, US), *1191* (K, PDA, US); 1 mi. SE Alutnuwara, *Worthington 1546* (BM). KANDY DISTRICT: Peradeniya Gardens, *Amaratunga 770* (PDA); Peradeniya, *Gamble 18748* (K); Hunnasgiriya, *Kostermans 25278* (K, L, PDA); above Peradeniya near University Grounds, Hantane, *Meijer 299* (K, PDA); Kandy, *Trimen s.n.* (K); Hantane and Peradeniya, *Gardner in C.P. 1630*, p.p. (PDA); Kadugannawa, Udawela, *Worthington 344* (BM), Allagalla Estate,

Worthington 818 (K), Marandawela, *Worthington 1328* (BM, K); "Attabagie", *Worthington 2848* (K). BADULLA DISTRICT: Pallewela, *Balakrishnan & Jayasuriya 807* (K, PDA, US); near Ella, *Kostermans 23362A, 23364* (BO, K etc.); Bibile to Mahiyangana, culvert 4/1, *Maxwell 1025* (PDA). AMPARAI DISTRICT: Gal Oya National Park, Inginiyagala, *Meijer & Balakrishnan 146* (PDA). RATNAPURA DISTRICT: Ratnapura, *Thwaites C.P. 1630*, p.p. (PDA). MONERAGALA DISTRICT: between Kandanketiya and Dambagalla, *Balakrishnan 395* (K, PDA, US); Ruhuna National Park Block 1, Menik Ganga, *Cooray 69073132* (K, US), Block 3, Veddangewadiya, Menik Ganga, *Wirawan 663* (K, PDA); 7 km N. of Wellawaya, *Hepper & de Silva 4771* (K, PDA); Bibile, Nogala, *Jayasuriya 1939* (K, PDA); Wellawaya, *Kostermans 23473* (K, L); Bibile, *Kostermans 24398* (K, L); mi. 137/4 W. of Wellawaya, *Meijer 201* (PDA); NE. of Bulupitiya (Bibile), *Worthington 4661* (BM, K). HAMBANTOTA DISTRICT: Ruhuna National Park, Block 2, about 1 mi. up stream from Kumana Road, *Cooray 68060604R* (K). LOCALITY UNKNOWN: *Gardner 45* (K); *Moon 588* (BM); *Thomson s.n.* (K); *Thwaites C.P. 1630* (BM, K).

2. Hydnocarpus octandra Thw., Hooker's J. Bot. Kew Gard. Misc. 7: 197. 1855; Thw., Enum. Pl. Zeyl. 19. 1858; Hook. f. & Thoms. in Hook. f., Fl. Br. Ind. 1: 197. 1872; Trimen, Handb. Fl. Ceylon 1: 76. 1893; Sleumer, Bot. Jahrb. Syst. 69: 59. 1938; Worthington, Ceylon Trees 21. 1959. Type: Ceylon, Ambagamuwa, *Thwaites C.P. 2640* (PDA, holotype; BM, K, isotypes).

Tree 8–25 m tall (see note); girth 0.7–1.2 m; bark brown, smooth, superficially cracked or finely fissured; live bark dark orange-brown, pinkish towards inside and smelling of almonds; young stems and foliage stellate-tomentose but soon glabrous. Young leaves thin and drying dark but soon coriaceous and green, ovate-oblong, 7.5–14 cm long, 3.3–6 cm wide, acuminate at the apex, rounded to cuneate at the base, ± unequal or oblique, entire or almost toothed, soon glabrous above, glandular-punctate and with stellate hairs beneath at first; petiole 1.3–2 cm long; stipules lanceolate, minute, deciduous. Inflorescences axillary, either fascicles or 2–8-flowered cymes; pedicels 0.5–1 cm long, very shortly yellowish puberulous. Sepals unequal, oblong, obtuse, concave, densely ferruginous stellate-puberulous outside, the outer ± 6 mm long, the inner a little shorter. Petals white, greenish or yellowish, round, 7–8 mm long, concave, ciliate with whitish hairs; scales round, about half the length of the petals, obtuse, silky. Male flowers: stamens 8, uniseriate, villous; rudimentary ovary minute. Female flowers: staminodes 8, the anthers sterile; ovary ovoid or ovoid-oblong, densely brown-hairy; style obsolete; stigma large, peltate, 4-lobed. Fruit globose, 4–7 cm diameter, velvety brown tomentose with thick woody wall fibrillated within. Seeds 4–12, rarely more, dark, oblong or compressed ellipsoid, 2 cm long, 1.2–1.5 cm wide, immersed in a soft pulp.

Distr. Endemic.

Ecol. Primary and secondary wet evergreen forest; 90–1200 m.

Uses. None recorded.

Vern. Wal dul (divul) (S).

Note. Waas (*2005*) gives the height as 80 m, bole 60 m and girth 1.2 m—perhaps at least the first two should have been feet? No one else records the tree reaching such dimensions.

Specimens Examined. KEGALLE DISTRICT: Pilakanda, *s. coll.* in Herb. Dehra Dun *35755* (DD). KANDY DISTRICT: 2 mi. W. of Double Cut Junction, *Jayasuriya & Sumithraarachchi 1559* (K, PDA, US); Ambagamuwa, *J.M. Silva s.n.* (PDA), *Thwaites C.P. 2640* (BM, K, PDA). KALUTARA DISTRICT: Morapitiya, *Cramer 4178* (K, PDA); Morapitiya Forest, *Kostermans 24976* (PDA). RATNAPURA DISTRICT: Sinharaja Forest, *Kostermans 26716* (PDA), Weddagale entrance, *Kostermans 27884* (K, L, PDA); above Balangoda, Rassagalle, *Kostermans 23579* (K, L); Warukandeniya to Sinharaja, *Waas 2005* (K, PDA, US); Rakwana, Aigburth Estate, *Worthington 2632* (BM); 59 mile Deniyaya, Panilkanda, *Worthington 6065* (BM, K, PDA). GALLE DISTRICT: Kanneliya, 4 mi. from Udugama road, *Cramer 4203* (K, PDA); Kottawa, 10 mi from Galle, *Worthington 2350* (BM); Naunkita Ela Forest Reserve, *Worthington 4139, 693* (BM, K), *3693* (BM). MATARA DISTRICT: S. Sinharaja forest, Enselwatte, *Waas 1480* (K); Akuressa, fide *Worthington.* HAMBANTOTA DISTRICT: Mandagala, fide *Thwaites*; Mandagala, *Trimen s.n.* (PDA). LOCALITY UNKNOWN: *Holtermann s.n.* (B, fide Sleumer).

3. TRICHADENIA

Thw., Hooker's J. Bot. Kew Gard. Misc. 7: 196, t. 5 (not 7 nor 8 as given in text). 1855; Gilg. in Pflanzenfam. ed. 2, 21: 409, fig. 179A–D. 1925; Sleumer in Fl. Males. 5: 39. 1954; Hutch., Gen. Fl. Pl. 2: 220. 1967. Type species: *T. zeylanica* Thw.

Leucocorema Ridley, Trans. Linn. Soc. London ser. 2, 9: 29. 1916. Type species: *L. latifolia* Ridley.

Dioecious trees. Leaves spirally arranged, petiolate, pinnately nerved, subentire to serrate, often lobed when young; stipules foliaceous, deciduous. Flowers in raceme-like axillary panicles. Calyx globose, calyptriform, splitting irregularly or transversely. Petals 5, imbricate, each with a lanceolate or narrowly oblong fleshy hairy scale adnate to inner side at the base. Male flowers: stamens 5, alternate with the petals; anthers with broad connective;

rudimentary ovary small or absent. Female flowers: staminodes absent; ovary sessile, 1-locular with 3 placentas each with 1(–2) ovules; styles 3, short; stigmas broad, reniform, crenate. Fruit globose, 1–3-seeded; pericarp crustaceous when dry. Seed with bony testa, oily fleshy endosperm and plicate-rugose foliaceous cotyledons.

Two species, the other occurring in the Philippines, E. Borneo, New Guinea and Melanesia.

Trichadenia zeylanica Thw., Hooker's J. Bot. Kew Gard. Misc. 7. 197, t. 5. 1855; Thw., Enum. Pl. Zeyl. 19. 1858; Hook. f. & Thoms. in Hook. f., Fl. Br. Ind. 1: 196. 1872; Trimen, Handb. Fl. Ceylon 1: 75. 1893; Alston in Trimen, Handb. Fl. Ceylon 6: 15. 1931; Lewis, Veg. Prod. Ceylon 9. 1934; Worthington, Ceylon Trees 19. 1959. Type: Ceylon, Central Province, *Thwaites* in *C.P. 2505* (PDA, syntypes; BM, K, isosyntypes).

Described as a large tree (only measurements seen are height 6–20 m and dbh 45 cm); bark thick, dull brown, roughish; trunk with large buttresses (fide Kostermans); branches contorted, ferruginous-tomentose, with prominent leaf-scars, the apices with imbricate concave tomentose stipules ± 7 mm long. Young leaves often large and deeply palmately 3–7-lobed, dentate and tomentose; mature leaves oblong, often broadest towards apex, 6.5–32 cm long, 3.5–24 cm wide, tapering to acuminate apex (especially in male, fide Trimen), rounded at base, sinuate (in males) or distantly serrate (particularly in females), or coarsely dentate, very tomentose beneath and on the venation above; venation very prominent beneath; petioles 2.5–17 cm long, tomentose. Inflorescences about 12 cm long; flowers pale green or yellow. Male flower: calyx pubescent on both surfaces; petals thin, elliptic, glabrous, 5 mm long, 3 mm wide; scale narrowly oblong, 4 mm long, 1.5 mm wide, adnate save at apex, densely pilose; filaments 2.5 mm long, linear-lanceolate, flat; anthers 2 mm long; rudimentary ovary lacking (Lewis refers to 5 curved stigmas in the male). Female flower: ovary ovoid, densely pubescent. Fruit globose, 3.8–4.5 cm diameter, 1–3-seeded. Seeds ovoid-globose, 2.5 cm long, 1.7 cm wide.

Distr. Endemic; already becoming scarce in Trimen's day. I have seen no recent wild material.

Ecol. Wet zone forests; 60–900 m.

Uses. Wood dull white, very soft and soon rotting; quite useless. Oil from the seeds used for children's skin complaints.

Vern. Hal-milla (Galle fide W. Ferguson); Keti Kesali, Tettigaha, Tettigas, Titta Eta, Tolol (S).

Specimens Examined. KEGALLE DISTRICT: Kegalle, *Lewis s.n.* (PDA); Dehiowita, *Worthington 2116* (BM). KANDY DISTRICT: Peradeniya

Botanic Gardens, *Kostermans 28155* (K, L); Hunnasgiriya, *Thwaites C.P. 2505*, p.p. (PDA); Madulkelle, Arratenna, *Worthington 2000* (K), Hatara Hora jungle, *Worthington 1983* (BM). RATNAPURA DISTRICT: Balangoda, Gilimale, Hangomuwa-Ganga and Rakwana, fide *Alston*; Gilimale Forest, *Meijer 928* (PDA). GALLE DISTRICT: Sinharaja Forest, Beverley, Deniyaya, *Worthington 2612*, (BM, K); Kanneliya Forest Reserve, *Jayasuriya & Kostermans 2350* (PDA); Nakiyadeniya, *Worthington 5241* (BM, K, PDA); Kottawa, *Ferguson s.n.* (PDA), *Worthington 623* (BM, K).

4. CHLOROCARPA

Alston in Trimen, Handb. Fl. Ceylon 6: 15. 1931; Hutch., Gen. Fl. Pl. 2: 218. 1969. Type species: *C. pentaschista* Alston.

Dioecious tree. Leaves alternate, oblong-lanceolate, obtuse, entire, pinnately nerved; stipules deciduous. Flowers in short axillary 1-sided inflorescences and sometimes also terminal at apices, the short rachis marked with circular scars. Sepals 4–5, imbricate, shortly united at the base. Petals 5 with a linear-lanceolate ciliate binerved scale on inner face. Stamens 5, free; ovary densely pubescent; styles not developed; stigmas 5. Fruit fleshy, ellipsoid, 5-angled and said to be tardily dehiscent into 5 valves (dehisced fruit not seen). Seeds numerous, oily, pilose.

A monotypic genus closely related to *Hydnocarpus* and restricted to Ceylon. Schaeffer (Blumea 20: 65–80. 1972) in a study of the pollen of *Hydnocarpus* states that the pollen of *Chlorocarpa* is very similar. I am very dubious indeed that the fruit ever dehisces; fruits on some sheets have broken under pressure.

Chlorocarpa pentaschista Alston in Trimen, Handb. Fl. Ceylon 6: 15. 1931; Worthington, Ceylon Trees 22. 1959, p.p.* Type: Ceylon, Pallegama, near Lagalla, *J.M. de Silva 147* (PDA, holotype; K, isotype).

Hydnocarpus alpina sensu Trimen, Handb. Fl. Ceylon 1: 76. 1893, non Wight.

Trees 4–15 m tall or sarmentose shrub 3–4 m tall (fide Bernardi); trunk short, 1.8 m tall, distorted, fluted, crown 15 m in large trees (fide Jayasuriya); d.b.h. 10–67 cm; bark smooth; young stems glabrous or with very sparse stellate hairs, densely lenticellate; outer bark pale brown, brittle; wood ochre-coloured, brown, hard. Leaves oblong, elliptic, elliptic-lanceolate or more rarely ovate-lanceolate, 4.5–18.5(–23) cm long, 1.7–8 cm wide, narrowed at the apex to a narrowly rounded tip, cuneate to ± rounded at the

* Worthington completely misunderstood that this was a new species and that *Hydnocarpus alpina* is not a synonym; some of the information he gives may therefore refer to the Indian species.

base, glabrous; venation raised and reticulate on both surfaces; petioles up to
1 cm long, wrinkled. Flowers in very characteristic axillary cymes, the rhachis
made up of very short thick reduced internodes, all the branches held on one
side, particularly in lower inflorescences; probably derived from a branched
system with all axes very reduced; pedicels 0.7–2.4 cm long. Sepals ovate
or elliptic, 6 mm long, 3.5–4.5 mm wide, rounded, tomentose or puberulous on both sides, joined at base for ± 3 mm. Petals white or greenish
white, drying dark, lanceolate, 7–9 mm long, 3.5 mm wide, glabrous; scale
5 mm long, 0.9 mm wide. Stamens white; anthers 2.5 mm long; filaments
very short. Ovary oblong, angled, 2 mm long, densely hairy. Stigma lobes
flat, 3–4 mm long, 0.8–1 mm wide, slightly bifid, reflexed, pubescent. Fruits
glaucous green, 2.7–3 cm long, 1.8–2.2 cm wide, 5-angled, said to be ultimately dehiscent, tomentose; stalk thick, 0.7–2.5 cm long. Seeds brown, ±
ovoid, 1–1.3 cm long, 8–9 mm wide, angled when compressed against others,
± reticulate and slightly tubercled.

Distr. Endemic.

Ecol. Riverine and streamside intermediate forest bordering 'parkland'
in dry zone; 45–350 m.

Uses. None recorded. Worthington mentions timber brown, straight-
grained and smooth but there is a slight doubt what he is referring to.

Vern. Makulla, Patma, Gomma (S); Attuchankular (T).

Specimens Examined. MATALE DISTRICT: Pallegama, Rana-
mure, along Halmini Oya, *Jayasuriya 336* (PDA, US); Pallegama, *de Silva
147* (K, PDA). BATTICALOA DISTRICT: Batticaloa, *s. coll. s.n.* (PDA).
KANDY DISTRICT: Peradeniya Botanic Gardens, *Kostermans 28207* (K, L);
Kadugannawa, Alagalla Estate, *Worthington 818* (K). AMPARAI DISTRICT:
Without exact locality, *Bernardi 15640* (PDA, US); Divulana Nuwaragala
Forest Reserve, *Jayasuriya 2083* (K, PDA), *2090* (PDA). RATNAPURA DIS-
TRICT: Karawita Kande, *s. coll. s.n.* (PDA). MONERAGALA DISTRICT:
Wellawaya road, *Kostermans 25432* (K, L, PDA); between Muppane and In-
digaswela, *de Silva s.n.* (PDA); bank of Kumbukkan oya, near Yala, *de Silva
s.n.* (PDA); Malwatuhela, a few mi. W. of Wellawaya, bank of Bibilehena
Ela, *Wirawan 735* (K, PDA, US); Bibile, Moneragala road, *Worthington 2977*
(BM, K, PDA); NE. of Bulupitya, Bibile, *Worthington 4661* (K); near Bibile,
mile 77/3, *Worthington 5310* (BM, K); Moneragala, 31 mi. post, Arugam
Bay road, *Worthington 5444* (BM, K), *5445* (PDA). GALLE DISTRICT:
trail to Godekanda off Hiniduma, *Jayasuriya & Kostermans 2329* (PDA).
BADULLA OR AMPARAI DISTRICT: Bintenne, *s. coll. s.n.* (PDA). LO-
CALITY UNKNOWN: *s. coll. C.P. 2918* (BM, PDA).

5. SCOLOPIA

Schreb., Gen. 335. 1789; Gilg in Pflanzenfam ed. 2, 21: 418. 1925; Sleumer, Blumea 20: 25–64. 1972 (revision), nom. cons. Type species: *S. schreberi* J.F. Gmel. (= *S. pusilla* (Gaertn.) Willd.).

Shrubs or trees, often with spines on the trunk and/or branches. Leaves alternate, persistent, subcoriaceous or coriaceous, entire or shallowly to rather deeply glandular-serrate-crenate, reddish or purplish when young, penninerved; stipules minute, caducous. Flowers small, hermaphrodite, rarely also male on the same plant, in axillary, mostly simple, sometimes compound false racemes, these rarely reduced to few-flowered fascicles or even to a solitary flower. Sepals 4–6, narrowly imbricate or subvalvate, ± connate at the base. Petals as many as sepals and similar to them. Receptacle flat, sometime set with hairs around the base of the ovary and the base of the filaments. Extrastaminal disk, if present, composed of one row of free, short, thick orange glands. Stamens indefinite in number, pluriseriate, exceeding the petals at full anthesis; anthers dorsifixed, the connective often produced beyond the thecae into an apicular appendage. Ovary sessile, with 2–4(5) few-ovuled placentas; style rather long, sometimes 3–4 partite distally; stigmas entire or slightly 2–3(–5) lobed. Berry subglobose or ovoid, (1–) 2–3(–6)-seeded, somewhat fleshy, with the withered sepals, petals and stamens at the base, crowned by the ± persistent style. Seeds with a hard testa without an aril.

A genus of about 37 species, 3 native in Ceylon, the rest in continental Africa, Madagascar, SE Asia, Malesia and Australia. *S. chinensis* has been treated in full since, although only cultivated, one of its synonyms was actually described from Ceylon. The species are in part poorly defined (see note at end of species 4).

KEY TO THE SPECIES

1 Leaves with 2 distinct orange (turning black when dry) glands at the base of the lamina . . .
. . . . **1. S. chinensis**
1 Leaves without such basal glands
  2 Leaves distinctly acuminate; extra-staminal disk-glands present or absent; fruit green (fide
    Trimen) . . . . **3. S. acuminata**
  2 Leaves not so distinctly acuminate (save in a few specimens)
    3 Pedicels ± slender; disk-glands absent or nearly so; fruits bright scarlet . . . . . **2. S. pusilla**
    3 Pedicels thick; disk-glands usually present as short thick lobes (but not always); fruits larger
      and green (fide Trimen) (not below 600 m and leaves never large) . . . . . . **4. S. crassipes**

1. Scolopia chinensis (Lour.) Clos, Ann. Sci. Nat. Bot. IV. 8: 249. 1857; Thw., Enum. Pl. Zeyl. 400. 1864; Gilg in Pflanzenfam. ed. 2, 21: 418. 1925; Merr., Trans. Amer. Philos. Soc. 24(2): 272. 1935; Sleumer, Blumea 20: 36. 1972.

Phoberos chinensis Lour., Fl. Cochinch. 318. 1790. Type: "China", *Loureiro s.n.* (P, syntype).

Phoberos cochinchinensis Lour., Fl. Cochinch. 318. 1790. Type: "Cochinchina" (BM, ? holotype).

Phoberos arnottianus Thw., Enum. Pl. Zeyl. 16. 1858. Type: Ceylon, Peradeniya Botanic Gardens, *Thwaites C.P. 3526* (PDA, holotype; BM, K, P, isotypes).

Scolopia crenata sensu Hook. f. & Thoms. in Hook. f., Fl. Br. Ind. 1: 191. 1872, p.p., non Clos.

Shrub or tree up to 5 m tall; branches usually with strong simple spines 1–6 cm long; young branchlets shortly spreading puberulous at first but soon glabrescent; bark grey to brownish with minute round reddish lenticels. Leaves elliptic to round or oblong-elliptic, 3–7(–10) cm long, 2–5.5 cm wide, shortly obtusely acuminate at the apex or almost rounded, cuneate to rounded at the base with 2 distinct basal glands, coriaceous, entire to subserrate, glabrous; petiole 3–5(–8) mm long, pubescent. Inflorescences few-flowered, often ± paniculate at tops of branchlets, spreading puberulous; pedicels 4–6(–10) mm long. Flowers (5–) 6(–7)-merous. Sepals ovate, 1.5–2 mm long, tomentose outside. Petals yellowish white, ovate, 2.5–3 mm long, ± glabrous outside but obscurely ciliolate. Receptacle hairy. Disk-glands (5–) 10. Stamens 40–60 with elongate hairy connective. Ovary glabrous; style slender, 2–3 mm long; stigma very shortly 3- or 4-lobed. Fruit dull purplish black, ellipsoid to subglobose, (6–) 8–10 mm in diameter, with 4 or 5 seeds.

Distr. Thailand, Laos, Vietnam and China, also naturalized around Djakarta, Java and cultivated at Calcutta and Peradeniya.

Note. Thwaites knew it was cultivated but not knowing its origin nor being able to identify it described it as new. He later discovered what it was and corrected the error.

Specimens Examined. KANDY DISTRICT: Peradeniya Botanic Gardens, *Thwaites C.P. 3526* (BM, K, P, PDA), June 1890, *s. coll. s.n.* (PDA).

2. Scolopia pusilla (Gaertn.) Willd., Sp. Pl. 2(2): 981. 1799; Moon, Cat. 39. 1824; Clos, Ann Sci. Nat. Bot. 4(8): 251. 1857; Gilg in Pflanzenfam. ed. 2, 21: 420. 1925; Sleumer, Blumea 20: 37. 1972.

Limonia pusilla Gaertn., Fruct. 1: 279, t. 58, fig 4. 1788. Type: seeds from Ceylon (L, syntype)* [Sleumer cites as *Koenig*, L Carp. Coll. 856, TUB ex L not seen].

Scolopia schreberi Gmel., Syst. Nat. ed. 13, 2(1): 793. 1791; Alston in Trimen, Handb. Fl. Ceylon 6: 14. 1931. Type: as for *Scolopia pusilla*;

---

* Gaertner also cites a Plukenet reference but with a query.

Schreber, Gen. Pl. ed. 8: 335. 1798 cites "Limonia Gaertn. 58", nom. illegit.

Phoberos gaertneri Thw., Enum. Pl. Zeyl. 17. 1858. Type: as for *S. pusilla*, nom illegit.

Phoberos gaertneri Thw., var. α *oblongifolius* Thw., Enum. Pl. Zeyl. 17. 1858. Type: Ceylon, Haragama, *s. coll. C.P. 2497* (PDA, holotype; BM, CAL, Fl, K, P, isotypes). (Sleumer gives Balangoda as locality).

Phoberous gaertneri Thw. var. β *cordifolius* Thw., Enum. Pl. Zeyl. 17. 1858. Type: Ceylon, Colombo, etc. *s. coll. C.P. 1076*, p.p. (PDA, syntypes; BM, CAL, FI, GOET, K, P, isosyntypes).

Phoberos gaertneri Thw. var. γ *lanceolatus* Thw., Enum. Pl. Zeyl. 17. 1858. Type: Ceylon, Galagama, *Thwaites C.P. 64, 181, 211* (PDA, syntypes; BM, CAL, FI, K, L, P, W, isosyntypes).

Scolopia gaertneri (Thw.) Thw., Enum. Pl. Zeyl. 400. 1864; Hook. f. & Thoms. in Hook. f., Fl. Br. Ind. 1: 191. 1872; Trimen, Handb. Fl. Ceylon 1: 71. 1893; Lewis, Veg. Prod. Ceylon 9. 1934.

Medium-sized tree or shrub 1–10(–20) m tall, up to 15–20(–45) cm dbh; trunk with formidable compound divaricate spines; bark greyish, smooth or ± rough, thin and hard; live bark light brown or very pale yellow; branches reddish-brown unarmed or with simple slender spines to 4 cm long. Leaves variable in shape, ± ovate to elliptic or oblong-elliptic, 1.2–7.5 cm long, 0.6–5 cm wide, rounded to broadly obtusely acuminate at the apex, attenuate to rounded at the base or sometimes subcordate, entire to slightly crenate, ± coriaceous, glabrous; venation reticulate; petiole 2–4 mm long. Inflorescences slender, lax, 4–8(–10)-flowered, 2–4 cm long, glabrous or spreading puberulous; pedicels slender, 1–1.3 cm long. Flowers scented, 4–5-merous. Sepals ovate-oblong, 1.5–2.5 mm long, ± 1.3 mm wide. Petals white, oblong, 1.5–3 mm long, ± 0.8 mm wide, thinner at margin, ciliate. Receptacle densely hairy; disk glands usually absent. Stamens 40–60; filaments 3–3.5 mm long. Ovary glabrous; style slender, 3–4 mm long; stigma obsoletely 3–4-lobed. Fruit yellow becoming scarlet when ripe, ovoid, 0.8–1.3 cm diameter, apiculate.

Distr. Endemic; Sleumer mentions Bourdillon recorded it from Travancore but no specimens have been seen.

Ecol. Dry and wet forests, submontane and montane forest, hillside scrub etc.; often on quartz; 150–1000 m (?–2140 dubious or sterile material).

Uses. Lewis describes the wood as dull red, close, hard and heavy; suitable for posts, wall plates and rafters and also used for arms to outriggers and walking-sticks.

Vern. Katu Keeree; Katu Kurundu; Katukenda (S).

Specimens Examined. PUTTALAM DISTRICT: Puttalam, *Gardner in C.P. 1076*, p.p. (PDA). MATALE DISTRICT: Laggala, *Balakrishnan & Dassanayake 1147* (PDA), *Trimen s.n.* (PDA); mi. 38 Rattota to Ilukkumbura road, *Huber 747* (PDA); 8 mi. E of Naula, *Huber 784* (PDA); Laggala to Ilukkumbura, *Jayasuriya 285* (K, PDA, US); Ilukkumbura, Karakolagastenna, *Jayasuriya et al. 417* (PDA); Ilukkumbura, *Jayasuriya & Bandaranayake 1777* (PDA); NE. Knuckles Mts., Rattota to Ilukkumbura, Dikpatana, *Kostermans 27953* (K, US), *Tirvengadum et al. 23* (PDA). COLOMBO DISTRICT: Botanic Gardens, Gampaha, *Amaratunga 667* (PDA); Makewita, *Amaratunga 1774* (PDA); Kimbulapitiya, *Simpson 8592* (BM); Colombo, *Macrae 146* (K), 19 Apr 1796, *Rottler s.n.* (K), *Thwaites C.P. 1076*, p.p. (PDA). KANDY DISTRICT: Knuckles, Corbet's Gap, *Ashton 2307* (PDA) (?); 8 mi. NE. of Hunnasgiriya, near mile post 29/10 along road to Mahiyangana, *Davidse & Jayasuriya 8405* (K); 9 mi. NE. on same road at 29/21, *Davidse & Jayasuriya 8423* (K); Hunnasgiriya-Madugoda road, marker 24/19, *Jayasuriya 358* (PDA); Hunnasgiriya, road marker 23/10, *Jayasuriya & Austin 2263* (K, PDA, US) (numerous small disk-glands present); Madugoda-Kaluwella road (Hunnasgiriya), *Kostermans 26726* (PDA); Madugoda-Mahiyangane road, *Kostermans 26769* (PDA), *26777* (PDA); Madugoda, *Simpson 8818* (BM, PDA), *Waas 1115* (K, PDA), *Worthington 1572* (BM); above Moray Estate, Fishing Hut, *Kostermans 27943* (K, PDA). RATNAPURA DISTRICT: Karawita Kanda, *J.M. Silva s.n.* (PDA); Galagama, *Thwaites C.P. 76, 181, 211* (BM, K, PDA); Kurulugala Kanda base, *Waas 406* (PDA); Belihuloya, *Worthington 448* (K); Petiagoda, Balangoda, *Worthington 3215* (BM, K). NUWARA ELIYA DISTRICT: planned Randenigala Reservoir area, E bank of Belihuloya, Pananewela, approach from Padiyapelella Pannala and Serasuntenna, *Jayasuriya & Balasubramaniam 3033* (PDA); W of Hewaheta, Rendapola, past Hope Estate, *Meijer 25* (PDA); Hakgala Peak, trail above Hakgala Botanic Garden, *Meijer 1839* (K). MONERAGALA DISTRICT: Maragala, Uva, *Alston s.n.* (BM); Kataragama Peak, Wedihitikanda, *Bernardi 15506* (G, K, PDA) (± acuminate leaves but no glands); Moneragala to Wellawaya, road marker 156/3, *Waas 77* (PDA); Moneragala road, Bibile, *Worthington 2976* (BM, K, PDA) (acuminate leaves). GALLE DISTRICT: Bona Vista, *Balakrishnan 943* (K, PDA, US), *946* (K, PDA, US), *972* (PDA), *975* (PDA), *Meijer 264* (K, PDA); Galle, Rumassala, *Jayasuriya & Balasubramaniam 3287* (PDA); Point de Galle, *Pierre s.n.* (K, P); Galle, *Wight s.n.* (K). KALUTARA DISTRICT: Pasdun Korale, ? *Trimen s.n.* (PDA); Katukurunda, *Kostermans 25720* (PDA). LOCALITY UNKNOWN: *Capt. Champion s.n.* (K); *Fraser 194* (BM); *Gardner 43* (K); *Graham 34* (K); *Kelaart s.n.* (K); *Mackenzie s.n.* (K); *Thwaites C.P. 1076* p.p. (BM, K, PDA); *C.P. 2497* p.p. (BM, K); *Col. Walker s.n.* (K).

3. Scolopia acuminata Clos, Ann. Sci. Nat. Bot. IV(8): 251. 1857; Thw., Enum. Pl. Zeyl. 400. 1864; Trimen, Handb. Fl. Ceylon 1: 70. 1893. Type: Ceylon, *Walker 34* (G, Herb. Delessert, syntype) & *Gardner 43* bis (Fl, Herb. Webb, syntype; BM, isosyntype); *s. coll. C.P. 1077* (PDA etc.) is presumably partly isosyntype material.

Phoberos acuminatus Gardn. ex Thw., Enum. Pl. Zeyl. 17. 1858. Type: Ceylon, Central Province, Hantane, 300–900 m, *Gardner* in *C.P. 1077* (PDA, syntypes; BM, BR, CAL, FI, K, P, W, isosyntypes).

Phoberos crenatus sensu Hook. f. & Thoms. in Hook. f., Fl. Br. Ind. 1: 191. 1872, p.p., non (Wight) Clos.

Tree 5–30 m tall, d.b.h. 20–40 cm, bole thorny; bark brown, smooth or ± rough; live bark light brown to light red or chocolate-coloured; wood hard, light brown; young branches with simple spines; puberulous. Leaves ovate to elliptic-lanceolate, 5–9(–12) cm long, 2–4 cm wide, distinctly acuminate at the apex, the actual tip ± obtuse, broadly cuneate at the base, subcoriaceous, ± entire to shallowly repand-dentate; venation finely densely reticulate; petiole 3–6 mm long. Inflorescences 6–12-flowered, solitary or panicled near apex, greyish-yellowish puberulous; rhachis slender, 1.5–5 cm long; pedicels 5–7 mm long, rather thickish. Flowers pinkish white or yellow, 5–6 merous, strongly scented. Sepals and petals ovate, 1.5–2 mm long, obtuse, glabrescent or pubescent outside, ciliolate. Receptacle shortly hairy; disk of few to ± numerous orange glands but sometimes ± absent. Stamens 80–100. Ovary glabrous; style slender, 3–4 mm long, the stigma obscurely 3-lobed. Fruit said to be green when mature, globose, 1–1.8 cm diameter, apiculate, the pedicel up to 1.3 cm long, 1 mm thick.

Distr. Endemic (does not occur in S. China as stated by Trimen and others).

Ecol. Wet evergreen forest and secondary forest in moist and dry zones; 150–960 m.

Uses. None recorded.

Vern. Katukenda, Katukurundu (S).

Specimens Examined. ANURADHAPURA DISTRICT: N. end of Kalawewa Bund, *Worthington 4256* (BM, K). MATALE DISTRICT: Dambulla, *Holtermann s.n.* (B), *Trimen s.n.* (PDA); E of Ambanganga, Clodagh, *Worthington 4796* (K). POLONNARUWA DISTRICT: Alut Oya, *Trimen s.n.* (PDA); Trinco Road, 35 m.p., *Worthington 985* (BM), *1042* (BM), *1084* (K). KEGALLE DISTRICT: Dickhena, *Waas 1663* (K, PDA); Koswatte Forest, *Waas 1678* (K, PDA, US). KANDY DISTRICT: Hantane, *Gardner 43* bis (K), *Gardner* in *C.P. 1077*, p.p. (PDA); Deltota, *Gardner* in *C.P. 1077*, p.p. (PDA); between Hasalaka and Madugoda, *Huber 463* (PDA); Maskeliya R. Valley,

Kostermans 24198 (PDA); Mahiyangana-Kandy Road, 31/4, *Tirvengadum & Jayasuriya 159* (PDA) (*crassipes/acuminata* intermediate); Peradeniya Botanic Gardens (cult.), *Trimen s.n.* (PDA); Madugoda, *Worthington 269* (K); Kadugannawa, Andiatenna, *Worthington 323* (K); Dolosbagie, *Worthington 1864* (K). RATNAPURA DISTRICT: Balangoda, *Holtermann s.n.* (B); above Balangoda, Rassagala, *Kostermans 23591* (BM, K, L); southern area of Sinharaja Forest from Deniyaya, Marathagala, *Waas 1745* (PDA); Rakwana, Suriyakanda, 95 m.p., *Worthington 2152* (BM, K). NUWARA ELIYA DISTRICT: Hanguranketa, *Gardner* in *C.P. 1077*, p.p. (PDA). GALLE DISTRICT: Hiniduma, *Kostermans 24712* (BM, K), *Cramer 3716* (PDA); Hinidumkanda, *Jayasuriya et al. 1791* (K, PDA, US), *Kostermans 25475* (K, L, PDA); Kanneliya forest, near Hiniduma, *Kostermans 28635* (PDA). LOCALITY UNKNOWN: *Macrae 334* (BM, K); *Thwaites 1077* (K).

4. Scolopia crassipes Clos, Ann. Sci. Nat. Bot. IV (8): 251. 1857; Thw., Enum. Pl. Zeyl. 400. 1864; Trimen, Handb. Fl. Ceylon 1: 71. 1893; Gilg in Pflanzenfam ed. 2. 21: 420. 1925; Sleumer, Blumea 20: 44. 1972. Type: "Oritur in insula Zeylanica (v.s. in Herb. Webb)" *Wight* (fide Sleumer) (FI, holotype) *Walker 58* in *Herb. Wight* at PDA is probably an isotype.

Phoberos hookerianus Wight ex Thw., Enum. Pl. Zeyl. 17. 1858. Type: Ceylon, Nuwara Eliya, *Gardner*, Deltota, ?*Thwaites* & Maturata, ?*Thwaites* in *C.P. 629*, p.p. (PDA, syntypes; BM, CAL, FI, L, P, W, isosyntypes) {K sheet not found).
Scolopia crenata sensu Hook. f. & Thoms. in Hook. f., Fl. Br. Ind. 1: 191. 1872, p.p. non (Wight) Clos.

Large shrub or tree 6–13.5 m tall with dense crown, d.b.h. ± 30 cm; bark light brown or blackish; living bark light red; young branches and sometimes base of bole with spines. Leaves elliptic, rounded-elliptic or less often elliptic-lanceolate, 3–5 cm long, 1.5–3 cm wide, obtuse to slightly emarginate or sometimes shortly acuminate at the apex, cuneate to ± rounded at the base, entire to slightly crenate or undulate, ± coriaceous, glabrous, shiny above; venation densely reticulate; petiole 2–4 mm long. Inflorescences axillary and terminal, ± few-flowered, grey-pubescent; rhachis rather thick, 1.5–4 cm long; pedicels ± thick, 4–5 mm long. Flowers whitish or yellowish, sweet-scented, 4–6-merous. Sepals ovate, 1.5–3 mm long, 1.5 mm wide. Petals similar, ovate-oblong, ± 2.5 mm long, 1 mm wide. Receptacle ± hairy. Disk-lobes ± 10–25 (10 fide Sleumer), ± thick, rarely absent. Stamens 50–80; filaments up to 6 mm long; connective elongate, glabrous. Ovary flask-shaped, 2–2.5 mm long, glabrous; style 4–5 mm long; stigma obscurely 2–3-lobed. Fruit bitter, said to be green when mature, ovoid-globose, up to 1.8 cm long, 1.3 cm wide, apiculate; pedicels 1.3 cm long, 1–1.3 mm thick.

Distr. Endemic.

Ecol. Montane and submontane wet forest, often on steep slopes, sometimes by streams or rivers; 600–1900 m.

Uses. None recorded.

Vern. None recorded.

Note. Trimen suggested this might be a montane form of *S. acuminata*. The specific delimitations between the native species of this genus seem very unsatisfactory. The main characters-acumination of the leaves, presence of disk-glands (character introduced by Sleumer not used by workers in Ceylon) and thickness of the pedicels seem rather arbitrary and variable. In *S. pusilla* the leaves are allowed to be very variable in shape and sometimes quite large; in *S. acuminata* Sleumer admits the disk-glands can be more or less absent and I have found them lacking in some specimens of *S. crassipes*; pedicel thickness is variable. Fruit colour I have been unable to assess but if it does remain green then that is certainly a good character. Fruit size varies considerably. I have kept Sleumer's arrangement as it is. Possibly *S. acuminata* is a mixture of varieties of *S. pusilla* and *S. crassipes* both with acuminate leaves. There are certainly specimens indistinguishable from *S. crassipes* which appear to have no disk-glands and it is best to take more note of the habit, usually more upland habitats, thicker pedicels and large fruits. The value of the disk-glands character needs a full field evaluation as do habitat differences. Despite the difficulties there is certainly more than one taxon.

Specimens Examined. KANDY DISTRICT: E of Madugoda near marker 31/4, *Jayasuriya & Tirvengadum 1009 & 1010* (K, PDA, US) (intermediate); Hunnasgiriya, road to Mahiyangane, *Kostermans 25147* (K); Deltota, fide *Trimen*; Dolosbage, *s. coll. s.n.* (PDA) (glands dubiously present); Kandy to Mahiyangane, 25/11 mile post, *Waas 212* (K, PDA, US) (doubtful, sterile); Dunally, Galaha, *Worthington 2500* (BM). BADULLA DISTRICT: Yalumala, Namunukula, *J.M. Silva 6* (PDA) (disk-glands absent?), ? *Silva s.n.* (PDA); Ohiya-Boralanda road, *Sohmer et al. 8554* (BM, PDA). RATNAPURA DISTRICT: Galagama, *Thwaites C.P. 629*, p.p. (PDA); Longford Division of Hayes Group, below Gongala, Cardamom Plantation, Field No. 14, *Jayasuriya & Balasubramaniam 3376* (PDA); Hayes-Lauderdale, *Worthington 6068* (K). NUWARA ELIYA DISTRICT: Nuwara Eliya, *Champion s.n.* (K) (a piece separated off looks exactly the same but has no data); *Gardner* in *C.P. 629*, p.p. (PDA); W of Hewaheta above Hope Estate, Upper Division, *Jayasuriya et al. 2857* (PDA); Kabaragala Estate, Cardamom Plantation, Field No. 4, *Jayasuriya & Gunatilleke 3444* (PDA); forest along Moon Plain, *Meijer 1900* (K); 1.5 mi below Hakgala, *A.M. Silva s.n.* (PDA) (disk-glands small); Hakgala ? *Silva s.n.* (PDA) (no disk-glands); Hakgala Forest Reserve

on trail to Hakgala Peak just beyond Hakgala Garden, *Sohmer et al. 8505*
(PDA); Maturata, ? *Thwaites C.P. 629*, p.p. (PDA); Hakgala Botanic Garden,
Waas 835 (PDA); Nuwara Eliya, *Worthington 2908* (BM); Glen Devon, Halgranoya, *Worthington 6210* (BM). LOCALITY UNKNOWN: *Thwaites C.P.
629*, p.p. (BM, CAL, FI, L, P, PDA); *Walker 58* in Herb. Wight (PDA); *Wight
s.n.* (FI).

6. HOMALIUM

Jacq., Enum. Pl. Carib. 5. 1760; Gilg in Pflanzenfam. ed. 2, 21: 425. 1925;
Sleumer, Bull. Jard. Bot. Etat. 43: 239. 1973. Type: *H. racemosum* Jacq.

Trees or shrubs. Leaves alternate, simple, entire or glandular-serrate-
crenate, penninerved, petiolate; stipules absent, or minute to large, sometimes auriculate, caducous or ± persistent. Flowers hermaphrodite, regular, (4–)5–8(–12)-merous, in axillary or subterminal false racemes, spikes
or panicles, solitary or fasciculate on the inflorescence-branches, sessile or
pedicelled, subtended by small caducous or persistent bracts. Calyx-tube (or
receptacle) adnate to the ovary. Sepals flat, usually narrow, persistent, often
accrescent and wing-like. Petals alternating with the sepals, persistent, sometimes accrescent. Stamens opposite the petals, solitary or in fascicles of 2–3;
filaments slender; anthers small, extrorse, dorsifixed. Disk represented by a
usually hairy gland opposite each sepal. Ovary conical and free in its upper
half, sunk in the receptacle in its lower half, 1-locular, with 2–3(–8) placentas,
each with 1-several ovules near the apex; styles 2–4(–7), free or connate into
a distinct column below (then only stigmatic arms free); stigmas punctiform.
Capsule half-inferior, coriaceous or woody, finally 2–3(–8)-valved from the
apex, or remaining indehiscent. Seeds 1-few, small, often not fully formed.

A genus of about 180 species occurring throughout the tropics, with 2
species in Ceylon.

KEY TO THE SPECIES

1 Flowers pedicellate; stamens 1 per petal .... **1. H. ceylanicum**
1 Flowers sessile; stamens 3 per petal .... **2. H. dewitii**

1. Homalium ceylanicum (Gardn.) Benth., J. Linn. Soc. Bot. 4. 35. 1859 (as
'zeylanicum'); Thw., Enum. Pl. Zeyl. 410. 1864; Clarke in Hook. f., Fl. Br.
Ind. 2: 596. 1879 (as *'zeylanicum'*); Trimen, Handb. Fl. Ceylon 2: 239. 1894
(as *'zeylanicum'*); Worthington, Ceylon Trees 286. 1959 (as *'zeylanicum'*).

Blackwellia ceylanica Gardn., Calcutta J. Nat. Hist. 7: 452. 1847; Thw.,
Enum. Pl. Zeyl. 79. 1858. Type: Ceylon, Central Province at 900 m,
Gardner (ubi?).

Blackwellia tetrandra Wight, Ic. Pl. Ind. Or. 5: (17) t. 1851. 1852. Type: India, Pulney Hills and other localities (ubi?).

Tree 4–22 m (–36 fide Worthington) tall, 10–40(–80) cm d.b.h.; bark grey, brown or blackish (mainly shades of pale grey, pink and yellow fide Worthington), smooth, flaking (Trimen states white, rather rough, breaking into irregular pieces); buttresses large and spreading; twigs lenticellate, glabrous. Young foliage red. Leaves elliptic to oblong-elliptic, 6–11.5 cm long, 3.3–6.5 cm wide, acuminate at the apex, the actual tip obtuse, cuneate at the base, crenate-dentate, rather thick, glabrous and shining; petiole 0.6–1.5 cm long. Inflorescences dark red (fide Kostermans) but others speak of white 'catkins', drooping, spike-like, interrupted, axillary or aggregated at the ends of the branchlets, 8–23 cm long with flowers fragrant, numerous in dense fascicles; pedicels 1–3 mm long; bracts minute. Calyx-tube red, 0.7–1.5 mm long, ± glabrous; sepals red, 4–6, oblong or spathulate, 0.8–1.1 mm long, 0.2–0.5 mm wide, obtuse, pubescent and densely ciliate. Petals pale green (but also described as yellow, violet or dirty white), 4–6, ± spathulate or obovate-oblong, 1–2 mm long, 0.5–1 mm wide, obtuse, pubescent and ciliate. Stamens 4–6, one opposite each petal; filaments filiform, 1.8 mm long, glabrous; anthers 0.2 mm long. Ovary with free part 0.7 mm long, densely hairy. Styles 3–6, filiform, spreading, often bent. Fruits not seen (see note).

Distr. Ceylon and S. India (Kerala).

Ecol. Remnants of tropical rain forest and secondary forest, sometimes on stony slopes; 150–1140 m.

Uses. Wood pale brown, the heart-wood darker, heavy, very hard, close-grained and durable; used for boats and building.

Vern. Eta-Heraliya, Liyang, Liyan (S).

Note. Trimen commented on not being able to find any fruiting specimens and there still seem to be none available which is extraordinary and indicates a field study to investigate fruit production.

Specimens Examined. MATALE DISTRICT: Matale, fide *Worthington*. KEGALLE DISTRICT: 12 mi. W. of Nawalapitiya, Kellie Estate Jungle, near Dolosbage, *Worthington 1916* (BM, K). KANDY DISTRICT: 2 mi. W. of Double Cut Junction, *Jayasuriya & Sumithraarachchi 1556* (K, PDA, USA); Gampola, *Thwaites C.P. 168, 388* (BM, K, PDA); Hantane, *Worthington 249* (BM); Udawela, Andiatenne, Kadugannawa, *Worthington 349* (K); W. of Peradeniya, Liyandeniya wela, Kadugannawa, *Worthington 651* (K); Kandy, Restharrow (planted), *Worthington 7072* (K, PDA). BADULLA DISTRICT: Tonacombe, Namunukula, *Worthington 6657* (K). RATNAPURA DISTRICT: Longford Division of Hayes Group, below Gongala, Cardamom

Plantation, Field No. 14, *Jayasuriya & Balasubramaniam 3389* (PDA); Sinharaja Forest, near Weddagala, *Kostermans 27158* (K, L, PDA), Weddagala Entrance, *Kostermans 27312* (K, L, PDA), *27471* (K, L, PDA); Pelmadulla, Kiribatgale, *Kostermans s.n.* (PDA); Maratenna Forest, *Waas 1732* (K, PDA, US); Balangoda Estate Jungle, *Worthington 774* (K); Rakwana, 95 m.p., Suriyakanda, *Worthington 2153* (BM, K); Balangoda Estate, *Childerstone in Worthington 6909* (K). RATNAPURA/MATARA DISTRICTS: between Bulutota Pass and Deniyaya, *Kostermans 28443* (K, L). NUWARA ELIYA DISTRICT: Ramboda, *Waas 1144* (K, PDA). GALLE DISTRICT: Kanneliya Forest, near Hiniduma, *Kostermans 24774* (PDA). MATARA DISTRICT: Near Deniyaya, *Kostermans 28166* (K, L, PDA); Deniyaya, Panilkanda, *Worthington 6066* (BM); Bengamuwa, Ratmale Forest, *Comanor 1204* (K, PDA). LOCALITY UNKNOWN: *Beckett 742* (BM); *Macrae 235* (BM); *Mrs. Col. Walker s.n.* (K).

2. Homalium dewitii Kostermans, Misc. Pap. Landbouwh. Wageningen 19: 220, fig. 5. 1980. Type: Ceylon, Amparai District, Nuwaragale Forest Reserve, base of Friars Hood, *Jayasuriya 2099* (G, holotype; K, PDA, US, isotypes).

Tree about 10 m tall, 20 cm d.b.h.; bark pale, the live part ochraceous, hard and gritty; branches grey, slender, glossy, smooth and glabrous. Leaves elliptic to oblong-elliptic, 5–9 cm long, 3 cm wide, acute to shortly acuminate at the apex, broadly cuneate at the base, serrate, rather thin, glabrous and glossy; venation reticulate; petiole slender, ± 1 cm long. Flowers numerous, sessile and densely packed in several spikes 5–11 cm long arranged in lax terminal inflorescences; each flower supported by 2 pairs of minute bracts. Calyx-tube obconic to obovoid, 1.5–2.2 mm long, ± ridged, densely pubescent, lobes 5–6, narrowly triangular, 1.3–1.5 mm long, acute. Petals greenish, 5–6, ± thick, ± spathulate, 2.5 mm long, 0.7 mm wide, densely grey-white hairy outside, the hairs forming a marginal fringe almost as wide as the petal, glabrous inside. Stamens 3 opposite each petal, the filaments ± 2 mm long, hairy. Ovary almost completely adnate to the calyx-tube, densely hairy; styles 5, linear, 2 mm long, joined at base. Fruits unknown.

Distr. Endemic; so far known only from one gathering.

Ecol. Intermediate zone forest; low altitude.

Uses. None recorded.

Vern. None recorded.

Note. As Kostermans points out this is very different from *Homalium ceylanicum* with which it had been identified; in its sessile flowers and bracts it approaches *Osmelia*. Kostermans describes the filaments as glabrous but his figure correctly shows them to be hairy.

Specimens Examined. AMPARAI DISTRICT: Nuwaragala Forest Reserve, at base of Friar's Hood, *Jayasuriya 2099* (G, K, PDA, US).

7. FLACOURTIA

Commers. ex L'Herit., Stirp. Nov. 3: 59, t. 30, 30/B. 1786; Gilg in Pflanzenfam. ed. 2, 21: 438, fig. 201. 1925. Type: *F. ramontchi* L'Herit. (= *F. indica* (Burm. f.) Merr.).

Usually dioecious shrubs or trees, sometimes with spiny branches and/or trunk. Leaves alternate, entire or serrate-crenate, penninerved, petiolate, exstipulate. Flowers unisexual, or rarely bisexual, small, in short axillary pseudoracemose cymes, sometimes reduced to a solitary flower. Sepals (3–) 4–5(–7), slightly connate at the base, imbricate in bud. Petals 0. Male flowers: stamens 15 to numerous, inserted on a receptacle which bears an extrastaminal annular disk usually broken into ± free glands; filaments filiform; anthers dorsifixed. Rudimentary ovary absent. Female flowers: sepals as in the male flowers. Receptacle with an entire, crenulate or lobed disk. Ovary sessile, incompletely (2–) 4–6(–8)-locular by false septa; placentas 4–8, pluri-ovulate, with 2 ovules per locule one above the other; styles as many as the locules, free or ± connate, persistent; stigmas small, inflated or shortly 2-lobed. Fruit a fleshy drupe with 4–16 seeds usually in pairs one above the other. Seeds obovoid-ellipsoid, somewhat flattened; testa crustaceous.

A palaeotropical genus with about 10 species, 2 in Africa. One species is native in Ceylon and another commonly cultivated and apparently sometimes self-sown. *F. jangomas* (Lour.) Rausch. (*F. cataphracta* Roxb. ex Willd.) has also been grown (Ratnapura District, Pussella, *s. coll. s.n.* (PDA)).

KEY TO THE SPECIES

1 Flowers hermaphrodite with stamens ± persistent at base of fruit; unarmed; leaves usually
  large (cultivated) .... **1. F. inermis**
1 Flowers unisexual; usually spiny
  2 Styles separate; leaves not usually very acuminate, often pubescent (wild) .... **2. F. indica**
  2 Styles entirely connate into a distinct column even in fruit; leaves distinctly long-acuminate,
    glabrous, thin and somewhat shiny (cultivated) .... **F. jangomas**

1. Flacourtia inermis Roxb., Pl. Corom. 3: 16, t. 222. 1811; Roxb., Hort. Beng. 73. 1814; Roxb., Fl. Ind. 3: 833. 1832; Trimen, Handb. Fl. Ceylon 1: 73. 1893; Heyne, Nutt. Pl. Indon. 1140. 1927; Sleumer in Fl. Males. 1, 5: 71. 1954 (extensive references); Allen, Malayan Fruits 57, fig. 21. 1967; Morton, Fruits of Warm Climates 317. 1987. Type: based on material grown in Calcutta Botanic Gardens from seeds obtained in the Moluccas by A. Berry, Roxburgh Drawing 568 (K, syntype) & India, *Roxburgh s.n.* (BM, syntypes; K, syntype ex Herb. Forsyth).

Unarmed tree 3–15 m tall described by Roxburgh as having a short trunk and being much branched above; bark brownish or grey, smooth; branchlets pubescent when young. Leaves reddish when young, elliptic, ovate-elliptic or ovate-oblong, 8–25 cm long, 4–12.5 cm wide, acuminate at the apex, cuneate to rounded at the base, coarsely crenate, glossy above, glabrous save for pubescent midrib; petiole 0.8–1.2 cm long, pubescent. Inflorescences 1–1.5 cm long, pubescent, ± 5-flowered; pedicels 4–10 mm long; bracts small, deciduous. Sepals (3)4–5, green, broadly ovate, 2–2.5 mm long, 1.5 mm wide, pubescent or ± glabrous. Stamens 15–25; disk fleshy, 6–8-lobed. Ovary ovoid; styles 4–5; stigma discoid or cuneate, bilobed. Fruit reddish orange or pink to cherry red, globose, 2–2.5 cm diameter, shallowly 5-lobed, usually with style remnants persistent. Seeds (4–) 8–10(–14), compressed ovoid, about 6 mm wide.

Distr. Widely cultivated throughout Malesia etc.; Sleumer describes two varieties, var. *rindjanica* (Sloot.) Sleumer and var. *moluccana* Sleumer, as wild in limited areas of W. Malesia; var. *inermis* he treats as only known in cultivation.

Ecol. Cultivated in Ceylon, probably more widely than the few specimens seen indicate. Worthington also states (*5014*) 'selfsown' so it is possibly naturalized in places; up to 600 m.

Uses. Widely cultivated for its fruit which is used in preserves, chutneys, pies, tarts, etc. but is usually too acid to eat raw. Sweet forms do exist.

Vern. Lobi-lobi, lovi-lovi (various Indonesian languages).

Specimens Examined. COLOMBO DISTRICT: planted commonly in village gardens in Colombo area, fide Worthington; Labugama waterworks bungalow, *Worthington 3495* (BM, K). KANDY DISTRICT: Kandy, *Worthington 4179* (BM, K), *Worthington 5014* (K) (naturalized); Peradeniya Botanic Gardens, *s. coll. s.n.* (PDA). KALUTARA DISTRICT: Matugama, *Balakrishnan 1012* (PDA).

2. Flacourtia indica (Burm. f.) Merr., Interp. Rumph. Herb. Amboin. 377. 1917; Gilg in Pflanzenfam. ed. 2, 21: 440, fig. 201. 1925; Alston in Trimen, Handb. Fl. Ceylon 6: 15. 1931; Sleumer in Fl. Males. 1, 5: 76, fig. 30/h, i. 1954; Sleumer in Fl. Trop. E. Africa, Flacourtiaceae 57, fig. 20. 1975.

Gmelina indica Burm. f., Fl. Ind. 132, t. 39, fig. 5. 1768. Type: Java, *s. coll,* Herb. Burman (G, holotype).

Flacourtia ramontchi L'Herit., Stirp. Nov. 59, t. 30 & 30/B. 1786; Hook. f. & Thoms. in Hook. f., Fl. Br. Ind. 1: 193. 1872; Trimen, Handb. Fl. Ceylon 1: 73. 1893; Lewis, Veg. Prod. Ceylon 9. 1934; Wealth of India 4: 43. 1956; Worthington, Ceylon Trees 18. 1959. Type: Madagascar, *Poivere, Commerson* (P, syntypes).

Flacourtia sepiaria Roxb., Pl. Corom. 1: 48, t. 68. 1796; Roxb., Fl. Ind. 3: 835. 1832; Thw., Enum. Pl. Zeyl. 17. 1858; Hook. f. & Thoms. in Hook. f., Fl. Br. Ind. 1: 194. 1872; Trimen, Handb. Fl. Ceylon 1: 73. 1893. Type: India, Roxburgh drawing 126 (K, lectotype).
Flacourtia sapida Roxb., Pl. Corom. 1: 49, t. 49. 1796; Roxb., Fl. Ind. 3: 835. 1832; Thw., Enum. Pl. Zeyl. 17. 1858. Type: Coromandel and Bengal, Roxburgh drawing 127 (K, lectotype).

Shrub or tree, usually dioecious, generally spiny, 2–10(–15) m tall; bark grey, speckled, rough; spines of the trunk sometimes branched, up to 12 cm long. Vegetative parts varying from glabrous to densely pubescent. Leaves red or pink when young, variable in shape and size, ovate or elliptic, sometimes suborbicular or obovate, apex obtusely acuminate, obtuse or rounded, base cuneate to rounded, membranous to almost coriaceous, serrulate-crenate, or more rarely subentire, 2.5–12(–16) cm long, 1.2–8 cm broad; lateral nerves 4–7 pairs, slightly prominent on both faces, as is the ± dense reticulation; petiole 0.3–2 cm long. Flowers unisexual, or occasionally bisexual (1 or several branches of a female specimen with perfect flowers, which, however, bear fewer stamens than in the males). Male flowers in axillary racemes 0.5–2 cm long; pedicels slender, ± pubescent, up to 1 cm long, the basal bracts minute and caducous. Sepals (4–) 5–6(–7), broadly ovate, apex acute to rounded, pubescent on both sides, 1.5–2.5 mm long and broad. Filaments 2–2.5 mm long; anthers 0.5 mm long. Disk lobulate. Female flowers in short racemes or solitary; pedicels up to 5 mm. Disk lobulate, clasping the base of the ovoid ovary; styles 4–8, central, connate at the base, spreading, up to 1.5 mm long; stigmas truncate. Fruit globular, reddish to reddish black or purple when ripe, fleshy, up to 2.5 cm across, with persistent styles, up to 10-seeded. Seeds 5–8, 8–10 mm long, 4–7 mm broad; testa rugose, pale brown.

Distr. Asia and Malesia, tropical and subtropical Africa, Madagascar, Mascarenes, Seychelles; also cultivated elsewhere for its fruit and as a hedge.

Ecol. Scrub, jungle, steep grassy roadsides, submontane forest; 260–810(–1200 cult.) m.

Uses. Cultivated for edible but usually acid fruits and for ornamental young foliage but can become an undesirable weed. The wood of the *ramontchi* form is brown or dull red, very close and hard with fine pores arranged in radial lines. A tough and durable timber for posts but liable to split.

Vern. Katukutundu, Uguressa, Ukkuressa, (S); Mulanninchil, Kutukali (T); Katai (Hindi).

Note. I have followed Sleumer in taking a wide view of *Flacourtia indica* but many Indian workers consider *Flacourtia ramontchi* to be a distinct species which is a larger tree in habit which does not bear flowers on the spines. In Africa several species were once separated including one which is a large rain forest tree; Sleumer has united all these under *indica*. The problem involves races adapted to rain forest and savanna conditions and would probably best be solved by using varietal names. Hooker f. and Thomson deal with some of the variation in Fl. Br. India (l. c.). *Flacourtia sepiaria* Roxb. (Mulanninchil, T) is a name given to the form bearing flowers and fruits on the spines. It must be admitted that 'Herbarium botanists' are woefully ignorant of differences in the field. The names are there for those who wish to keep what they think are different plants separate.

Specimens Examined. JAFFNA DISTRICT: Jaffna, *Gardner* in *C.P. 1650*, p.p. (PDA); Pallawarayankaddu, ?*Trimen s.n.* (PDA). ANURADHAPURA DISTRICT: Mihintale Hill, *Balakrishnan & Jayasuriya 1118* (K, US). PUTTALAM DISTRICT: Puttalam, *Kundu & Balakrishnan 384* (PDA); Wilpattu National Park, Malimaduwa, *Wirawan et al. 1105* (K, US, PDA). TRINCOMALEE DISTRICT: Foul Point, *Jayasuriya et al. 666* (PDA). MATALE DISTRICT: Rattota to Illukkumbura road mi 28, *Huber 755* (PDA); Leloya, *Jayasuriya 332* (K, PDA, US); Dikpatana, *Jayasuriya & Faden 2409* (K, PDA). COLOMBO DISTRICT: 7 mi. from Colombo, *Ferguson s.n.* (PDA); Colombo, *Gardner* in *C.P. 2583*, p.p. (PDA). KANDY DISTRICT: Peradeniya, *Gardner 42* (K), in *C.P. 2583*, p.p. (PDA), in 1832, *s. coll.* in *C.P. 2583*, p.p. (PDA); Hunnasgiriya-Mahiyangana road, *Kostermans 25153* (PDA); Aniewatte road towards river, *Meijer 694* (K, PDA); village near Peradeniya, *Meijer 1993* (K); Kandy, *Moon 294* (BM); Kandy-Mahiyangana road mi. 28, *Sohmer 8269* (PDA); Peradeniya, Botanic Garden, *Thomson s.n.* (K); Kandy Jungle, *Worthington 700* (BM, K); Andiatenna, Kadugannawa, *Worthington 1323* (BM); Madugoda, *Worthington 1570* (BM, K); Kandy, Royston, *Worthington 1944* (K); Madulkele, *Worthington 1948* (BM, K); Kandy, Hillcrest, *Worthington 6337* (K); Peradeniya, Botanic Garden, Old Gravel Pit, *s. coll. s.n.* (PDA); wild part of Garden, *s. coll. s.n.* (PDA). BADULLA DISTRICT: Erabedde, Forest nursery, *Worthington 2925* (K); Bandarawela, *Worthington 583* (BM); Keppitipola, *Meijer 1756* (PDA). MONERAGALA DISTRICT: Kataragama, *Cooray 69090413R* (PDA); Bibile to Nilgala, *Trimen s.n.* (PDA); Bibile, fide *Worthington*. HAMBANTOTA DISTRICT: Ruhuna National Park, Block 1 at Rakinawala, plot R33, *Cooray 68102218R* (PDA). MATARA DISTRICT: Akuressa, *Worthington 2524* (K). WITHOUT LOCALITY: *Kelaart* (K); *Thwaites C.P. 46* (K); *C.P. 1650* (K, PDA); *C.P. 2583* (BM, K); *Col. Walker* (K).

8. DOVYALIS

Arn. in Hooker's J. Bot. Kew Gard. Misc. 3: 251. 1841; Gilg in Pflanzenfam. ed. 2, 21: 440. 1925; Sleumer, Bot. Jahrb. Syst. 92: 64. 1972. Type species:

D. zizyphoides Mey. & Arn., nom. illeg. (= *Flacourtia rhamnoides* Burchell ex DC.).

Aberia Hochst., Flora 27, Bes. Beil. 2. 1844; Oliver in Fl. Trop. Africa 1: 121. 1868. Type species: *A. verrucosa* Hochst. (= *Dovyalis verrucosa* (Hochst.) Warb.).

Shrubs or trees, unarmed, or often with simple or compound spines on trunk and branches, and simple axillary spines on branchlets, dioecious or the male plant bearing occasionally a few bisexual flowers and fruits. Leaves alternate, sometimes fascicled on much reduced lateral shoots, generally persistent, rarely deciduous, entire, denticulate or crenate, exstipulate, petiolate, with 1 or 2 basal or suprabasal and a few upper spreading pairs of nerves, with fine pellucid points visible against strong light in a number of species. Flowers greenish to yellowish, rarely white, solitary or in short racemes, these generally reduced to rather few-flowered fascicles; pedicels with small basal bracts. Calyx-lobes (3–) 4–6(–7, very rarely–9), free almost to the base, subimbricate to practically valvate, variously pubescent on both faces, sometimes with sessile or stipitate marginal glands, sometimes accrescent in later stages. Male flowers: stamens 10–50(–80), inserted on a rather fleshy receptacle, with several irregularly arranged small disk-glands between; filaments filiform; anthers dorsifixed. Rudiment of ovary absent. Female flowers: calyx-lobes as many as (rarely–12) and similar to those of the male, often slightly larger, persistent and ± recurved at fruiting time. Staminodes rarely, stamens very rarely present. Ovary surrounded at the base by a cupular or annular disk (lobed according to the number of the calyx-lobes), unilocular or incompletely 2–4(–8, rarely more)-locular; placentas 2–4(–8, rarely more), each with 1 or 2(–6) ovules; styles 2–8(–20, rarely –40), divergent, channelled; stigmas ± lobed and papillose. Fruit a fleshy berry. Seeds few, or occasionally up to 12, ellipsoid, rather flattish, embedded in a pulp; testa coriaceous, glabrous, hairy, or sometimes woolly.

About 15 species, all African save for the one from Ceylon described below. Apart from this the Kei apple, *Dovyalis caffra* (Hook. f. & Harv.) Hook. f., a native of southern Africa, has been grown for its fruits and as a hedge plant, its very spiny branches making it impenetrable. Only one specimen has actually been seen (Hakgala Botanic Gardens, ?*Trimen s.n.* (PDA)).

KEY TO THE SPECIES

1 Ovary minutely or very shortly pubescent; fruit ± glabrous (cult.) .... **D. caffra**
1 Ovary velutinous: fruit velvety .... **D. hebecarpa**

Dovyalis hebecarpa (Gardn.) Warb. in Pflanzenfam. III, 6a: 44. 1893; Gilg in Pflanzenfam, ed. 2, 21: 441. 1925; Alston in Trimen, Handb. Fl. Ceylon 6: 15. 1931; Burkhill, Dict. Econ. Prod. Malay Pen. 857. 1935.

Roumea hebecarpa Gardn., Calcutta J. Nat. Hist. 7: 449. 1857; Thw., Enum.
Pl. Zeyl. 18. 1858. Type: "Jungles of the Central Province, as at Cundasalle", *Gardner* in *Thwaites C.P. 1075*, p.p. (PDA, holotype), *Gardner
44* (BM, isotype; K, 4 isotypes).
Aberia gardneri Clos, Ann. Sci. Nat. Bot. IV, 8: 236. 1857; Thw., Enum.
Pl. Zeyl. 400. 1864; Hook. f. & Thoms. in Hook. f., Fl. Br. Ind. 1: 195.
1872; Trimen, Handb. Fl. Ceylon 1: 74, t. 7. 1893, nom. illegit. (based
on *Roumea hebecarpa*).
Aberia hebecarpa (Gardn.) Kuntze, Rev. Gen. Pl. 1: 43. 1891.

Tree or shrub 2.5–6 m tall; stem unarmed. Branchlets brownish straw
or purplish brown, wrinkled, pale rusty pubescent at tips, older parts
soon glabrous, sometimes with slender axillary spines up to 3 cm long.
Leaves ovate-oblong to oblong-lanceolate, 2.5–12 cm long, 1.3–5 cm wide,
shortly acuminate at the apex, the tip subacute, cuneate to rounded at base,
thinly chartaceous, sparsely pubescent and finally glabrescent above, short-
tomentose beneath especially along midrib and nerves, entire or minutely
glandular-dentate or -serrate; lateral nerves ± 6 pairs, rather steeply curved-
ascending, slightly raised beneath, reticulation obscure; petiole 0.5–1.2 cm
long, hairy. Male flowers 8–10(–20) in almost sessile umbellate clusters
often in leafless axils, pale rusty or yellowish-tomentellous; pedicels slender,
up to 8 mm long; tepals 5(–7), ovate-acuminate, 3–4 mm long, 1.5–2 mm
wide; stamens 50–60; disk glands small, hairy. Female flowers solitary or
2–3 in a fascicle at leafy or leafless axils, yellowish-tomentellous; pedicels
stoutish, 3–4 mm; tepals as in the male flowers, though slightly larger; disk
consisting of c. 10 subglobular basally connate tomentose glands; ovary
densely spreading yellow-brown hairy; styles 5–7, hairy at base. Fruit purple
to bronze-maroon, globular, 2.5–4 cm in diam., velvety; seeds c. 10, flattened
ellipsoid, 7 mm long, 5 mm wide, densely hairy.

Distr. Endemic; quite widely cultivated in the tropics, including Ceylon.

Ecol. Roadside scrub, 450–780 m.

Uses. The acid fruits are used in curries, preserves, etc.

Vern. Ketambilla (S); Ceylon gooseberry (E).

Note. The *Gardner* type specimen and Thwaites's Hanguranketa specimens are mounted on one sheet as *C.P. 1075* but it is unclear which is which;
Kew sheets are labelled *Gardner 44* and I have treated them as isotypes but
all could equally be treated as syntypes. Gardner presumably sent material
direct to Hooker. There are no *C.P. 1075* sheets at Kew.

Specimens Examined. MATALE DISTRICT: N. of Matale,
Gammaduwa, *s. coll. s.n.* (PDA). KANDY DISTRICT: Kandy, Kundasale
["Cundasalle"], *Gardner 44* (K), *Gardner* in *Thwaites C.P. 1075*, p.p.

(PDA); Peradeniya, *Macrae 455* (K); Kandy, Graylands, *Worthington 2030* (K); King's Pavillion, *Worthington 4263* (K). BADULLA DISTRICT: near Palugama, below Hakgala, *Fosberg & Sachet 53208* (K, US); between Nuwara Eliya and Palugama, *Balakrishnan 1059* (K, PDA); below Hakgala, Keppitipola Village, *Meijer 1758* (K). NUWARA ELIYA DISTRICT: Maturata, fide *Thwaites*; Hanguranketa, *Thwaites* in *C.P. 1075*, p.p. (PDA). LOCALITY UNKNOWN: *Macrae 214* (BM); *Thwaites 1075*, p.p. (BM); *Col. Walker s.n., 203* (K).

9. OSMELIA

Thw., Enum. Pl. Zeyl. 20-Nov. 1858; Sleumer in Fl. Males. ser. 1, 5: 77. 1954; Hutch., Gen. Fl. Pl. 2: 228. 1967. Type: *O. gardneri* Thw.
Stachycrater Turcz., Bull. Soc. Imp. Naturlistes Moscou 31(2): 464. 1858 (see note). Type: *S. philippinus* Turcz.

Dioecious trees or flowers hermaphrodite in Ceylon species (fide Thwaites). Leaves spirally arranged, ovate to oblong-lanceolate, petiolate, penninerved and often plainly 3-nerved from base, mostly entire; stipules small or larger and leafy, deciduous or somewhat persistent. Flowers small, ± sessile, numerous in long terminal axillary simple or panicled ± interrupted drooping spike-like inflorescences; each flower supported by a cupuliform involucel and also by 1–2 minute bracteoles. Calyx-tube very short with imbricate lobes. Corolla absent. Male flowers: stamens 8–10, ± exserted, 4–5 alternating with the sepals, the other 4–5 opposite the sepals and inserted between the lobes of the staminodes: anthers small; staminodes opposite the sepals and joined to their bases, flattened, deeply bilobed, densely hairy; ovary rudiment villous with 3 short styles. Female flowers similar but stamens shorter, not exserted and without pollen. Hermaphrodite flowers as male with ovary developed. Ovary ovoid or oblong, sessile, unilocular, densely hairy; styles 3, shortly curved; stigmas capitellate or ± reniform and distinctly bilobed; placentas 3 with 2–3 or few ovules. Capsule subglobose or oblong-trigonous, 3-valved, very shortly stalked, 1–4-seeded, tomentose, the pericarp subcoriaceous. Seeds ellipsoid to subglobose with red or yellow fleshy aril.

Four species, one in Ceylon, the others from Malay Peninsula to New Guinea (Hutchinson's mention of 12 species is presumably a slip).

Osmelia gardneri Thw., Enum. Pl. Zeyl. 20. 1858; Benth., J. Linn. Soc. Bot. 5 (suppl.): 89. 1861; Trimen, Handb. Fl. Ceylon 2: 238. 1894. Type: Ceylon, Hantane, *Gardner 193* and *Thwaites C.P. 1246* (PDA, syntypes; BM, K, isosyntypes).

Stachycrater zeylanicus Turcz., Bull. Soc. Imp. Naturlistes Moscou 36(1): 609. 1863. Type: Ceylon, *Gardner 193* (?KW, holotype; K, PDA, isotypes).

Osmelia zeylanica Clarke in Hook. f., Fl. Br. Ind. 2: 595. 1879 (sphalm. pro *gardneri*). Type: as for *O. gardneri* (see note).

Osmelia paniculata Warb. in Pflanzenfam. 3, 6a: 49, fig. 18J. 1893; Gilg in Pflanzenfam. ed. 2, 21: 450, fig. 206J. 1925; Hutch., Gen. Fl. Pl. 2: 229. 1967, nom. illeg. (based on *Casearia paniculata* Gardner MS published only in synonymy by Thwaites).

Tree 6–15 m tall, d.b.h. 10–40 cm, fluted at the base, the buttresses rounded, symmetrical; bark grey, smooth, scaly but also has been described as brown and rough; inner bark pale ochre. Leaves elliptic or oblong-elliptic, 4–14.5 cm long, 1.3–5.5 cm wide, narrowed into a distinct acumen up to 1.5 cm long, cuneate at the base, entire, glabrous; venation reticulate and raised; petiole 0.5–1.5 cm long; stipules minute, deciduous. Flowers purplish or greenish white tinged pink, numerous in drooping spike-like inflorescences 10.5–20 cm long, one from each ± 6 upper axils and also one apparently terminal as well, sometimes pubescent; pedicel and cupuliform involucel together about 1 mm long, also minute triangular aristate bracteoles 0.5 mm long. Sepals convex, oblong, 1.7–2 mm long, 0.8–1.2 mm wide, with very few hairs inside. Staminal filaments 2 mm long; staminodes ± 1 mm long. Styles thickened and joined at base, rather flattened above, about 0.8 mm long, the stigmas subcapitate. Fruit pink or purplish, subglobose or ellipsoid, 0.8–1.2 cm long, 1–4-seeded, subtomentose and glandular. Seeds subspherical, ± 5 mm diameter, enclosed in a bright red aril.

Distr. Endemic.

Ecol. Primary wet evergreen forest; 300–1200 m.

Uses. None mentioned but Thwaites describes it as a very beautiful tree so is presumably worth cultivating.

Vern. None recorded.

Note. Clarke's use of the epithet *zeylanica* is a slip; he attributes it to Thwaites and does not mention Turczaninow's *Stachycrater zeylanicus*. It seems fairly certain that *Stachycrater* is an older name by a month or so than *Osmelia*. Stafleu and Cowan TL 2 state "censor Vol. 7 Sept. 1858" which is presumably the date the volume passed the Russian censors. It surely was actually published before November but Bentham (tom. cit. 88) states that *Osmelia* definitely has right of priority and he would have had access to Russian literature as published. Neither the copy at Kew nor at the British Museum (Nat. Hist.) offers any clues. Until further definite information is available I have retained *Osmelia*.

Sleumer says of the genus "dioecious trees" but Thwaites definitely states the Ceylon species has hermaphrodite flowers and Warburg figures one as such; flowers I have dissected undoubtedly have pollen and ovules in the same flower.

Specimens Examined. KEGALLE DISTRICT: Alagalla fide *Trimen*; Koswatte Forest, *Waas 1682* (K, PDA). KANDY DISTRICT: Hantane, *Gardner 193* (K, PDA). RATNAPURA DISTRICT: Sinharaja forest, Weddagale Entrance, *Kostermans 28733* (K, L); Adams Peak Sanctuary, *Meijer 505* (K, PDA); Potupitiya, fide *Trimen*. LOCALITY UNKNOWN: *Thwaites C.P. 1246* (K).

10. CASEARIA

Jacq., Enum. Pl. Carib. 4, 21. 1760; Gilg in Pflanzenf. ed. 2, 21: 451. 1925; Hutch., Gen. Fl. Pl. 2: 22. 1967. Type species: *C. nitida* (L.) Jacq.

Shrubs or mostly trees, sometimes up to 40 m tall. Leaves alternate, distichous, mostly coriaceous, entire, wavy or subserrate-crenulate especially when young, usually marked with pellucid lines or dots at least in the young state, penninerved; lateral veins arcuate, or rarely nearly straight from the midrib; stipules small, caducous. Flowers hermaphrodite, small, solitary or usually few to numerous, clustered in axillary or slightly supra-axillary fascicles or glomerules, each from a cushion formed by a few to numerous small scale-like bracts; pedicels articulated at the base. Receptacle cupular or funnel-shaped, ± perigynous, ± deeply 5-lobed; calyx-lobes slightly imbricate, persistent. Petals 0. Stamens 8–10(–12), equal or alternate ones with longer filaments, alternating with as many flattened or clavate, usually densely hairy appendages (or staminodes) and connate with them below into a ± perigynous tube; filaments filiform; anthers small, often apiculate by the slightly protruding connective. Ovary ovoid, attenuate to a short style; stigma capitate to almost disk-like. Capsule coriaceous, hard or succulent in its outer part, yellow to orange-red, ovoid-globose to ellipsoid, slightly 3–4(–6)-angled when fresh, often 4–6-ribbed when dry, splitting from above into (2–) 3–4 valves. Seeds few to numerous, partly enveloped in a soft membranous usually fimbriate aril which becomes reddish when exposed to the air.

A pantropical genus of about 160 species, but only 3 in Ceylon.

KEY TO THE SPECIES

1 Leaves (beneath) and young shoots, etc. ± pubescent .... 3. C. tomentosa
1 Leaves and young shoots practically glabrous
  2 Leaves lanceolate or elliptic .... 1. C. zeylanica
  2 Leaves obovate; usually more coriaceous .... 2. C. thwaitesii

1. Casearia zeylanica (Gaertn.) Thw., Enum. Pl. Zeyl. 19. 1858; Alston in Trimen, Handb. Fl. Ceylon 6: 131. 1931.

Vareca zeylanica Gaertn., Fruct. 1: 290, t. 60, fig. 6. 1788. Type: Seeds from Leiden Botanic Garden (L or TUB?, holotype).

Casearia esculenta Roxb., Fl. Ind. 2: 422. 1832; Clarke in Hook. f., Fl. Br. Ind. 2: 592. 1879; Trimen, Handb. Fl. Ceylon 2: 237. 1894; Lewis, Veg. Prod. Ceylon 208. 1934; Worthington, Ceylon Trees 285. 1959; Anon., Drawings Indian Pl., fasc. 7, t. 24. 1976. Type: India, Circar Mts., Roxburgh drawing 148 (CAL, lectotype).

Casearia championi Thw., Enum. Pl. Zeyl. 19. 1858. Type: Ceylon, Hewaheta, *Thwaites C.P. 2608* (PDA, holotype; BM, K, isotypes).

Casearia varians Thw., Enum. Pl. Zeyl. 19. 1858. Type: Ceylon, Ratnapura District, Palabaddala; Kandy District, Hunnasgiriya and Palagalla, *Thwaites C.P. 2604, 2657* (PDA, syntypes; K, isosyntypes).

Casearia varians var. α *ovata* Thw., Enum. Pl. Zeyl. 19. 1858. Type as for last (PDA, syntypes; K, isosyntypes).

Casearia varians var. β *minor* Thw., Enum. Pl. Zeyl. 19. 1858. Type: Ceylon, Central Province, without locality and Kandy District, Raxawa, *Thwaites C.P. 3365* (PDA, syntypes).

Casearia esculenta Roxb. var. *angusta* Clarke in Hook. f., Fl. Br. Ind. 2: 592. 1879. Type: Ceylon, *Thwaites C.P. 2603* (K, holotype; PDA, isotype).

Shrub or tree 2–10(–30) m tall, d.b.h. 10–70 cm; bole slender, fluted, sparsely branched, the branches slender and sagging, glabrous or with slight adpressed pubescence, somewhat coffee-like; bark whitish to grey, brown or yellowish-grey, smooth to roughish, faintly cracked, soft, sometimes with small raised dark patches; under-bark light brown, hard. Leaves broadly elliptic to oblong-elliptic or ovate, 4–15.5 cm long, 2.7–7.3 cm wide, shortly acuminate, obtuse or subacute at apex, cuneate at the base, entire or slightly serrate, subcoriaceous, glabrous; petiole ± 1.3 cm long; stipules leafy, persistent. Inflorescences numerous on raised axillary bosses or at leafless nodes, few- to many-flowered; buds yellowish, pubescent, pedicels 0.5 mm long, articulate at base and with many very small bracts. Calyx-lobes green, ± round, 1.5 mm diam. Stamens 6–8; filaments ± 1 mm long. Staminodes oblong, about as long as filaments, ciliate or hairy. Fruit yellow or orange-yellow, ovoid, 1.2 cm (–2 fide Trimen) long, ± 1 cm wide, distinctly 3-ribbed, glabrous or pubescent, finally dehiscing by (2)3 thick valves; fruiting pedicels (3–) 7–8 mm long. Seeds several, almost entirely covered by a large fleshy lacerate dark scarlet aril.

Distr. Ceylon, S. India and Malay Peninsula.

Ecol. Open scrub, low windswept forest, dry zone, secondary forest, mangrove swamp margins, secondary submontane forest, coconut plantations, sand dunes and sand spits with jungle, gallery forest along sandy river banks; 0–1500 m.

Uses. Wood yellow or honey—or biscuit-coloured, fine-grained and with fine rays, moderately hard and heavy, durable; medicinal uses; fruit sweet and eaten. Lewis states rarely more than 12 ft and not big enough for timber.

Vern. Walwaraka (S); Kakkapalai, Kakapelar, Kakkaipalai, Tey pala (T) (Gaertner gives the name walwareka presumably the origin of his name *Vareca*).

Specimens Examined. VAVUNIYA DISTRICT: Mullaitivu, *Alston 1001* (PDA); Chandakulam Tank, 18 mi. NE. of Puliyankulam, *Fosberg & Balakrishnan 53486* (PDA, US); Periyapantrichurichchan, *Jayasuriya et al. 978* (PDA). PUTTALAM DISTRICT: Wilpattu National Park, entrance of road to Malikaran Villu from Kali Villu, *Cooray & Balakrishnan 69050119R* (K, PDA); Puttalam, *Gardner* in *C.P. 415*, p.p. (PDA); 3 mi. W. of Wilpattu National Park Office, *Meijer 352* (PDA); W. of Wilpattu National Park, between Periya and Periya Naga Villus, *Mueller Dombois et al. 69043008* (K), N. of Maraivillu on Puttalam-Mannar road, close to plot W 39, *Wirawan et al. 890* (K, PDA), Kokkare Villu, *Wirawan 1066* (K, US), Kuda Patessa, *Wirawan et al. 1147* (K, PDA). ANURADHAPURA DISTRICT: Wilpattu National Park, entrance to Timbiri Wila from Mahapatassa, *Mueller-Dombois 69042615* (K, PDA, US); Wilpattu National Park, without exact locality, *Waas & Tirvengadum 782* (K, US). TRINCOMALEE DISTRICT: near Trincomalee, Periakulam, *Bernardi 15290* (PDA) (slight indumentum); Palaiottu, beside Blue Lagoon, *Cramer 4375* (K); Trincomalee to Kuchchaveli road, *Kostermans 24807* (BM, K, L, PDA); road to Trincomalee, sea coast, *Kostermans 24808* (PDA); Villa to Kuchchaveli road, *Nowicke & Jayasuriya 291* (K, PDA, US), *292* (PDA); Foul Point, *Simpson 9677* (BM); Trincomalee, Deadman's Cove, *Worthington 482* (K); Muttur, *Worthington 1651* (BM, K). KURUNEGALA DISTRICT: Belligama Estate, *Worthington 493* (K); Belligama Bungalow, *Worthington 955* (BM, K). MATALE DISTRICT: 9 mi. SE. of Kandalama, Dambulla, *Sumithraarachchi 470, 478* (PDA); Galewela, *Worthington 4249* (BM, K); Etabendiwewa, Dambulla, *Worthington 4811* (K); Dambulla, *Gardner* in *C.P. 415* (BM, K, PDA). POLONNARUWA DISTRICT: 1 mi NE. of Elahera along the Amban Ganga, *Davidse 7366* (K, PDA); Polonnaruwa Sacred Area, *Dittus 71090502* (PDA), along N-S trail, P270, *Hladik 921* (PDA), *1057* (PDA), Sect. 4, *Ripley 75* (PDA), *208* (PDA) (shoot from stump, traces of indumentum on young leaves); Polonnaruwa Sanctuary, *Huber 430* (PDA); Polonnaruwa Jungle, Thunmodera, banks of Mahaweli Ganga, *Waas 528* (PDA). BATTICALOA DISTRICT: Batticaloa-Trincomalee road, mile-marker 30/1, *Balakrishnan 364* (K, PDA); Vantharumoolai, *Cramer & Jayaratnam 5900, 5978* (K); 4 mi NW. of Batticaloa near mile-post 4/6 on the Trincomalee road, *Davidse & Sumithraarachchi 8989* (K, PDA); Kalkudah, *Somanader* in *Worthington 7095* (K); *Somanader & Charavanapavan* in *Worthington 7097* (K, PDA); Keelikudah, *Waas 2127* (K, PDA, US); Mankeni, *Worthington 6704* (K). KEGALLE DISTRICT: Kegalle, collector (illegible) in *Worthington 7098*

(K, PDA). KANDY DISTRICT: Hantane, *Champion s.n.* (K); Peradeniya Botanic Gardens, *Charavanapavan* in *Worthington 7087* (K), *Kostermans 24584* (K, L, PDA); Rangala to Corbet's Gap, *Kostermans 23486* (K, L); Knuckles, Madulkelle, *Kostermans 25035* (K, L); Madulkelle area, above Lebanon Estate, *Kostermans 27188* (PDA), Perera Cardamom Estate, trail from Lebanon Estate, *Kostermans 27561* (K, L); Knuckles Mts., near summit of Gambiya Ridge, N. of Madulkelle, *Kostermans 28528* (K, L, PDA); Knuckles, *Kostermans s.n.* (K, L); Hunnasgiriya, *Thwaites 2604*, p.p. (PDA); Dungolla off Madugoda, *Waas 1105* (K, US); Kellie Estate, Dolosbage, *Worthington 1882* (BM, K). BADULLA DISTRICT: road to Dunhinda Falls, *Balakrishnan & Jayasuriya 875* (K, PDA); Haputale, *? Trimen s.n.* (PDA). AMPARAI DISTRICT: Mt. Wadinagala, *Bernardi 15626* (G, PDA); Gallodai near Padiyatalawa, *Huber 473* (PDA); Lahugula Tank, *Muller-Dombois & Comanor 67072519* (PDA). RATNAPURA DISTRICT: Longford Division of Hayes Group, below Gongala Cardamom Plantation Field 14, *Jayasuriya & Balasubramaniam 3221* (PDA); Palabaddala, *Thwaites* in *C.P. 2604*, p.p. (PDA); Ratnapura, *Thwaites C.P. 2657* (K); Gongala, *Waas 1379* (PDA). MONERAGALA DISTRICT: N. of Kataragama, 1 mi. S. of Vaddangewadiya at Menik Ganga, *Mueller-Dombois 68102312* (PDA). GALLE DISTRICT: Mt. Godekanda, E. of Hiniduma, *Bernardi 15427* (K, PDA) (? abnormal leaves to 22 × 8 cm); Galle Jetty, *Capt. Champion s.n.* (K); Galle, *Gardner s.n.* (K, PDA); Kanneliya Forest Reserve, *Jayasuriya 3394* (PDA); Kanneliya Forest, near Hiniduma, *Kostermans 28632* (K, L); Bonavista, *Kundu & Balakrishnan 503* (PDA). LOCALITY UNKNOWN: *Champion s.n.* (K); *Dyke s.n.* (K); *Macrae 413* (BM); *Thwaites C.P. 2603, 2604* bis, *2608* (BM, K); "*Mrs. Col. Walker, Col. Walker 245, Mrs. Gen. Walker 245*" (K).

2. **Casearia thwaitesii** Briq., Annuaire Conserv. Jard. Bot. Gèneve 2: 62. 1898; Alston in Trimen, Handb. Fl. Ceylon 6: 131. 1931. Type: as for *C. coriacea*.

Casearia coriacea Thw., Enum. Pl. Zeyl. 20. 1858; Clarke in Hook. f., Fl. Br. Ind. 2: 592. 1879; Trimen, Handb. Fl. Ceylon 2: 237. 1894; Gamble, Fl. Pres. Madras (3): 521. 1919; Fyson, Fl. South Ind. Hill Stations 1: 238. 1932; Worthington, Ceylon Trees 284. 1959. Type: Ceylon, Adam's Peak, *Thwaites C.P. 465* (PDA, holotype; BM, isotype, K, not found), non Vent.

Casearia varians Thw. var. *obovata* Thw., Enum. Pl. Zeyl. 20. 1858. Type: Ceylon, Nuwara Eliya, 2100–2400 m, *Thwaites C.P. 1217, 1247* (PDA, syntypes; BM, K, isosyntypes).

Tree 4–12 m or shrub 2–3 m tall with dense canopy; d.b.h. up to 60 cm; bark dark, smooth or very rough (fide Worthington); living bark yellowish;

young branches glabrous. Leaves ± congested, red-purple when young, obovate or some shortly elliptic, 1.5–7.7 cm long, 1.3–5.2 cm wide, rounded to emarginate at the apex, cuneate at the base, entire, mostly distinctly coriaceous, the margins often recurved, glabrous; petiole 4–8 mm long; stipules rounded ovate, 1.5 mm long. Flowers in few-flowered axillary fascicles, pinkish white; pedicels 2–4 mm long. Calyx lobes rounded or broadly ovate, 1.5–2.8 mm long. Stamens 6–8, filaments 0.7 mm long; anthers 0.5 mm long; staminodes 8, squarish, 0.8 mm long, the upper margin ciliate. Ovary ovoid, ± 2 mm tall, glabrous; stigma discoid, subsessile, 0.8 mm wide. Fruit orange or yellow, ovoid, 1.5–2 cm long, ± 1 cm wide, apiculate, glabrous, 2–3–valved.

Distr. Ceylon and also India (Tamilnadu)—see note.

Ecol. Primary and secondary montane forest edges, cloud-forest, jungle, scrub and thicket edges; 1500–2300 m.

Uses. None recorded.

Vern. None recorded.

Note. Gamble has annotated material from the Pulney, Nilgiri and Annamally Hills as this species and there are drawings of his floral dissections on the sheets. Although the leaves attain a larger size and are more elliptic I think he is correct to consider it conspecific with the Ceylon material.

Specimens Examined. KANDY DISTRICT: Slopes of Adam's Peak, along SE trail from Moray Group tea estates, *Davidse & Sumithraarachchi 8648* (PDA). BADULLA DISTRICT: Namunukala, *Willis s.n.* (PDA). RATNAPURA DISTRICT: Carney to Adam's Peak, *Bernardi 15783* (GE, K); Adam's Peak, top cone on N side, *B, & K. Bremer 990* (PDA, S, US); Ascent to Adam's Peak on the southern slope above Sri Palabaddala, *Jayasuriya & Gunatilleke 3138* (PDA); Adam's Peak, *Thwaites C.P. 465* (PDA). NUWARA ELIYA DISTRICT: Horton Plains, World's End Trail, *Balakrishnan 1045* (K, PDA, US), *1048* (K, PDA, US), *1205* (K, US), *Cramer 4343* (K), road from Ohiya to Farr Inn, *Huber 662* (PDA), Farr Inn to Big World's End, jungle patch, *Sohmer & Sumithraarachchi 10047* (K), midway Pattipola to Horton Plains, *Tirvengadum & Cramer 113* (K), road to World's End, *Tirvengadum & Cramer 275* (K); slopes of Mt. Pidurutalagala, directly N of Nuwara Eliya, *Davidse & Sumithraarachchi 8073* (K); Nuwara Eliya, *Gardner C.P. 1217* (PDA), *Gardner 192* (BM, K, PDA); Hakgala jungle, *Kostermans 24190* (PDA); near Ambawela, Elk Plains, *Kostermans s.n.* (K, L); road from Horton Plains to Ohiya, *Nowicke & Jayasuriya 272* (K, US); Hakgala, *A.M. Silva s.n.* (PDA); Sita Eliya Forest Reserve on N side of Nuwara Eliya to Hakgala road near 32 mi marker, *Sohmer et al. 8459* (PDA); Nuwara Eliya, Bambarakelle, *Worthington 6774* (K, PDA), *s. coll.*

s.n. (K). LOCALITY UNKNOWN: *Thwaites C.P. 1217* (BM, PDA), *s. coll. C.P. 1247* (K); *Mrs. Col. Walker 218* (K); *s. coll. s.n.* Herb. Wight (PDA).

3. Casearia tomentosa Roxb., Fl. Ind. 2: 421. 1832; Clarke in Hook. f., Fl. Br. Ind. 2: 593. 1879; Anon., Drawings Indian Pl., Fasc. 7, t. 25. 1976. Type: India, Circar Mts., Roxburgh drawing 147 (K, syntype), *Roxburgh s.n.* (BM, syntype).

Anavinga lanceolata Lam., Enc. 1: 146. 1783. Type: 'India orientali', *Sonnerat s.n.* (P-LAM, holotype), non *Casearia lanceolata* Miq.
Casearia elliptica Willd., Sp. Pl. 2: 628. 1799; Matthew & Brito, Fl. Tamilnadu Carnatic 2. Illustr. t. 40. 1982, 3. Fl. (2): 58. 1983. Type: as for *Anavinga lanceolata*, nom. illegit.

Tree (2–)12–16 m tall, 20–35 cm d.b.h.; bark brownish grey, rough, irregularly cracked, scaling in rounded patches; living bark beefy red outside, dirty white or red brown inside; branches graceful, the young shoots usually sparsely to densely pubescent. Leaves narrowly elliptic to narrowly oblong-elliptic, 3–12.5(–17.5) cm long, 1.7–2.7(–5) cm wide, acute, obtuse or acuminate at the apex, cuneate to rounded at the base, entire or slightly toothed, glabrous to finely velvety above, densely velvety to finely pubescent or sometimes almost glabrous beneath; petiole 0.8–1.5 cm long. Flowers whitish or pale green in dense or sometimes few-flowered axillary clusters; pedicels jointed at the base, ± 4 mm long. Sepals ovate or lanceolate, 2.2–4 mm long, pubescent, persistent. Stamens 6–8, connate at base to staminodes; filaments 2 mm long; staminodes 6, clavate, 1.5 mm long, densely pubescent at apex. Ovary ovoid-globose, 1.5–2 mm long, glabrous save at apex; style 1.5 mm long; stigma cupular or disk-like. Fruit green, ± globose or ellipsoid, up to 2.5 × 2 cm, shining, glabrous, ribbed, 3–4-valved.

Distr. Widespread in India; former indications of its occurrence in Malesia and Australia actually refer to other species.

Subsp. **reducta** Verdc., subsp. nov.* Type: Ceylon, Jaffna, *Gardner* in *C.P. 1248*, p.p. (K, holotype; PDA, isotype).
Casearia tomentosa sensu Thw., Enum. Pl. Zeyl. 19. 1858; Trimen, Handb. Fl. Ceylon 2: 238. 1894; Lewis, Veg. Prod. Ceylon 208 adnot. 1934, non Roxb. sensu stricto.
Casearia elliptica sensu Alston in Trimen, Handb. Fl. Ceylon 6: 131. 1931, non Willd. sensu stricto.

---

* **Casearia tomentosa** Roxb. subsp. **reducta** Verdc., a subsp. *tomentosa* ramulis juvenilibus pagina inferiore folii inflorescentiisque minus pubescentibus, foliis basi magis cuneatis, cymis paucifloris differt. Typus: Ceylon, Jaffna, *Gardner* in *C.P. 1248* p.p. (K, holotype; PDA, isotype).

Young stems, lower surfaces of leaves and inflorescences much less thickly hairy; leaves more cuneate at the base, inflorescences fewer-flowered, than subsp. *tomentosa*.

Distr. N and S Ceylon.

Ecol. Banks of watercourses including riparian woodland, etc., flood plain undergrowth, evergreen forest on alluvial plains; 3–150 m.

Uses. Lewis mentions the hard wood is used for combs.

Vern. Kiri Makulu (?).

Notes. Willdenow based his name on the earlier name of Lamarck so it is illegitimate. There was no bar to using the epithet *lanceolata* in *Casearia* at that date. There is no doubt Ceylon material referred to this species is not the same as the Indian material with woolly indumentum and many-flowered inflorescences. Although I have treated it as a variant of *C. tomentosa* it seems to me to differ little from *C. zeylanica* which is not always entirely glabrous. No one has previously queried the identification but there is very little material. A number of explanations might be possible; a once widespread population of *C. tomentosa* may have been virtually eliminated by introgression with *C. zeylanica*. The more distinctly hairy northern populations appear to be extinct; the southern ones are virtually indistinguishable from *C. zeylanica*. The subspecific name used is a device for dealing with a situation not yet understood.

Specimens Examined. JAFFNA DISTRICT: Jaffna, *Dyke s.n.* (K); *Gardner C.P.1248* (K, PDA). MATALE DISTRICT: Sigiriya, *Gardner in C.P. 1248*, p.p. (PDA). RATNAPURA DISTRICT: Uggalkaltota on Walawe Oya, *Worthington 3317* (K). HAMBANTOTA DISTRICT: Ruhuna National Park, *Bernardi 15551* (G, K, PDA), along the Menik Ganga NW of Yala, *Davidse 7789* (PDA), Block 1, Rugamtota on Menik Ganga, (Plot 31), *Fosberg & Mueller Dombois 50160* (PDA), Menik Ganga, 1 mi. above Yala Bungalow, *Fosberg et al. 51054* (K, PDA, US), along Menik Ganga, *Meijer 220* (K), Block 1 near Yala Bungalow, *Wirawan 779* (K).

HIPPOCRATEACEAE
(by B.M. Wadhwa*)

Juss., Ann. Mus. Natl. Hist. Nat. 18: 486. 1811 ("Hippocraticeae"). Type genus: *Hippocratea* L.

Lianes, stragglers or erect shrubs. Leaves usually opposite, simple, coriaceous, mostly entire, pinnately veined; stipules small or absent. Inflorescence axillary, flowers in fascicles or dichotomous or panicled cymes; bracts triangular-ovate; bracteoles sometimes in clusters. Flowers bisexual, regular, usually 5-merous. Calyx lobes minute, imbricate, connate below, persistent. Petals free, inserted under the disk, spreading or erect, imbricate. Disk usually prominent, fleshy, thick, cupular, apex entire to crenulate. Stamens usually 3, inserted on or within the top of the disk; filaments flat, dilated below, recurved during anthesis; anthers usually dehiscing transversely. Ovary superior, more or less embedded in the disk, 3-celled; ovules 2 or more per cell on an axile placenta; style short, stout, mostly 3-fid; stigma simple. Fruit a berry or a capsule with 3 divergent flattened follicles. Seeds usually many, often winged, sometimes not; albumen 0; embryo straight.

About 8 genera, distributed to tropical areas of New World, Africa, Madagascar, Malaysia, India, Burma, S. China, Australia and Sri Lanka; 3 genera and 8 species in Sri Lanka.

KEY TO THE GENERA

1 Flowers in axillary fascicles; fruits drupaceous or baccate, with a terminal, persistent style or its scar or with a depressed cavity at the upper end, indehiscent; seeds not winged . . . .
. . . . 3. **Salacia**
1 Flowers in axillary, dichotomous or panicled cymes, with or without supplementary branchlets in the dichotomy; fruits capsular, dehiscent; seeds winged
  2 Leaves sub-entire or crenate; inflorescence without short supplementary branchlets in the dichotomy; flowers laxly arranged; disk conspicuous . . . . 1. **Loeseneriella**
  2 Leaves entire or serrate; inflorescence with short supplementary branchlets in the dichotomy or in the axil of the branches; flowers densely arranged; disk inconspicuous . . . .
. . . . 2. **Reissantia**

* Royal Botanic Gardens, Kew.

1. LOESENERIELLA

A.C. Smith, Amer. J. Bot. 28: 438. 1941; Halle, Mem. Inst. Franc. Afrique Noire n. 64: 103. 1962. Type species: *L. macrantha* (Korth.) A.C. Smith (= *Hippocratea macrantha* Korth.).

Lianes or stragglers, rarely erect shrubs. Leaves decussate, simple, coriaceous, glabrous, glossy above; stipules interpetiolar. Inflorescence axillary, panicled cymes or dichotomously cymose, sometimes umbellate clusters. Flowers bisexual, 5-merous. Calyx deeply 5-lobed, imbricate, persistent. Petals usually thick, subcoriaceous to coriaceous when dry, margins slightly overlapping, acuminate, entire, puberulous without, sometimes glabrescent. Disk prominent, fleshy, simple, annular-pulvinate, rarely double and then inner part forming a kind of receptacular androgynophore, glabrous or distinctly strigose. Stamens 3, inserted at the base of the free part of the pistil; filaments linear, reflexed at anthesis; anthers extrorse, dehiscing transversely. Ovary superior or half immersed, 3-celled; style distinct; stigma obscure; ovules 2–12 in each cell on axile placenta. Fruit capsular, consisting of 3 separate, divergent, dorsiventrally flattened follicles, each dehiscing along inconspicuous median suture into 2 valves. Seeds usually with a basal wing, the wing usually membranous; endosperm 0; cotyledons completely free, sometimes partially united.

About 26 species distributed in tropical Africa, S. China, S.E. Asia, Indo-Malaya, New Hebrides and Australia; 3 species in Sri Lanka.

KEY TO THE SPECIES

1 Petals on upper 1/2 to 2/3 of inner surface and top of disk distinctly pilose; sutural margins of pericarp slightly reflexed after dehiscence .... **1. L. macrantha**
1 Petals on inner surface and the entire disk completely glabrous; sutural margins of pericarp spreading after dehiscence
  2 Petals orbicular with a narrow, short claw, imbricate; leaves elliptic, upto 15 cm long; follicles oblanceolate, upto 8 cm long .... **2. L. arnottiana**
  2 Petals triangular-lanceolate, without claw, valvate; leaves elliptic-oblong or ovate-lanceolate, up to 10 cm long; follicles obovoid-oblong, not exceeding 5 cm in length ....
  .... **3. L. africana**

1. Loeseneriella macrantha (Korth.) A.C. Smith, Amer. J. Bot. 28: 439. 1941; Ding Hou in Fl. Males. ser.1. 6(3): 398. 1964.

Hippocratea macrantha Korth., Kruidk. Verh. Nat. Gescheid. Neder. Ind. 187. t. 39. 1842; Miq., Fl. Ind. Bat. 1(2): 599. 1859; Alston in Trimen, Handb. Fl. Ceylon 6: 49. 1931, p.p.; Loes. in Pflanzenfam. ed. 2, 20b: 213. 1942; Amshoff, Blumea 5: 517. 1945; Loes., Bot. Jahrb. Syst. 63: 274. 1970. Type: from Borneo, *Korthals s.n.* (K).

238 HIPPOCRATEACEAE

Lianes, up to 20 m; young branches divaricate, glabrous. Leaves elliptic to elliptic-lanceolate, sometimes broadly elliptic, 5.5–9.5 (–12) × 3.2–5.5 (–6.5) cm, chartaceous to thinly coriaceous, base obtuse-subcuneate, apex acute to shortly acuminate, margin sub-entire to slightly crenulate, nerves 6–7 pairs; petioles 5–7 mm long, glabrous. Inflorescence dichotomously cymose, up to 8 cm long, 4–5 times branched, sometimes flowers on young, axillary short shoots. Flowers bisexual, yellowish-green or green; bracts sparsely brown pubescent; pedicels 5–7 mm long, the central one usually longer, up to 9 mm. Calyx lobes deltoid, c.1 mm long, brown puberulous without. Petals ovate-lanceolate. 4.5–5.5 × 1.5–2.0 mm, laxly brown pilose along the margin and upper 1/2–2/3 within. Disk annular-pulvinate, 1.5–2.5 mm broad, with the basal part slightly extended outwards, pilose at the top. Stamens 3, c. 2 mm long; filaments linear, adnate to inner surface of the disk; anthers extrorse, dehiscing transversely. Ovary 3-celled, half-immersed, c. 1.5 mm emerging from the disk; ovules 4–6 in each cell on axile placenta. Follicles elliptic-oblong, 5–7 × 2–3 cm, obtuse at apex, dorsiventrally compressed. Seeds obovate-oblong with membranous wings, 3.5–4.5 × 0.7–0.9 cm (incl. wings).

Distr. New Ireland, Solomon Islands, New Hebrides, Malaysia and Sri Lanka.

Ecol. In lowland forests, rare.

Vern. Diyakirindiwel (S).

Note. The bark surface is yellowish, bright or orange-yellow beneath and inner bark chocolate brown and wood whitish.

Specimens Examined. RATNAPURA DISTRICT: Sinharaja Forest-Waturawa main trail, 26 Oct 1987, *Jayasuriya et al.* 4125 (PDA), Tundole Ele, 25 Nov 1992, *Wadhwa & Weerasooriya 338* (PDA). LOCALITY UNKNOWN: *Col. Walker s.n.* (K); *s.coll. s.n.* (PDA).

2. Loeseneriella arnottiana (Wight) A.C. Smith, J. Arnold Arbor. 26:174. 1945: Manilal & Sivarajan, Fl. Calicut 66. 1982; Ramchand. & Nair, Fl. Cannannore 97. 1988.

Hippocratea arnottiana Wight, Ill. Ind. Bot. 1:133. tt. 46, 47A. 1839; Lawson in Hook. f., Fl. Br. Ind. 1: 624. 1875; Trimen, Handb. Fl. Ceylon 1: 275. 1893; Gamble, Fl. Pres. Madras 1: 213. 1918. Type: India, Malabar, Herb. *Wight 2445* (K).
Salacia terminalis Thw., Enum. Pl. Zeyl. 407. 1864. Type: Sri Lanka, Galle Dist., "Hikkadoowe" Oct 1861, *s. coll. C.P. 3737* (K).
Prionostemma arnottiana (Wight) Halle, Bull. Mus. Hist. Nat. (Paris) sect. B, Adansonia ser. 4. 3(1):7. 1981.

A large woody climber; branchlets glabrous. Leaves broadly elliptic or elliptic-ovate, 6.2–14.0 × 2.8–6.5 cm, glabrous, very coriaceous, attenuate at base, apex acute to subacuminate, margin entire to faintly crenate; veins 5–6 pairs; petioles 0.8–1.4 cm long, glabrous. Flowers c. 8 mm in diam., in axillary and terminal, lax, divaricate, ± 15 cm long paniculate cymes exceeding the length of the leaves, yellow; pedicels 4–8 mm long, glabrous. Calyx lobes 5, orbicular-ovate, c. 1.5 mm long, entire, glabrous, reflexed. Petals orbicular, 3.5–4.0 mm wide, with a narrow short claw, glabrous, imbricate, very concave, margin involute. Disk thick, cup-shaped, papillose. Stamens 3, inserted within the disk; filaments connate with the ovary; anthers extrorse, transversely dehiscent. Ovary 3-locular, sunk in the disk; style trigonous, subulate; Fruits capsular; follicles oblanceolate or obovate-oblong, 6.5–7.8 × 1.4–1.8 cm, convex. Seeds 8, 4.4–4.9 × 1.4–1.6 cm (incl. membranous wing), apex swollen; cotyledons convex, united.

Distr. India and Sri Lanka.

Ecol. In moist zone forests at lower elevations; rare.

Specimens Examined. KALUTARA DISTRICT: Yagirala forest range, 1 June 1990, *Jayasuriya 5195* (PDA). GALLE DISTRICT: "Hikkadoowe", Oct 1861, *s. coll. C.P. 3737* (CAL, PDA—*Isotypes* of *S. terminalis* Thw.). MATARA DISTRICT: Kekunadura forest range, 19 Oct 1990, *Jayasuriya 5824* (PDA).

3. Loeseneriella africana (Willd.) Wilczek, Fl. Congo Belge & Ruanda-Urundi 9:154. 1960; Halle, Mem. Inst. Franc. Afrique Noire n. 64:104. 1962.

Tonsella africana Willd., Sp. Pl. 1:194. 1797. Type: from W. Africa, Guinea (B-W).

Hippocratea obtusifolia Roxb. [Hort. Beng. 5. 1814, nom. nud.], Fl. Ind. 1:170. 1820; Roxb., Fl. Ind. ed. Carey 1:166. 1832; Wight & Arn., Prod. 104. 1834; Wight, Ic. Pl. Ind. Or. t. 963. 1845; Thw., Enum. Pl. Zeyl. 52. 1858, p.p.; Lawson in Hook. f., Fl. Br. Ind. 1: 623. 1875; Trimen, Handb. Fl. Ceylon 1: 275. 1893, p.p.; Gamble, Fl. Pres. Madras 1: 213. 1918. Type: India, Coromandel coast.

Hippocratea viridiflora Moon, Cat. 5. 1824, nom. nud.

Loeseneriella obtusifolia (Roxb.) A.C. Smith, Amer. J. Bot. 28: 440. 1941; Matthew, Fl. Tamilnadu Carnatic 1: 262. 1983.

A climbing shrub, 8(–12) m high; young branchlets sometimes cirrhose, glabrous to slightly pubescent. Leaves elliptic-oblong or ovate-lanceolate, 4–9.5 × 2.4–5.2 cm, coriaceous, glabrous, glossy above, base obtuse to subacute, apex acute-subacuminate, margin entire to faintly crenate, nerves prominent below; petioles 5–8 mm long, glabrous. Flowers c. 7 mm across, in axillary, lax, spreading, paniculate cymes; peduncles up to 5 cm long; bracts triangular, 1.5 mm long; pedicels 4–5 mm long, glabrous. Calyx lobes 5, ovate, c. 1 mm long, pubescent without. Petals 5, triangular-lanceolate,

3.5–4 mm long, spreading, yellowish-green or green, glabrous within. Disk cupular. Stamens 3, filaments linear, 2 mm long, green; anthers reflexed. Ovary 3-celled, with 6 ovules per cell; style 2 mm long; stigma 3-lobed. Fruits capsular; follicles obovoid-oblong, 5.0 × 3.0 cm, obtuse, emarginate, dorsiventrally compressed. Seeds 6 per cell, subfalcate, 1.5 × 0.5 cm; wing membranous, 3.5 × 1.5 cm.

Distr. Tropical Africa, India, Australia and Sri Lanka.

Ecol. In lowland dense forests on alluvial flats in wet zone; common.

Note. On the specimen *C.P. 3158* in PDA, "Caltura" (Kalutara), *Moon s.n.*, Nov 1854 is recorded in pencil.

The bark surface is greyish; inner bark chocolate-brown and the wood is whitish with chocolate-brown patches/streaks.

Specimens Examined. KANDY DISTRICT: 1.5 miles S.E. of Teldeniya. 80° 47′ E. 7° 17′ N., 450 m, 6 Jan 1972, *Jayasuriya, Dassanayake & Balasubramaniam 474* (K, PDA). RATNAPURA DISTRICT: Sinharaja forest-Waturawa main trail, 22 Oct 1987, *Jayasuriya et al. 4118* (PDA). HAMBANTOTA DISTRICT: Ruhuna National Park, c. 1 mile above mouth of Menik Ganga river, 4 April 1968, *Fosberg 50215* (K), Block-I, at Rugamtota, 24 Feb 1968, *Mueller-Dombois & Cooray 68022405* (PDA). LOCALITY UNKNOWN: *s. coll. C.P. 3158* (BM, K, PDA).

2. REISSANTIA

Halle, Bull. Mus. Hist. Nat. (Paris) ser. 2.30: 466. 1958; Halle, Mem. Inst. Franc. Afrique Noire n. 64: 84. 1962. Type species: *R. astericantha* Halle.

Lianes or stragglers, sometimes erect shrubs. Leaves decussate, coriaceous; stipules present, caducous. Inflorescence axillary, dichotomously cymose, sometimes paniculate, usually with supplementary branchlets in the dichotomy or in the axils of branchlets. Flowers small. Calyx lobes 5, imbricate. Petals 5, imbricate, entire or laciniate. Disk extrastaminal, inconspicuous, most of it usually united with the ovary, the uppermost part slightly extended like a rim. Stamens 3, inserted on top of the disk; filaments erect or recurved, base dilated, flat; anthers extrorse, transversely dehiscent. Ovary 3-celled, ovules 2 or more per cell; style short; stigma simple. Fruit capsular, consisting of 3 divergent, separate follicles, dehiscing along an inconspicuous median suture into 2 valves. Seeds with a basal ± transparent, membranous wing; endosperm 0; cotyledons free.

About 7 species in the Old World tropics: Central and West Africa and Indo-Malaysia; 1 in Sri Lanka.

Reissantia indica (Willd.) Halle [Bull. Mus. Hist. Nat. (Paris) ser. 2, 30: 466. 1958, nom. illeg.], Mem. Inst. Franc. Afrique Noire n. 64: 85. 1962:

Ding Hou, Blumea 12: 33. 1963; Ding Hou in Fl. Males. ser. 1. 6(3): 401. 1964; Matthew, Fl. Tamilnadu Carnatic 1: 264. 1983.

Hippocratea indica Willd., Sp. Pl. 1: 193. 1797; Roxb., Pl. Corom. 2: 16. t. 130. 1798; Roxb., Fl. Ind. ed. Wall. 1: 169. 1820; DC., Prod. 1: 568. 1824; Wight & Arn., Prod. 104. 1834; Thw. Enum. Pl. Zeyl. 52. 1858; Lawson in Hook. f., Fl. Br. Ind. 1: 624. 1875; Kurz, J. Asiat. Soc. Bengal 44(2): 164. 1875; Trimen, Handb. Fl. Ceylon 1: 276. 1893; Gamble, Fl. Pres. Madras 1: 213. 1918. Type: from India.

Pristimera indica (Willd.) A.C. Smith, Amer. J. Bot. 28: 440. 1941; A.C. Smith, J. Arnold Arbor. 26: 175. 1945.

Lianes or straggling shrubs, 4(6) m high; branchlets glabrous. Leaves broadly ovate or elliptic-oblong, rarely obovate, 3.0–9.0 × 1.8–5.5 cm, thinly coriaceous, glabrous, base attenuate or cuneate, apex abruptly acuminate or shortly apiculate, margin finely serrate; nerves obscure above, prominent below, 5–7 pairs; petioles 5–10 mm long. Inflorescence dichotomously cymose, 4–8 cm long, usually with supplementary branchlets in the dichotomy; peduncles usually short, sometimes up to 3.5 cm; bracts triangular, c. 1 mm long, laciniate. Flowers 1 mm across, pale yellow or greenish-yellow; pedicels 1–2 mm long. Calyx lobes 5, triangular, c. 1 mm long, papillose on both surfaces. Petals 5, oblong, 1 mm long, papillose on both surfaces. Disk cupular, thick. Stamens 3, filament subulate, flat, 0.5 mm long. Ovary flask–like, 3-celled, with 2 ovules per cell; style stout; stigma obscure. Follicles elliptic or obovate-oblong, 2.3–3.4 × 0.7–1.1 cm, much compressed, striate, sutural margin slightly spreading after dehiscence. Seeds 2, 2.1–2.4 × 0.5–0.7 cm (incl. wing); wing emarginate, with a central vein.

Distr. India, Burma, Thailand, Indo-China, S. China, Malaysia and Sri Lanka.

Ecol. In dry region and lowland wet region forests, up to 800 m, common; fls & frts June–Dec.

Specimens Examined. JAFFNA DISTRICT: Vaddakkachchi, Paranthan-Mullaitivu road, 17 Nov 1970, *Kundu & Balakrishnan 708* (PDA). MANNAR DISTRICT: Madhu-Tiruketiswaram jungle road, 22 June 1975, *Sumithraarachchi & Sumithraarachchi 770* (K, PDA). PUTTALAM DISTRICT: N. of Puttalam, N.W. coast, 24 July 1974, *Kostermans 25256* (K, PDA); Palavi abandoned airport, Puttalam-Kalpitiya road, 15 Nov 1992, *Wadhwa 280, 286* (PDA). ANURADHAPURA DISTRICT: Kekirawa-Anuradhapura road, 62/3 mark, 12 Jan 1973, *Tirvengadum & Jayasuriya 251* (PDA); Wilpattu National Park, 15 Oct 1975, *Bernardi 15365* (PDA), at Dangaha Uraniya, Plot W32, 28 April 1969, *Mueller-Dombois, Wirawan, Cooray & Balakrishnan 64042815* (K, PDA), between Marandanmaduwa

& Kumbukwewa, 9 July 1969, *Wirawan, Cooray & Balakrishnan 1075* (K, PDA), close to Plot W39, 30 June 1969, *Wirawan, Cooray, & Balakrishnan 893* (K, PDA). TRINCOMALEE DISTRICT: Niroddumunai, littoral jungle, Trincomalee, 1 Dec 1932, *Simpson 8498* (BM). MATALE DISTRICT: Dambulla-Kanarewa road, 86 km mark, 13 Nov 1992, *Wadhwa 221* (PDA). BATTICALOA DISTRICT: Batticaloa, NW of Unichchai tank, 14 Aug 1967, *Mueller-Dombois 67081421* (PDA); Arugam Bay, c. 5 m, 30 Oct 1975, *Bernardi 15562* (PDA). COLOMBO DISTRICT: Matalana, Feb 1889, *Nevill s.n.* (PDA). KANDY DISTRICT: Madugoda, 1000 m, 13 Oct 1977, *Nooteboom 3371* (PDA); Madugoda to Mahiyangana road, 800 m, 8 June 1978, *Kostermans 26618, 26627* (PDA); between Hasalaka and Madugoda, 700 m, 16 Oct 1977, *Huber 464* (PDA); Mount Kokegala, Mahiyangana, 730 m, 11 Nov 1973, *Bernardi 15689* (PDA); 2 miles beyond Madugoda, 835 m, 80°54' E, 7°20' N, 27 Oct 1971, *Jayasuriya 366* (K, PDA); Kandy, Lady Macarthy's Drive, 12 June 1926, *Alston 258* (K, PDA). HAMBANTOTA DISTRICT: Ruhuna National Park, between Komawawewa & Talgasmankada, 1 Dec 1969, *Cooray 69121017* R (K, PDA), Block-I, NS boundary, S. of Sithulpahuwa road in Plot R4, 20 Oct 1968, *Mueller-Dombois 68102003* (K, PDA); Tissamaharama, Dec 1882, *s. coll. s.n.* (PDA); "Yala Reserve", 2 Oct 1991, *Jayasuriya 5788* (PDA). LOCALITY UNKNOWN: *Thwaites C.P. 1164* (BM, CAL, K, PDA); *Kostermans 25152* (PDA).

3. SALACIA

L., Mant. ed. 2, 159, 293. 1771; A.C. Smith, Brittonia 3: 423. 1940; Loes. in Pflanzenfam. ed. 2, 20b: 217. 1942. Type: *S. chinensis* L.
Johnia Roxb., Fl. Ind. ed. Wall. 1: 172. 1820; Roxb., Fl. Ind. ed. Carey 1:168. 1832. Type species: not designated.

Lianes, scandent or sometimes erect shrubs. Leaves opposite, sometimes subopposite. Flowers in axillary or extra-axillary fascicles or in cymes, thyrsiform or paniculiform, usually 5-merous, sometimes 4-merous. Calyx 5 (–4)-lobed, rarely not lobed, imbricate. Petals 5 or 4, imbricate. Disk large, intrastaminal, fleshy, annular-pulvinate, sometimes truncate-conical or flattened. Stamens usually 3, sometimes 2, inserted at the base of the free part of the pistil, usually reflexed at anthesis; filaments subulate, broadened towards the base; anthers usually transversely dehiscent. Ovary partly or completely immersed in the disk, the free part conical, 3 (or 2)-celled, with 2–8 ovules in each cell; style short, simple; stigma simple. Fruit drupaceous or baccate, globose or subglobose, usually 3-celled, indehiscent; pericarp coriaceous when dry. Seeds 1–8, embedded in mucilaginous pulp, angular, not winged; cotyledons massive, free or united.

About 200 species, distributed in tropical parts of India, Burma, Sri Lanka, Malaysia, Solomon Islands, Africa and New World; 4 species in Sri Lanka.

KEY TO THE SPECIES

1 Flowers in axillary, very short pedunculate heads; stamens dehiscing longitudinally; ovary much exserted from the disk; leaves yellowish when dry. .... **4. S. oblonga**
1 Flowers in fascicles or on very short axillary, bracteate, tubercles or peduncles; stamens dehiscing transversely; ovary nearly embedded in the disk or only a small part emerging; leaves not yellowish when dry
  2 Flowers tetramerous; stamens 2; ovary 2-locular. .... **3. S. diandra**
  2 Flowers pentamerous; stamens 3; ovary 3-locular
    3 Petals orbicular or obovate, always clawed; fruits globose, not exceeding 1.5 cm in diam., smooth; leaves without prominent reticulate veins beneath. .... **1. S. chinensis**
    3 Petals ovate or oblong, without claws; fruit globose, up to 3.5 cm in diam., tuberculated; leaves with prominent reticulate veins beneath. .... **2. S. reticulata**

1. Salacia chinensis L., Mant. ed. 2, 293. 1771; Ding Hou in Fl. Males. ser. 1. 6(3):419. 1964; Matthew, Fl. Tamilnadu Carnatic 1:265. 1983. Type: from China.

Tonsella prinoides Willd., Ges. Naturf. Freunde Berlin, Neue Schriften 4:184. 1803. Type: from India (B-W).
Tonsella chinensis (L.) Spreng., Syst. 1:177. 1824.
Salacia prinoides (Willd.) DC., Prod. 1:571. 1824; Wight & Arn., Prod. 1:105. 1834; Thw., Enum. Pl. Zeyl. 53. 1858; Lawson in Hook. f., Fl. Br. Ind. 1:626. 1875; Kurz, J. Asiat. Soc. Bengal 44(2):163. 1875; Trimen, Handb. Fl. Ceylon 1:276. 1893; Gamble, Fl. Pres. Madras 1:214. 1918.

A large woody climber; branches divaricate, glabrous. Leaves opposite, thinly coriaceous, broadly elliptic or elliptic-lanceolate, 4.2–10.5 × 2.2–4.0 cm, glabrous, slightly glossy above, base subacute to obtuse, apex acute to shortly acuminate, margin entire or shallowly crenate-serrate; petioles 5–8 mm long. Flowers few to many in fascicles on axillary bracteate tubercles, 6–7 mm in diam., yellow or yellowish-green; pedicels 5–7 mm long; bracts triangular. Calyx shallowly 5-lobed; lobes broadly ovate, c. 1 mm long, obtuse, puberulous, persistent. Petals 5, orbicular or obovate, 3.0–4.0 × 2.5–3.0 mm, shortly clawed, spreading. Disk intrastaminal, annular-pulvinate, thick, 1–1.5 mm wide, slightly lobed, base papillose. Stamens 3, inserted on the rim of the disk, 1.5 mm long; filaments linear, subulate; anthers transversely dehiscent. Ovary triangular or ovoid, partly emerging from the disk, 3-celled, with 2 or more ovules per cell; style short, attenuate; stigma simple. Fruit a drupe, globose, 1.0–1.5 cm in diam., 1-celled, 1-seeded, supported on the disk and somewhat enlarged, persistent calyx smooth, on maturity red or orange-red. Seeds globose, up to 8 mm in diam.

Distr. India, Burma, Thailand, Indo-China, S. China, Malaysia, New Britain, Solomon Islands. Fiji and Sri Lanka.

Ecol. In forests along river banks and lowland areas, chiefly in dry region; fairly common.

Vern. Hinhimbutuwel (S).

Note. Plate No. 426 based on *C.P. 1165* (habit only) and named as *Salacia prinoides* DC. in PDA belongs to this species.

Specimens Examined. JAFFNA DISTRICT: near Pooneryn, 5 m, 17 March 1973, *Bernardi 14289* (K, PDA); Ampan Chempiyanpattu, 15/3 mile mark, alt. low, 12 Oct 1974, *Tirvengadum 496* (K, PDA); Ampan, alt. sea level, 12 Oct 1974, *Cramer 4336* (K, PDA); c. 8 miles from Elephant Pass, low tree Parkland, 4 May 1931, *Simpson 8026* (BM); Pallavarayan kaddu, Mannar-Pooneryn road, 13 Nov 1970, *Kundu & Balakrishnan 640* (PDA). VAVUNIYA DISTRICT: Near 113 mile post, Medawachchiya-Mannar, 80°16′ E, 8°41′ N, alt. low, 12 Jan 1972, *Jayasuriya, Balakrishnan, Dassanayake & Balasubramaniam 586* (K, PDA). ANURADHAPURA DISTRICT: Ritigala Strict Natural Reserve, 14 Nov 1992, *Wadhwa 241* (PDA); Wilpattu National Park, Kuruttu Pandi Villu, Plot W5, 10 Oct 1968, *Cooray 68091009* (K), 9 July 1969, *Wirawan, Cooray & Balakrishnan 1060* (K, PDA), Maduru odai, Mannar-Puttalam road, junction to Periya Naga villu, 30 June 1969, *Wirawan, Cooray & Balakrishnan 928* (K, PDA), Periya Naga villu, near Plot W38, 30 April 1969, *Mueller-Dombois, Wirawan, Cooray & Balakrishnan 69043005* (K, PDA), Manikapola Uttu, Plot W31, 9 July 1969, *Wirawan 1071* (K, PDA). MATALE DISTRICT: Dambulla, IFS-Popham Arboretum, *Cramer 6053* (PDA); Dambulla, alt. low, 16 May 1973, *Kostermans 24796* (K). TRINCOMALEE DISTRICT: Kanniya, near Hot wells, 24 June 1975, *Sumithraarachchi & Sumithraarachchi 848* (PDA); Horowapatana-Trincomalee, 28 March 1963, *Amaratunga 560* (PDA). KANDY DISTRICT: Hantane, 915 m, *Gardner 1201* (K, BM). KALUTARA DISTRICT: Kalutara, *Macrae 98* (K). RATNAPURA DISTRICT: Horagal Kanda-Sinharaja, 300 m, 25 Feb 1977, *Waas 2014* (K, PDA). HAMBANTOTA DISTRICT: Ruhuna National Park East, Kumana villu, alt. low, 2 May 1975, *Jayasuriya 2013* (K, PDA); Patanagala area, 5 March 1968, *Comanor 1953* (K). LOCALITY UNKNOWN: *Col. Walker s.n.* (K); *Thwaites C.P. 1165* (BM, K, CAL); *Thwaites s.n.* in *C.P. 3148* (BM); *Kurz s.n.* (CAL).

2. Salacia reticulata Wight, Ill. Ind. Bot. 1:134. 1839; Thw., Enum. Pl. Zeyl. 53. 1858; Lawson in Hook. f., Fl. Br. Ind. 1:627. 1875 excl. var. *diandra*; Trimen, Handb. Fl. Ceylon 1:277. 1893 excl. var. *diandra*; Gamble, Fl. Pres. Madras 1:214. 1918. Lectotype: Sri Lanka: no precise locality, *Col. Walker s.n.* (K).

Woody climber or scandent shrub, dichotomously branched; stem relatively tough; young parts glabrous. Leaves opposite, coriaceous, elliptic or oval, 4.2–11.6 × 2.2–5.4 cm, glabrous, narrowed at base, apex obtuse or with a short point, margin entire or very shallowly crenate-serrate, shining above, paler with fine, prominent, reticulate veins beneath; petioles 0.7–1.2 cm long. Flowers 2–8 on axillary tubercles or peduncles, 6–7 mm in diam., greenish-yellow; pedicels 4–5 mm long, glabrous. Calyx entire or scarcely lobed, glabrous. Petals 5, ovate or oblong, 3–3.5 mm long, obtuse, spreading. Disk intrastaminal, annular-pulvinate, thick, 1 mm wide, slightly lobed. Stamens 3, inserted on the rim of the disk, 1.5 mm long; filaments broad at the base; anthers transversely dehiscent. Ovary ovoid, nearly embedded in the disk, 3-locular, with 2–6 ovules per cell; style short. Fruits drupaceous, globose, 1.6–3.8 cm in diam., tuberculate, bright pink-orange when ripe. Seeds 1–4, almond-like; testa membranous, yellow.

Distr. India (Malabar), and Sri Lanka.

Ecol. In moist zone in wet secondary forests, up to 1500 m, rather common.

Vern. Himbutuwel, Kotala Himbutu (S).

Specimens Examined. ANURADHAPURA DISTRICT: Ritigala Strict Natural Reserve, Weweltenna Plain, 530–700 m, Apr–May 1974, *Jayasuriya 2155, 2156, 2157, 2158* (PDA), *1672* (K, PDA), ± 600 m, 28 Sept 1972, *Jayasuriya 922* (PDA); Ritigala Strict Natural Reserve, 14 Nov 1992, *Wadhwa 240* (PDA); Wilpattu National Park, N.E. of Kuruttu Pandi villu, 24 March 1968, *Koyama 13401B* (K), before Borupan villu, 23 March 1968, *Koyama, Herat & Cooray 13365* (PDA), *Koyama 13359* (PDA), Kuruttu Pandi villu, Plot W5, 10 Oct 1968, *Cooray 68091009* (K, PDA). KURUNEGALA DISTRICT: Kumbalpola forest, 17 Nov 1992, *Wadhwa 329* (PDA). KANDY DISTRICT: Mirisketiya, 1000 m, 19 Jan 1977, *Nooteboom 3407* (PDA); Summit of Hantane ridge, 750-800 m, *Kostermans 28087, 28269* (PDA); Madugoda-Mahiyangana, 600 m, 6 Aug 1978, *Kostermans 26782* (PDA); Mount Kokegala at Mahiyangana, 600 m, 11 Nov 1975, *Bernardi 15691* (PDA); Nawalapitiya, 17 April 1939, *Worthington 126* (BM); Hantane, Apr 1864, *s. coll. C.P. 741* (K); Royal Botanic Garden, Peradeniya, *s. coll. s.n.* (BM); Guru-oya, 80°47' E, 7°17' N, 460 m, 30 April 1975, *Jayasuriya 1916* (K, PDA). BADULLA DISTRICT: Ratkarawwa, Boralanda, 80°54' E, 6°48' N, 1300 m, 26 April 1972, *Jayasuriya & Cramer 775* (K, PDA); Boralanda, ± 1300 m, 28 April 1972, *Cramer & Jayasuriya 3745* (PDA); Palugama to Boralanda, mile marker 1/8, 1500 m, 5 Nov 1971, *Balakrishnan 1057* (K, PDA). KALUTARA DISTRICT: Kalutara, Himbutuwela, March 1821, *Moon s.n.* (BM); Kalutara, *Macrae 99* (K). RATNAPURA DISTRICT: Gilimale forest reserve, 3 March 1987, *Jayasuriya 3780* (PDA); Belihuloya, 155/5 mark along

Ratnapura road, 4 March 1987, *Jayasuriya 3787, 3789* (PDA); Sinharaja forest, Sinhagala,. 600 m, 13 July 1978, *Kostermans 26798* (PDA); Horagalkanda-Sinharaja forest, 300 m, 25 Feb 1977, *Waas 2014* (PDA). GALLE DISTRICT: Haycock (Hinidumkande), 500 m, 31 Aug 1974, *Kostermans 25476* (PDA); Kottawa, 3 July 1990, *Jayasuriya 5397* (PDA); Weligama, sea coast, Redcliff rock, 20 m, 25 June 1980, *Kostermans 28561* (PDA); Hiniduma forest reserve, alt. low, 6 May 1973, *Kostermans 24708* (K, PDA); Kanneliya forest, near Hiniduma, 200 m, 19 Aug 1980, *Kostermans 28634* (PDA). MONERAGALA DISTRICT: Rambukkanoya, Mullegama, alt. low, 4 May 1975, *Jayasuriya 2053* (K, PDA). LOCALITY UNKNOWN: South coast, no precise locality, Jan 1846, *Thwaites C.P. 658* (BM, K, CAL); *Gardner 269* (BM, K).

3. Salacia diandra Thw., Enum. Pl. Zeyl. 53. 1858. Type: Sri Lanka, *C.P. 2720* (K).

Salacia reticulata Wight var. *diandra* (Thw.) Lawson in Hook. f., Fl. Br. Ind. 1:627. 1875; Trimen, Handb. Fl. Ceylon 1:277. 1893.
Salacia acuminatissima Kostermans, Reinwardtia 11(1): 53. 1992. Type: Sri Lanka: Sinharaja forest reserve, 200 m, *Gunatilleke s.n.* (PDA).

Scandent shrub or woody climber; stem flexible; young branches angular, glabrous, black. Leaves opposite, ovate-oblong, 5.2–11.0 × 2.2–4.5 cm, subcoriaceous, glabrous, narrowed at base, apex abruptly shortly acuminate or apiculate, margin entire, black above, paler with very prominent reticulate veins beneath; petioles 3–5 mm long, glabrous. Flowers 1–2 on axillary peduncles, 4 mm in diam., greenish-yellow, 4-merous; pedicels 4–5 mm long, glabrous. Calyx lobes 4, triangular, 1 mm long, glabrous. Petals 4, rotundate-obovate, shortly clawed, 2.5–3 mm long (incl. claw), spreading at anthesis. Disk intrastaminal, pulvinate, thick, 1.5 mm. Stamens 2, inserted on the rim of the disk, 1.5 mm long; filaments short, c. 1 mm long; anthers extrorse, dehiscing transversely. Ovary 2–locular, with 2 ovules in each cell; style short. Fruit a drupe, ovoid, 1.4–1.8 cm in diam., 1-celled, tuberculate.

Distr. Endemic.

Ecol. In wet zone forests, upto 600 m, rare.

Note. Thwaites (l.c.) has given Central Province and "Saffragam" District (Sabaragamuwa) as the area of distribution while Trimen (l.c.) has mentioned Ekneligoda and Sabaragamuwa as two localities.

Plate No. 427 A in PDA based on *C.P. 2720* (habit & dissected flower) belongs to this species.

Specimens Examined. RATNAPURA DISTRICT: W. Ratnapura, Mar 1853, *s. coll. s.n.* (K); Sinharaja forest reserve, 200 m, *Gunatilleke s.n.* (Type of *S. acuminatissima* Kosterm., PDA); Sinharaja forest, Tundole ela,

25 Nov 1992, *Wadhwa & Weerasooriya 333* (PDA). GALLE DISTRICT: Nakiyadeniya forest reserve, 5 Aug 1991, *Jayasuriya 6604* (PDA); Kanneliya forest reserve, near river, close to old forest Bungalow, 30 Nov 1992, *Wadhwa & Weerasooriya 376* (PDA); Hiniduma forest, 30 Nov 1992, *Wadhwa & Weerasooriya 371* (PDA). MATARA DISTRICT: Badullekele forest reserve, 22 Oct 1991, *Jayasuriya et al. 5868* (PDA). LOCALITY UNKNOWN: *s. coll. C.P. 2720* (K, type; BM, isotype).

4. Salacia oblonga Wall. ex Wight & Arn., Prod. 1:106. 1834; Wight, Ill. Ind. Bot. t. 47B. 1839; Thw., Enum. Pl. Zeyl. 53. 1853; Lawson in Hook. f., Fl. Br. Ind. 1:628. 1975; Trimen, Handb. Fl. Ceylon 1:277. 1893; Talbot, For. Fl. Bombay & Sind 1:287. 1909; Gamble, Fl. Pres. Madras 1:214. 1918. Type: India, *Wallich* n. *4226.*

A large woody climbing shrub; young branches terete, densely sprinkled with lenticels, glabrous. Leaves elliptic-oblong or oblong-lanceolate, 4.8–11.2 × 2.2–5.2 cm, glabrous, subcoriaceous, turning yellow when dry, apex obtuse or shortly acuminate, margin distantly serrate, veins prominent beneath; petioles 5–7 mm long, glabrous. Flowers axillary, 1–3, in short pedunculate heads, 4–5 mm across, dull yellow; peduncles 2 mm long. Calyx lobes 5, rotund, entire, glabrous. Petals 5, broadly elliptic, 2–2.5 mm long, obtuse, slightly toothed along the margin, erect. Disk flat, with slight elevation in the middle. Stamens 3; filaments much dilated at the base; anthers 2-celled, dehiscing longitudinally. Ovary elongate-conical, much exserted from the disk, 3-locular, with 4 ovules in each cell; style short. Fruits baccate, globose or subglobose, 2.0–4.8 cm in diam, supported on persistent calyx, tuberculate, bright orange-red when ripe. Seeds 1–8, angular, 2.7 × 1.5 cm.

Distr. India (Concan southwards) and Sri Lanka.

Ecol. In secondary open forests, occasional.

Vern. Chundan (T).

Note. Plate No. 428 in PDA, based on *C.P. 1061* (habit & dissected flower), represents this species.

Specimens Examined. ANURADHAPURA DISTRICT: Anuradhapura road from Trincomalee, 200 m, 13 Aug 1950, *Worthington 4880* (PDA). TRINCOMALEE DISTRICT: Trincomalee, Sandy Bay, alt. low, 23 May 1975, *Waas 1252* (K, PDA). KURUNEGALA DISTRICT: Kumbalpola forest, 17 Nov 1992, *Wadhwa 330* (PDA). MATALE DISTRICT: Galboda, off Naula-Elahera road, 2000 m, 5 Jan 1971, *Jayasuriya 306* (K); near Wewala tank, 5 Jan 1974, *Sumithraarachchi 483* (K); c. 8 miles SE of Dambulla, 270 m, 11 Oct 1974, *Davidse 7410* (PDA); Between Kibissa and Sigiriya, 250 m, 15 Oct 1977, *Huber 460* (PDA). BADULLA DISTRICT: Uma oya,

S. of Badulla, 200 m, 30 June 1973, *Kostermans 25185* (PDA); 13th post on Bibile-Mahiyangana road, alt. low, 9 Sept 1978, *Huber 931* (PDA). NUWARA ELIYA DISTRICT: Randenigala reservoir area, 7° 11′ N, 80° 51′ E, 230 m, 20 Nov 1984, *Jayasuriya 3038* (PDA). MONERAGALA DISTRICT: Ruhuna National Park, Suandana Ara, 8 km NE of Galge, 23 Sept 1992, *Jayasuriya 6735, 6737* (PDA), Kosgasmankada, 1 km. N. of Yala Bungalow, 22 Sept 1992, *Jayasuriya 6728* (PDA), Katagamuwa, 4 Oct 1991, *Jayasuriya 5807* (PDA), Tammitagala, 14 km NW of Galge, 24 Jan 1992, *Jayasuriya 6751* (PDA), Block 2, on way to Plot 16, 30 Aug 1967, *Mueller-Dombois 67083014* (PDA). HAMBANTOTA DISTRICT: Kumbukkan oya, c. 2 miles above mouth, Meegahakanda Meda wewa, Block 2, 6 July 1969, *Fosberg, Mueller-Dombois, Wirawan, Cooray & Balakrishnan 51111* (K). LOCALITY UNKNOWN: *Thwaites C.P. 3148* (K, BM); *Thwaites C.P. 1061* (K).

ICACINACEAE
(by W. Meijer* and B. Verdcourt**)

Miers, Ann. Mag. Nat. Hist. Ser. 2, 8: 174. 1851. Type genus: *Icacina* Juss.

Trees, shrubs or climbers. Leaves simple, alternate or less often opposite, usually entire; stipules absent. Inflorescences usually axillary, sometimes terminal spikes, false racemes, cymes or few-flowered panicles, or flowers arranged in fascicles or heads. Flowers (3–) 4–6-merous sometimes with only petals present (in *Pyrenacantha* and female flowers of *Gomphandra*). Calyx lobes imbricate or less often valvate; petals free or partly connate, valvate; disk occasionally present, annular or cup-like. Stamens in a single whorl, opposite the sepals, free or inserted on the corolla tube; filaments often hairy above. Anthers opening by slits or in one genus (not in Ceylon) by operculate pores. Ovary mostly unilocular, less often 2–5-locular; ovules usually 2, rarely 1 by abortion, apical, pendulous, anatropous, with one integument; style 1 or obsolete. Drupes ellipsoid or globose with fleshy exocarp and thin crustaceous to woody or fibrous endocarp. Seed 1, without an aril, usually with abundant endosperm; embryo straight.

About 56 genera with about 300 species in the tropics and subtropics of both hemispheres.

There is no general agreement as to the taxonomic position of this family. It is often considered to be closely related to Celastraceae and Aquifoliaceae (Celastrales) (e.g. Stebbins, Flowering Plants 352. 1974) or recently evolved from Olacaceae (Santalales, Stebbins, 1974) which would suggest a close relationship with the archaic Proteaceae. According to Takhtajan the primitive representatives of Santalales are very near the primitive families of the Celastrales especially the Icacinaceae. it might be possible to extend the definition of Icacinaceae so wide that it could not be kept distinct from Olacaceae.

The definition of the family and concept of the genera follows Sleumer's revision for Flora Malesiana (Blumea 17: 181–264. 1969) and account for Flora Malesiana (I, 7: 1–87. 1971) which results in an interpretation of the names *Stemonurus* and *Gomphandra* differing from that of Howard (J. Arnold

* University of Kentucky, Lexington, Kentucky, U.S.A.
** Royal Botanic Gardens, Kew (account based partly on W. Meijer's notes including all his records of PDA material none of which has been seen by B.V.

Arbor. 21: 461–489. 1940) and from the checklist of Abeywickrama (Ceylon J. Sci., Biol. Sci. 2: 188. 1959).

Gonocaryum litorale (Blume) Sleumer (= *G. pyriforme* Scheff.) was grown in the Botanic Garden, Peradeniya, last century but is not listed in recent guides. It might persist somewhere so is included in the key.

KEY TO THE GENERA

1 Climbing plants; leaves mostly with distant teeth tipped with hydathodes; milky latex present in stem; flowers 4-merous . . . . **5. Pyrenacantha**
1 Trees or shrubs
  2 Petals connate beneath with filaments adnate to the cupular part of corolla; flowers unisexual (cult.) . . . . **Gonocaryum**
  2 Without these characters combined (petals connate and flowers unisexual in *Gomphandra*)
    3 Flowers in axillary inflorescences
      4 Flowers in paniculate clusters or almost solitary, unisexual; fruits 1–5 cm long without fibrous outer layer; leaves coriaceous in one species but not so thick . . . .
      . . . . **1. Gomphandra**
      4 Flowers in pedunculate heads, hermaphrodite; drupes 4–5 cm long, with fibrous outer part; leaves coriaceous and very thick . . . . **4. Stemonurus**
    3 Flowers in terminal or subterminal branched inflorescences
      5 Calyx and corolla ± glabrous; fruits asymmetrical with flattish shield-like appendage at base, the style finally lateral to almost basal; leaves up to 15 × 8 cm, often small, glabrous
      . . . . **2. Apodytes**
      5 Calyx and corolla tomentose; fruit without appendage, the style terminal; leaves often large, up to 6–28 × 2–19 cm, very sparsely hairy to densely velvety beneath . . . .
      . . . . **3. Nothapodytes**

1. GOMPHANDRA

Wall. ex Lindl., Nat. Syst. ed. 2: 439. 1836; Sleumer, Blumea 17: 189. 1969 (full synonymy and references); Sleumer, Fl. Males. I, 7: 21–35. 1971. Type species: *Gomphandra tetrandra* (Wall.) Sleumer.

Dioecious trees. Leaves alternate, entire. Flowers in axillary or leaf-opposed cymes, the males more numerous than the females; bracts small; pedicels thickened and jointed below the calyx. Calyx short, saucer-shaped, 4–5-toothed or entire. Petals 4–5, thin, valvate, united into a short tube at the base, reflexed after flowering. Stamens 4–5, free; filaments glabrous or with club-shaped hairs below the anthers; pollen lacking in functionally female flowers. Disk in male flowers cushion-shaped, united with base of rudimentary ovary, often absent in female flowers. Ovary conical to cylindric, as long as the filaments, 1-locular; ovules 2, pendulous; stigmas sessile or disciform. Drupes ovoid-ellipsoid, crowned by the ± excentric stigma; mesocarp thin, fleshy; endocarp crustaceous-woody, more or less asymmetrically 10-ribbed. Seed 1; raphe elongate; embryo small in fleshy endosperm.

About 33 species in tropical and subtropical Asia, Malesia to New Guinea, Solomon Is. and St. Cruz Is., also NE. Australia and Melanesia. Two species occur in Ceylon.

KEY TO THE SPECIES

1 Leaves thin, drying green with venation more evident beneath; petioles often (but by no means always) shorter, 0.5–0.8 (–1.2) cm long; inflorescences axillary, 4–10 mm long and with peduncles 2–10 mm long (in Ceylon); stamens conspicuously hairy ....
.... **1. G. tetrandra**
1 Leaves thicker, usually drying more blackish and rugulose, the venation between lateral nerves not evident; petioles often longer; inflorescences leaf-opposed or axillary, 1–1.4 cm long with the peduncles 0.6–2 cm long (in Ceylon); staminal hairs short or even absent ....
.... **2. G. coriacea**

1. Gomphandra tetrandra (Wall.) Sleumer, Notizbl. Bot. Gart. Berlin 15: 238. 1940; Sleumer, Blumea 17: 204. 1969 (extensive synonymy).

Lasianthera? tetrandra Wall. in Roxb., Fl. Ind. ed. Carey & Wall.2:328. 1824. Type: India, Silhet, *M.R. Smith* (K-Wall. 3718, lectotype; BM, isolectotype) (Wallich was dubious if it belonged to the genus not that it was a distinct species).

Gomphandra polymorpha Wight, Ill. Ind. Bot. 1: 103. 1840 (including five varieties); Wight, Ic. Pl. Ind. Or. 3(3): 3, t. 954. 1844; Masters in Hook. f., Fl. Br. Ind. 1: 586. 1875 (including the five varieties). Type: India, Courtallum, Herb Wight prop. and Herb. Wight *428* (K, lectotype & isolectotype) (It seems clear from Sleumer that he considered var. *angustifolia* Wight to be what would now be considered the typical variety and the type of that is the type of the species but Wight gives no hint; Sleumer has by this action lectotypified the species).

Stemonurus axillaris Miers, Ann. Mag. Nat. Hist. II, 10: 41. 1852, nom illegit. (based on *G. axillaris* Wall. ined. but *L. tetrandra* cited in synonymy).

Stemonurus ceylanicus Miers, Ann. Mag. Nat. Hist. II, 10: 42. 1852. Type: Ceylon, *Macrae 428* (CGE (LINDLEY), K (HOOK.), syntypes).

Stemonurus heyneanus Wall. ex Miers, Ann. Mag. Nat. Hist. II, 10: 40. 1852. Type: India Orient. *Heyne* in *Wallich 6780* (K-WALL., K-HOOK., syntypes; BM, isosyntype) & Ceylon, *Gardner 102* (BM, syntype; K, isosyntype) (on page 38 Miers refers this same number to *S. polymorphus* var. *longifolius*).

Stemonurus polymorphus (Wight) Miers, Ann. Mag. Nat. Hist. II, 10: 37. 1852. (including the five varieties).

Platea axillaris (Miers) Thw., Enum., Pl. Zeyl. 44. 1858.

Gomphandra axillaris Wall. ex Bedd., Fl. Sylv, For. Man. LXI. 1871; Masters in Hook. f., fl. Br. Ind. 586. 1875; Trimen, Handb. Fl. Ceylon 1: 261. 1893, nom illegit. (several earlier names cited in synonymy).

Stemonurus tetrandrus (Wall.) Alston in Trimen, Handb. Fl. Ceylon 6: 48. 1931.

Shrub or small tree 2–9 m tall; d.b.h. 2–19 cm; bark grey, rough; young branches puberulous. Leaves thin, variable, broadly elliptic to narrowly oblong or narrowly obovate to oblanceolate, 2.7–14 cm long, 0.8–5(–7) cm wide, acuminate to caudate at the apex, the tip obtuse, cuneate or tapering to base, entire, glabrous, dying ocherous beneath; lateral nerves 5–6 pairs, ± prominent beneath, the venation evident between; petiole 0.5–0.8 (1.2) cm long. Flowers 2–10 in small axillary clusters up to 8 mm long; peduncle 0.2–1.5 cm long, puberulous; buds very blunt. Calyx cupular, 0.5 mm long, toothed. Corolla pale green or yellowish white, ± 3 mm long including lobes up to 0.5 mm. Stamens exserted with conspicuous hairs (gold-tipped fide Cramer). Drupes white, oblong, 1.2–2.5 cm long, 5–8 mm wide, blunt at both ends, smooth but wrinkled when dry with persistent style disk at apex.

Distr. Ceylon, India, Burma, Indo-China, Thailand and China (including Hainan).

Ecol. Primary and secondary wet evergreen forest; 75–1200 m.

Uses. None recorded.

Vern. None recorded.

Note. The fruit has been described as ribbed or wrinkled; *Davidse* (8331) describes the fruit as red with dark brown 'seeds' others as fruit translucent with furrowed white 'seed' (endocarp).

Specimens Examined. KURUNEGALA DISTRICT: Weudakanda, *Waas 1644* (K, PDA, US). MATALE DISTRICT: Rattota—Illukkumbura road, *Cramer 4869* (K); Rattota, Bambaragala, road to Midlands, *Cramer 4994* (K); Rattota, Midlands, *Tirvengadum et al., 1* (K). COLOMBO DISTRICT: Lubugama, *Waas 1237* (K, PDA, US). KEGALLE DISTRICT: Kitulgala, *Comanor 541* (K, US); Kitulgala, across Kelani R. from Rest House, *Sohmer & Waas 10550* (K). KANDY DISTRICT: Gammaduwa, *Alston 666* (PDA); Hantane, 2000–3000 ft, *Gardner 102* (BM, K), *Gardner in C.P. 251*, p.p. (PDA), *Thwaites C.P. 762* (K); Kadugannawa, *Kostermans 24946* (K, PDA, US); Kandy, *Macrae 10* (K); Dolosbage, *Trimen s.n.* (PDA); Madulkelle, Hatale Estate, *Worthington 1980* (BM, K); Adam's Peak, *Thwaites C.P. 251*, p.p. (K); Pitakanda, *J.M. de Silva s.n.* (PDA). KALUTARA DISTRICT: Badureliya, Maguru Ganga, *Cramer 4166* (K, PDA, US); Morapitiya, 1.5 mi s of logging camp, *Meijer 2057* (K, PDA, US); Morapitiya, *Waas 919* (K, US); Hewessa, *s. coll. s.n.* (PDA). RATNAPURA DISTRICT: SW. base of Peak Sanctuary, *Jayasuriya & Robyns 126* (K, PDA); Rassagala, above Balangoda, *Kostermans 23578* (K, L); 2 mi from Rassagala, slopes of Dotalugala, *Sohmer & Waas 10518* (K); Gilimale Forest

Reserve, *Tirvengadum & Cramer 178* (PDA, US); Kuruwita Korale, ? *Trimen s.n.* (PDA); Palabaddala, *Waas 258* (PDA, US); Mandagala Oya Forest, along trail to Maskeliya from Malibode, *Waas 1687* (K, PDA, US); Marathagala, southern area of Sinharaja from Deniyaya, *Waas 1748* (K, PDA, US); Horagulkanda to Sinharaja, *Waas 2022* (K, PDA, US). GALLE DISTRICT: Hiniduma, Haycock Mt., *Davidse 7840* (K, PDA, US); Galle, *Gardner 102 bis* (K); Kanneliya Forest Reserve, near Udugama, Hiniduma, *Kostermans 24739* (BM, K, L); trail to Haycock Hill near Hiniduma, *Meijer 565* (K); Kanneliya Forest Reserve, *Waas 1342* (PDA, US, K); Foot of Titaweralawa, *Lewis & J.M. de Silva s.n.* (PDA); foot of Tittaweraluwa Kotha, *Lewis & J.M. de Silva s.n.* (PDA); LOCALITY UNKNOWN: *s. coll. C.P. 251* (BM, K); *Macrae 10, 422 & 426* (BM); *Walker 144* (K).

2. Gomphandra coriacea Wight, Ill. Ind. Bot. 1: 103. 1840; Trimen, Handb. Fl. Ceylon 1: 261. 1893; Sleumer, Blumea 17: 206. 1969. Type: India, Pulney Hills, *Wight s.n.* (K, Herb. Wight prop., holotype; Herb. Wight *427 & 433*, isotypes).

Gomphandra polymorpha sensu Wight, Ic. Ind. Or. 3(3): 3, t. 953. 1844; sensu
   Masters in Hook., Fl. Br. Ind. 1: 586. 1875, p.p., non Wight sensu stricto.
Stemonurus gardneri Miers, Ann. Mag. Nat. Hist. II, 10: 38. 1852; Miers,
   Contr. Bot. 1: 89, t. 13/2. 1852. Type: India, Nilgiri Hills, *Gardner & Wight* (BM, holotype; K, isotypes) (Sleumer speaks of the lectotype being at Kew but Miers' own herbarium is at BM and moreover Sleumer has annotated the BM specimen ?holotype).
Stemonurus coriaceus (Wight) Miers, Ann. Mag. Nat. Hist. II, 10: 37. 1852;
   Miers, Contr. Bot. 1: 87. 1852; Alston in Trimen, Handb. Fl. Ceylon 6: 48. 1931.
Stemonurus walkeri Miers, Ann. Mag. Nat. Hist. II, 43. 1852; Miers, Contr.
   Bot. 1: 93, t. 14/2. 1852. Type: Ceylon, *Col. Walker s.n.* (male), Ramboda, *Gardner 101* (female) (K-HOOK, syntypes; BM, fragments from Hooker sheets) (Sleumer has chosen *Gardner 101* as lectotype).
Platea coriacea (Wight) Thw., Enum. Pl. Zeyl. 44. 1858.

Shrub or small to medium-sized tree 2–15 m tall; d.b.h. 15–32 cm; bark grey or light brown, smooth; live bark yellowish or pinkish, fibrous. Leaves ± coriaceous, elliptic, oblong-elliptic or ± obovate, 3–12 cm long, 1.3–5.5 cm wide, rounded to bluntly acuminate or even emarginate at the apex, cuneate at the base, glabrous, minutely rugulose when dry; lateral nerves in 3–4 pairs; petiole (0.5–) 1 cm long. Inflorescences leaf-opposed or axillary (sometimes said to be extra-axillary but probably a misinterpretation of terminal leaf and its opposed inflorescence), 1.5–4 cm long, dichotomously branched; peduncle 1–2 cm long; secondary branches 1.5 cm long; pedicels 3–4 mm long, together with axes ± ferruginous adpressed pubescent. Calyx 0.5 cm long, flattened, shallowly toothed. Corolla white or greenish yellow, tube 2.5 mm long, the lobes ± triangular, 1.2 mm long, glabrous. Stamens slightly exserted with no or short hairs. Drupes green turning white, ellipsoid, 1.5–2 cm long, 7–9 mm wide, tipped with sessile stigma; endocarp with longitudinal lines.

Distr. Ceylon, India (Pulney and Nilgiri Hills).

Ecol. Primary and secondary intermediate and montane forest, disturbed forest, jungle; 600–1800 m.

Uses. None recorded.

Vern. None recorded.

Note. The fruits have also been described as orange-yellow. Sleumer has suggested this is possibly only a montane variant of the previous species but nevertheless maintains them distinct. The altitudinal ranges overlap considerably although *G. coriacea* is not found in the lowlands as *G. tetrandra* is nor is the latter ever found really high up. I have found it relatively easy to tell the two apart from dried material and Trimen, Gamble and others who knew them in the field clearly did not doubt their distinct status.

Specimens Examined. KANDY DISTRICT: Rangala, Corbet's Gap, *Alston 921* (K, PDA); Rangala to Knuckles, *Balakrishnan 200* (US), *205* (K, PDA, US); Rangala, *Balakrishnan 603* (K, PDA, US), *s. coll. s.n.* (PDA); Rangala-Hunasgiriya road, *Cramer et al. 3928* (PDA); Hantane, *Gardner in C.P. 251*, p.p. (PDA); Rangala to Corbet's Gap, *Kostermans 23075A* (BM, K, L); Knuckles, Madulkelle, *Kostermans 25028* (K, PDA, US), *25075* (PDA, US); Hunasgiriya Hill above Madulkelle Estate, *Sohmer & Jayasuriya 10623* (K); Ambagamuwa, ? *Thwaites C.P. 375*, p.p. (PDA); Hunasgiriya, close to Rangala-Hunasgiriya road, *Tirvengadum et al., 71* (PDA, US); Madulkele, Nilloomally, ? *Trimen s.n.* (PDA); Dolosbagie, *Trimen s.n.* (PDA); Corbet's Gap to Knuckles, *Waas 187* (K, US); Hunasgiriya, *Waas 974* (K, US), *978* (K, US); Kobonilgala, *Waas 1067* (K, PDA, US); Adam's Peak, *Thwaites C.P. 370*, p.p. (BM, K), *Balakrishnan 284* (PDA, US); Moray Estate, *Sohmer & Sumithraarachchi 8703, 9848* (PDA, US). BADULLA DISTRICT: Above Haputale, Thotulagala Estate, *Sumithraarachchi 900* (K). RATNAPURA DISTRICT: Pinnawala, *Balakrishnan 546* (PDA, US); Dotulagala, *Waas 1034* (K, US), *1037* (K, PDA, US), *1637* (US); Maratenna, *Waas 1783* (K, US); S of Adam's Peak, Mukkuwatta Forest, *Waas 1813* (K, PDA, US). NUWARA ELIYA DISTRICT: Dimbula, *Thwaites C.P. 375*, p.p. (BM, PDA); Maturata, ?*Trimen s.n.* (PDA); Mool Oya, *Waas 1130* (PDA, US); Bogawantalawa, *Worthington 781* (K). LOCALITY UNKNOWN: *Gardner s.n.* (PDA).

2. APODYTES

E. Meyer ex Arn. in Hooker's J. Bot. Kew Gard. Misc. 3: 155. 1840; Sleumer, Blumea 17: 184. 1969; Sleumer in Fl. Males. I, 7: 47. 1971. Type species: *A. dimidiata* E. Meyer ex Arn.

Trees or shrubs. Leaves alternate, spirally arranged, entire, pinnately nerved, often shiny and blackening on drying. Flowers hermaphrodite, small, in terminal and axillary corymbs. Calyx cupular, small, (4–) 5-toothed, persistent. Petals (4–) 5, free or slightly coherent at base, valvate, inflexed at apex, glabrous. Stamens 5, alternating with and almost as long as the petals, slightly coherent with them at the base; anthers linear-oblong, introrse, sagittate at base, medifixed, opening by lateral slits. Ovary oblique with a lateral swelling at the base, 1-locular; ovules 2, pendulous; style thick, ± excentric; stigma small. Drupe obliquely ellipsoid or rounded, ± unilaterally developed, ± compressed, the style finally lateral to almost basal and the lateral appendage large and succulent; endocarp crustaceous or woody; seed 1, compressed, ovate-reniform; embryo small within the apex of the fleshy endosperm.

Formerly considered to comprise about 17 species. Sleumer admits only 2, one in Queensland, the other extending from SE Asia and Malesia to Africa and occurring in Ceylon.

Apodytes dimidiata E. Meyer ex Arn. in Hooker's J. Bot. Kew Gard. Misc. 3: 155. 1840; Benth., Trans. Linn. Soc. London 18: 680, 683, t. 41. 1841; Sleumer, Blumea 17: 185. 1969 (very full synonymy); Sleumer in Fl. Males. I, 7: 48, fig. 17. 1971; Matthew, Fl. Tamilnadu Carnatic, Illustr., t. 139. 1982, Fl. (1): 250. 1983. Type: S. Africa, Uitenhage, Krakakamma Forest, *Zeyher 673* (E (ex GL), syntype; K, isosyntype) & S. Africa, Durban (Port Natal), *Drege s.n.* (E (ex GL), syntype; K (ex Benth.), isosyntype).

Apodytes benthamiana Wight, Ic. Pl. Ind. Or. 3: (14), t. 1153. 1846; Bedd., Fl. Sylv. 3: LXII, t. 140 var. A & B. 1871; Masters in Hook. f., Fl. Br. Ind. 1: 588. 1875; Gamble, Fl. Pres. Madras 1: 195. 1915. Type: India, Nilgiri Hills, woods near top behind Avalanche Bungalow, *Wight s.n.* (K, holotype).

Apodytes gardneriana Miers, Ann. Mag. Nat. Hist. II 9: 389. 1852; Miers, Contr. Bot. 1: 58, t. 5. 1852, p.p.; Thw., Enum. Pl. Zeyl. 42. 1858; Masters in Hook. f., Fl. Br. Ind. 1: 588. 1875; Trimen, Handb. Fl. Ceylon 1: 262. 1893. Type: Ceylon, Nuwara Eliya at 6000 ft, *Gardner 189* (BM, holotype; CGE, G, K, isotypes).

Nothapodytes zeylanica Kostermans, Acta Bot. Neerl. 31: 127. 1982. Type: Ceylon, Matale District, Ratotta-Illukkumbura road, mile 37, *Huber 712* (L, holotype; PDA, isotype).

Shrub or spreading tree, mostly (2) 15–30 m tall, the largest specimens with fluted trunks (6–) 20–70 cm diam.; crown high and lax; bark smooth or rough, dark grey to brown; live bark straw-coloured; young branchlets glabrous to sparsely pubescent; older branches with pale lenticels. Leaves very variable, ovate, elliptic, narrowly oblong or obovate-oblanceolate, 2–15 cm long, 1–8 cm wide, shortly acutely acuminate to obtuse at the apex, cuneate to obtuse at the base, sometimes unequal, somewhat decurrent on the petiole, ± thin to coriaceous, dark green and shining in life, drying blackish-brown, sometimes with firm ± adpressed yellowish hairs on midrib above, otherwise ± glabrous; nerves raised or plane beneath; petiole 0.5–3 cm long. Flowers numerous, sweet-scented, shortly pedicellate or sessile in terminal or less often axillary ± pubescent corymbs 3–8 cm long; peduncles 1–3 cm long, buds ellipsoid-oblong; bracts minute or absent. Calyx cupular, 0.5–0.7 mm long with 5 short triangular teeth, puberulous. Petals 5, white or yellowish, drying black, narrowly oblong, (2–)–6 mm long, 1 mm wide. Anthers yellow, 2–3.5 mm long; filaments 1.5–3 mm long. Ovary ovoid, 0.7 mm long, densely pubescent; style white, 2–2.5 mm long. Fruit dark purple becoming black and shining with appendage green turning scarlet, asymmetrically oblong-ovoid, oblique, ± 5–11 mm long, 5–9 mm wide, laterally compressed, 3–4 mm thick, glabrous or pubescent.

Distr. Widely distributed in Africa, Ceylon, S. India, Assam, Burma, Thailand, Indo-China, Yunnan, Hainan, and Sumatra to the Moluccas.

Ecol. Montane forest including primary and distrubed jungle, windswept dense low forest and scrub, sometimes in valleys and by streams, often on rocky slopes; 200–2100 m.

Uses. None reported.

Vern. None reported.

Note. Montane populations mostly have smaller leaves with recurved margins.

Specimens Examined. MATALE DISTRICT: Rattota to Illukkumbura road mi. 37, *Huber 712* (L, PDA); Laggala to Illukkumbura, mi. 37, *Jayasuriya 280* (K, PDA, US); near Laggala, *Trimen s.n.* (PDA); Rattota to Illukkumbura road, *Kostermans 27992* (K, L), *27993* (G, L), same road steep slope down to Illukkumbura above Midcar Estate, *Kostermans 28083* (K, L). KANDY DISTRICT: Rangala to Corbet's Gap, *Kostermans 23489* (K, PDA, L); Knuckles, *J.M. de Silva 12* (PDA); Rangala, ?*Trimen s.n.* (PDA); Nilloomaly, Madulkelle ? *Trimen s.n.* (PDA); Hunasgiriya, *Waas 982* (K, US); Knuckles, Madulkele, *Worthington 1952* (K); Corbet's Gap, Rangala, *Worthington 5483* (K, PDA); Ambagamuwa, *s. coll. C.P. 204* p.p. (PDA); Gartmore, Rajamally, *J.M. de Silva s.n.* (PDA). RATNAPURA DISTRICT:

Warnagala, Adam's Peak, *Cramer 4691* (K); Galagama, *Thwaites C.P. 204* (BM, K), *C.P. 495* (K); Maskeliya-Maliboda trail, Mandegale Oya Forest, *Waas 1691* (K, PDA, US). NUWARA ELIYA DISTRICT: Moon Plains, *Balakrishnan & Dassanayake 1131* (K, PDA, US); Hakgala-Nuwara Eliya road, Culvert 56/8, *Cramer 4247* (PDA, US); Ragala, Uda Pussellawa, *W.F. s.n.* (PDA); Nuwara Eliya, *Herb Gardner 98, Herb Wight* (PDA), *Gardner 189* (BM, CGE, G, K, PDA), Hakgala, *A.M. Silva s.n.* (PDA); Maturata, ?*Thwaites in C.P. 204*, p.p. (PDA); Pattipola, approach to Horton Plains, *Tirvengadum & Cramer 97* (PDA); Pattipola, Horton Plains, *s. coll. s.n.* (PDA). BADULLA DISTRICT: Namunukula, *J.M. Silva s.n.* (PDA); on way to Fort Macdonald, *J.M. Silva s.n.* (PDA). LOCALITY UNKNOWN: *Gardner in C.P. 204*, p.p. (PDA); *Thwaites C.P. 1218 & 1219* (K).

3. NOTHAPODYTES

Blume, Ann. Mus. Bot. Lugduno Batavum 1: 243. 1850; Sleumer, Notizbl. Bot. Gart. Berlin 15: 246. 1940; Sleumer in Pflanzenfam. ed. 2, 20b: 274. 1942; Sleumer, Blumea 17: 232. 1969; Sleumer in Fl. Males. I, 7: 53. 1971. Type species: *N. montena* Blume.

Neoleretia Baehni, Compt. Rend. Seances Soc. Phys. Geneve 53: 1, 33. 1936; Baehni, Candollea 7: 177. 1936. Type species not designated.

Trees or shrubs with alternate or almost opposite mostly oblong to broadly elliptic, entire, glabrous to densely hairy leaves. Flowers hermaphrodite or polygamous, small, in lax terminal cymes; pedicels articulated beneath the calyx. Calyx campanulate or saucer-shaped. 5-lobed, persistent. Petals 5, linear, valvate, hairy within, somewhat connivent at the base. Stamens 5; filaments thickened at the base; anthers ovoid to oblong, dorsifixed. Ovary oblong-ovoid, unilocular, hairy; ovules 2, pendulous from apex of locule; style terminal, filiform to conical; stigma capitate, concave or 2-lobed. Disk leafy, annular, 5–10-toothed or -lobed. Drupe ellipsoid, 1-seeded. Seed with copious endosperm, the cotyledons thin, leafy, almost as long as the seed.

Four species in W. China, Ryu-Kyu Is., Indo-China, Thailand, S. India, Ceylon, Burma, Malaysia, Sumatra and Philippine Is.

Kosterman's supposed new species of *Nothapodytes* appears to be no more than *Apodytes dimidiata* q.v.

Nothapodytes foetida (Wight) Sleumer, Notizbl. Konigl. Bot. Gart. Berlin 15: 247. 1940; Howard, J. Arnold Arbor. 23: 70. 1942; Sleumer, Blumea 17: 232, 1969 (very extensive synonymy); Sleumer in Fl. Males. I, 7: 55, fig. 26. 1971.

Stemonourus foetidus Wight, Ic. Pl. Ind. Or. 3(3): 4, t. 955. 1845 (as 'eaetidus' in text). Type: India, Nilgiri Hills, *Herb. Wight 431* (K, holotype; BM, ?fragment).

Mappia championiana Miers, Ann. Mag. Nat. Hist. II, 9: 397. 1852. Type: Ceylon, *Col. Walker, Maj. Champion s.n.* (K, syntypes; BM, fragment).

Mappia foetida (Wight) Miers, Ann. Mag. Nat. Hist. II, 9: 395. 1852; Thw., Enum. Pl. Zeyl. 43. 1858; Bedd., Fl. Sylv. 3, t. 141. 1871; Masters in Hook. f., Fl. Br. Ind. 1: 589. 1875.

Mappia gardneriana Miers, Ann. Mag. Nat. Hist. II, 9: 396. 1852; Miers, Contr. Bot. 1: 66. 1852. Type: Ceylon, Nuwara Eliya, *Gardner 98* (BM, K, syntypes) & Galagama, *Thwaites C.P. 492* (BM, K, syntypes; PDA etc., isosyntypes).

Mappia oblonga Miers, Ann. Mag. Nat. Hist. II, 9: 396. 1852; Miers, Contr. Bot. 1: 65. 1852; Masters in Hook. f., Fl. Br. Ind. 1: 589. 1875. Type: India, Bombay, 'Ghauts' *Dalzell s.n.* (K, holotype).

Mappia ovata Miers, Ann. Mag. Nat. Hist. II, 9: 396. 1852; Masters in Hook. f., Fl. Br. Ind. 1: 589. 1875; Trimen,* Handb. Fl. Ceylon 1: 262. 1893. Type: Ceylon, Hantane, *Gardner 99* (BM, K, syntypes).

Mappia sp. Thw., Hooker's J. Bot. Kew Gard. Misc. 7: 212. 1855.

Mappia foetida (Wight) Miers var. *championiana* (Miers) Thw., Enum. Pl. Zeyl. 43. 1858.

Mappia foetida (Wight) Miers var. *gardneriana* (Miers) Thw., Enum. Pl. Zeyl. 43. 1858.

Mappia ovata Miers var. *championiana* (Miers) Trimen, Handb. Fl. Ceylon 1: 263. 1893.

Neoleretia foetida (Wight) Baehni, Candollea 7: 177, pl. 4, f. 2/D-F. 1936.

Nothapodytes gardneriana (Miers) Kostermans, Acta Bot. Neerl. 31: 127. 1982.

Small tree or shrub 1.5–5 m tall; d.b.h. 3–12.5 cm; bark whitish or grey, smooth; inner bark yellowish, soft; twigs drying yellowish, wrinkled, lenticellate; buds yellowish, pubescent. Leaves with ± foetid smell, ovate or ovate-oblong, 3.5–28 cm long, 1.5–19 cm wide, acute or shortly acuminate at the apex, broadly cuneate to rounded at the base, entire; lateral nerves ± 6–8 pairs usually with a tuft of hairs in axils; venation reticulate; petiole 0.8–6 cm long; cymes often foetid, 2.5–9 cm long, up to 15 cm wide; peduncle short, 0.7–5 cm long; buds ovoid or ± globular, hairy. Calyx 0.5 mm long, pubescent. Petals yellow or occasionally white, lanceolate, 5 mm long, acute, pubescent on both sides. Ovary flask-shaped, hairy. Drupes green turning

---

* Trimen considered *M. foetida* distinct from *M. ovata,* so he cited *M. foetida* as sensu Thw. non·Miers in the synonymy of *M. ovata.*

green and purple, dull red and yellow, finally bright red or purple, oblong-ovoid, minutely apiculate, smooth or wrinkled when dry. Seeds purplish.

Distr. Ceylon, India, Burma, Thailand, Indo-China, China and Malesia.

Ecol. primary and secondary lowland wet forest, intermediate and montane forest, scrub forest by small rocky streams; 100–1800 m (to 2100 fide Trimen).

Uses. None recorded.

Vern. Gandapana (S).

Specimens Examined. MATALE DISTRICT: Campbells Lane, *Waas 948, 965* (K, US). KANDY DISTRICT: Kandy jungle, *Alston 102* (PDA); Haloluwa-Getambe road, *Cooray 69100903R* (PDA, US); slopes of Adam's Peak, along the SE. trail from the Moray Group Tea Estates, *Davidse & Sumithraarachchi 8627* (K, PDA, US); Hantane, *Gardner 99* (K); E. of Madugoda, *Jayasuriya 1922* (K, PDA, US); Kandy, *Moon 112* (BM), *310* (BM). BADULLA DISTRICT: Namunukula, *s. coll. s.n.* (PDA). RATNAPURA DISTRICT: Galagama, *Thwaites C.P. 492*, p.p. (K); above Aberfoyle Tea Estate, Bulutota, above Bulutota Pass, *Kostermans 28123* (K, L); Horagulkanda to Sinharaja, *Waas 2028* (K, PDA, US); Belihuloya, *Worthington 425* (K, PDA). NUWARA ELIYA DISTRICT: Forest opposite Hakgala Gardens, *Balakrishnan 1062* (PDA, US); Pundaluoya, *Cramer 4068* (PDA, US); Hakgala-Keppetipola road, near culvert 86/8, *Cramer 4477* (PDA, US); Nuwara Eliya at 6000 ft, *Gardner 981* (K); Elk Plains, near Ambawela, *Kostermans 23083* (K, L); Hakgala, near MAB Reserve, trail to summit of Hakgala Mt., *Meijer 1753* (K); 0.6 mi. W. of Ambawela station, *Mueller-Dombois 67071003* (PDA); Hakgala, *A.W.S. 83/66* (PDA), *Simpson 9504* (BM); Hakgala Trail from Garden to first summit of the Hakgala Mts, *Sohmer & Sumithraarachchi 10084* (K, PDA, US); Nuwara Eliya, *Thwaites in C.P. 492* p.p. (BM, PDA); Totupola, Horton Plains, *?Trimen s.n.* (PDA); Hakgala Nature Reserve, *Wirawan s.n.* (PDA). MONERAGALA DISTRICT: Badulawela, *Jayasuriya 2064* (PDA, US). GALLE DISTRICT: Kanneliya logging site, *Waas 1516* (PDA, US). LOCALITY UNKNOWN: *Champion s.n.* (K); *Macrae 239* (BM); *Walker s.n.* (K).

Note. I have followed Sleumer in his wide interpretation of this species but some populations, e.g. densely tomentose specimens from Burma, deserve varietal names.

4. STEMONURUS

Blume, Bijdr. 468. 1826; Sleumer, Blumea 17: 255. 1969. Type species: *Stemonurus secundiflorus* Blume.

Urandra Thw. in Hooker's J. Bot. Kew Gard. Misc. 7: 211. 1855. Type species: *Urandra apicalis* Thw. (= *Stemonurus apicalis* (Thw.) Miers). *Lasianthera* sensu Miq., Fl. Ind. Bat. 1: 790. 1856 p.p., non P. Beauv.

Trees, sometimes buttressed, glabrous save for some parts of the inflorescence. Leaves spirally arranged, ± coriaceous and thick, entire; petioles thick. Flowers hermaphrodite, sessile in axillary usually solitary umbel-like inflorescences or heads supported by ± conspicuous often persistent bracts. Calyx cupular, truncate or ± 5-lobed. Petals (4–) 5, valvate, joined at base, inflexed at apex. Stamens (4–) 5; filaments fleshy, flattened, shortly bearded below anthers on ventral side, long-penicillate dorsally, the pencil of hairs inflexed in bud but erect and exserted in open flower. Disk rim- or cup-like. Ovary ovoid-conic, unilocular with 2 pendulous ovules; style short, slender, with punctiform stigma. Drupe ovoid, ellipsoid or fusiform-oblong, 1-seeded, usually bicoloured, the lower part red or purple, the upper part cream or greenish; endocarp coriaceous to woody, the outer part fibrous.

Twelve species extending from Ceylon (one), Indo-China and Malesia to the Solomon and Palau Islands.

Stemonurus apicalis (Thw.) Miers, Trans. Linn. Soc. London 22: 110. 1856; Thw., Enum. Pl. Zeyl. 43. 1858; Miers, Contr. Bot. 1: 305. 1861; Becc., Malesia 1, t. 4, fig. 14 & 15. 1877.

Urandra apicalis Thw., Hooker's J. Bot. Kew Gard. Misc. 7: 211. 1855. Type: Ceylon, Central Prov., Alagalle, 1000–2000 ft, *Thwaites* in *C.P. 2569* (PDA, holotype; A, BM, CGE, FI, G, K, L, P, isotypes).
Lasianthera apicalis (Thw.) Thw.,* Enum. Pl. Zeyl. 405. 1864; Bedd., Fl. Sylv., t. 139, For. Man. 61. 1871 (as *apicaulis*); Masters in Hook. f., Fl. Br. Ind. 1: 584. 1875, p.p.; Trimen, Handb. Fl. Ceylon 1: 260. 1893; Lewis, Veg. Prod. Ceylon 101. 1934.

Tree 6–20 (–70 fide *Waas 2031*) m tall with dense pyramidal crown; d.b.h. 10–40 (45 fide *Waas 2031*) cm; bark grey, greenish grey, pale straw or yellowish, thin, smooth, lenticellate; inner bark dirty white or straw-coloured, laminated, gritty; lower branches somewhat drooping; young parts glabrous, dark, varnished, rugulose when dry. Scars prominent, pale. Leaves oblong-elliptic to obovate-oblong, 5–18 cm long, 2–7.5 cm wide; ± cuneate at the base, ± obtuse to abruptly and shortly acuminate at the apex, entire, thick and coriaceous; midrib very prominent but the 10–12 pairs of lateral nerves hardly visible; petiole 0.7–2 cm long. Flowers green or reddish in bud, ± crowded into heads 0.5–1 cm long; peduncle stout, 0.7–1 cm long; bracts

---

* Always attributed to Benth. & Hook. f., Gen. pl.1: 350. 1862 but combination not actually made there.

4, ovate, 2.5–4 mm long, 2–2.5 mm wide, pubescent. Calyx ± 3 mm long, lobed for about a third to half its length, the lobes ± acute or obtuse with slight pubescence persistent. Petals reddish or white, oblong, 4–5 mm long, 1.5 mm wide. Stamens with hairs exserted beyond the petals, completely concealing the anthers from the outside. Disk slightly 5-lobed. Drupe at first with upper part grey-green and lower dark green becoming yellow-green but finally white above and red-brown or purplish green below, the two colours sharply demarcated, oblong-ovoid, 4–5 cm long, ± 2.3 cm wide, somewhat acute; pulp scanty; endocarp thick, woody with large cavity filled with soft tissue, the outer part with vertical fibrous woody bands.

Distr. Endemic. Trimen's report of it from Borneo is a result of a misidentification in Fl. Br. Ind. of another species.

Ecol. Primary wet evergreen forest; 45–700 m.

Uses. A resinous wax exudes from the bark of old trees which is inflammable and slightly scented when burned.

Vern. Urukanu, Uruhonda (S).

Specimens Examined. COLOMBO DISTRICT: Kottawa, *Worthington 2353 (2352?* —specimen not now at K), *5272* (K); Indikada Mukulana, Waga, *Worthington 3542* (K). KEGALLE DISTRICT: Kitulgala, *Lewis s.n.* (PDA), *Worthington 393* (BM). KANDY DISTRICT: Maskeliya Valley Road, *Kostermans 24199* (BM, K, L, PDA); Kadugannawa, Udawela, Andiakande, *Worthington 359* (K); Udawela, Kadugannawa, *Worthington 1435* (BM), *Worthington 2834* (K); Alagalle, *Thwaites 2569* (A, BM, CGE, FI, G, K, L, P, PDA). KALUTARA/RATNAPURA DISTRICT: Kaluganga Valley, *Lewis s.n.* (PDA). KALUTARA DISTRICT: Pasdun Korale, *?Trimen s.n.* (PDA). RATNAPURA DISTRICT: Gilimale Forest Reserve, *Jayasuriya & Bandaranayake 1874* (K, PDA, US), *Meijer 413* (K, PDA), *Tirvengadum et al. 166* (PDA, US); Ratnapura-Carney road, *Kostermans 23297* (BO, K); Tumbagoda, N. of Rassagalla-Ratnapura road, *Kostermans 24647* (BM, PDA), *24677* (K, L); Morapitiya Forest Reserve, *Meijer 2067* (K); Horagulkanda to Sinharaja, *Waas 2031* (K, PDA, US). GALLE DISTRICT: Hiniduma, near summit of Godekande Peak, *Cramer 3696* (PDA, US); Kanneliya Forest Reserve, *Jayasuriya & Kostermans 2336* (K, PDA, US); Hiniduma Kanda, *Jayasuriya et al. 789* (K, PDA, US); Kanneliya Forest near Hiniduma, *Kostermans 24980* (BM, K, L, PDA), *25538* (K, L); Kanneliya Forest Reserve, mile 17.5 logging road, *Meijer 986* (PDA), *Meijer 1018* (PDA); Kanneliya, *Waas 1337* (K, PDA, US), *Waas & Peeris 553* (K, PDA, US); Hiniduma, Kuneli Mookalana, *Worthington 2252* (BM); Homodola Estate, Udugama, *Worthington 4125* (BM). MATARA DISTRICT: Dediyagala Forest, *Worthington 2534*

(BM). LOCALITY UNKNOWN: *Thwaites C.P. 1396* (K); *s. coll. C.P. 2569* (BM, K, PDA); Herb. Wight, *s. coll. s.n.* (K).

5. PYRENACANTHA

Hook.,* Bot. Misc. 2: 107, suppl. tt. 9 & 10. 1830; Sleumer, Blumea 17: 249. 1969; Sleumer in Fl. Males. I, 7: 76. 1971, nom. cons. Type species: *Pyrenacantha volubilis* Hook.

Scandent dioecious or less often monoecious shrubs or climbers, often from tuberous roots or massive partly buried swollen stems. Leaves alternate, entire to deeply lobed, penni- or palmati-nerved; hydathodes sometimes present; petiole sometimes prehensile. Flowers in ± axillary spikes or racemes. Calyx absent. Petals (3-) 4-5(-6), united at the base, valvate, persistent. Male flowers: stamens the same number as and alternating with the petals; filaments usually short; rudimentary ovary represented by a few bristle-like hairs. Female flowers: staminodes sometimes present, alternating with the petals; ovary ovoid, unilocular; stigma sessile, discoid or variously divided; ovules 2, pendulous. Fruit a 1-seeded drupe with fleshy exocarp; endocarp woody bearing peg-like protuberances on inner surface deeply penetrating the endosperm.

About 20 species in Asia, tropical and subtropical Africa and Madagascar; only 1 in Ceylon.

Pyrenacantha volubilis Hook., Bot. Misc. 2: 107, suppl. tt. 9, 10. 1830; Thw., Enum. Pl. Zeyl. 290. 1861; Trimen, Handb. Fl. Ceylon 1: 263. 1893; Gamble, Fl. Pres. Madras 1: 198. 1915. Type: India, ? Madras, Vellengamy, *Wight s.n.* (K, holotype) (see note).

A climber or scandent shrub about 2 m long partly climbing by means of petioles; stems slender with milky juice, young parts slightly rough with setulose adpressed pubescence. Leaves elliptic, oblong-elliptic or narrowly oblong-lanceolate, 2–12 cm long, 1.2–7.5 cm wide, acute and mucronate at the apex, narrowing to a truncate base with a gland on either side, entire, crenate or with shallow distinct angular teeth tipped by a hydathode, glabrous above, rough with stiff adpressed hairs beneath; lateral nerves in ± 4 pairs; petioles flexuous, 0.3–1.8 cm long, glabrous. Male flowers in filiform axillary spikes 1–4 cm long with peduncle ± 1.5 cm long, puberulous; bracts minute, lanceolate, ± 0.5 mm long; petals ovate, 0.7 mm long, 0.3 mm wide; filaments shorter than anthers. Female flowers in more compact ± 8–10-flowered head-like spikes up to 1.5 cm long with peduncles 2–4 cm long; ovary hairy. Drupes orange-red, compressed, ovoid or ellipsoid, 1.1–1.5 cm long, 8–9 mm

---

* Often attributed to Wight but he sent it to Hooker without a name; although he recognised it was a new genus he can hardly be treated as an author (see note).

wide, obtuse, puberulous, irregularly rib-veined at least when dry. Seeds compressed, oblong, 9.5 mm long, 5 mm wide, densely perforated.

Distr. India ('Madras'), Ceylon, Indo-China, Hainan.

Ecol. Intermediate forest at low altitudes.

Uses. None recorded.

Vern. None recorded.

Specimens Examined. MANNAR DISTRICT: Madhu-Tirukiteswaram road, *Sumithraarachchi 772* (PDA, US). VAVUNIYA DISTRICT: 10 mi. from Mullaitivu towards Puthukkudiyiriruppu, *Kundu & Balakrishnan 718* (PDA, US). PUTTALAM DISTRICT: *Gardner* in *C.P. 1024*, p.p. (PDA). KURUNEGALA DISTRICT: Dumkanda, Arankele, *Jayasuriya & Balasubramaniam 521* (K, PDA, US); Kurunegala, *Gardner* in *C.P. 1024*, p.p. (BM, K, PDA). POLONNARUWA DISTRICT: Manampitiya, *Kundu 231* (PDA, US). LOCALITY UNKNOWN: *Col. Walker s.n.* (K); *s. coll. s.n.* (K).

Note. The specimen cited as holotype is one from the herbarium of W. Hooker which bears a note by C. Fischer that it is almost certainly the original Wight specimen from Vellengamy; another label in Mrs. Graham's handwriting (fide J.D. Hooker) bears the number 142 and states that it is a new genus and that the specimen had been sent to Hooker. Another sheet, *Wight in Wallich 8581* is from a locality not cited by Wight (Tulassipatnam and the sheet in the Wallich Herbarium also bears a Rottler label Aug 1814). The citation as *Pyrenacantha volubilis* Wight in Wallich's numerical list dates from 1847/8. The genus was omitted from the Flora of British India.

LOGANIACEAE
(by D. Philcox[*])

Mart., Nov. Gen. Sp. Pl. 2: 133. 1827 ("Loganieae"). Type genus: *Logania* R.Br., nom. cons.

Trees, shrubs, woody climbers or herbs. Leaves usually opposite, rarely ternate, entire, pinnately- or more rarely 3–7-veined, stipulate or not; stipules where present usually interpetiolar, at times reduced to stipular line. Flowers actinomorphic, bisexual, usually arranged in simple or compound, terminal or axillary cymes, or solitary. Calyx 5-, rarely 4-lobed, lobes free or rarely (in our material) united. Corolla tubular, 5-, rarely 4-lobed, lobes valvate or imbricate. Stamens 5 (4), inserted on corolla tube, alternating with lobes; anthers 2-celled, usually basifixed, dehiscing laterally. Ovary superior, 2-celled; ovules 1-many in each cell. Style simple. Stigma capitate or 2-lobed. Fruit (in our material) baccate, rarely capsular. Seeds 1-many.

Family of some 28 genera occurring in the tropics of both Old and New Worlds, and represented in our area by only three or four which contain eleven species. The genus in doubt is *Spigelia* which was recorded as occurring in Ceylon by Abeywickrama (1959), but to date no material has been found. Many earlier authors have also included *Gaertnera* in Loganiaceae, but most now agree that it is more correctly placed in Rubiaceae, where it may now be found.

KEY TO THE GENERA

1 Annual herb; fruit capsular
  2 Calyx and corolla 4-lobed; stamens 4 .... **1. Mitrasacme**
  2 Calyx and corolla 5-lobed; stamens 5 .... **2. Spigelia**
1 Trees or shrubs; fruit baccate
   3 Corolla lobes valvate in bud .... **3. Strychnos**
   3 Corolla lobes imbricate in bud .... **4. Fagraea**

1. MITRASACME

Labill., Nov. Holl. Pl. Spec. 1: 35, tab. 49. 1804. Type species: *Mitrasacme pilosa* Labill.

---

[*] Royal Botanic Gardens, Kew.

Herbs, annual or perennial. Leaves opposite, sessile or subsessile, connate at base. Flowers solitary-axillary, rarely terminal. Calyx 4-lobed, lobes connate to about halfway, campanulate. Corolla 4-lobed, campanulate, valvate in bud. Stamens 4, inserted on lower half of corolla tube, mostly included; filaments usually filiform; anthers 2-celled. Ovary 2-celled; ovules many in each cell. Styles 2, eventually distinct. Fruit capsular, dehiscing by apical slits, 2-horned, with horns terminated by styles, which may be either separate or partly connate. Seeds many, minute, ovoid, minutely lobed.

Genus of about 40 species mainly from Australia, New Zealand, New Caledonia, and Malaysia, and from India and Ceylon to Korea and central Japan.

Mitrasacme indica Wight, Ic. Pl. Ind. Or. 4, 4: 15, tab. 1601. 1850; Thw., Enum. Pl. Zeyl. 200. 1860; Gamble, Fl. Pres. Madras 5: 864. 1923; Leenhouts in Fl. Males. 6: 384. 1962. Type: from India.

Mitrasacme pusilla Dalz., Hooker's J. Bot. Kew Gard. Misc. 2: 136. 1850. Type: from India.

Mitrasacme alsinoides sensu Clarke in Hook. f., Fl. Br. Ind. 4: 80. 1883, non R. Br. 1810; Trimen, Handb. Fl. Ceylon 3: 170. 1895.

Annual herb, erect, up to 12 cm tall, slender, branched, clearly 2–4-angled to shortly 2–4-winged, glabrous, or less frequently subglabrous. Leaves 2.5–7 × 0.8–1 mm, linear-lanceolate to narrowly ovate, acute, glabrous, patent at right-angles to stem. Flowers solitary in upper axils of stem or branch, pedicellate. Pedicels 3.5–11 mm long, filamentous, glabrous to subglabrous. Calyx c. 2 mm long, lobes c. 1 mm wide, ovate-lanceolate, acute, glabrous, connate about halfway. Corolla white, 3–4 mm long, campanulate, sparsely bearded at mouth, lobes c. 1.5 × 0.8 mm, rounded at apex. Ovary c. 0.5 mm diameter. Styles distinct at base, connate only at apex. Stigma truncate, shortly bilobed. Capsule c. 2 mm diameter, vertically compressed-globose, glabrous, with remains of stigmas appearing as two horns capping capsule, frequently continuing connate at apex. Seeds many, c. 0.2 mm diameter, ovoid, angular, densely reticulate, pale brown.

Distr. southern India and Ceylon; Central China, Japan, Hainan through to Malaysia and the Philippines and New Guinea to northern and eastern Australia.

Ecol. Dry to marshy, sandy soils up to 500 metres.

Specimen Examined. JAFFNA DISTRICT: Jaffna, *Gardner s.n.* in *C.P. 1677*, p.p. (PDA), Feb 1890, *s. coll. s.n.* (PDA). PUTTALAM DISTRICT: Marai Villu, 1 Feb 1970, *Cooray 70020102R* (PDA). MATALE DISTRICT: Dikpatana, W of Illukkumbura, 825 m, 9 Jan 1984, *Jayasuriya*

3063 (PDA); Laggala, c. 900 m, 14 Feb 1978, *Cramer 5144* (K). POLON-
NARUWA DISTRICT: Polonnaruwa, Mar 1853, *Thwaites s.n.* in *C.P. 1677*,
p.p. (PDA); Habarana, top of Bahirawagala, c. 90 m, 14 Jan 1978, *Cramer
5091* (K, PDA). KANDY DISTRICT: Dolosbage, Apr 1882, *s. coll. s.n.*
(PDA). AMPARAI DISTRICT: Lahugala, sea-level, 11 Jan 1978, *Cramer
5048* (K, PDA). MONERAGALA DISTRICT: Nannapura, 18 July 1972,
Hepper & de Silva 4727 (K). LOCALITY UNKNOWN: *Thwaites s.n.* in
C.P. 1677, p.p. (BM, K).

2. SPIGELIA

L., Sp. Pl. 149. 1753; L., Gen. Pl. ed. 5: 74. 1754. Type species: *Spigelia
anthelmia* L.

Annual or perennial herbs or undershrubs, glabrous or pubescent, hairs
simple or stellate. Stems simple or branched. Leaves simple, opposite or
whorled usually at apex of stem, connate at base or with interpetiolar stipules.
Inflorescence terminal or almost so. Sepals free or connate, equal or unequal.
Corolla 5-lobed, regular, valvate in bud, lobes somewhat triangular, acute.
Stamens 5, alternating with corolla lobes, inserted on corolla tube; filaments
free; anthers 2-celled, dehiscing by longitudinal split. Ovary 2-celled, superior, with several ovules in each cell. Style 1, deciduous in upper half. Capsule
2-lobed, 4-valved, valves deciduous leaving cupular base in persistent calyx.
Seeds obliquely ellipsoid or ovoid, reticulate, verrucose or tuberculate.

Genus of about 50 species from tropical or subtropical America, with
one species, *S. anthelmia* L., naturalized in the Old World.

Spigelia anthelmia L., Sp. Pl. 149. 1753; DC., Prod. 9: 7. 1845; Leenh. in
Fl. Males. 6: 377, fig. 38. 1962. Type: from Brazil.

Annual herb, 12–50 cm tall, stem simple or less frequently stoutly
branched below, especially towards base, glabrous. Leaves occasionally
1, or 2 remote pairs on stem, but each stem or branch terminated by
whorl or pseudowhorl of 4 leaves in 2 pairs, often each pair of different
sizes, connected at base by broadly triangular, interpetiolar stipules; lamina
2.5–9 (–10.5) × 0.3–3 (–4) cm, ovate-oblong to lanceolate, acute apex,
base cuneate, entire, glabrous, slightly scabrid above; petiole 0–1 cm long.
Inflorescences 4–12 cm long, usually in axils of terminal leaves. Flowers
white, through pink and lilac to mauve and purple, subsessile, bracteate;
bracts 1.5–3.5 mm long, linear-lanceolate, acuminate. Calyx 2–3.5 mm long,
sepals free, somewhat unequal, glabrous. Corolla infundibuliform, tube c.
8 mm long, lobes 1–1.5 × 1–1.5 mm, broadly triangular. Stamens glabrous,
filaments filiform, c. 1 mm long, inserted below midway on corolla tube.
Ovary subglobose, glabrous. Capsule c. 3.5–4.5 × 4.5–5 mm, squamate or

tuberculate. Seeds 2 × 1.5 mm, (?immature) obliquely ellipsoid, pale brown with orange-brown tubercles.

Distr. Native of New World from Florida and Mexico to Brazil and Peru; naturalised in tropical Africa and Asia.

Ecol. Weed of roadsides, gardens and waste places.

Note. The description above was made from extra-Ceylon material.

Specimen Examined. Currently no material seen from Ceylon, but cited in Ceylon J. Sci. 12 as occurring there.

3. STRYCHNOS

L., Sp. Pl. 189. 1753; L., Gen. Pl. ed. 5: 86. 1754; A.W. Hill, Bull. Misc. Inform. 1917: 121. 1917. Lectotype species: *Strychnos nux-vomica* L. (vide Hitchcock, Prop. Brit. Bot. 133. 1929).

Trees or, more usually, scandent shrubs provided with short, paired, axillary tendrils, or at times axillary spines; where tendrils occur in subterminal axils with both subtending leaves and terminal portion of branch suppressed, false impression given of tendrils occurring terminally. Leaves opposite, estipulate, 3–5-veined. Inflorescence cymose, terminal or axillary, with scale-like bracts. Flowers small, white or yellowish. Calyx 4–5-lobed, small, lobes usually shortly ciliate. Corolla 4–5-lobed, tube long or short, lobes valvate in bud, spreading to reflexed when open. Stamens 5, inserted on corolla-tube, exserted; filaments very short. Ovary 2-celled; ovules several in each cell. Style long or short or almost lacking; stigma capitate or obscurely 2-lobed. Fruit globose or subglobose berry with thin pericarp. Seeds large, subcompressed, (1–) 2–several, immersed in pulp.

Genus of about 60 species from both Old and New World tropics.

KEY TO THE SPECIES

1 Erect trees
  2 Leaves petiolate
    3 Corolla tube longer than, and up to twice as long as, lobes .... 1. S. nux-vomica
    3 Corolla tube half as long as lobes .... 2. S. tetragona
  2 Leaves sessile or subsessile; corolla tube and lobes subequal .... 3. S. potatorum
1 Shrubs, scandent or suberect
    4 Calyx and corolla 5-lobed
      5 Corolla tube 7.5–12 mm long, more than twice as long as lobes .... 4. S. wallichiana
      5 Corolla tube 1–3 mm long, shorter than lobes
        6 Stem glabrous; corolla lobes 3–4 times longer than tube .... 5. S. minor
        6 Stem pubescent, at least when young; corolla lobes only slightly longer than tube
          7 Woody climber; leaves rhomboid, acute at base, 5-veined; anthers hirsute at base, sessile .... 6. S. trichocalyx
          7 Erect shrub; leaves ovate, obtuse-rounded at base, 3-veined; anthers glabrous, filaments at least 1 mm long .... 7. S. coriacea

4 Calyx and corolla 4-lobed .... **8. S. benthamii**

1. Strychnos nux-vomica L., Sp. Pl. 189. 1753; Roxb., Fl. Ind. 2: 261. 1824; Thw., Enum. Pl. Zeyl. 201. 1860; Clarke in Hook. f., Fl. Br. Ind. 4: 90. 1883; Trimen, Handb. Fl. Ceylon 3: 175. 1895; A.W. Hill, Bull. Misc. Inform. 1917: 183, fig. 1917; Bisset, Philcox et al., Lloydia 36: 189. 1973. Type: Ceylon: Colombo, *Hermann s.n.* in Herb. Hermann 4: fol. 33 (BM, holotype).

Tree, 8–15 m tall, erect; bark smooth, yellow-grey; stems compressed-cylindric, dichotomously branched, swollen at nodes, glabrous. Leaves petiolate; lamina 3.5–12 × 1.75–7 (–10) cm, broadly ovate to subrotund, apex obtuse, rounded, shortly acuminate, frequently uneven at obtuse or acute base, thin, glabrous, often somewhat shiny, especially above, prominently 5-veined with outer pair often obscure, dense intermediate venation; petiole 5–10 mm long, glabrous. Flowers many, greenish, slender-pedicellate in short-pedunculate, minutely pubescent, terminal cymes. Peduncle (1–) 1.4–3.5 cm long, terminated below first branches by pair of lanceolate bracts, up to 5 mm long. Pedicel up to 1.75 mm long or absent. Calyx 1 × 0.5 mm, ovate-triangular, acute. Corolla 8–11 mm long overall, tube 5–7.5 mm long, slightly pubescent without, hairy in throat with few long hairs, lobes 5, 2.5–3 × c. 1 mm, oblong to oblong-lanceolate, obtuse or subacute. Anthers 1.3–1.5 mm long, oblong, glabrous. Ovary glabrous; style 9–10 mm long, slender, glabrous; stigma capitate. Fruits c. 3.5–5.5 cm diameter (? immature), globose, yellow-orange, orange to orange-red with thin pericarp. Seeds (not seen) 1–6, c. 1.5–1.8 cm diameter, somewhat discoid, immersed in pinkish pulp, drying shiny, silver-grey, very finely appressed-pubescent.

D i s t r. India and Ceylon; Thailand, Laos, Cambodia and South Vietnam.

E c o l. Dry region forests.

V e r n. Godakaduru (S), Eddi, Kanchurai (T).

S p e c i m e n E x a m i n e d. MANNAR DISTRICT: Radamagama Veva, near Madhu Road, 2 Apr 1932, *Simpson 9409* (BM). JAFFNA DISTRICT: Milepost 17.5 on road from Mankulam to Mullaitivu, 10 m, 14 Dec 1970, *Ripley 391* (PDA), *410* (K, PDA). VAVUNIYA DISTRICT: c. 3 km from Mullaitivu on road to Mankulam, 24 Jun 1975, *Sumithraarachchi & Sumithraarachchi 802* (PDA). PUTTALAM DISTRICT: Wilpattu, sea-level, 21 may 1976, *Cramer 4678* (E); Wilpattu National Park, Kali Villu, 11 Sept 1968, *Mueller-Dombois 68091110* (K, PDA); Chilaw, 3 m, 8 Dec 1953, *Worthington 6519* (K, PDA). ANURADHAPURA DISTRICT: Anuradhapura, *Moon s.n.* in *C.P. 2839*, p.p. (PDA), c. 0.25 km E of Wilpattu turn-off, on road from Anuradhapura to Puttalam, 0–100 m, 30 Mar 1971, *Robyns 7344* (K, PDA);

mile-marker 65/4, Anùradhapura to Trincomalee road, 5 Oct 1973, *Sohmer 8142* (PDA); Maradankadawala, 4 Feb 1971, *Amaratunga 2197* (PDA); c. 12.8 km SW of Anuradhapura on road to Puttalam, 90 m, 31 Oct 1974, *Davidse & Sumithraarachchi 8168* (K); Ganewalpola, 23 Oct 1973, *Waas 234* (PDA). TRINCOMALEE DISTRICT: Trincomalee, c. 6 m, 5 May 1939, *Worthington 183* (BM), c. 12 m, 5 Oct 1947, *Worthington 3092* (BM), Aug 1859, *Glenie s.n.* in *C.P. 2839*, p.p. (PDA), *s. coll.* in *C.P. 2839*, p.p. (PDA), Kandy Road, c. 8 m, 24 Nov 1949, *Worthington 4397* (BM); Plantain Point, c. 13 m, 5 Oct 1947, *Worthington 3092* (K, PDA); Shell Bay, Foul Point, 3 m, 17 Apr 1940, *Worthington 878* (K). KURUNEGALA DISTRICT: Kurunegala, *s. coll.* in *C.P. 2839*, p.p. (PDA); Nikaweratiya, 14 Sept 1970, *Jayasuriya & Lazarides 2* (PDA). MATALE DISTRICT: Dambulla, SE of Kandalama Tank, 17 Sept 1974, *Sumithraarachchi 451* (PDA). POLONNARUWA DISTRICT: Kekirihena, Kandakaduwa, 8 Jun 1974, *Waas 626* (K, PDA). BATTICALOA DISTRICT: c. 47 km N of Batticaloa, 2 m, 19 Aug 1950, *Worthington 4899* (K, PDA); c. 21 km NE of Maha Oya, 50 m, 16 Apr 1973, *Stone 11174* (PDA). KANDY DISTRICT: Peradeniya, Royal Botanical Gardens, 15 May 1985, *Jayasuriya & Bandara 3325* (PDA). AMPARAI DISTRICT: near Kalmunai, 30 May 1971, *Kostermans 24329* (BM, K, PDA), 30 Jun 1968, *Amaratunga 1617* (PDA); Bibile to Batticaloa road, culvert 50/7, c. 152 m, 17 Jun 1953, *Worthington 6310* (K, PDA). MONERAGALA DISTRICT: Siyambalanduwa, 27 Jun 1970, *Meijer 167* (K, PDA). LOCALITY UNKNOWN: *Gardner s.n.* (K); *Thwaites s.n.* in *C.P. 2839* (BM); *Walker 1419* (E); *Walker s.n.* (K).

2. Strychnos tetragona A.W. Hill, Bull. Misc. Inform. 1917: 140. 1917; Alston in Trimen, Handb. Fl. Ceylon 6: 196. 1931; Bisset, Philcox et al., Lloydia 36: 193. 1973. Type: Ceylon: without locality, *Thwaites s.n.* in *C.P. 3720B* (K, holotype).

Shrub or tree up to 8 m tall; branches 4-angled when young, furrowed, pubescent. Leaves subcoriaceous, petiolate; lamina 6–8.5 × 3.25–4 cm, ovate or ovate-elliptic, long-acuminate, base rounded or rounded-cuneate, glabrous above, minutely pubescent on major veins beneath, 5-veined, veins prominent, markedly so beneath; petiole c. 5 mm long, minutely pubescent. Flowers many, pedicellate, in axillary panicles 2.5–5 cm long, inflorescence branches 4-angled, furrowed, minutely pubescent. Calyx lobes 0.75 mm long, triangular-ovate, obtuse, glabrous, ciliate. Corolla 2.5 mm long, tube c. 0.75 mm long, campanulate, lobes 1.75 mm long, glabrous without, pilose-lanate within especially at throat. Anthers c. 0.8 mm long, lightly hirsute to subglabrous at base. Filaments up to c. 1 mm long, shortly, sparsely hirsute. Ovary with style 2.25 mm long, glabrous. Fruit 1.5 cm diameter, globose, drying glossy, dark-brown. Seeds 1.25 cm diameter, discoid.

Distr. Endemic.

Ecol. Wet forests.

Note. *Thwaites 3720B* housed at Kew and cited above, is clearly this species while the other two specimens there numbered *3720A* and considered by A.W. Hill to represent *S. micrantha* Thw., a species here considered to be synonymous with *S. minor* Dennst. (see there).

Specimens Examined. PUTTALAM DISTRICT: between Kokkare Villu and Kurutu Pandi Villu, 29 Jun 1969, *Wirawan et al. 856* (E, K). KALUTARA DISTRICT: Morapitiya, Arambakanda logging area, c. 300 m, 30 Aug 1976, *Waas 1908* (E, K). RATNAPURA DISTRICT: Rassagala, Balangoda, 6 Sept 1895, *Trimen's Collector 45* (PDA). LOCALITY UNKNOWN: *Thwaites s.n.* in *C. P. 3720B* (K, holotype).

3. Strychnos potatorum L. f., Suppl. Pl. 148. 1781; Thw., Enum. Pl. Zeyl. 201. 1860, and 425. 1864; Clarke in Hook. f., Fl. Br. Ind. 4: 90. 1883; Trimen, Handb. Fl. Ceylon 3: 176. 1895; A.W. Hill, Bull. Misc. Inform. 1917: 154. 1917; Alston in Trimen, Handb. Fl. Ceylon 6: 197. 1931; Bisset, Philcox et al., Lloydia 36: 191. 1973. Type: from India.

Tree, 5–15 m tall, erect; bark corky, dark-grey, furrowed; stems smooth, thickened at nodes, glabrous. Leaves sessile or subsessile; lamina 4–9.5 (–12) × 2.5–6.25 cm, ovate to ovate-elliptic or ovate-lanceolate, apex acute or more rarely subacute to obtuse, base acute to rounded or obscurely subcordate, glabrous, somewhat glossy, 3–5-veined with veins arising up to 1 cm or more above base of midrib, midrib pinnately veined with 5–10 pairs of lateral veins, these anastomosing, forming, sometimes obscurely, 2 intramarginal costae. Flowers greenish-yellow, slender pedicelled in small, compact, axillary, small-bracteate cymes. Pedicel 3–4.5 mm long, subtended by broadly ovate bracteole c. 0.5 mm long. Calyx up to c. 1.2 × 0.6 mm, ovate, glabrous. Corolla glabrous without, tube 3.5–4 mm long, lobes 3–3.5 mm long, ovate-lanceolate, subdensely pilose within. Stamens inserted towards mouth of tube; anthers 1–1.25 mm long, glabrous. Stigma small, capitate, style c. 4–4.5 mm long, slender, glabrous, exserted. Fruit 1.25–2 cm diameter, ripening blackish. Seeds 1 or 2, c. 1 cm diameter.

Distr. India, Ceylon and Burma; also in Zaire, from Tanzania to Botswana and northern Transvaal; Madagascar.

Ecol. Low altitude, dry-zone forest on well-drained soils, or bordering mangrove swamps.

Vern. Ingini (S), Tetta (T).

Uses. Pulverized seed used to make turbid water clear.

Specimens Examined. VAVUNIYA DISTRICT: Kolavil, 28 Jun 1931, *Simpson 8279* (BM). ANURADHAPURA DISTRICT: Wilpattu National Park, Maradan Maduwa, 14 Jun 1968, *Ripley 52* (PDA); between Anuradhapura and Mihintale, 29 Apr 1921, *Silva s.n.* (PDA). TRINCOMALEE DISTRICT: Trincomalee, *Glenie s.n.* in *C.P. 3719* (BM, K, PDA); Shell Bay, Foul Point, 3 m, 20 Nov 1939, *Worthington 662* (K), 1 Jan 1944, *Worthington 1361* (K). MATALE DISTRICT: Dambulla, Nov 1858, *s. coll.* in *C.P. 3719*, p.p. (PDA), 18 Jun 1932, *Simpson 9824* (PDA), behind Rest House, 15 May 1969, *Kostermans 23543* (K, PDA); Nalanda, 305 m, 3 Nov 1949, *Worthington 4329* (K). POLONNARUWA DISTRICT: Polonnaruwa Rest House, 7 Apr 1939, *Worthington 80* (K); Sacred Area, 61 m, 27 Aug 1971, *Ripley 460* (K, PDA), 25 May 1969, *Hladik 807* (PDA), 24 Nov 1975, *Sohmer & Sumithraarachchi 10823* (PDA), 11 Jan 1970, *Dittus 70011101* (PDA), 7 Sept 1971, *Dittus 71090706* (PDA), 10 Sept 1974, *Tirvengadum 466* (PDA); Kalkudah, 42 m, 18 Apr 1950, *Worthington 5850* (K). KURUNEGALA DISTRICT: Ganewatta Medicinal Plant Garden, 100 m, 25 May 1993, *Dhanesekara 4* (PDA). BATTICALOA DISTRICT: near to Ottamavedi railway bridge, on Batticaloa to Polonnaruwa road, 10 Sept 1974, *Tirvengadum & Waas 465* (K); Mankeni, 2 m, 17 Dec 1954, *Worthington 6704B* (K). AMPARAI DISTRICT: Mullaitivu road, 13 Mar 1932, *Simpson 9290* (BM). NUWARA ELIYA DISTRICT: E bank of Belihuloya, Randenigala planned reservoir, 220 m, 20 Nov 1984, *Jayasuriya 3034* (PDA). MONERAGALA DISTRICT: Bibile, 21 Feb 1951, *Worthington 5108* (K), 23 Feb 1951, *Worthington 5124*, 25 Oct 1925, *Silva s.n.* (PDA); between Ekiriyankumbura and Bibile, 28 Oct 1971, *Jayasuriya 384* (K, PDA); Wirawila, 28 Apr 1951, *Worthington 5215* (K). HAMBANTOTA DISTRICT: Buttawa Beach area, 3–4 m, 30 Jan 1968, *Comanor 898* (K, PDA); Katagamuwa Tank, 26 Aug 1967, *Mueller-Dombois 67082614* (PDA); Ruhuna National Park, Andunoruwawewa, S of mouth of Menik Ganga, below 5 m, 14 Nov 1977, *Huber 620* (PDA), Block I, Yala Camp, 8 Dec 1967, *Mueller-Dombois et al., 67120821* (PDA); Patanagala Camp, 19 Oct 1968, *Cooray 68101901R* (K, PDA), 13 Sept 1969, *Cooray 69091301R* (K, PDA); behind Smithsonian [Patanagala] Camp, 28 Oct 1968, *Wirawan 682C* (K, PDA), 2–3 m, 6 Apr 1968, *Fosberg 50356* (PDA); Jamburagala, 7 Mar 1969, *Ripley 109* (PDA); Block II, Kumbukkan Oya, 400 m, W of river, 2 m, 1 Oct 1967, *Mueller-Dombois et. al. 67100109* (PDA). BADULLA DISTRICT: Ooma (Uma) Oya, Jun 1881, *s. coll. s.n.* (PDA). LOCALITY UNKNOWN: *s. coll.* in *C.P. 3719*, p.p. (PDA); 5 Sept 1927, *Jayawardene 1* (PDA).

4. Strychnos wallichiana Steud. ex DC., Prod. 9: 13. 1845; A. W. Hill, Bull. Misc. Inform. 1917: 198. 1917; Bisset & Philcox, Taxon 20: 543. 1971; Bisset, Philcox et al., Lloydia 36: 195. 1973. Type: from India.

Strychnos cinnamomifolia Thw., Enum. Pl. Zeyl. 201. 1860; Clarke in Hook. f., Fl. Br. Ind. 4: 89. 1883; Trimen, Handb. Fl. Ceylon 3: 174. 1895; A.W. Hill, Bull. Misc. Inform. 1917: 194. 1917. Type. Ceylon: Hantane, *Thwaites s.n.* in *C.P. 1867* (PDA, holotype; BM, K, isotypes). *Strychnos cinnamomea* Thw., Enum. Pl. Zeyl. 425. 1864, err. typ.

Large scandent, woody climber; stems up to 7 cm diameter or more; bark smooth, yellowish-grey; branches glabrous or minutely pubescent when young, glabrescent with age; tendrils present, single, axillary, appearing paired when occurring in each of opposite axils. Leaves shortly petiolate; lamina 4–9.5 × 1.75–6.5 cm, broadly elliptic, ovate-elliptic to ovate-lanceolate, apex acute, shortly acuminate, broadly cuneate at base, entire, glabrous, glossy especially above, markedly 3-veined with veins arising from or very near base, occasionally obscurely 5-veined with outer pair of veins running parallel to and just within margin; petiole 5–8 mm long, slender, glabrous. Flowers pedicellate, in terminal, many-flowered cymes, on minutely pubescent, 1–2 cm long peduncle, individual branches of cyme subtended by lanceolate bracts up to 3.5–4 mm long. Pedicel 0.5–2.5 mm long, bracteoles 0.5–0.75 mm long, ovate. Calyx 5-lobed, c. 2 mm long, lobes c. 1 × 0.8 mm, ovate, obtuse, minutely pubescent, shortly ciliate or not. Corolla 5-lobed, minutely pubescent without, tube 7.5–12 mm long, lobes 3.5–5 × 1.5 mm, ovate to oblong, obtuse, glabrous within, throat also. Stamens inserted just within throat of corolla; anthers c. 1.5 mm long, glabrous. Ovary glabrous. Stigma capitate; style up to 13.5 mm long, slender, exserted. Fruit reportedly c. 4 cm diameter (not seen). Seeds c. 2.5 × 1.75 cm, very compressed ovoid, pale fawn with fine, felt-like indumentum.

Distr. India and Ceylon; North Vietnam, China (Yunnan) and the Andaman Islands.

Ecol. Moist low country, to about 950 metres.

Vern. Etakirindiwel, Welbeli (S).

Specimens Examined. TRINCOMALEE DISTRICT: Ostenburg, 9 Aug 1939, *Worthington 190* (K, PDA). COLOMBO DISTRICT: Henaratgoda, 24 Jan 1918, *Petch s.n.* (K). KANDY DISTRICT: Peradeniya, Gannoruwa Hill, 27 Feb 1917, *s. coll. s.n.* (K); Hantane, in 1859, *Thwaites s.n.* in *C.P. 1867*, p.p. (PDA, holotype of *S. cinnamomifolia* Thw.), *Gardner s.n.* in *C.P. 1867*, p.p. (PDA). GALLE DISTRICT: Galle, *Champion s.n.* in *C.P. 1867*, p.p. (PDA). LOCALITY UNKNOWN: *Gardner 578* (K); *Thwaites s.n.* in *C.P. 1867*, p.p. (BM, K), 610 m, in 1860, *Thwaites s.n.* in *C.P. 1867*, p.p. (K), 915 m, in 1861, *Thwaites s.n.* in *G.P. 1867*, p.p. (K).

5. Strychnos minor Dennst., Schluss. Hort. Ind. Malab. 33. 1818; Bisset & Philcox, Taxon 20: 542. 1971; Leenh. in Fl. Males. 6: 958. 1972. Type: Rheede Ic. in Hort. Malab. 7: 9, t. 5. 1688.

Strychnos colubrina Stokes, Bot. Mat. Med. 1. 414. 1812, p.p., non L., 1753. Type: Rheede Ic. in Hort malab. 7: 9, t. 5. 1688.

Strychnos colubrina auct.: A.W. Hill, Bull. Misc. Inform. 1917: 157. 1917; Leenh. in Fl. Males. 6: 355. 1962, non L., 1753.

Strychnos micrantha Thw., Enum. Pl. Zeyl. 425. 1864; Clarke in Hook. f., Fl. Br. Ind. 4: 86. 1883; Trimen, Handb. Fl. Ceylon 3: 172. 1895; Bisset, Philcox et al., Lloydia 36: 187. 1973. Type: Ceylon: Trincomalee, *Thwaites s.n.* in *C.P. 3720A* (K, lectotype; P, isolectotype).

Strychnos lenticellata A.W. Hill, Bull. Misc. Inform. 1917: 159, fig. 1917. Type: from India.

Shrub, climbing or scrambling up to 15 m; stems glabrous, smooth, slightly thickened at nodes; tendrils paired. Leaves subcoriaceous, petiolate; lamina 4.5–11.5 × 2–5.5 cm, lanceolate to ovate, acuminate or rounded at apex, base obtuse to subacute, glabrous, somewhat shiny above, apparently 3-veined, but with further obscure pair of veins anastomosing and finely intramarginal; petiole 4–12 mm long, glabrous. Flowers many in much-branched, axillary or terminal, minutely pubescent cymes. Calyx 5-lobed, 0.5–0.75 mm long, broadly-ovate to suborbicular, obtuse, subglabrous, shortly ciliate. Corolla 5-lobed; tube c. 2 mm long, lobes c. 2 mm long, elliptic-lanceolate, subacute, sparsely hairy without. Stamens inserted in mouth of tube; anthers 0.75–c. 1 mm long, ovate-lanceolate to oblong, glabrous or subglabrous; filaments c. 1 mm long, glabrous. Ovary with style c. 3 mm long, overall hirsute with lax, spreading hairs, style often glabrous. Mature fruit and seeds not seen.

Distr. Widely spread from India and Ceylon, through Burma, Thailand and Vietnam, throughout Malesia to the Solomon Islands to Australia (Queensland).

Ecol. Low country.

Vern. Kaduru (S), Kachchalkodi (T).

Specimens Examined. PUTTALAM DISTRICT: Panirendawa Forest Reserve, 25 Jan 1974, *Jayasuriya et al. 1444* (K, PDA). ANURADHA-PURA DISTRICT: Habarane, 3 May 1927, *Alston 511* (PDA). TRINCOMA-LEE DISTRICT: Trincomalee, up to 610 m, *Thwaites s.n.* in *C.P. 3720A* (K); Sober Island, Mar 1892, *Nevill s.n.* (PDA). COLOMBO DISTRICT: Henaratgoda, 7 Aug 1926, *Alston 838* (PDA). KANDY DISTRICT: Gannoruwa Hill, Peradeniya, 800 m, 6 Oct 1968, *Wirawan 625* (K, PDA). MONERAGALA DISTRICT: Bibile, *s. coll.* in *C.P. 3540*, p.p. (PDA). GALLE DISTRICT:

Galle, *s. coll.* in *C.P. 3540*, p.p. (PDA). LOCALITY UNKNOWN: *Thwaites s.n.* in *C.P. 3540* (BM, K), *Thwaites s.n.* in *C.P. 1866* (K, PDA); *s. coll.* in *C.P. 3720*, p.p. (PDA).

6. Strychnos trichocalyx A.W. Hill, Bull. Misc. Inform. 1917: 174, fig. 1917; Alston in Trimen, Handb. Fl. Ceylon 6: 197. 1931; Bisset, Philcox et al., Lloydia 36: 194. 1973. Type: Ceylon: without locality, *Mrs General Walker s.n.* in Herb. Hook. (K, lectotype).

Strychnos minor Dennst. var. *nitida* Benth., J. Linn. Soc. Bot. 1:101. 1856. Syntypes: Ceylon: without locality, *Walker s.n.*, *Thwaites s.n.* in *C.P. 2516* (K).

Strychnos colubrina sensu Thw., Enum. Pl. Zeyl. 201. 1860.

Strychnos colubrina var. *zeylanica* Clarke in Hook. f., Fl. Br. Ind. 4: 87. 1883; Trimen, Handb. Fl. Ceylon 3: 173. 1895. Type: Ceylon: without locality, *Thwaites s.n.* in *C.P. 2516* (BM, K).

Large, scandent, woody climber; bark grey, smooth, heavily lenticelled; stems minutely pubescent when young; tendrils strongly circinnate. Leaves coriaceous, shortly petiolate; lamina 3.5–8 × 1.25–5 cm, rhomboid, cuneate, acute at both ends or obovate, rounded at apex, margin slightly recurved, shiny above, less so beneath, glabrous except occasionally minutely pubescent towards base beneath, 5-veined with veins arising from base, running parallel for up to 0.5 cm before spreading, central 3 veins prominent beneath, outer pair somewhat obscure; petiole 3–5 mm long, glabrous or minutely pubescent. Flowers white, pedicellate, in many-flowered, subcompact, minutely pubescent, axillary cymes. Peduncle (0.1–) 0.4–0.6 cm long, bearing bracts up to 2.5 mm long at points of insertion of inflorescence branches. Pedicel 0.1–1.5(–2.5) mm long, subtended by bracteoles up to 1.25 mm long, lanceolate, densely pubescent within and without. Calyx (4–) 5-lobed, lobes c. 1 × 0.5–0.6 mm, ovate to subrotund, pubescent without, erect-hirsute towards base within, shortly ciliate. Corolla 5-lobed, c. 2.5 mm long, lobes connate for 1 mm at base, c. 0.5 mm wide, oblong, long-hirsute within above midway. Stamens arising at base of sinus of corolla lobes; anthers 0.6–0.7 mm long, sessile, hirsute at point of attachment to corolla. Ovary with style attached up to 1.75–2 mm long, glabrous. Fruit 1.2–1.35 cm diameter, ovoid-globose, shiny, olivaceous-brown. Seeds 1 or 2 in each fruit, c. 0.9 × 1.2 cm, compressed, ellipsoid.

Distr. Endemic.

Ecol. Low country.

Vern. Kaduru, Gonakaramba (S).

Note. One specimen housed at Kew, that collected from Ceylon by Col. Walker and originally from Wight's herbarium, was identified, named and cited by A.W. Hill as this species, but clearly has 4-merous flowers. This character has been allowed for in the above description.

Unfortunately, much of the material cited below was collected at a very late stage of fruiting, leaving only old calyces along with the leaves as an aid to identification. However, the author is confidant that the affinities in every case are with the species under which they are included.

Specimens Examined. JAFFNA DISTRICT: Pallavarayankaddu to Mankulum road, 13 Nov 1970, *Kundu & Balakrishnan 642* (PDA). MANNAR DISTRICT: c. 6 km S of Madhu Road, 9 July 1971, *Meijer 805* (PDA). PUTTALAM DISTRICT: Wilpattu National Park, near Manikepola Uttu, 28 Apr 1969, *Mueller-Dombois 69042805* (K, PDA); between Dangaha Uraniya and Occapu Kallu, 11 July 1969, *Wirawan et al. 1096* (PDA). ANURADHAPURA DISTRICT: Anuradhapura, *Moon s.n.* in *C.P. 2516*, p.p. (PDA); Ritigala Strict Natural Reserve, 610 m, 4 Aug 1972, *Jayasuriya et al. 829* (K, PDA); Weweltenna plain, c. 580 m, 2 Jun 1974, *Jayasuriya 2142* (PDA); Ritigala, c. 750 m, *Cramer 3832* (K, PDA); Wilpattu National Park, c. 3 km from Maradanmaduwa to Kokkare villu, 29 Aug 1974, *Tirvengadum & Waas 424* (K); Habarane, 3 May 1927, *Alston 499* (PDA). KURUNEGALA DISTRICT: Doluwa Kanda, 305 m, 14 Feb 1975, *Sumithraarachchi 641* (PDA). MATALE DISTRICT: Dambulla, July 1887, *s. coll. s.n.* (PDA); Galagama, *Gardner s.n.* in *C.P. 2516*, p.p. (PDA); milepost 37, between Laggala and Illukkumbura, 1000 m, 4 Oct 1971, *Jayasuriya 270* (K, PDA); milepost 38 on Rattota to Illukkumbura road, 900 m, 13 Aug 1978, *Huber 723* (PDA); Ridiella, 16 km N of Naula-Elahera road, off milepost 5, 4 Oct 1971, *Jayasuriya 314* (PDA). KANDY DISTRICT: below Madugoda, road to Mahiyangane, 800 m, 21 Aug 1974, *Kostermans 25427* (K, PDA). Kandy, 8 Mar 1819, *Moon 346* (BM); Udawatte, 950 m, 14 Dec 1971, *Jayasuriya & Balasubramaniam 467 1/2* (PDA). MATARA DISTRICT: Galagama, ?Feb 1846, *Thwaites s.n.* in *C.P. 330* (K). BADULLA DISTRICT: Katalaywela, Bintenna, 26 Apr 1923, *Silva 38* (PDA); Uma Oya, Kuruminiya, Kandura, 8 Dec 1927, *Silva 267* (PDA); Nilgala Hill, Jan 1888, *s. coll. s.n.* (PDA). LOCALITY UNKNOWN: *Macrae 197* (BM); *Thwaites s.n.* in *C.P. 2516*, p.p. (K, PDA); *Walker s.n.* (K).

7. Strychnos coriacea Thw., Enum. Pl. Zeyl. 425. 1864; A.W. Hill, Bull. Misc. Inform. 1917: 155. 1917; Alston in Trimen, Handb. Fl. Ceylon 6: 196. 1931; Bisset, Philcox et al., Lloydia 36: 183. 1973. Type: Ceylon: Central Province, without further locality, *Thwaites s.n.* in *C.P. 3367* (PDA, holotype; BM, K, isotypes).

Strychnos beddomei Clarke var. *coriacea* (Thw.) Clarke in Hook. f., Fl. Br. Ind. 4: 89. 1883; Trimen, Handb. Fl. Ceylon 3: 173. 1895.

Shrub; young branches pubescent. Leaves coriaceous, shortly petiolate; lamina 5–6.5 × 2–3 cm, ovate- or elliptic-lanceolate, apex acute or shortly acuminate, base rounded-obtuse, margin slightly reflexed, 3-veined, veins prominent, especially beneath; petiole c. 2 mm long, hirsute. Flowers cymose, axillary, cymes 3.5–4 cm long, peduncles and pedicels pubescent. Calyx 5-lobed, 1.5 mm long, lobes c. 1 mm wide, broadly ovate, obtuse, ciliate. Corolla 5-lobed, 5 mm long, sparsely pubescent without, lobes slightly longer than tube, lobes and throat densely villous within, tube glabrous within. Anthers 1 mm long, elliptic, glabrous. Filaments 1.2 mm long, somewhat stocky. Ovary including style 4.5 mm long, erect pilose. Fruit and seeds not seen.

Distr. Endemic; known only from the type collection.

Specimens Examined. LOCALITY UNKNOWN: Feb 1855, *Thwaites s.n.* in *C.P. 3367* (holotype, PDA; BM, K, isotypes).

8. Strychnos benthamii Clarke in Hook. f., Fl. Br. Ind. 4: 87. 1883; Trimen, Handb. Fl. Ceylon 3: 174. 1895; A.W. Hill, Bull. Misc. Inform. 1917: 165. 1917; Bisset, Philcox et al., Lloydia 36: 182. 1973. Type: Ceylon, Colombo, in 1860, *Thwaites s.n.* in *C.P. 187*, p.p. (K, lectotype; P, isolectotype (vide Bisset & Philcox, 1973)).

Strychnos minor Dennst. var. *angustior* Benth., J. Linn. Soc. Bot. 1: 101. 1856. Type: Ceylon: Ratnapura District, Balangoda, Feb 1846, *Thwaites s.n.* in *C.P. 187*, p.p. (K, lectotype; PDA, isolectotype).
Strychnos minor var. *ovata* Benth., J. Linn. Soc. Bot. 1: 101. 1856. Type: Ceylon: without locality, *Kelaart s.n.* (K, holotype).
Strychnos minor var. *parvifolia* Benth., J. Linn. Soc. Bot. 1: 101. 1856. Type: Ceylon: without locality, *Gardner 580* (K, holotype, BM, isotype).
Strychnos benthamii var. *angustior* (Benth.) A.W. Hill, Bull. Misc. Inform. 1917: 166. 1917.
Strychnos benthamii var. *parvifolia* (Benth.) Clarke in Hook. f., Fl. Br. Ind. 4: 87. 1883; A.W. Hill, Bull. Misc. Inform. 1917: 166. 1917; Trimen, Handb. Fl. Ceylon 3: 174. 1895.

Shrub or small treelet, 2–4.5 m tall, mostly erect, rarely scandent; branches many, divaricate, occasionally bearing small, solitary tendrils, when young frequently with patent, axillary spines of lengths up to 1.3 cm. Leaves very varied in· size, shortly petiolate to subsessile; lamina 1.25–1.75–3.75–9.5 × 0.6–0.75–1.75–4.5 cm, lanceolate-ovate to lanceolate-elliptic, apex acute-acuminate or obtuse, acute or obtuse at base, glabrous above, membranous or subcoriaceous, where membranous frequently patent, ferruginous, short-hirsute, especially on major veins beneath, where subcoriaceous margin slightly revolute, glabrous throughout, markedly 3-, obscurely 5-veined; petiole 2–3 mm long, minutely pubescent.

Flowers greenish-yellow to white, shortly pedicellate in few-flowered, axillary cymes. Pedicels 1–2.5 mm long, with peduncles minutely pubescent to subglabrous. Calyx 4-lobed, lobes c. 0.5 mm long, broadly triangular, shortly acute, subglabrous without, shortly ciliate. Corolla 4-lobed, rarely 5, 2–2.25 mm long, lanceolate-oblong, subacute, very minutely pubescent without, long, white villous within with hairs up to 0.6 mm long. Anthers c. 0.6 mm long, ovate, shortly apiculate, hairy at base. Ovary with style 1.5 mm long, glabrous, glossy. Fruit c. 8 mm diameter, globose, drying black. Seeds c. 5 mm diameter, discoid, off-white.

Distr. Endemic.

Ecol. Somewhat rare in moist, lowland areas, but the small-leaved state is commoner in the mountains.

Specimens Examined. ANURADHAPURA DISTRICT: Ritigala Strict Natural Reserve, Weweltenna plain, SE edge, 31 May 1974, *Jayasuriya 1713* (PDA). MATALE DISTRICT: Dambulla, Apr 1856, *Thwaites s.n.* in *C.P. 341*, p.p. (PDA); milepost 37, between Laggala and Illukkumbura, 1000 m, 4 Oct 1971, *Jayasuriya 274* (K, PDA); Dikpatana, W of Illukkumbura, 825 m, *Jayasuriya 3062* (PDA), 900 m, 1 Dec 1971, *Jayasuriya et al. 432* (PDA). COLOMBO DISTRICT: Colombo, in 1860, *Thwaites s.n.* in *C.P. 187*, p.p. (BM, K, PDA); Dewalapola, 7 Mar 1971, *Amaratunga 2243* (PDA); Ekala, 28 Mar 1971, *Amaratunga 2265* (PDA); Danowita, 5 Jan 1963, *Amaratunga 451* (PDA); Pasyala, 16 Feb 1967, *Amaratunga 1241* (PDA). KANDY DISTRICT: Kandy, 17 Feb 1819, *Moon 203* (BM); Kandy to Mahiyangane road, 28 mile-marker, 870 m, 10 Oct 1973, *Sohmer et al. 8274* (PDA); Kaluntenne, 975 m, 3 Feb 1975, *Waas 1092* (PDA); Hunasgiriya, 1000 m, 6 Jun 1971, *Kostermans 24422* (K), 27 Jun 1973, *Kostermans 25150* (K, PDA), 1386 m, 19 Jan 1975, *Waas 991* (K, PDA); NE of Hunasgiriya, near milepost 29/21 on road to Mahiyangana, 810 m, 14 Nov 1974, *Davidse & Jayasuriya 8421* (K, PDA); Madugoda to Mahiyangane, 800 m, 8 Jun 1978, *Kostermans 26616* (PDA); between Hunasgiriya and Weragamtota, 830 m, 30 May 1972, *Maxwell & Jayasuriya 724* (PDA); Corbet's Gap, 1220 m, 27 Feb 1948, *Worthington 3591* (PDA); Kovilmada, mile-marker 27/17 on Mahiyangana road, 690 m, 9 Sept 1974, *Waas & Tirvengadum 796* (PDA); Madugoda, 8 Nov 1931, *Simpson 8780* (BM); Moray Estate, 15 Nov 1973, *Sohmer 8734* (PDA), *Nooteboom & Huber 3128* (PDA), *Sohmer & Waas 8737* (PDA), *Balakrishnan et al. 927* (PDA), Maskeliya Oya, 1300 m, 24 May 1971, *Kostermans 24266A* (PDA); near Fishing Hut, 27 Mar 1974, *Sumithraarachchi & Jayasuriya 172* (PDA), *Fosberg 57995* (E); Adams Peak, Mar 1846, *Thwaites s.n.* in *C.P. 341* (BM, K); Maskeliya, 29 Feb 1926, *Alston 26* (PDA); Wariagala, Hantane, Dec 1889, *s. coll. s.n.* (PDA); Nilumally, Madulkelle, Oct 1887, *s. coll. s.n.* (PDA); summit of Rangala Hill, Sept 1888, *s. coll. s.n.* (PDA); Nitre Cave, Sept 1888, *s. coll. s.n.* (PDA). BADULLA DISTRICT: Haputale, 24 May 1906, *s. coll. s.n.* (PDA). KALUTARA DISTRICT: Kalutara ("Caltura"), *Gardner s.n.* in *C.P. 187*, p.p. (PDA); Arambakanda logging area, c. 300 m, 30 Aug 1976, *Waas 1908* (PDA). RATNAPURA DISTRICT: Pedigalla, c. 250 m, 28 Aug 1976, *Waas 1865* (E, PDA); Balangoda, Feb 1846, *Thwaites s.n.* in *C.P. 187*, p.p. (K), *Gardner s.n.* in *C.P. 187*, p.p. (PDA). NUWARA ELIYA DISTRICT: Elephant plains, c. 1800 m, *Gardner 580* (BM); Nuwara Eliya, *s. coll.* in *C.P. 341*, p.p. (HAK); Hakgala, Mar 1922, *de Alwis s.n.* (PDA), 27 May 1926, *Alston 1295* (PDA), 9 Jan 1932, *Simpson 9068* (BM); trail to top of Hakgala rock, c. 2000 m, 6 may 1976, *Waas 1633* (E, K, PDA); path from Hakgala to Fort Macdonald, 25 Apr 1906, *s. coll. s.n.* (PDA); Maturata, 15 Sept 1851, *s. coll. 122* in *C.P. 341*, p.p. (PDA), Nov 1851, *Thwaites s.n.* in *C.P. 341*, p.p. (PDA); mt Galaha, near Hakgala, 1800 m, 9 Dec 1975, *Bernardi s.n.* (PDA); Moon Plains, 1525 m, 12 July 1978, *Meijer 1888* (K). LOCALITY UNKNOWN: *Gardner 580* (K); *Kelaart s.n.* (K); *Walker 244* (K); *Walker s.n.* (E, K).

4. FAGRAEA

Thunb., Kongl. Vetensk. Acad. Handl. 3: 132, tab. 4. 1782. Type species: *Fagraea ceilanica* Thunb.

Trees, shrubs or woody climbers. Leaves opposite, large, coriaceous, petiolate or sessile, appearing estipulate, but with initially connate stipules or sheathing petiole bases which split interpetiolarly, leaving 2 axillary or interpetiolar scales usually adnate to base of petiole. Flowers large, in 3-many-flowered terminal, trichotomous cymes. Bracts small, scale-like. Bracteoles, when present similar to bracts but smaller. Calyx deeply 5-lobed, lobes fleshy, rounded, imbricate. Corolla 5-lobed, tube long, funnel-shaped, lobes rounded, turning to left in bud. Stamens 5, inserted on corolla-tube. Ovary ellipsoid, 2-celled. Style filiform. Stigma capitate. Ovules many in each cell. Fruit 1–2-celled berry, indehiscent. Seeds many, embedded in pulp.

Genus of about 35 species from southern India and Ceylon to China, Hainan and Formosa, thence through Malaysia to northern Australia and the Pacific islands.

Fagraea ceilanica Thunb., Kongl. Vetensk. Acad. Handl. 3: 132, tab. 4. 1782; Thunb., Nov. Gen. Pl. 35. 1782; Thw., Enum. Pl. Zeyl. 200. 1860 ('zeylanica'); Clarke in Hook. f., Fl. Br. Ind. 4: 83. 1883 ('zeylanica'); Trimen, Handbk. Fl. Ceylon 3: 170. 1895 ('zeylanica'); Leenh. in Fl. Males. 6: 315, figs. 13–15. 1962. Type from Ceylon (not seen).

Fagraea obovata Wall. in Roxb., Fl. Ind. 2: 33. 1824; Thw., Enum. Pl. Zeyl. 200. 1860; Clarke in Hook. f., Fl. Br. Ind. 4: 83. 1883; Trimen, Handb. Fl. Ceylon 3: 171. 1895. Type: from India.

Fagraea gardneri Thw., Enum. Pl. Zeyl. 200. 1860. Type: Ceylon: Hantane, *Gardner* in *C.P. 1826*, p.p. (PDA, holotype).
Fagraea obovata var. *gardneri* (Thw.) Clarke in Hook. f., Fl. Br. Ind. 4: 84. 1883; Trimen, Handb. Fl. Ceylon 3: 171. 1895.

Epiphytic or terrestrial shrub or small tree, up to 10 m or more tall. Leaves fleshy or coriaceous; lamina (7–)10.5–18(–22) × 4–9.5 cm, obovate to obovate-oblong, obtuse apex, cuneate at base into petiole or subsessile, entire, glabrous; indistinctly 4–6 pairs of veins, frequently prominent beneath; petiole robust to somewhat slender, 0.5–5.5 cm long, axillary scale appressed to base. Flowers very large, pale yellow to white on short, thick pedicels, sweet-scented, arranged in dense to subdense, glabrous, dichasial cymes. Pedicels 0.5–4 cm long. Calyx 1–1.5 cm long, coriaceous, lobes connate 1/4 to more than 1/2 their length, rounded, glabrous. Corolla somewhat fleshy, narrowly funnel-shaped, tube 6.5–9.5 cm long, glabrous, lobes 2.5–3.5 × 1–1.75 cm, broadly obovate-oblong to oval, obtuse, glabrous. Stamens inserted about 1/3 way down tube, exserted. Ovary oblong. Style up to c. 10 cm long, exceeding stamens. Fruit 3–5.5 cm long, ovoid or ellipsoid, slightly or strongly beaked, drying glossy black, calyx lobes appressed, markedly lenticellate. Seeds numerous, ovoid, tuberculate, brown.

Distr. Southern India and Ceylon; Burma, Thailand and throughout Malaysia; China, Hainan, Hongkong and Formosa.

Ecol. In and along borders of both primary and secondary forests in both dry and marshy areas, from sea-level up to 1500 m or more in our area.

Vern. Etamburu (S).

Note. Collections of *Fagraea fragrans* Roxb. are occasionally found in various herbaria made from trees planted as ornamentals in Ceylon. Additionally, specimens do occur purportedly from wild sources, but these are almost certainly from trees which have escaped from cultivation. As such, this species is not treated further in this work.

Specimens Examined. KURUNEGALA DISTRICT: Arankale, 14 Feb 1975, *Sumithraarachchi 650* (K, PDA); Doluwakande Hill, 600 m, 5 Jan 1977, *Faden & Faden 77/43* (K, PDA); Mt Kokegala, E of Mahiyangana, 600 m, 11 Nov 1975, *Bernardi 15699* (PDA). ANURADHAPURA DISTRICT: Ritigala summit, 15 Mar 1939, *Worthington 287* (BM), 24 Mar 1905, *Willis* (?) *s.n.* (PDA); Ritigala Strict Natural Reserve, summit, 732 m, 30 Sept 1972, *Jayasuriya 900* (PDA), 4 Aug 1972, *Jayasuriya et al. 831* (PDA). KEGALLE DISTRICT: Udabage, Kelani Valley, 275 m, 12 Sept 1946, *Worthington 2096* (K). KANDY DISTRICT: Corbet's Gap, 13 Aug 1970, *Meijer & Dassanayake 664* (PDA); between Rangala and Corbet's Gap, 610 m, 28 Mar 1970, *Balakrishnan 202* (PDA); Rangala Forest, along Rangala to Corbet's Gap road, 14 Nov 1975, *Sohmer & Jayasuriya 10647* (K, PDA); E of Corbet's Gap, 1300 m, 17 Nov 1974, *Davidse 8494* (PDA); Peradeniya, Gannoruwa, behind Royal Botanic Gardens, 600 m, 30 May 1973, *Kostermans 24922* (E, K); E Hoolankande Estate, N of Madulkele, 1400 m, 20 Aug 1978, *Huber 793* (PDA); Hantane, *Champion s.n.* (K), c. 750 m, *Gardner 583* (BM), *Gardner s.n.* in *C.P. 1824* (PDA), *C.P. 1826* (PDA); Bible Rock, c. 800 m, 19 Mar 1974, *Sumithraarachchi & Fernando 129* (PDA); Nawanagala, 1400 m, 31 Jan 1975, *Waas 1021* (K, PDA); Kadugannawa, 27 Nov 1969, *Amaratunga 1995* (PDA); Urugala, 610 m, 24 Apr 1958, *Worthington s.n.* (K); Madulkelle, below Knuckles, c. 1310 m, 23 July 1946, *Worthington 1968* (K); Knuckles, 1300 m, *Nooteboom 3415* (PDA), from Rangala to Looloowatte, c. 1300 m, 11 Sept 1977, *Nooteboom 3080* (PDA); Nillambe, c. 1035 m, 13 July 1947, *Worthington 2876* (K), Jan 1891, *s. coll. s.n.* (PDA); Loolewatte, 17 Apr 1932, *Simpson 9439* (BM); Kiriwanna Eliya, Laxapana, 28 Mar 1974, *Sumithraarachchi & Jayasuriya 207* (PDA); Laxapana to Maskeliya, 900 m, 12 May 1971, *Kostermans 24090* (PDA); Moray Estate, Maskeliya, 1500 m, 16 May 1971, *Kostermans 24150* (K, PDA); Laxapana to Maskeliya road, 900 m, 12 May 1971, *Kostermans 24090* (K). BADULLA DISTRICT: Thotulagalle Estate above Haputale, 1700 m, 18 Apr 1969, *Kostermans 23208* (K); Namunukula, 25 Apr 1974, *Waas 461* (PDA). AMPARA DISTRICT: near Arugam Bay, on road to Panama, 30 m, 1 Nov 1975, *Bernardi 15565* (K, PDA); 196 mile-marker on Potuvil to Moneragala road, 29 Jun 1975, *Sumithraarachchi & Sumithraarachchi 892* (PDA). RATNAPURA DISTRICT: Ela Uda, c. 600 m, 23 July 1976, *Cramer 4716* (E, K, PDA); Karawita kanda, 17 Feb 1974, *Waas 391* (PDA); c. 21 km NE of Deniyaya on Highway A17 to Ratnapura at milepost 64, 1050 m, 22 Oct 1974, *Davidse 7913* (PDA) above Belihul Oya, 600 m, 10 may 1969, *Kostermans 23461* (K); Bambarabotuwa Forest Reserve, between Pelmadulla and Rassagala, 810 m, 30 Oct 1977, *Huber 506* (PDA). NUWARA ELIYA DISTRICT: "Rambodde" (= Ramboda), *s. coll.* in *C.P. 757*, p.p. (HAK), Oct 1845, *Thwaites s.n.* in *C.P. 757* (K, PDA); Meeriyatenne, Hanguranketa, 1280 m, 4 Feb 1975, *Waas 1124* (K, PDA); Pundaluoya, c. 925 m, 28 Feb 1978, *Cramer 5149* (E, K, PDA); Coneygar, Halgranoya, 1615 m, 1 Oct 1940, *Kershaw s.n.* (PDA). MONERAGALA DISTRICT: Moneragala, 11 Jun 1974, *Waas 688* (K, PDA); Uda Walawe, 17 Oct 1971, *Balakrishnan & Jayasuriya 885* (K, PDA); Diyaluma Oya, c. 190 m, 23 Sept 1976, *Cramer 4742* (E, K, PDA). GALLE DISTRICT: Yatalamatta Forest, 18 Nov 1968, *Huber 59* (BM, K, PDA); Hiniduma, 23 Mar 1970, *Balakrishnan 616* (K, PDA); Bambarawana, c. 75 m, 2 Apr 1978, *Cramer 5217* (K); Kanneliya Forest Reserve, 28 July 1971, *Meijer 1062* (PDA); about milepost 5, on Dediyagala road, Kanneliya Forest, 1 Sept 1974, *Jayasuriya & Bandaranayake 1805* (K, PDA); Kanneliya, 150 m, 25 July

1976, *Jayasuriya & Kostermans 2358* (PDA); Kottawa, 2 Oct 1973, *Waas & Peeris 534* (PDA). LOCALITY UNKNOWN: *Gardner 583* (K), *584* (K); in 1861, *Thwaites* in *C.P. 1824* (BM), *C.P. 757* (BM); *Worthington 7166* (K), *7177* (K).

MONIMIACEAE
(by M.D. Dassanayake*)

Juss., Ann. Mus. Natl. Hist. Nat. 14: 133. 1809 ("Monimieae"); Perkins and Gilg in Pflanzenr. 4(4): 1–122. 1959. Type genus: *Monimia* Thouars.

Trees or shrubs, rarely climbers. Leaves almost always opposite, simple, pinnately veined, exstipulate, often containing aromatic oils or resin. Flowers mostly in axillary or rarely terminal cymose inflorescences, rarely solitary, small or rarely medium-sized, bisexual or unisexual on monoecious or dioecious plants, usually perigynous, regular. Bracts small or absent. Floral axis forming a cup-shaped or very deeply hollowed hypanthium, with the inner surface often nectariferous. Perianth 4-many, all sepaloid when parts are few, inner segments petaloid when many, free or connate, rarely absent. Stamens mostly many, in one or two series; staminodes sometimes present. Filaments sometimes with two lateral nectariferous appendages at the base. Anthers basifixed, extrorse or introrse, opening by longitudinal slits, or by 2 valves. Ovary mostly with many free carpels, rarely unicarpellary. Styles long and filiform or short, or absent, stigma terminal, small. Ovule single in each carpel, usually anatropous, bitegmic or unitegmic, pendulous or erect. Fruit of 1-many free achenes, drupelets or nutlets standing on the floral axis or enclosed by the enlarged hypanthium. Seeds with copious and oily endosperm and small or medium-sized embryo.

About 39 genera, tropical and subtropical, predominantly in the Southern Hemisphere: Oceania, Polynesia, Australia, Malesia, Madagascar, South America; very rare in Africa, absent from India.

HORTONIA

Wight ex Arn., Mag. Zool. Bot. 2: 545. 1838. Type species: *Hortonia floribunda* Wight ex Arn.

Shrub or small tree; young parts with peltate scales. Leaves opposite or subopposite. Flowers few, bisexual, regular. Perianth segments many, imbricate, in many series, the outer sepaloid, changing gradually inwards to 16–20 peltoid segments. Stamens 4–12, shorter than the perianth segments, attached

---

* Flora of Ceylon Project, Peradeniya.

to the margin of the hypanthium in one or two series. Filaments linear, somewhat short, bent outwards at apex, with 2 large turbinate appendages (glands) at the base on the abaxial side. Anthers extrorse. Carpels 8–10, free, sessile in centre of hypanthium, ovules pendulous, anatropous. Drupelets ovoid, shortly stalked, with a hard endocarp. Seed broadly ovoid, compressed.

3 species, closely similar, considered by Thwaites (Enum. Pl. Zeyl. 11. 1858) as varieties of one species.

KEY TO THE SPECIES

1 Leaves elliptic, rounded at base and apex . . . . .2. **H. ovalifolia**
1 Leaves lanceolate to narrowly ovate, base and apex acute
  2 Leaves lanceolate to narrowly ovate, major veins arched, peduncles stout . . . .
  . . . . .1. **H. floribunda**
  2 Leaves narrowly lanceolate to narrowly elliptic, major veins parallel to margin, peduncles
  slender . . . . .3. **H. angustifolia**

1. Hortonia floribunda Wight ex Arn., Mag. Zool. Bot. 2: 546. 1838; Wight, Ic. Pl. Ind. Or. 6: t. 1997. 1853; A. DC. in DC., Prod. 16(2): 672. 1868; Hook. f., Fl. Br. Ind. 5: 115. 1886; Trimen, Handb. Fl. Ceylon 3: 436. 1895. Type: Sri Lanka, vicinity of Pussellawa & Ramboda, *Walker s.n.*

Hortonia acuminata Wight, Ic. Pl. Ind. Or. 6: t. 1998. 1853.
Hortonia floribunda Wight ex Arn. var. *acuminata* (Wight) Hook. f. and
  Thoms., Fl. Ind. 1: 166. 1855; Thw., Enum. Pl. Zeyl. 11. 1858.

Tree, to c. 15 m or more high, trunk to c. 50 cm diam. at base. Leaf 7–17 × 2–7.5 cm, lanceolate to narrowly ovate; acute at base and apex, acuminate, coriaceous, lamina sometimes curved downwards, with the apex pointing down, margins often reflexed, lateral veins arching. Petiole 6–20 mm long, more or less flattened on upper surface and more than 2 mm broad. Peduncle 10–15 mm long. Bracts 2–5 mm long, caducous. Pedicel 10–15 mm long, flower 15–25 mm across. Sepals 4–6, 3–6 mm long, ovate. Petals 16–20, 8–11 × 2–3 mm, pale greenish yellow or pale yellow. Stamens 8–10, filaments 1.5–2 mm long; appendages yellowish, thick, with concave top about 1 mm wide, on a stalk about 0.5 mm long; anthers about 1.5 mm long, broadly elliptic, spreading. Carpels 8–9. Fruit of 1–7 separate drupelets, each 10–18 mm long, about 10 mm across, green, red, then purplish black when ripe; seed 8–15 mm long.

Distr. Endemic.

Ecol. Montane zone, about 1300 m and higher, in forest, locally common, in shade, often near streams, flowering March to November.

Vern. Wawiya (S).

Specimens Examined. MATALE DISTRICT: Kabaragala, off Madulkele-Matale road, *Dassanayake 589* (PDA). KANDY DISTRICT: Peak Wilderness, near Rajamally Estate, *Dassanayake 546* (PDA); Dolosbage, 6 Feb 1969, *Hoogland 11418* (PDA); Hantane and Hunasgiriya, in 1851, *s. coll.* in *C.P. 1027*, p.p. (PDA); Ramboda, in 1848, *Gardner* in *C.P. 1027*, p.p. (PDA); Maskeliya-Double Cut, May 1971, *Kostermans 24074A* (PDA), Maskeliya-New Dam site, 13 May 1971, *Kostermans 24102* (PDA), Maskeliya, 30 Apr 1926, *J.M. Silva s.n.* (PDA); Rangala-Corbet's gap, 8 Sep 1927, *Alston 1512* (PDA); Hewaheta jungle, 19 Apr 1927, *Alston 1008* (PDA); Madulkele, Oct 1887, *s. coll. s.n.* (PDA). BADULLA DISTRICT: Madawelgama, 15 Oct 1971, *Balakrishnan 817* (PDA); Namunukula, 19 Apr 1924, *J.M. Silva s.n.* (PDA). RATNAPURA DISTRICT: Kitulgala/Maskeliya, 12 May 1971, *Kostermans 24083* (PDA). NUWARA ELIYA DISTRICT: Hakgala, 11 Feb 1970, *Cooray 70021118R* (PDA), 9 Aug 1970, *Meijer 606* (PDA), 28 Jun 1971, *Balakrishnan 525* (PDA), 22 May 1911, *J.M. Silva s.n.* (PDA); Maturata, Jun 1848, *Gardner* in *C.P. 1027*, p.p. (PDA); Pidurutalagala, 21 Jun 1970, *Meijer 77* (PDA).

2. Hortonia ovalifolia Wight, Ic. Pl. Ind. Or. 6: t. 1998. 1853; A. DC. in DC., Prod. 16(2): 672. 1868. Lectotype: Sri Lanka, Adam's Peak, *s. coll. C.P. 159*.

Hortonia floribunda var. *ovalifolia* (Wight) Hook. f. and Thoms., Fl. Ind. 1: 166. 1855; Thw., Enum. Pl. Zeyl. 12. 1858; Hook. f., Fl. Br. Ind. 5: 115. 1886; Trimen, Handb. Fl. Ceylon 3: 437. 1895.

Small tree, about 7 m tall, with trunk about 50 cm diam. Leaves 5–20 × 3–9 cm, narrowly elliptic, broadly elliptic or narrowly ovate, obtuse or acute at apex, obtuse or rounded, rarely acute at base, often apiculate, thick, coriaceous, with more or less strongly reflexed margins. Petiole 6–22 mm long, subcylindric. Peduncle 5–25 mm long. Bracts 4–5 mm long, caducous. Pedicel about 12 mm long, reddish. Flower 20–25 mm in diameter. Sepals 3–6, broadly ovate, reddish green, 2–6 mm long. Petals about 20, orange-yellow or greenish yellow, 5–10 mm long. Stamens 8–10, filaments 1-1.5 mm long, with appendages as in *H. floribunda*. Carpels 8–9. Fruit as in *H. floribunda*.

Distr. Endemic.

Ecol. Adam's Peak, about 1600 m and higher. Flowering March to August.

Specimens Examined. KANDY DISTRICT: Foot of Adam's Peak, May 1891, *s. coll. s.n.* (PDA), Dalhousie Estate, foot of Adam's Peak, *Worthington 2727* (PDA), Adam's Peak, March 1846, *s. coll. C.P. 161* (PDA), Apr 1852, *s. coll. C.P. 159* (PDA), 21 Sept 1969, *ven Beusekom 1551* (PDA), Jun 1971, *Balakrishnan 522* (PDA).

3. Hortonia angustifolia (Thw.) Trimen, Cat. 75. 1885; Trimen, Handb. Fl. Ceylon 3: 437, t. 78. 1895.

Hortonia floribunda Wight ex Arn. var. *angustifolia* Thw., Enum. Pl. Zeyl. 12. 1858; Hook. f., Fl. Br. Ind. 5: 116. 1886. Lectotype: Sri Lanka, between Galle and Hiniduma, Dec 1852, *s. coll.* in *C.P. 1026* (PDA). *Hortonia acuminata* A. DC. in DC., Prod. 16(2): 672. 1898, non Wight 1853.

Tree, to about 5 m. Twigs standing horizontal or somewhat drooping. Leaves 8–14 × 2–6 cm, narrowly lanceolate, lanceolate, narrowly elliptic or very narrowly elliptic, base acute, sometimes cuneate or asymmetric, attenuate at apex, long-acuminate or caudate, 3-veined from above base; veins pinnate, major veins parallel to the margin, prominent beneath, lamina flat or bent down at apex, margins sometimes reflexed or slightly repand. Petiole 10–15 mm long, cylindric, channelled above, 1–1.5 mm thick. Peduncle 8–30 mm long. Bracts 2–4 mm long. Pedicel 6–10 mm long, flower about 10 mm in diameter. Sepals 4–5, 1.5–3 mm long; very broadly deltoid, acute, obtuse or rounded. Petals 10–13, 4–8 × 2 mm, margins ciliate, with hairs in tufts, somewhat repand, pale yellow. Stamens 4–6; filaments 1–1.5 mm long, appendages thick, brownish, concave at top, about 0.5 mm wide, with stalk about 0.5 mm long. Carpels 8–10. Fruit of 3–8 drupelets, about 12 mm long, dark purplish crimson. Seed about 10 mm long.

Distr. Endemic.

Ecol. Moist lowlands, at elevations to about 700 m; Galle, Kalutara, Colombo and Ratnapura Districts; locally somewhat common.

Ecol. At edges of streams. Flowering July to November. Flowers pale yellow.

Specimens Examined. COLOMBO DISTRICT: Kalatuwawa, below reservoir, *Dassanayake 488* (PDA); Panangala Estate, near Waga, *Dassanayake 487* (PDA). KALUTARA DISTRICT: Hewessa, Pelawatte, *Worthington 6597* (PDA). RATNAPURA DISTRICT: Diganda, 6 Aug 1971, *Meijer 1083* (PDA). GALLE DISTRICT: Dediyagala forest, *Worthington 2540* (PDA); Tawalama, 3 Apr 1970, *Balakrishnan 214* (PDA); Hiniduma, 1 July 1971, *Balakrishnan 526* (PDA); Galle, Dec 1853, *s. coll.* in *C.P. 1026*, p.p. (PDA); between Galle and Hiniduma, Dec 1853, *s. coll.* in *C.P. 1026*, p.p. (PDA). MATARA DISTRICT: near Morawaka, 5 Mar 1881, *s. coll. s.n.* (PDA). LOCALITY UNKNOWN: *Moon* in *C.P. 1026*, p.p. (PDA).

NELUMBONACEAE
(by M.D. Dassanayake[*])

Dumort., Anal. Fam. Pl. 53. 1829 ("Nelumboneae"). Type genus: *Nelumbo* Adans.

Aquatic herbs with branching rhizomes and fibrous roots. Leaves arising from the rhizome on the upper side, each leaf accompanied by 2 scale leaves and an ochrea-like stipule forming a sheath around the terminal bud at its base. Petiole long. Lamina large, suborbicular, peltate, usually standing above the water. Flowers large, solitary, from axil of upper scale-leaf, standing above the water, regular, bisexual, hypogynous, pollinated by beetles. Perianth segments many, free, spirally arranged, few outer greenish and sepaloid, inner ones progressively petaloid, caducous. Stamens many, free, spirally arranged. Filaments slender or dilated. Anthers linear, with 4 pollen sacs, introrse, the connective prolonged beyond the anther as a terminal appendage. Pollen tricolpate. Ovary of many free carpels sunk individually in the large obconic flat-topped spongy receptacle, with the stigmas protruding. Ovule solitary, pendulous from the top of the loculus, anatropous, bitegmic. Fruit of separate nuts loosely embedded in the enlarged receptacle. Seed solitary, without endosperm or perisperm, with large embryo.

One genus.

NELUMBO

Adans., Fam. Pl. 2: 76, 582. 1763. Type species: *N. nucifera* Gaertn.
Nelumbium Juss., Gen. Pl. 68. 1789, ed. 2, 76. 1791, orth. var.

Characters of the family.

2 species.

Nelumbo nucifera Gaertn., Fruct. 1: 73. t. 19, f. 2. 1788; Nicolson et al., Interp. Hort. Mal. 197. 1988. Type: from India.

Nymphaea nelumbo L., Sp. Pl. 511. 1753. Type: from India.
Nelumbium speciosum Willd., Sp. Pl. 2: 1258. 1799; Trimen, Handb. Fl. Ceylon 1: 51. 1893, nom. superfl.

---

[*] Flora of Ceylon Project, Peradeniya.

Rhizome creeping, to c. 5 cm broad, white, brown-spotted, with internodes to 50 cm or more long. Roots fibrous, white, arising all round the rhizome and confined to a zone just behind a node. Lower scale leaf 5–6.5 cm long, upper 12–15 cm long. Lamina 30–60 cm across, suborbicular, flat and floating with relatively weak petiole or more or less concave or cup-like and raised above the water on a rigid petiole, glabrous, pergameneous, dark green and glaucous above, light green beneath, hyaline at the edge; veins prominent beneath, 20 or 21 veins radiating out from the petiole apart from the midrib and bifurcating 2 or 3 times, ending in loops just within the leaf margin, secondary veins forming a net-work; midrib with 3 or 4 secondary veins on either side, and ending in a slight mucro at the leaf apex, where the margin is slightly depressed. Petiole 10–12 mm broad, light green, scabrous, with distant prickles. Stipule 3–4 cm long, ribbed, brownish. Pedicel erect, rigid, scabrous with short prickles, c. 10 mm broad. Bud to c. 9 cm long, ovoid, tapering and acute at end. Outer tepals 4 or 5, 1.5–2 × 1.2–1.5 cm, more or less deltoid, acute, green or greenish purple, caducous, leaving more or less prominent scars. Inner tepals 4.5–12 × 2–8 cm, broadly elliptic, narrowly ovate, ovate or obovate, pink, purplish pink, pinkish white or white, truncate at base, very concave, acute or obtuse at apex, mucronate or cuspidate, pergameneous, thick, with prominent veins on both surfaces; innermost ones narrower and shorter, some divided or lobed. Filaments 3–18 mm long, white or cream-coloured. Anthers 10–14 mm long, c. 1 mm wide, yellow, the connective with a white spathulate terminal appendage 2–7 mm long. Outer stamens transitional to inner tepals, with progressively dilated filaments and reduced anthers and appendages. Carpels 16–24; stylar region c. 0.25 mm long; stigma exserted, disk-like, c. 2 mm broad, somewhat depressed at the centre, yellow, sticky. Ripe receptacle to c. 10 cm across at top, spongy, black and light when dry, becoming detached and floating on the surface, with the cavities enlarged, allowing nuts to fall out. Nuts c. 18 mm long, ovoid, brownish black, glaucous. Seed filling the nut, with thin, brown testa, embryo large, cotyledons thick and fleshy.

Distr. Warmer parts of Asia and northeastern Australia.

Ecol. In ponds and reservoirs, particularly in the dry region, often cultivated. Flowering throughout the year.

Uses. The flowers are a favoured temple offering. Rhizomes are eaten cooked and used in medicine. Seeds are said to be used as food in Kashmir, etc.

Vern. Nelum (S), Lotus (E).

Note. The flower colour varies from white to purplish pink.

Specimens Examined. VAVUNIYA DISTRICT: "Moolaitivu", *Gardner* in *C.P. 1022* (PDA). ANURADHAPURA DISTRICT: near Rambwewa, "flowers white, pinkish outside, leaves reddish below", 28 March 1927, *Alston 1116* (PDA); Anuradhapura, Raja Pokuna, 30 Nov 1973, *Sohmer 8956* (PDA); NE of Mihintale, Mahakandara River, 20 March 1969, *Robyns 6972* (PDA). MATALE DISTRICT: Tank, Dambulla-Sigiriya Road, 22 Sept 1971, *Dassanayake 522* (PDA). COLOMBO DISTRICT: Jaela, marshy rice-fields, "flowers rose-coloured", 3 Jan 1952, *Senaratne s.n.* (PDA). HAMBANTOTA DISTRICT: Tissamaharama Tank, "flowers white with pinkish tinge, leaves red below", Jan 1926, *Alston 117* (PDA).

NYMPHAEACEAE
(by M.D. Dassanayake*)

Salisb., Ann. Bot. (König & Sims) 2: 70. 1805 ("Nymphaeeae"). Type genus: *Nymphaea* L., nom. cons.

Aquatic rhizomatous herbs. Leaves with long petioles arising from the rhizome, spirally arranged, stipulate or exstipulate; lamina ovate to orbicular, peltate with a basal sinus, usually floating. Flowers usually large, solitary, bisexual, regular, hypogynous to epigynous, entomophilous. Sepals 4–6 or more, free, sometimes more or less petaloid. Petals 8 to many, free, the inner often passing into the stamens. Stamens many, spirally arranged, inner and outer ones often reduced to staminodes. Filaments often broad, with 3 veins. Anthers linear, introrse, with longitudinal slits. Ovary syncarpous, of 5-many carpels and 5-many loculi, superior to inferior, with many bitegmic anatropous or orthotropous ovules scattered over the partitions. Fruit ripening under water, spongy, berry-like, indehiscent or irregularly dehiscent by swelling of mucilage within. Seeds small, often arillate, with scanty endosperm, copious perisperm and small embryo with one or two cotyledons.

6 genera and c. 60 species, cosmopolitan.

NYMPHAEA

L., Sp. Pl. 1: 510. 1753. Type species: *N. alba* L. (typ. cons.).

Rhizomes erect. Leaves large or medium-sized, usually floating, stipulate. Lamina deeply cordate, ovate to orbicular, palmately veined. Sepals 4, imbricate. Petals many, spirally arranged, outer 4 alternating with sepals and next 4 with those. Stamens many, spirally arranged, with or without an appendage formed by the connective. Filaments dilated at the base, petaloid. Ovary sunk in and adnate to the fleshy receptacle, petals and filaments inserted on the side of the receptacle and near its apex. Carpels many, loculi many, each with many anatropous ovules. Stigmas many, sessile, linear, radiating over the upper surface of the ovary, with appendages at the margin. Fruit a large fleshy berry. Seed surrounded by an aril open at the top.

* Flora of Ceylon Project, Peradeniya.

c. 35 species, cosmopolitan. The water lilies, 2 species in Sri Lanka. Several exotic species and hybrids are cultivated as ornamentals.

KEY TO THE SPECIES

1 Leaves pubescent beneath with many short hairs, margin sharply dentate-mucronate. Stamens without a tongue-shaped appendage beyond the anther or appendage very short. Flowers red or white .... **1. N. pubescens**
1 Leaves glabrous, margin entire to dentate with blunt teeth. Stamens with a tongue-shaped appendage beyond the anther to 5 mm long. Flowers blue or white .... **2. N. nouchali**

1. Nymphaea pubescens Willd., Sp. Pl. 2: 1154. 1799; Gamble, Fl. Pres. Madras 34. 1915; Nicolson et al., Interp. Hort. Mal. 198. 1988. Type: From India.

Nymphaea lotus L. var. *pubescens* (Willd.) Hook. f. & Thoms., Fl. Ind. 1: 241. 1855; Trimen, Handb. Fl. Ceylon 1: 50. 1893.
Nymphaea nouchali sensu Alston in Trimen, Handb. Fl. Ceylon 6: 8. 1931, non Burm. f. 1768.

Petioles to c. 15 mm thick, terete, scabrous, red-brown. Stipules c. 2 cm long, connate to petiole base below and free for 3–5 mm at their apices. Lamina to c. 50 cm broad, floating on the surface, coriaceous, smooth, glossy dark green and glabrous above, dark purplish green and velvety, with a felt of fine short purplish hairs beneath, somewhat peltate, orbicular, incised-cordate at base, margin repand, dentate-mucronate; veins very prominent beneath. Pedicel erect, stout, with short prickles. Sepals 5–10 × 2–3 cm, lanceolate or narrowly oblong, with a broad base, tapering and acute or obtuse at apex, green or purplish green abaxially with 7–10 prominent lighter-coloured veins, white or purplish pink and smooth within. Petals 12–22, 5–10 × 1.6–3.5 cm, inserted spirally on the axis, oblanceolate, narrowly elliptic or narrowly obovate, acute or obtuse at the apex, white, purplish pink or red, inner petals smaller. Stamens to c. 70, inner ones much reduced. Filaments 5–22 mm long, 2.5–10 mm broad, much dilated at the base, tapering upwards, yellowish white becoming deeper yellow distally, or pale purplish pink to crimson. Anthers 6–33 mm long, linear, with a 1–2 mm long hook-like prolongation of the connective, yellow. Ovary 1.5–2.5 cm broad, 1–2.5 cm high, the top sloping down to the centre, with a central knob c. 2 mm high. Carpels c. 20. Ovules with long funicles, with collar at apex. Stigmatic rays c. 3 mm broad, with appendages incurved over the ovary. Fruit 3–9 cm broad, depressed globose, fleshy, dark green, bearing the now stout persistent filaments at the sides and the remnants of sepals at the base. Seeds many, subglobose, 1.5–2 mm broad, pale black-brown, with tubercles in longitudinal rows and transversely striate with reticulations. Aril white, sac-like.

Distr. Tropical Asia, India to New Guinea.

Ecol. Lowlands, particularly in the dry region, in ponds and reservoirs. Very common. Flowering throughout the year. Flowers fragrant, opening in the evening, closing late next morning.

Uses. The flowers are a favoured offering in temples. The seeds are eaten boiled. The rhizome, flower, fruits and seeds are used in medicine.

Vern. Olu, Etolu (S).

Note. The flower colour varies from nearly white to red.

Specimens Examined. JAFFNA DISTRICT: Jaffna, in 1846, *Gardner* in *C.P. 1019*, p.p. (PDA); Jaffna, temple tank, in 1848, *Gardner* in *C.P. 1021* (PDA). ANURADHAPURA DISTRICT: tank near Kalawewa, Feb 1888, "flowers deep rich rose colour", *s. coll. s.n.* (PDA); Mavala tank, near Kekirawa, 22 Jan 1982, *Dassanayake 616* (PDA); near Kawakkulama, 17 July 1972, *Hepper & Jayasuriya 4663* (PDA); near Rambewa, "flowers white, pinkish outside, leaves reddish below", 28 March 1927, *Alston 1116* (PDA); Anuradhapura, Raja Pokuna, 30 Dec 1973, *Sohmer 8956* (PDA); NE of Mihintale, Mahakanadara River, 20 March 1969, "blade 50 cm wide, flowers white". *Robyns 6972* (PDA). POLONNARUWA DISTRICT: Nikawewa tank, 2 April 1932, *Simpson 9413* (PDA). MATALE DISTRICT: 6.5 miles N of Dambulla, 8 Dec 1970, *Fosberg & Balakrishnan 729* (PDA); Dambulla-Sigiriya Road, 22 Sept 1971, *Dassanayake 522* (PDA). MATALE & ANURADHAPURA DISTRICTS: Matale, Oct 1848, *Gardner* and Anuradhapura, *Brodie*, in *C.P. 1019*, p.p. (PDA). KURUNEGALA DISTRICT: estate near Bolana, Feb 1853, *s. coll. C.P. 1019*, p.p. (PDA). AMPARAI DISTRICT: Lahugala, 13 Feb 1972, *Jayasuriya, Cramer & Balasubramaniam 729* (PDA). HAMBANTOTA DISTRICT: Ruhuna National Park, Palugaswela, 23 March 1970, *Cooray 70032314* (PDA); Buttuwa-Patanagala, 5 Jan 1969, *Fosberg, Mueller-Dombois, Wirawan, Cooray & Balakrishnan 51040* (PDA); Tissamaharama tank, "flowers white, with pink tinge, leaves red below", 1 Jan 1926, *Alston 117* (PDA).

2. Nymphaea nouchali Burm. f., Fl. Ind. 120. 1768; Nicolson et al., Interp. Hort. Mal. 198. 1988. Type: From India.

Nymphaea stellata Willd., Sp. Pl. 2: 1153. 1799; Trimen, Handb. Fl. Ceylon 1: 50. 1893. Type: from Malabar, India.

Petioles to c. 8 mm thick, terete, smooth, purplish green. Stipules fused to petiole base for c. 5 mm, free for c. 2 cm, free part falcate, membranous, purplish. Lamina 10–30 × 10–26 cm, floating on the surface, coriaceous, smooth, glossy bright green above, dark purplish green and glabrous beneath, with green veins, somewhat peltate, rotund, orbicular or suborbicular, incised-cordate at base; veins prominent beneath. Pedicel erect, stout, smooth,

c. 12 mm broad, pale red. Bud narrowly conical. Sepals 4–5.5 × 1–2 cm, lanceolate, acute and rounded at apex, cuspidate, glabrous, coriaceous, bright green with purple dots and narrow streaks and 9–12 lighter-coloured veins abaxially, greenish white and smooth within. Petals 9–16, 4–5.5 × 1–2 cm, outer somewhat similar to sepals, inner smaller and more petaloid, lanceolate or narrowly elliptic, acute or subobtuse at apex, with a broad base, pale violet or pale blue fading to a dull blue, yellowish at base. Stamens 16–30, the outer longer. Filaments flat, pale yellowish, to c. 1.5 cm long, c. 5 mm broad, inner c. 4 mm long, 1 mm broad. Anthers 4–10 mm long, yellow, with a tongue-shaped appendage beyond, this in the outer stamens faintly violet, 5–8 mm long, in the innermost yellow, c. 1 mm long. Ovary 10–12 mm broad, yellow. Carpels 10–16. Ovules embedded in mucilage, funicle with a swollen end. Stigmas bright yellow, rays acute, curved upwards at ends. Fruit subglobose, to c. 5 cm broad. Seeds many, narrowly ovoid, 1–1.5 mm long, greyish white, marked with longitudinal ridges.

Distr. Also India to New Guinea.

Ecol. Lowlands, particularly in the dry region, in ponds and tanks, often cultivated. Common. Flowering all the year. Flowers slightly fragrant. Open from about sunrise till early afternoon.

Uses. The flowers are offered in temples. The starchy rhizomes are eaten roasted. Rhizomes and seeds used in medicine.

Vern. Manel (S).

Specimens Examined. JAFFNA DISTRICT: Kaddaikadu, SE of Point Pedro, 18 Mar 1973, *Bernardi 14268* (PDA). ANURADHAPURA DISTRICT: Mavala Tank, near Kekirawa, 22 Jan 1972, *Dassanayake 619* (PDA). PUTTALAM DISTRICT: Tabbowa, 21 Mar 1973, *Cramer 4069* (PDA). COLOMBO DISTRICT: Seeduwa, cultivated, 1 Sept 1967, *Jayasuriya 2669* (PDA). BATTICALOA DISTRICT: Mankeni, 11 Feb 1972, *Jayasuriya, Cramer & Balasubramaniam 697* (PDA).

OLACACEAE

(by B. Verdcourt*)

Mirbel ex. DC., Prod. 1: 531. 1824 ("Olacineae"). Type genus: *Olax* L.

Trees, shrubs or less often lianes, occasionally dioecious (one African genus). Leaves usually alternate, simple, entire, exstipulate. Flowers bisexual or rarely unisexual, borne in axillary cymes or apparent racemes either pedunculate or reduced to fascicles, sometimes solitary. Sepals 3–7, often small, usually partly united or forming a cupuliform structure with entire or lobed margin, often accrescent in fruit, sometimes adnate to the ovary. Petals 3–7, free or sometimes joined, valvate. Stamens 4–12(–15), free or adnate to the petals when same number as them; some occasionally sterile (staminodes); anthers dehiscing longitudinally or in a few genera by means of apical pores. Disk usually present. Ovary superior or ± immersed in the disk or occasionally inferior, 1–5(–7)-locular, sometimes several-locular below but 1-locular above; ovules usually anatropous, 1 per locule, pendulous or in 1-locular ovaries with 2–7 ovules from the apex of a free central placenta. Fruit a berry or drupe, often surrounded by the accrescent free or adnate calyx. Seeds with copious endosperm.

About 28 genera and 170 species in the tropics of both Old and New Worlds. Less conservative estimates put the genera at 29 and up to 250 species. There is no doubt that the family is related to Santalaceae, Loranthaceae, Icacinaceae and Opiliaceae and in older treatments the latter two were usually included in Olacaceae. Some genera (e.g. *Olax* and *Ximenia*) have partially parasitic species which are not host-specific.

KEY TO THE GENERA

1 Stamens (including staminodes if present) more numerous than the petals
  2 Petals densely hairy within; staminodes absent; calyx not cupuliform nor accrescent; plant usually spiny .... **1. Ximenia**
  2 Petals glabrous within; staminodes present, bifid; calyx cupuliform, accrescent in fruit; plant not spiny save in *O. scandens* .... **2. Olax**
1 Stamens and petals both 5; flowers fasciculate; calyx accrescent but not cupuliform, adnate to the ovary .... **3. Strombosia**

* Royal Botanic Gardens, Kew.

1. XIMENIA

L., Sp. Pl. 1193. 1753; L., Gen. Pl. ed. 5: 500. 1754; Sleumer in Pflanzenfam. ed. 2, 16B: 22. 1935. Lectotype species: *X. americana* L.

Usually spiny trees and shrubs, often with very short shoots (brachyblasts) bearing the leaves and inflorescences. Leaves alternate. Flowers 4–5-merous in axillary fascicles, pedunculate cymes or sometimes solitary, rarely functionally unisexual. Calyx small, 4–5-lobed, not accrescent. Petals 4–5, linear-oblong, densely pilose on inner surface, glabrous or sparsely pubescent outside. Stamens 8(–10), free, half opposite the petals the other half alternating with them; filaments filiform; anthers erect, dehiscing longitudinally. Disk absent. Ovary superior, 3(–4)-locular; style ± slender; stigma capitate. Drupes ovoid to globose.

About 8 species (10–15 fide A.S. George) in the tropics and subtropics of both Old and New World, the single Ceylon species also pantropical.

Ximenia americana L., Sp. Pl. 1193. 1753; Masters in Hook. f., Fl. Br. Ind. 1: 574. 1875; Trimen, Handb. Fl. Ceylon 1: 255. 1893; Sleumer in Pflanzenfam. ed. 2, 16B, fig. 11. 1935; Lucas in Fl. Trop. E. Africa, Olacaceae: 3, fig. 2/1–5. 1968; De Filipps, Bol. Soc. Brot. II, 43: 194. 1969; Matthew, Fl. Tamilnadu Carnatic 2 (Illustr.): 136. 1982 & 3(1); 245. 1983; Sleumer in Fl. Males. 10: 11, fig. 4. 1984 (numerous references); A.S. George in Fl. Austral. 22: 15. 1984; A.C. Smith, Fl. Vitiensis Nova 3: 732, fig. 184. 1985. Type: Herb. Clifford 483, *Ximenia* No. 1 (BM, lectotype).

Spiny (rarely not) often straggling glabrous (save for flowers) shrub or tree to 10.5 m (usually ± 3–7 m) tall. Leaves narrowly to broadly elliptic, ovate, obovate, almost round or lanceolate, 1.3–10 cm long, 0.8–6 cm wide, retuse, obtuse or acute, sometimes mucronate, membranous to ± coriaceous, sometimes glaucous; petiole 0.2–1.2 cm long. Inflorescences often umbel-like, 2–10-flowered or flowers solitary; peduncles 0.1–1.5 cm long or absent; pedicels 0.2–1.2 cm long; without bracteoles save in solitary-flowered inflorescences where pedicels are 2–4-bracteolate near the middle. Flowers fragrant, usually hermaphrodite. Sepals 4(–5), 0.5–1.4 mm long, sometimes ciliate. Petals white or greenish, 4.5–8.5(–12) mm long, 1–2.5 mm wide, apiculate, densely white-bearded within to 1–3 mm below the apex, recurved when mature. Stamens 8(–10); filaments 1–4 mm long. Ovary ovoid-conic, 2.4–3.7 mm long; style (0.2–)0.5–4.3(–5.5) mm long, rarely absent. Drupes mostly yellow or sometimes orange to red, ovoid or ± globose, 1–3.5 cm long, 1–3 cm wide; seeds ellipsoid, 1.5–2.5 cm long, 1–2 cm wide.

var. **americana**

Leaves 2–10 cm long, 1–6 cm wide, not glaucous. Inflorescences 3–10-flowered; pedicels without bracteoles; filaments 2.5–4 mm long; style (1–)2.5–4.3 mm long.

Distr. Widespread in tropical America, Africa and Asia extending to N. Queensland, Fiji, New Caledonia, Tuamotu Is. etc.

Ecol. Dry areas at low altitudes. No recent material has been seen of this species and it has apparently always been rare. Capt. Walker's specimen was collected in Sept. 1885. Thwaites did not mention it in his enumeration nor does he list *C.P. 2382* in his index of numbers. Perhaps he failed to identify this gathering.

Uses. The fruit is edible, sour but refreshing, much eaten in countries where the species is common; an oil has been extracted from the seeds and used for dressing leather in Africa. The bark has been used medicinally in Africa and presumably elsewhere.

Vern. Chiru Illantai (Siru Illanthai fide Walker) (T).

Specimens Examined. TRINCOMALEE DISTRICT: Trincomalee, *s. coll. C.P. 2382* (K, PDA). BATTICALOA DISTRICT: Batticaloa, *Walker 204* (PDA).

Note. De Filipps (l.c.) has made a reasonable case to separate the African material into two varieties, var. *africana* and var. *microphylla* Welw. ex Oliv.

2. OLAX

L., Sp. Pl. 34. 1753; L., Gen. Pl. ed. 5: 20. 1754; Sleumer in Pflanzenfam. ed. 2.16B: 24. 1935. Type: *O. zeylanica* L.

Small to medium-sized trees or shrubs with mostly glabrous or slightly scabrid somewhat ridged branchlets. Leaves alternate, distichous, often yellowish green. Flowers hermaphrodite or unisexual, solitary or in fascicles scattered along the axis or axes of simple or branched inflorescences which are often simplified to produce apparent racemes; pedicels articulate at the base. Calyx cupuliform, truncate or slightly undulate or lobed at the margin, often accrescent in fruit. Petals 3 or 5 (–6), free or connate (see note below), linear-spathulate, apiculate, glabrous inside. Stamens 3–6, placed between the petals; staminodes 3–6, ± opposite and adnate to petals; antherodes bifid, the lobes filiform usually longer than the anthers. Ovary superior, 1-locular or imperfectly 3-locular (below); disk circular; style ± short with unlobed or 3–4-lobed capitate stigma. Drupes oblong, ovoid or ± globose, usually completely or half-included in the accrescent membranous calyx which is free to base inside (as in Ceylon species) or (fide A.S. George) with succulent exocarp formed by the enlarged hypanthium with calyx persistent as an apical collar. About 45–50 (60 fide A.S. George, 11 in Australia) species in the tropics of the Old World; three in Ceylon. The petals are sometimes considered to be cleft (often to the middle, or joined in pairs which has led to the same number being variously described as 3 or 6 or a tube with 5–6 lobes.

KEY TO THE SPECIES

1 Climbing shrubs; or if rarely a small tree then branches pubescent; drupes larger, up to 2.5 cm long, 2/3 to almost completely covered by the enlarged membranous calyx; petals appearing to be 3, deeply bifid or 3–5 free to base or variously connate
  2 Branches pubescent; old stems with slightly curved spines; flowers ± 6.5–8 mm long, rarely more .... **1. O. scandens**
  2 Branches glabrous; spines absent; flowers about 1 cm long or more ..... **2. O. imbricata**
1 Small tree with glabrous branches; drupes smaller, not over 1 cm long, less than half-covered by the enlarged membranous calyx; petals appearing to be 3 entire .... **3. O. zeylanica**

1. Olax scandens Roxb., Pl. Corom. 2: 2, t. 102. 1798; Roxb., Fl. Ind. ed. Carey 1: 163. 1832; Wight & Arn., Prod. 1: 89. 1834; Thw., Enum. Pl. Zeyl. 42. 1858; Masters in Hook. f., Fl. Br. Ind. 1: 575. 1875; Trimen, Handb. Fl. Ceylon 1: 256. 1893; Gamble, Fl. Pres. Madras 1: 16. 1915; Sleumer in Fl. Males. I, 10: 7, fig. 1. 1982. Type: 'India orientali, *Roxburgh*' (BM, lectotype).

Olax psittacorum sensu Vahl, Enum. Pl. 2: 33. 1805 p.p.; Almeida, J. Bombay Nat. Hist. Soc. 81: 742. 1985; Van Steenis & De Wilde in Fl. Males. I, 10: 717. 1989, non (Willd.) Vahl, sensu stricto.

Liane 2–20 m long or sometimes a straggling shrub or small tree to 5 m with pendent branches; older branches with strong slightly curved spines; young branchlets ± densely pubescent but soon glabrescent; stems with ± smooth grey bark. Leaves yellowish green, ovate, elliptic or oblong-elliptic or ± round, (1.8–) 3.5–8(–9.5) cm long, (0.3–) 0.8–4.5 cm wide, ± narrowed to rounded at the apex, attenuate to rounded at the base, thinly coriaceous, glabrous above, at first shortly pubescent on midrib beneath; petiole 4–7(–10) mm long, shortly pubescent. Inflorescences sweetly scented, 1–3 per axil, simple or branched, many-flowered, 0.5–3.5 cm long, densely shortly hairy, bracteate at the base; peduncles 0–2 mm long (sometimes appearing much longer on leafless short shoots which simulate peduncles); bracteoles 2–3 mm long, obtuse, keeled, pubescent; pedicels 1–5(–2) mm long, glabrous. Calyx 0.5–1 mm long, truncate. Petals white, 3, 2–3 split about halfway, linear-oblong, (4–) 7(–9) mm long, 1.5 mm wide, lobes acute at the apex, incurved, glabrous. Stamens 3, reaching to base of sinus between petal lobes in long-styled form and to 2–2.5 mm beyond it in short-styled form: anthers oblong; staminodes with very narrow, deeply bifid antherodes 1.5 mm long. Ovary ellipsoid or ovoid. Style 5–6 mm long in long-styled flowers, 1.5–2.5 mm in short-styled form; stigma obscurely 3-lobed. Drupe yellow to orange or pink, juicy, ovoid or subglobose, 0.8–1.8 cm long, 0.6–1.2 cm wide, about (1/2–)2/3–4/5-covered by the accrescent membranous calyx. Seed 7 mm long, 5 mm wide.

Distr. India from W. Himalaya to Ceylon, Burma, Indo-China, Thailand, Malay Peninsula, Java.

Ecol. Secondary lowland forest, rocky slopes, dry regions.

Uses. None recorded.

Vern. None recorded.

Specimens Examined. MANNAR DISTRICT: near Madhu, *Kostermans 25127* (K, L). PUTTALAM DISTRICT: Mi Oya 3 mi. E of Puttalam on A 12, *Maxwell & Jayasuriya 811* (PDA); N. of Chilaw, Karukupone, *Simpson 8563* (BM); (Puttalam fide *Trimen*). ANURADHAPURA DISTRICT: Anuradhapura, *Trimen s.n.* (PDA). TRINCOMALEE DISTRICT: (Trincomalee fide *Trimen*); Koddiar ("Koddiyar"), *Trimen s.n.* (PDA). LOCALITY UNKNOWN: *Thwaites C.P. 1216* (BM, K); *Col. Walker 39* (K).

Note. As Wight and Arnott and Trimen noted long ago *Olax psittacorum* is endemic to the Mascarene Islands and not conspecific with *O. scandens* as claimed by Almeida. *Fissilia psittacorum* Willd., Sp. Pl. 1: 194. 1797 is based entirely on '*Fissilia*' Lam., Enc. t. 26. 1791 which is the Mauritian plant. The fact that Vahl widened the description to include a Ceylon plant is of no consequence; his new combination is based only on the type. Neither is Roxburgh's name *O. scandens* invalidated by his including *Fissilia psittacorum* in synonymy in 1832; he did not do so in 1798 when he described the plant. It is unfortunate that this illconceived attempt at name changing was uncritically accepted in the Flora Malesiana corrigenda.

2. Olax imbricata Roxb., Fl. Ind. ed. Carey & Wall. 1: 169. 1820, ed. Carey 1: 164. 1832; Masters in Hook. f., Fl. Br. Ind. 1: 575. 1875; Sleumer, Blumea 26: 156. 1980 (detailed synonymy); Matthew, Fl. Tamilnadu Carnatic 2 (Illustr.), t. 135. 1982, 3 (Fl. (i)): 245. 1983; Sleumer in Fl. Males. I, 10: 8. 1984. Type: Bangladesh, Chittagong, *Roxburgh s.n.* (CAL, lectotype) (Sleumer cites this specimen which he had not seen and presumably this must be accepted as a lectotypification).

Olax wightiana Wall. ex Wight & Arn, Prod. 1: 89. 1834; Thw., Enum. Pl. Zeyl. 42. 1858; Masters in Hook. f., Fl. Br. Ind. 1: 575. 1875; Trimen, Handb. Fl. Ceylon 1: 256. 1893; Alston in Trimen, Handb. Fl. Ceylon 6: 48. 1931. Type: India, Courtallum, *Wallich 6779* (ex Herb. Wight) (K, lectotype; K-WALL, K, L, P, isotypes).

Liane or rarely a shrub with unarmed branches; branchlets striate, glabrous or slightly puberulous, dark red-brown with pale lenticels. Leaves ovate to elliptic-oblong, 4–15(–18) cm long, 2–7.5 cm wide, ± acuminate, acute or obtuse at the apex, cuneate to rounded at the base, said to be ± shining above (not when dry), glabrous; petiole 0.5–1 cm long. Inflorescences branched from the base, many-flowered, 1–3(–5) cm long; peduncles obsolete; pedicels 2.5–4 mm long; bracts ovate, imbricate, 1.5–3 mm long. Calyx ± 0.7 mm tall, slightly undulate. Petals white or pinkish, 3–5, linear-oblong, 1–1.2 cm long, free to base or variously connate. Stamens 3; staminodes 5–6, bifid. Drupes orange, subglobose, oblong or obovoid, (1.4–)1.7–2.5 cm long, almost covered by the thin accrescent calyx.

Distr. India, Ceylon, Burma, Andaman and Nicobar Is., Indo-China, Hainan, Sumatra to Solomon Is., Taiwan and Micronesia (Palau).

Ecol. Disturbed primary forest by streamside, rocky slopes, scrub jungle; low altitude.

Uses. None recorded. Drupes are "semi-sweet".

Vern. Telatiya (S).

Specimens Examined. ANURADHAPURA DISTRICT: Ganewalpola, track to Ritigala, *Jayasuriya & Sumithraarachchi 1611* (K, PDA, US); Kalawewa, *Trimen s.n.* (PDA); Mahailluppallama, ?*Willis s.n.* (PDA). KANDY DISTRICT: Kandy, *Moon 297* (BM). BADULLA DISTRICT: Bank of Uma ("Ooma") Oya, *de Silva s.n.* (PDA). KALUTARA DISTRICT: Kalutara ("Caltura"), *Macrae 264* 'Herb. Soc. Hort. Lond.' (K); Madola, *Trimen s.n.* (PDA). MONERAGALA DISTRICT: Beduwela, *Jayasuriya 2060* (K); Bibile-Nilgala road, 3rd mi. marker, Nagala, *Waas 654* (K, US). GALLE DISTRICT: Hiniduma, ?*Trimen s.n.* (PDA). LOCALITY UNKNOWN: *Gardner s.n.* (K); *Thwaites 1215* (BM, K); *Col. Walker s.n.* (K, PDA).

Note. Wight & Arnott suggest that one of the Ceylon specimens Vahl included under *O. psittacorum* is actually this species which is probably correct since *O. imbricata* resembles *O. psittacorum* more closely than does *O. scandens.*

3. Olax zeylanica L., Sp. Pl. 34. 1753; Wight & Arn., Prod. 1: 88. 1834; Thw., Enum. Pl. Zeyl. 42. 1858; Masters in Hook. f., Fl. Br. Ind. 1: 576. 1875; Trimen, Handb. Fl. Ceylon 1: 257. 1893; Gamble, Fl. Pres. Madras 1: 190. 1915; Alston in Trimen, Handb. Fl. Ceylon 6: 48. 1931; Lewis, Veg. Prod. Ceylon 100. 1934; Worthington, Ceylon Trees 131. 1959. Type: Ceylon, *Hermann s.n.* (Herb. Hermann 1: 76, No. 34) (BM, lectotype).

Shrub or small tree 4–8 m tall, d.b.h. ± 22 cm, unarmed; branchlets yellowish or bronze-brown, angled-ridged, characteristically finely transversely wrinkled, glabrous; bark yellowish or brown, ridged. Leaves yellow-bronze when young, later sage-green, elliptic to ovate-lanceolate, 2.5–8 cm long, 1–3.4 cm wide, acute to obtuse at the apex, rounded then abruptly cuneate at the base, somewhat coriaceous, drying with crinkly margins and midrib impressed above, glabrous, said to be shining in life but hardly so when dry; petiole very short, scarcely 2 mm long, ± winged and channelled. Inflorescences raceme-like, 1-few-flowered, 1–1.3 cm long, glabrous; peduncles obsolete; pedicels ± 2 mm long; bracts ± 0.8 mm long. Calyx a minute rim 0.3 mm tall, subcrenulate. Petals greenish, 3, linear-oblong, 4.5–5 mm long, 1.2 mm wide, not divided. Stamens 3, adnate to middle of lobes; staminodes 5–6, often one on each side of stamens, bifid, rather shorter than the petals but overtopping anthers. Ovary ovoid; style 1.5–3 mm long. Drupe scarlet, broadly ellipsoid, ± 1 cm long, slightly apiculate, less than half-covered by the ± 3 mm long thin cup-shaped accrescent calyx.

Distr. Ceylon. Also recorded for Madras—Herb Wight *302*—but no other material seen.

Ecol. Moist low forest, secondary rain forest, submontane forest, often by steamsides, also recorded for dry evergreen forest; 30–800 m.

Uses. Leaves eaten as a salad.

Vern. Maila, Malla, Mella (S).

Specimens Examined. ANURADHAPURA DISTRICT: Ritigala Strict Natural Reserve, *Jayasuriya 1691* (K, US); top of Ritigala, ?*Willis s.n.* (PDA). COLOMBO DISTRICT: near Colombo, *Thwaites 1214*, p.p. (K, PDA). KANDY DISTRICT: E. of Madugoda, *Jayasuriya 2119* (K), *2121* (K). KALUTARA DISTRICT: (Kalutara, fide *Lewis*); Hewesse, *Trimen s.n.* (PDA); E. Kalugala Forest, *Waas 1546* (K, US); Hewesse, Pelawatte (Pellewatta), *Worthington 6598* (K). NUWARA ELIYA DISTRICT: Above Belihuĺoya, dirt road to Horton Plains, *Kostermans 23456* (BO, K, etc.). GALLE DISTRICT: Kaneliya (Kanneliya), Naunkita ela, *Worthington 3684* (BM, K). NOT PLACED: Galagama (several places of same name), *C.P. 1214*, p.p. (PDA). LOCALITY UNKNOWN: *Gardner s.n.* (K, PDA); *Hermann s.n.* (BM); *Koenig s.n.* (BM); *Thwaites C.P. 1214*, p.p. (K, PDA) (& *Pierre* drawing (BM, P).

3. STROMBOSIA

Blume, Bijdr. 1154. 1826–1827; Sleumer in Pflanzenfam. ed. 2, 16B: 21. 1935. Type species: *S. javanica* Blume.

Trees or shrubs. Leaves distinctly petiolate, glabrous, sometimes finely pellucid-dotted, often coriaceous; lateral nerves prominent beneath, usually with rather faint but characteristic tertiary venation, the veins ± parallel and

± at right angles to the nerves. Flowers 5-merous, borne in axillary fascicles, pedicellate. Calyx small, cupular, 5-lobed, adnate to the ovary, accrescent. Petals with fine to dense indumentum on the inner surface usually confined to the upper half. Stamens epipetalous with filaments free for only a short distance. Ovary initially ± superior, finally semi-inferior to inferior, 3–5–(6)-locular below, borne on or partially immersed in the prominent lobed disk; style short to elongate with 3–5(–6)-lobed stigma. Drupe bearing a circular depression at the apex, the rim of which is top of adnate calyx (hypanthium) and from centre of which the style-base persists. About 16 species in the Old world tropics, only two of which occur in Ceylon. I have not agreed with Sleumer's circumscription of the species (see note).

KEY TO THE SPECIES

1 Leaves larger, mostly 10–17.5 cm long, 5–7.2 cm wide; lateral nerves mostly leaving midrib at an angle of 30° and fine parallel venation at ± right angles visible between them; bark "hammered", peeling in big pieces; up to 25 m tall .... **1. S. ceylanica**
1 Leaves smaller, 3.5–6.5(–10.5) cm long, 1.5–4 cm wide; lateral nerves mostly leaving midrib at about 45–60°, hardly visible with intermediate venation ± invisible; bark finely longitudinally fissured or finely cracked; about 6 m tall .... **2. S. nana**

1. Strombosia ceylanica Gardner, Calcutta J. Nat. Hist. 6: 350. 1846; Masters in Hook. f., Fl. Br. Ind. 1: 579. 1875; Trimen, Handb. Fl. Ceylon 1: 257. 1893 (as *zeylanica*); Gamble, Fl. Pres. Madras 1: 191. 1915 (as *zeylanica*); Lewis, Veg. Prod. Ceylon 100. 1934 (as *zeylanica*); Worthington, Ceylon Trees 132. 1959 (as *zeylanica*); Sleumer in Fl. Males. I, 10: 22. 1984, p.p. (see note). Type: Ceylon, near Kandy, Hantane, virgin forest on mountain, *Gardner 97* (K, syntypes; PDA, syntypes; BM, isosyntypes (ex Herb. Miers)).

Sphaerocarya leprosa Dalz. in Hooker's J. Bot. Kew Gard. Misc. 3: 34. 1851; A.DC. in DC., Prod. 14: 629. 1857. Type: India, Bombay, *Dalzell s.n.* (K, syntypes) (see note).
Stombosia javanica sensu Thw., Enum. Pl. Zeyl. 42. 1858, non Blume.
Levallea ceylanica (Gardner) Baill., Adansonia 2: 361. 1862 (as *zeylanica*).

Tree (8–) 15–30 m tall, d.b.h. 15–38 cm, with compressed slightly twisted trunk and pyramidal crown; branches ridged, glabrous; bark grey or pinkish brown, speckled with small lenticels, scaly, fairly smooth but "hammered" and flaking off in very large plaques leaving pale spots; inner bark ochraceous, laminated. Leaves oblong to ovate-oblong, (5–)10–17.5 cm long, (2.5–)5–7.2 cm wide, rounded or very bluntly to narrowly acuminate at the apex, the actual tip rounded, ± cuneate at the base, glabrous; lateral nerves at an angle of 30° to midrib and a fine parallel tertiary venation evident between them; petiole 0.8–1.7 cm long, thickened towards the apex, channelled above. Flowers in 6–15-flowered subsessile axillary and extra-axillary fascicles; pedicels 1–2 mm long, bearing a few small reddish bracts. Calyx shallowly concave, 0.5 mm tall; lobes 5, rounded, 1 mm tall, persistent. Petals greenish white or yellow, oblong-lanceolate, 2–3 mm long, 0.8 mm wide, hairy inside around anthers. Filaments adhering to the petals for half their length or more. Ovary completely immersed in the yellow conical fleshy entire disk, ± 2.5 mm long including style, 5-locular at the base, 1-locular above; ovules 4–6, pendulous from apex of the central placenta; style short, slender; stigma obsoletely 5-lobed. Fruit black, somewhat fleshy or ± woody, ellipsoid-obovoid, 1.6–2.5 cm long including stipe, 1.5–1.7 cm wide, rugulose or ± smooth, 1-seeded, impressed around style-base; seed subglobose, ± 1 cm long, wrinkled when dry.

Distr. India (Bombay, Madras); Sleumer gives the distribution of his extended *S. ceylanica* as Ceylon and SW. India (Western Ghats from Kanara southward); Sumatra, Malay Peninsula, Anambas Is., W. & Central Java and Borneo. Is also cultivated at Peradeniya.

Ecol. Montane forest, Upper Zone of moist low country; 200–900 m.

Uses. Woody yellow-brown, soft and heavy, used for poles (Kostermans says 'wood moderately hard').

Vern. Pub-beriya (S) (fide Lewis but crossed out in Worthington's copy of the book so possibly incorrect).

Specimens Examined. ANURADHAPURA DISTRICT: Ritigala Strict Natural Reserve, E. slope towards Unakanda, *Jayasuriya 1096* (PDA), Weweltenna. Plain, *Jayasuriya 2145* (PDA). KURUNEGALA DISTRICT: Dunkanda, Arankele, *Jayasuriya & Balasubramaniam 520* (K, US); near Wewahene, Doolukande Hill, *Meijer 375* (PDA). MATALE DISTRICT: (Nalanda, Tumpane Valley, fide *Trimen*); Nalanda, *Worthington 1038, 3447* (BM); 32 mi. post, Nalanda, *Worthington 2415* (BM), *2416* (BM). KANDY DISTRICT: Hantane, *Gardner 97* in *C.P. 1237*, p.p. (BM, K, PDA); Peradeniya (cult.), *Kostermans 27416* (K, L), *27635* (K, L); Kadugannawa-Gampola, *Worthington 553* (BM); Kadugannawa, Udawela, *Worthington 1432* (BM). RATNAPURA DISTRICT: Mi. 10 on Ratnapura to Carney road, *Kostermans 23307* (BO, K, etc.); (Gilimale, fide *Trimen*). NUWARA ELIYA DISTRICT: Watagoda ("Watugoddi, Wattegodde"), *s. coll. C.P. 1237*, p.p. (PDA). MONERAGALA DISTRICT: Galoya National Park, Sellaka Oya Sanctuary, *Jayasuriya 2069* (PDA); MATARA DISTRICT: Morawaka Kanda, *Waas 1503* (PDA).

Note. Sleumer takes a very wide view of *S. ceylanica* in his Flora Malesiana treatment giving its distribution as from India to Borneo and including under it many populations from individual islands and restricted areas formerly considered to belong to distinct species; in many cases it appears to me that the original authors were correct to consider them distinct taxa differing in leaf-size, texture and venation and fruit shape. I am not convinced *S. ceylanica* occurs outside India and Ceylon. Treatments of populations from so many different islands such as is so often necessary in Flora Malesiana accounts is a matter of great difficulty but differenct taxa, at least subspecies, can often be expected to occur on different groups of islands. Worthington records the species from Ceylon only but not I think due to any belief the Indian populations were different (although even here they are not identical).

Index Nominum Genericorum lists *Sphaerocarya* as Dalz. ex A.DC. non Wall. with the type species *S. leprosa* Dalz. ex A.DC. but this is wrong. Dalzell gives a long description of the species and both he and A. De Candolle used the genus *Sphaerocarya* in the sense of Wallich (in the Santalaceae) so I do not see any reason for the "non Wall."

2. Strombosia nana Kostermans, Misc. Pap. Landbouwh. Wageningen 19: 228. 1980. Type: Ceylon, Galle District, Kanneliya Forest, Hiniduma, *Kostermans 24987* (G, holotype; K, L, PDA, US, isotypes).

Tree to 6 m tall, d.b.h. 10–20 cm, glabrous; bark grey or light to dark brown, smooth, very finely longitudinally fissured or finely cracked; underbark white to straw-coloured. Leaves spirally arranged, subcoriaceous, ovate to ovate-elliptic, 3.5–6.5 (–10.5) cm long, 1.5–4 cm wide, obtuse to acuminate at the apex, the actual tip obtuse, cuneate at the base, dull, minutely pustular above; lateral nerves at up to 45–60° to the midrib hardly visible and intermediate venation even more obscure or even invisible, certainly not fine and closely parallel; petiole slender, 0.5–1 cm long. Flowers 1–5 in axillary and extra-axillary inflorescences borne on very short tubercles; peduncles short, up to 5 mm long; pedicels minute, merging with fleshy calyx; bracts and bracteoles minute. Calyx-tube obconic, 0.5 mm long; lobes 5, oblate, 0.5–1 mm long, rounded, ciliate at the margins. Petals lanceolate, 1.5–2 mm long. Disk obscure, flat. Staminal filaments adnate to the petals for most of their length. Ovary almost superior; stigma 0.5 mm long. Fruit obovate-ellipsoid or clavate, 1.5–2 cm long, 1–1.4 cm wide; style-base persistent.

Distr. SW. Ceylon. Endemic.

Ecol. Lowland rain forest with *Shorea* etc.; 150–300 m.

Vern. None recorded.

Uses. None recorded.

Specimens Examined. KALUTARA DISTRICT: Denihena, *Waas 1874* (K, US). RATNAPURA DISTRICT: Sinharaja Forest, Weddagala Entrance, *Gunatilleke 3306 B* (PDA), *3502* (G, L), *Kostermans 27277* (G, K,

L, PDA) (K sheet not found), *27730* (K, L). GALLE DISTRICT: Kanneliya Forest, *Kostermans 24987* (G, K, L, PDA, US).

Note. Sleumer does not mention this species in his synonymy of *S. ceylanica* in Flora Malesiana but had so annotated the Kew sheets of *Kostermans 24987, 27730* and *Waas 1874* as that species in 1979–1980. *S. nana* resembles populations from the Malay Peninsula previously called *S. maingayi* (Masters) Whitmore (*S. rotundifolia* King) more than it does the type of *S. ceylanica*. I have preferred to keep it separate.

PASSIFLORACEAE

(by B.M. Wadhwa* and A. Weerasooriya**)

Juss. ex Kunth in H.B.K., Nov. Gen. Sp. 2: ed. fol. 100, ed. qu. 126. 1817 ('Passifloreae'). Type genus: *Passiflora* L.

Herbaceous or shrubby climbers, often climbing by means of axillary tendrils, sometimes erect shrubs or trees (as in tribe Paropsieae), glabrous or hairy. Leaves alternate, simple or compound, entire or lobed, often with glands on petiole and blade. Stipules small, 2, persistent, Sometimes caducous. Flowers regular, hermaphrodite or unisexual, axillary, solitary, racemose or cymose-paniculate, often showy; bracteoles usually 3, minute and scattered or foliaceous, forming an epicalyx, rarely absent. Hypanthium saucer-shaped to tubiform. Sepals 4–5 (–6), imbricate, free or partially connate (*Adenia* in part), often persistent. Petals (3–) 4–5 (–6), imbricate, free or connate. Corona extrastaminal, inserted on the hypanthium, various, composed of hairs or of 1 or more whorls of thread-like processes or filaments or tubiform or cup-shaped or absent (*Adenia* in part). Disk mostly extrastaminal, annular or of 5 mostly strap-shaped parts (as in *Adenia*) or absent. Stamens 4-many, inserted on the hypanthium or on an androgynophore, free or partially connate; anthers 2-celled, basifixed or versatile, introrsely dehiscent. Ovary superior, on a gynophore or subsessile, 1-locular, 3–5 (–6)-carpellate, with 3–5 (–6) parietal placentas; ovules many; styles 1 or 3, very short to distinct, free or partially connate; stigma capitate to subglobose, sometimes much divided (as in *Adenia*). Fruit a loculicidally 3–5-valved capsule, or berry-like. Seeds numerous, anatropous, mostly compressed, often pitted, with a fleshy aril; albumen fleshy; testa crustaceous; embryo large with foliaceous cotyledons.

A pantropical family of about 500 species, comprising 17 genera; represented by 2 genera and 11 species in Sri Lanka.

KEY TO THE GENERA

1 Flowers hermaphrodite; and androgynophore distinct, much longer than the ovary; corona conspicuous .... **1. Passiflora**

* Royal Botanic Gardens, Kew.
** Flora of Ceylon Project, Peradeniya.

1 Flowers mostly unisexual; androgynophore shorter than the ovary or absent; corona inconspicuous, at most composed of short hairs or absent .... **2. Adenia**

1. PASSIFLORA

L., Sp. Pl. 955. 1753; L., Gen. Pl. ed. 5, 410. 1754; Benth & Hook. f., Gen. Pl. 810. 1867; Masters in Hook. f., Fl. Br. Ind. 3: 599. 1879; Harms in Pflanzenfam. ed. 2, 21: 495. 1925; Alston in Trimen, Handb. Fl. Ceylon 6: 132. 1931; Killip, Field Mus. Nat. Hist., Bot. Ser. 19: 1–613. 1938; Chakravarty, Bull. Bot. Soc. Bengal 3 (1): 47. 1951; de Wilde in Fl. Males. ser. 1, 7: 407. 1972; de Wilde in Polhill, Fl. Trop. E. Africa, Passifloraceae. 11. 1975. Lectotype species: *P. incarnata* L. (vide N.L. Britton & A. Brown, Ill. Fl. N. U.S. ed. 2, 2: 565. 1913).

Perennial climbing herbs to large lianas, rarely shrubs or trees, glabrous or hairy, usually climbing by tendrils. Leaves mostly alternate, simple or compound, lobed or unlobed, palminerved or pinnatinerved, petiolate, margin mostly dentate, often with small gland-teeth; petiole often glandular. Stipules minute to large. Inflorescences sessile or peduncled, 1-many-flowered, with or without a simple tendril; bracts and bracteoles small to large, forming a conspicuous involucre or not. Flowers hermaphrodite, 5-merous, often showy. Hypanthium saucer-shaped to cylindrical. Sepals 5, free, fleshy or membranous, often dorsally corniculate or aristate below the apex. Petals 5, membranous, mostly resembling sepals, alternate with the sepals, sometimes absent. Corona extrastaminal, variously shaped, simple or mostly composed of a usually complicated outer corona consisting of threads/ filaments and flat or plicate inner coronas, sometimes with a nectary ring or annulus. Androgynophore mostly distinct, 3 mm or more. Stamens 5 (–8), free (in Asian species partly connate), reflexed in older flowers; anthers dorsifixed, versatile, linear-ovate or oblong, 2-celled. Gynophore absent or sometimes up to 7 mm; ovary globose, ovoid or fusiform, triplacentiferous; ovules many; styles usually 3, free or connate at base; stigmas capitate, or reniform. Fruit usually indehiscent, ± baccate, globose, ellipsoid or ovoid, rarely fusiform, often with coriaceous exocarp, pulp mucilaginous. Seeds many, ± compressed, reticulate.

About 370 species, c. 350 in the Americas and 20 in S-E continental Asia, Indo-Australia and the West Pacific; represented in Sri Lanka by 9 species. A number of species, introduced as ornamentals, now naturalised. Two species, *P. foetida* L. and *P. suberosa* L., are locally established weeds in many tropical countries. Another species, *Passiflora vitifolia* H.B.K., with red flowers is introduced in the Royal Botanic Gardens, Peradeniya (Herbarium specimen, *Amaratunga 795* (PDA)).

The genus is subdivided by Harms (1925) into 21 sections; Killip (1938) accepts 22 subgenera and many sections and series for the American species.

KEY TO THE SPECIES

1 Bracts and bracteoles minute, inconspicuous, not forming an involucre; stipules ± linear,
  sometimes deciduous; flowers small, up to 1.5 cm across, apetalous, pale green . . . .
  . . . . **1. P. suberosa**
1 Bracts and bracteoles conspicuous, forming an involucre
  2 Involucral bracts finely and deeply divided, 2–4-pinnatisect, segments gland-tipped; flowers
    2.5–5 cm in diam.; petals white; corona filaments ± purplish; fruits subglobose-ellipsoid,
    1.5–2.5 cm in diam., yellow to orange . . . . **2. P. foetida**
  2 Involucral bracts not divided
    3 Involucral bracts free or shortly connate at base; hypanthium less than 3 (–4) cm
      4 Pedicels very long, up to 35 cm; flowers red or pink-red; fruits fusiform, longitudinally
        ribbed, to c. 12 cm long; leaves strigillose on the nerves beneath; lobes lanceolate . . .
        . . . . **3. P. antioquiensis**
      4 Pedicels much shorter
        5 Leaves entire, not lobed, pinnately nerved
          6 Stem 4-angled or winged; petioles with 3 pairs of nearly sessile glands; stipules ovate
            or ovate-lanceolate, more than 1 cm wide; flowers 8–12 cm in diam., purple-red;
            corona filament banded, purple-pink; fruits 12–30 cm long . . . .
            . . . . **4. P. quadrangularis**
          6 Stem terete, sub-angular, not winged, glands on petioles 2–3 pairs, thread-like or long
            clavate, 3–10 mm long; stipules foliaceous; flowers 6–9 cm in diam., white or pale-
            pink; corona filaments banded; fruits 6–8 cm long . . . . **5. P. ligularis**
        5 Leaves mostly lobed or partite, palmately nerved
          7 Stipules foliaceous; involucral bracts subentire
            8 Stipules large, 1.5–4 cm, asymmetrical ovate, entire; flowers 4–5 cm in diam., white;
              sepals horned; fruits to 4 cm in diam., with thick pericarp . . . . . **6. P. subpeltata**
            8 Stipules large, c. 2 cm long, subreniform, mostly dentate; leaves usually 5-(rarely
              3 or 7) lobed nearly to the base; flowers 6–10 cm in diam., greenish-white; corona
              purplish or bluish; fruits ovoid to subglobose . . . . **7. P. caerulea**
          7 Stipules lanceolate or linear-subulate, c. 1 cm long; involucral bracts glandular serrate-
            denticulate, 1.5–2.5 cm; flowers 4–7 cm in diam.; petals white; corona filaments white,
            with purple base; fruits globose to ellipsoid, 4–5 cm in diam., purple, sometimes yellow
            . . . . **8. P. edulis**
    3 Involucral bracts to 4 cm long, partially connate, up to halfway in a tube; flowers ± pink,
      with hypanthium 6–9 cm long; fruits ellipsoid, 7–12 cm . . . . **9. P. mollissima**

1. Passiflora suberosa L., Sp. Pl. 958. 1753; Masters, Trans. Linn. Soc.
London 27: 630. 1871; Trimen, Handb. Fl. Ceylon 2: 241. 1894; Alston in
Trimen, Handb. Fl. Ceylon 6: 133. 1931. Killip, Field Mus. Nat. Hist., Bot.
ser. 19: 88. 1938; Chakravarty, Bull. Bot. Soc. Bengal 3 (1): 54. 1951; G.
Cusset, Adansonia ser. 2, 7 (3): 376. 1967; de Wilde in Fl. Males. ser. 1, 7:
407. 1972. Type: from Hispaniola (Dominica); Specimen grown in Uppsala
Botanic Garden, Linnean Herbarium 1070. 21 (LINN, syn.).

Passiflora walkeri Wight, Ill. Ind. Bot. 2:˙39. t. 108. 1831. Type: from Sri
Lanka, s. loc., *Mrs. Walker 1416* (K).

Climber to 6 m, glabrous to densely pubescent; stem corky when older. Leaves polymorphous (subcircular to ovate or oblong), entire to deeply 3-lobed, 4–10 × 4–14 cm, base rounded or cordate, occasionally peltate, membranous or subcoriaceous, margin entire; lobes linear, triangular to broadly ovate, suberect or widely divergent, acute or obtuse; petioles 0.5–4 cm long, with 2 stipitate glands above the middle. Stipules linear-subulate, 5–8 mm long. Flowers 1–2 cm in diameter, solitary or in pairs in the axils of the leaves or occasionally in leafy, axillary racemes; bracts minute, setaceous, c. 1 mm long, caducous. Hypanthium saucer-shaped, 3–5 mm wide. Sepals ovate to lanceolate, 5–10 mm long, subobtuse, pale greenish-yellow. Petals absent. Corona filaments in 2 series; operculum (inner corona) plicate, minutely fimbriate, white, margin incurved. Androgynophore 2–4 mm; filaments ± subulate, 2–3 mm long. Ovary subglobose or ovoid, 1–2 mm, glabrous; styles 2–3 mm long. Fruit a berry, subglobose or ovoid, 0.8–1.5 cm in diameter, glabrous, purple-blackish, glaucous when young. Seeds many, subovoid, 3–4 × 2 mm, flattened, abruptly acuminate at apex, tapering at base, coarsely reticulate.

Distr. A native of tropical America, introduced and naturalized throughout the tropics.

Ecol. Naturalised, to 2,000 m, very common; fls & frts June–Dec.

Specimens Examined. PUTTALAM DISTRICT: Ambakela Coconut Reserach Centre, 13 Oct 1994, *Weerasooriya 139* (PDA). ANURADHAPURA DISTRICT: Ritigala Hill, base jungle, 6 Apr 1974, *Sumithraarachchi & Jayasuriya 223* (PDA); Ritigala Strict Natural Reserve, 3 Dec 1994, *Wadhwa, Weerasooriya & Samarsinghe 502* (K, PDA). MATALE DISTRICT: Illukkumbura-Rattota Road, 16 Nov 1994, *Weerasooriya, Clayton & Samarasinghe 153* (K, PDA); 2 miles E. of Illukkumbura, 600 m, 4 Oct 1971, *Jayasuriya 287* (K, PDA); KEGALLE DISTRICT: Galpitamada, 22 Aug 1966, *Amaratunga 1143* (PDA). KANDY DISTRICT: Gannoruwa hills, south slope, 6 Oct 1968, *Wirawan 609* (K, PDA); Peradeniya, 25 Jan 1927, *Livera s.n.* (PDA); Royal Botanic Gardens, Peradeniya, 15 June 1910, *s. coll. s.n.* (PDA); Peradeniya Univ. Campus, 500 m, 16 May 1967, *Comanor 324* (K); Kandy, 7 Nov 1994, *Wadhwa, Weerasooriya & Samarasinghe 479* (K, PDA). BADULLA DISTRICT: Road between Bandarawala & Haputale, below Kahagalla Tea Factory, 2 Apr 1985, *Koyama, Sumithraarachchi, Mii, Strudwick & Hanashiro 16035, 16036* (PDA); Badulla-Mahiyangana road, 23 Oct 1994, *Wadhwa, Weerasooriya & Samarasinghe 459* (K, PDA); Along Bandarawala-Badulla Road, near 15/6 road mark, 24 Apr 1974, *Sumithraarachchi & Waas 254* (PDA); Mirahawatte, 20 Nov 1994, *Wadhwa, Weerasooriya & Samarasinghe 483* (K, PDA). RATNAPURA DISTRICT: 12th mile post, between Panamure & Kolonne, 17 Sept 1971, *Balakrishnan & Jayasuriya 911* (K, PDA). NUWARA ELIYA DISTRICT: Botanic Garden, Hakgala, 25 Jan 1993,

Philcox & Weerasooriya 16441 (PDA); Hakgala forest, below garden, 21 Oct 1994, *Wadhwa, Weerasooriya & Samarasinghe 458* (K, PDA); Rendapola-Ambewale road, 19 Oct 1994, *Wadhwa, Weerasooriya & Samarasinghe 424* (K, PDA). GALLE DISTRICT: Bentota Beach, 25 Apr 1972, *Amaratunga 2489* (PDA); Bonavista Hill, 6 Nov 1994, *Wadhwa, Weerasooriya & Samarasinghe 474* (K, PDA). LOCALITY UNKNOWN: *Col. Walker s.n.* (K, PDA); *Mrs. Walker 1416* (K); *s. coll. s.n.* (K); cultivated, *s. coll. 1239* (K).

2. Passiflora foetida L., Sp. Pl. 959. 1753; Masters, Trans. Linn. Soc. London 27: 631. 1871; Trimen, Handb. Fl. Ceylon 2: 242. 1894; Alston in Trimen, Handb. Fl. Ceylon 6: 133. 1931; Gamble, Fl. Pres. Madras 1: 524. 1919; Killip, Field Mus. Nat. Hist., Bot. ser. 19: 474. 1938; Chakravarty, Bull. Bot. Soc. Bengal 3 (1): 57. 1951; G. Cusset, Adansonia ser. 2, 7 (3): 377. 1967; de Wilde in Polhill, Fl. Trop. E. Africa—Passifloraceae 13. 1975; Sriniv. in Nair & Henry, Fl. Tamilnadu 1: 169. 1983; Saldanha, Fl. Karnataka 1: 169. 1984. Type: Lesser Antilles, Linnean Herbarium 1070. 24 (LINN, lectotype).

Herbaceous climber to 3 m, with foetid smell; stem terete, glabrous or with variable degrees of whitish or yellowish-green hairs or indumentum. Leaves suborbicular to ovate, 4–11 × 3–10 cm, base cordate, membranous, usually 3–5-lobed up to halfway, sometimes entire, 3–5-nerved from near the base; lobes up to 4 cm, usually acute-acuminate, margin entire or subentire, with coarse gland-tipped hairs; petioles 1.5–6 cm long, eglandular. Stipules semi-auriculate, 5–10 mm, deeply cleft into filiform, occasionally pinnatisect, gland-tipped segments. Inflorescences sessile, 1 (–2)-flowered; peduncles 2–6 cm long, inserted beside a simple tendril, 5–15 cm long. Bracts and bracteoles (1–) 2–4 cm long, deeply 2–4-pinnatifid or pinnatisect; segments filiform, gland-tipped, forming an involucre just below and enveloping the flowers. Flowers 2.5–5 cm in diameter, pink, lilac or purplish, sometimes white. Hypanthium short, saucer-shaped. Sepals ovate-lanceolate or ovate-oblong, 1.5–2 cm long, 5–7 mm broad, awn dorsally just below the apex. Petals oblong, oblong-spathulate or oblong-lanceolate, slightly shorter than the sepals. Corona filaments in several series, those of the 2 outer series filiform, c. 1 cm long, several inner series of capillary threads 1–2 mm long; operculum membranous, ± erect, denticulate; disk annular, conspicuous. Androgynophore 4–6 mm; filaments flattened, 5–6 mm long. Ovary globose to ellipsoid, 2.5–3 mm, usually glabrous; styles 4–5 mm long. Fruit a dry berry, subglobose to ellipsoid, 1.5–2.5 cm diam., glabrous, yellowish to red, ± enveloped by persistent involucre. Seeds many, ovoid to cuneiform, 5 × 2.5 mm, obscurely tridentate at apex, coarsely reticulate in the middle.

Distr. A tropical American species, is introduced in other tropical countries as well, where it has settled down comfortably.

Ecol. Introduced and naturalised, very common; fls & frts Aug–Mar.

Specimens Examined. PUTTALAM DISTRICT: Lunoya, Iranavillu Scheme, 07 Oct 1972, *Amaratunga 2603* (PDA). ANURADHAPURA DISTRICT: 4 mile N. of Anuradhapura, 14 July 1972, *Hepper & Jayasuriya 4659* (K); Anuradhapura-Rajangane Highway, 28A, c. 220 m, 24 March 1985, *Koyama, Mii, Jayasuriya, Strudwick & Harashiro 15972* (PDA). TRINCOMALEE DISTRICT: China Bay, *Worthington 1203* (K); Kantalai Tank, Aug 1885, *s. coll. s.n.* (PDA). KURUNEGALA DISTRICT: Dodangaslande, 27 May 1967, *Amaratunga 1314* (PDA). MATALE DISTRICT: Sigiriya Wewa, 11 March 1973, *Townsend 73/204* (K). COLOMBO DISTRICT: Delgode, 24 March 1919, *Lewis & J.M. Silva s.n.* (PDA). KEGALLE DISTRICT: Danowita, 02 Aug 1967, *Amaratunga 1395* (PDA). KANDY DISTRICT: Off railroad, behind Botany Building, Peradeniya Univ., 508 m, 2 June 1967, *Comanor 336* (PDA, K); Randenigala, 23 Oct 1994, *Wadhwa, Weerasooriya & Samarasinghe 462* (K, PDA). KALUTARA DISTRICT: East of A–2 Highway, c. 5 miles S. of Panadura, c. 10 m, 17 Dec 1970, *Theobald & Krahulik 2777* (PDA). MONERAGALA DISTRICT: Wellawaya to Ella Road, 3 Apr 1985, *Koyama, Mii, Sumithraarachchi, Strudwick & Hanashiro 16071* (PDA); Pelmadulla-Ambilipitiya Road, near Udawelawe Reservoir, 07 Oct 1994, *Wadhwa, Weerasooriya & Samarasinghe 407* (K, PDA). HAMBANTOTA DISTRICT: Ruhuna National Park, Yala Bungalow, 28 Oct 1968, *Wirawan 688* (PDA). LOCALITY UNKNOWN: in 1849, Fraser 168 (BM).

3. Passiflora antioquiensis H. Karst., Linnaea 30: 162. 1859; Killip, Field Mus. Nat. Hist., Bot. ser. 19: 302. 1938; Chakravarty, Bull. Bot. Soc. Bengal 3 (1): 56. 1951; G. Cusset, Adansonia ser. 2, 7 (3): 377. 1967; Green, Kew Bull. 26 (3): 554. 1972; Sriniv. in Nair & Henry, Fl. Tamilnadu 1: 169. 1983. Type: from Colombia (cultivated at Bogota).

Tacsonia van-volxemii Lem. Ill. Hort. 10: t. 381. 1863; Hook., Bot. Mag. ser. 3, 22: t. 5571. 1866; Masters, Trans. Linn. Soc. London 27: 628. 1871; Matthew, Rec. Bot. Surv. India 20 (1): 123. 1969. Type: from Colombia (Cultivated in Europe).
Passiflora van-volxemii (Lem.) Triana & Planch., Ann. Sci. Nat. Bot. 17: 141. 1873.

Climber; stem terete, younger parts subangulate, hirsute to tomentose. Leaves dimorphic, entire to 3-lobed to within 1 cm of base, ovate or ovate-lanceolate, 6–15 × 3.5–8 cm, rounded or subcordate at base, unevenly and sharply serrate, puberulous on nerves above and pilose or strigillose on nerves beneath and slightly tomentellous elsewhere beneath; lobes lanceolate or elliptic lanceolate, up to 3 cm wide; petioles 2–4 cm long, stout, indistinctly glandular. Stipules subulate, 5–8 mm long. Bracts distinct to base. Flowers red to pink-red; pedicels 20–35 cm long. Calyx tube cylindric, 2.5–4 cm long, glabrous, ventricose at base. Sepals oblong-lanceolate, 5–7 × 1.5–2.5 cm, obtuse, short-awned. Petals similar to the sepals, obtuse, clawed at base. Corona filaments 3-seriate, the outer 2 c. 2 mm apart, the outermost minutely tuberculate, the inner varying from minutely tuberculate to filamentose, often in same flower, the third series, located at c. 1 cm above base of tube, filamentose, the filaments 4–6 mm long; operculum membranous, deflexed, margin recurved, almost entire to minutely denticulate. Ovary linear-narrowly ellipsoid, glabrous or puberulent. Fruit ± fusiform, to c. 12 cm long, longitudinally ribbed.

Distr. A native of tropical America, introduced and naturalised in mountainous regions of Sri Lanka and India.

Ecol. Montane region, to 2,220 m; fls & frts Oct–Dec.

Specimens Examined. NUWARA ELIYA DISTRICT: Nuwara Eliya, surrounding Hills, c. 1850 m, 17 Sept 1969, *C.F. & R.J. van Beusekom 1403* (PDA); St. Andrew's Drive, trail to Pidurutalagala, 1850 m, 6 Nov 1994, *Jayasekera & Scotland 374* (PDA); Road to Nanuoya, 1800 m, 20 Nov 1994, *Wadhwa, Weerasooriya & Samarasinghe 480* (K, PDA); Kandapola Forest Reserve, along loop road, 3 Oct 1973, *Sohmer, Jayasuriya & Eleizer 8361* (PDA).

4. Passiflora quadrangularis L., Syst. Nat. ed. 10, 2: 1248. 1759; DC., Prod. 3: 328. 1828; Masters, Tran. Linn. Soc. London 27: 635. 1871; Killip, Field Mus. Nat. Hist., Bot. ser. 19: 335. 1938; Chakravarty, Bull. Bot. Soc. Bengal 3 (1): 64. 1951; G. Cusset, Adansonia ser. 2, 7 (3): 379. 1967; Green, Kew Bull. 26 (3): 557. 1972; de Wilde in Polhill, Fl. Trop. E. Africa—Passifloraceae 15. 1875; Sriniv. in Nair & Henry, Fl. Tamilnadu 1: 170. 1983. Type: from Jamaica.

Granadilla quadrangularis (L.) Medik., Malvenfam. 97. 1787.

Climber to ± 12 m, glabrous throughout; stem stout, 4-angled, distinctly winged. Leaves entire, broadly ovate, 8–20 × 6–15 cm, base rounded to subcordate, apex abruptly acuminate, margin entire, penninerved, membranous; petioles 2.5 cm long, stout, canaliculate on upper side, with 3 pairs of wart-like, nearly sessile glands. Stipules ovate or ovate-lanceolate, 2–4.5 × 1–2 cm, acute to mucronate, narrowed at base, membranous. Inflorescences 1-flowered; peduncles 1.5–3.5 cm long, 3-angled, inserted beside a simple 10–20 cm long tendril. Bracts and bracteoles distinct, cordate-ovate, 3–5.5 × 1.5–4 cm, acute-acuminate, entire, thin-membranous, forming an involucre. Flowers 10–12 cm diam., whitish or pinkish. Hypanthium shallowly cup-shaped or campanulate. Sepals ovate or ovate-oblong, 3–4 × 1.5–2.5 cm, concave, cucullate, corniculate, green or greenish-red without, white, violet or pink within. Petals ovate-oblong or oblong-lanceolate, 3–4.5 × 1–2 cm, obtuse, white, deeply pink-tinged. Corona 5-seriate, 2 outer ones filamentose, subequal, banded with reddish-purple and mottled with pink-blue, third series tubercular, fourth filamentous and innermost membranous, 3–7 mm long, curved inward, lacerate; operculum membranous, 4–6 mm long, curved inward, denticulate; disk annular, fleshy. Androgynophore 12–14 mm, stout, with 2 annular processes below; filaments ± flattened, 7–8 mm long. Ovary ellipsoid-ovoid, 8–10 mm, glabrous; styles 10–12 mm long. Fruit oblong-ovoid, 20–30 cm × 12–15 cm, terete or longitudinally 3-grooved. Seeds many, broadly obcordate, 7–10 mm long, centrally reticulate, margin striate.

Distr. A tropical American species, cultivated throughout tropical America and other tropical countries, and escaped in some countries.

Ecol. Cultivated and naturalized; fls & frts Sept–Dec.

Uses. Fruits edible.

Vern. Tun-Tun, Desi-Puhul (S).

Note. The fruits are flavoured and edible.

Specimens Examined. KANDY DISTRICT: Royal Botanic Gardens, Peradeniya, 3 Nov 1921, *J.M. Silva s.n.* (PDA). RATNAPURA DISTRICT: Sinharaja Forest, near village, cultivated, 5 Nov 1994, *Wadhwa, Weerasooriya & Samarasinghe 468* (K, PDA); Sinharaja Forest, 3 Sept 1994, *Weerasooriya & Samarasinghe 129* (PDA).

5. Passiflora ligularis Juss., Ann. Mus. Natl. Hist. Nat. 6: 133. t. 40. 1805; Killip, Field Mus. Nat. Hist., Bot. ser. 19: 344. 1938. Type: from Peru.

Climber, glabrous throughout; stem terete. Leaves broadly ovate, 8–15 × 6–13 cm, abruptly acuminate, deeply cordate, entire, penninerved; petioles 4–10 cm long, with 4–6 thread-like or long clavate glands 3–4 mm long. Stipules oblong-lanceolate or ovate-lanceolate, 1–2.5 × 1–1.3 cm, acute or acuminate, narrowed at base, entire or serrulate. Bracts 2–3.5 × 1–1.5 cm, connate one-fifth to one-third their length, free part ± ovate, acute, entire, tomentose at margin within. Flowers 6–9 cm in diameter, solitary or in pairs, white or pale-pink; pedicels 2–4 cm long; Hypanthium short-campanulate. Sepals ovate-oblong, 2.5–3.5 × 1–1.5 cm, acute, white within, green without. Petals oblong, c. 3 × 1 cm, white or pinkish white. Corona 5–7-seriate, the filaments of 2 outer rows as long as the petals, radiate, terete, blue at apex, banded with white, reddish purple below, the inner rows closely approximate, filaments c. 2 mm long, dilated above middle; operculum membranous, slightly incurved, denticulate, white, margin red-purple; disk cupuliform, surrounding base of gynophore. Ovary ovoid. Fruit ovoid, 6–8 × 4–5 cm, the pericarp yellow or purple, parchment-like, with white pulp. Seeds narrowly obcordate, 6 × 4 mm, apex minutely tridentate, middle tooth the largest.

Distr. Native of tropical America, naturalised in many other tropical countries.

Ecol. Montane region, to 2,000 m, fairly common; fls & frts Oct–Dec.

Note. The fruits are edible, and tasty.

Specimens Examined. NUWARA ELIYA DISTRICT: Nuwara Eliya-Hakgala Road, 19 Oct 1994, *Wadhwa, Weerasooriya & Samarasinghe 421* (K, PDA); Rendapola-Ambewela road, 19 Oct 1994, *Wadhwa, Weerasooriya & Samarasinghe 436* (K, PDA); Queen's Cottage Garden, Nuwara Eliya, c. 2,000 m, cult, 1 Apr 1985, *Koyama, Sumithraarachchi, Mii, Strudwick, Hanashiro 16025* (PDA); Nuwara Eliya, Aug 1985, *Jayasuriya 3367* (PDA); Ohiya, 2,400 m, 4 Nov 1971, *Balakrishnan 1051, 1051A* (PDA).

6. Passiflora subpeltata Ortega, Nov. Rar. Pl. Hort. Matrit. 6: 78. 1798; Killip, Field Mus. Nat. Hist., Bot. ser. 19: 436. 1938; Chakravarty, Bull. Bot. Soc. Bengal 3 (1): 59. 1951; G. Cusset, Adansonia ser. 2, 7 (3): 379. 1967; de Wilde in Polhill, Fl. Trop. E. Africa—Passifloraceae 16. 1975. Type: from Mexico.

Passiflora stipulata sensu Griseb., Bonplandia 6: 7. 1858; Masters, Trans. Linn. Soc. London 27: 638. 1871; Trimen, Handb. Fl. Ceylon 2: 242. 1894; Alston in Trimen, Handb. Fl. Ceylon 6: 133. 1931, non Aubl., 1775.

Climber or herbaceous creeper, glabrous; stem terete, striate. Leaves 3-lobed to about half-way, suborbicular, 4–9 × 5–12 cm; base rounded or cordate, often subpeltate, glabrous or slightly puberulous above; lobes elliptic to oblong, up to 5 cm, apex obtuse or acute, ± 1 mm mucronate, margin entire except a few glandular teeth in or near the lobe-sinuses; petioles 3–6 cm long, slender, with 2–5 scattered, minute c. 1 mm long glands above the middle. Stipules ovate-oblong or asymmetrical ovate, 1.5–4 × 0.5–2 cm, mucronulate, entire, except glandular-crenulate at base. Inflorescences 1-flowered; peduncles 3–6 cm long, inserted beside a simple, 4–12 cm long tendril. Bracts and bracteoles ovate-oblong, 1–1.5 × 1.0 cm, ± acute, entire or with a few glandular teeth at base, forming an involucre. Flowers 4–5 cm in diameter. Hypanthium broadly cup-shaped, 7–10 mm wide. Sepals oblong, 2.0–2.5 cm long, obtuse, green without, white within, carinate, the keel terminating in a green, foliaceous c. 1 cm long horn. Petals linear-oblong, 1.5–2 cm long, acute. Corona filaments white, 5-seriate, two outer up to 1 (–1.5) cm long, those of inner series 2–6 mm long; operculum subplicate, c. 2.5 mm high, with a fringe of inward curved dentiform processes; disk annular, limen with lobulate edge, closely surrounding androgynophore, margin reflexed, crenulate. Androgynophore 10–12 mm; filaments 5–6 mm long, dilated. Ovary ovoid or ellipsoid, 3–4 mm across, glabrous; styles 8–10 mm long. Fruit ellipsoid or subglobose, 3–4 cm in diam., ± leathery, greenish turning yellow. Seeds many, obovate, c. 5 × 3 mm, flattened, the beak slightly curved, finely reticulate.

Distr. A native of tropical America, cultivated as an ornamental, freely escaped and naturalized in submontane regions of tropical countries of Africa and Asia.

Ecol. Sub-montane region, 600–1,500 m, very common; fls & frts Aug-Mar.

Specimens Examined. MATALE DISTRICT: Near Midlands, Road marker 37/10, past Laggala, 26 June 1973, *Nowicke & Jayasuriya 206* (K, PDA); Elkaduwa, 800 m, 14 Feb 1982, *Veldkamp 7850* (K). KANDY DISTRICT: Madugoda, 8 Nov 1931, *Simpson 8821* (BM); Kadugannawa, Aug 1884, *W. Ferguson s.n.* (PDA); Madugoda-Mahiyangana Road, 600 m, 6 Aug 1978, *Kostermans 26754* (PDA). BADULLA DISTRICT: Welimada, 1,800 m, 12 March 1971, *Balakrishnan 443* (K), Pagoda to Jangula, 26 June 1931, *Simpson 8255* (BM). NUWARA ELIYA DISTRICT: Nuwara Eliya, Mar 1884, *s. coll. s.n.* (PDA); Road to Nuwara Eliya, 21 March 1906, *A.M. Silva s.n.* [Acc. No. 000223 (HAK)]; Hakgala, 12 March 1971, *Balakrishnan 428* (K); Namunukula, 1,400 m, 22 Sept 1976, *Cramer 4724* (K); Nuwara Eliya, Hakgala Road, 19 Oct 1994, *Wadhwa, Weerasooriya & Samarasinghe 423* (K, PDA); Hakgala forest, 21 Oct 1994, *Wadhwa, Weerasooriya & Samarasinghe 455* (K, PDA); Diagama, 3 March 1971, *T. & M. Koyama & Balakrishnan 14094* (PDA); Horton Plains, Farr Inn-Ohiya road, 29 Jan 1974, *Sumithraarachchi, H.N. & A.L. Moldenke & Jayasuriya 64* (K); Ambewala-Rendapola Road, 19 Oct 1994, *Wadhwa, Weerasooriya & Samarasinghe 435* (K, PDA). RATNAPURA DISTRICT: Pinnawala, Balangoda road, 1,000 m, 19 March 1968, *Comanor 1089* (K).

7. Passiflora caerulea L., Sp. Pl. 959. 1753 (non Lour., 1790); L., Amoen. Acad. 1: 231. t. 10. f. 20. 1787; Masters, Trans. Linn. Soc. London 27. 638. 1871; Killip, Field Mus. Nat. Hist., Bot. ser. 19: 423. 1938; Chakravarty, Bull. Bot. Soc. Bengal 3 (1): 56. 1951; G. Cusset, Adansonia ser. 2, 7 (3): 378. 1967; Matthew, Rec. Bot. Surv. India 20 (1): 124. 1969; Green, Kew Bull. 26 (3): 554. 1972; Sriniv. in Nair & Henry, Fl. Tamilnadu 1: 169. 1983. Type: from Brazil (LINN. Herb.).

Granadilla caerulea (L.) Medik., Malvenfam. 91. 1787.

Climbers, glabrous, often glaucous; stems subangular, striate, grooved. Leaves palmately 5 (occasionally 3, 7 or 9)-lobed, usually nearly to base, cordate, membranous; lobes nearly ovate-oblong or linear-oblong, c. 10 × 0.5–2.5 cm, obtuse or emarginate, rarely acute or mucronulate, entire, 2–4-glandular in sinuses; petioles 1.5–5 cm long, ± pilosulous, with 2–4 stipitate glands. Stipules semiovate or subreniform, 1–2 × 0.5–1 cm, aristate or acuminate, mostly denticulate, sometimes subentire. Inflorescences 1-flowered; peduncles 3–7 cm long, usually stout, sometimes slender. Bracts and bracteoles broadly ovate to ovate-oblong, 1.5–2.5 × 1–1.5 cm, rounded at apex, thin membranous. Flowers 6–10 cm in diameter. Hypanthium cupuliform. Sepals oblong or lanceolate-oblong, 1.5–2 × 1–1.5 cm, obtuse, subcoriaceous, green without, white or pinkish within, keeled; keel terminating in a slender awn. Petals oblong, 1.5–2.5 × 1–1.5 cm, obtuse, membranous, white or pinkish. Corona filaments 4-seriate, 2 outer series c. 6 mm long, sometimes as long as the petals, very slender, inner 2 series, 1–2 mm long, capitellate, erect, white, purplish at apex; operculum membranous below, filamentose above, the filaments 3–4 mm long, dark purple, nectar-ring fleshy; disk cup-shaped, closely surrounding base of androgynophore, crenulate. Ovary ovoid or subglobose, pruinose. Fruit ovoid to subglobose, 6 × 4 cm, orange or yellow. Seeds obcordate, 5 × 3.5–4 mm, coarsely reticulate.

Distr. East tropical South America; central America; Mascarene Islands; South Africa, India and Sri Lanka; cultivated in China.

Ecol. Introduced and naturalised in montane region, 1,800–2,000 m, not common; fls & frts Oct–Dec.

Vern. Blue passion fruit (E).

Specimens Examined. NUWARA ELIYA DISTRICT: Hakgala forest, below Botanic Garden, 19 Oct 1994, *Wadhwa, Weerasooriya & Samarasinghe 432* (K, PDA); Nuwara Eliya, near lake, 20 Nov 1994, *Wadhwa, Weerasooriya & Samarasinghe 481* (K, PDA).

8. Passiflora edulis Sims, Bot. Mag. 45: t. 1989. 1818; Masters, Trans. Linn. Soc. London 27: 637. 1871; Trimen, Handb. Fl. Ceylon 2: 242. 1894; Alston in Trimen, Handb. Fl. Ceylon 6: 133. 1931; Gamble, Fl. Pres. Madras 1:524. 1919; Killip, Field Mus. Nat. Hist., Bot. ser. 19: 393. 1938; Chakravarty, Bull. Bot. Soc. Bengal 3 (1): 61. 1951; G. Cusset, Adansonia ser. 2, 7 (3): 378. 1967; Green, Kew Bull. 26 (3): 554. 1972; de Wilde in Polhill, Fl. Trop. E. Africa—Passifloraceae 15. 1975; Sriniv. in Nair & Henry, Fl. Tamilnadu 1: 169. 1983. Type: Plant cultivated in Europe (probably originally from Brazil).

Climbers, to 12 m, glabrous, rarely pilosulus. Leaves 3-lobed to three-fourths, suborbicular to broadly ovate in outline, 5–11 × 4–10 cm, base rounded or shallowly cordate, subcoriaceous, serrate, 3-nerved from base; lobes elliptic to oblong, up to 8 cm long, acute or acuminate; petioles 1–4 cm long, biglandular at apex, the glands sessile or short-stipitate. Stipules lanceolate or linear-lanceolate, 10 × 1 mm, entire or minutely glandular-serrate. Inflorescence 1-flowered; peduncle 2–6 cm long, inserted beside a simple, 5–20 cm long tendril. Bracts and bracteoles ovate, 1.5–2.5 × 1–1.5 cm, obtuse or acute, glandular serrate-denticulate, forming an involucre. Flowers 4–7 cm in diameter, white. Hypanthium cup-shaped, ± 1 × 1–1.5 cm. Sepals oblong, 2.5–3 × 1 cm, corniculate. Petals oblong, ± 3 cm long, 5–7 mm wide, obtuse. Corona filaments 4–5-seriate, the outer 2 series filiform, 1.5–2.5 cm long, crispate at apex, white, purple at base, the inner ones shorter, 2–2.5 mm long; operculum membranous, incurved, crenulate or short-fimbriate; disk cupuliform, entire or crenulate. Androgynophore 6–8 mm, thickened at base; filaments ± subulate, 6–8 mm long. Ovary subglobose to ellipsoid, 3–5 mm, shortly pubescent or glabrous; styles 10–12 mm long. Fruits berry-like, globose to ellipsoid, with coriaceous pericarp, 4–5 cm in diameter, glabrous, yellow or purplish. Seeds many, ellipsoid or oval, 5–6 × 3–4 mm, minutely reticulate.

Distr. Probably native of Brazil, extensively cultivated elsewhere and naturalised in many tropical countries.

Ecol. Introduced for the flavoured fruits, now naturalised along forest edges, thickets and disturbed places, to 2,200 m, very common; fls & frts Sept–Apr.

Specimens Examined. TRINCOMALEE DISTRICT: Hood's Tower, Trincomalee, cultivated, 16 June 1943, *Worthington 1279* (K). MATALE DISTRICT: Dikpatana, Laggala, 700 m, 21 Oct 1974, *Jayasuriya & Bandaranayake 1835* (K, PDA). KANDY DISTRICT: Near Hewaheta, 1 Nov 1931, *Simpson 8767* (BM, PDA); Madulkele, 1,310 m, 23 July 1946, *Worthington 1974* (K). Kandy, 7 Nov 1994, *Wadhwa, Weerasooriya & Samarasinghe 478* (K, PDA); Dolosbage, Kelvin Estate, 850 m, 12 June 1946, *Worthington 1909* (K). BADULLA DISTRICT: Bandarawela-Haputale, below Kahagalla Tea Factory, 2 Apr 1985, *Koyama, Mii, Sumithraarachchi, Strudwick & Hanashiro 16037* (PDA). RATNAPURA DISTRICT: Sinharaja Forest, 5 Nov 1994, *Wadhwa, Weerasooriya & Samarasinghe 470* (K, PDA). NUWARA ELIYA DISTRICT: Botanic Gardens, Hakgala, June 1893, *s. coll. s.n.* (PDA); Hakgala, in 1898, *s. coll. s.n.* (PDA); Queen's Cottage Garden, Nuwara Eliya, c. 2,000 m, 1 Apr 1985, *Koyama, Mii, Sumithraarachchi, Strudwick, & Hanashiro 16025* (PDA); Forest edge at Fairyland Tea estate, 29 Oct 1975, *Sohmer & Sumithraarachchi 10220* (PDA); Ambewala, roadside, 26 March 1906, *A.M. Silva s.n.* (Acc. No. 1000234 HAK); Hakgala, 21 Oct 1994, *Wadhwa, Weerasooriya & Samarasinghe 456* (K, PDA); Ambewale-Kandapola Road, 19 Oct 1994, *Wadhwa, Weerasooriya & Samarasinghe 384* (PDA).

9. Passiflora mollissima (H.B.K.) Bailey, Rhodora 18: 156. 1916; Killip, Field Mus. Nat. Hist., Bot. ser. 19: 291. 1938; Chakravarty, Bull. Bot. Soc. Bengal 3 (1): 55. 1951; G. Cusset, Adansonia ser. 2, 7 (3): 378. 1967; de Wilde in Polhill, Fl. Trop. E. Africa—Passifloraceae 14. 1975; Sriniv. in Nair

& Henry, Fl. Tamilnadu 1:170. 1983. Green, Kew Bull. 11 (4): 183–186. 1994.

Tacsonia mollissima H.B.K., Nov. Gen. Sp. 2: 144. 1817; Hook., Bot. Mag. ser. 3, 1: t. 4187. 1845; Masters, Trans. Linn. Soc. London 27: 629. 1871; Fyson, Fl. S. India Hill Stns 2: t. 188. 1932. Type: from Colombia: Monserrate, near Bogota.

Passiflora tomentosa Lam. var. *mollissima* (H.B.K.) Triana & Planch., Ann. Sci. Nat. Bot. 17: 131. 1873.

Climber to 20 m; perennial, densely and softly villous; stem terete, striate. Leaves 3-lobed, 5–10 × 6–12 cm, divided to about two-thirds, subcordate, membranous to thinly coriaceous, subglabrous to softly pubescent above, tomentose beneath, sharply glandular serrate-dentate; lobes elliptic to lanceolate, 3–6 cm long, middle lobe 2–4 cm broad, acute or acuminate; petioles 1.5–5 cm long, with 4–6 paired, sessile or somewhat stipitate, glands. Stipules subreniform, 5–9 × 3–4 mm, aristate, finely glandular-serrate or subentire. Flowers axillary, solitary; peduncles 2–6 cm long, inserted beside a simple, 5–18 cm long tendril. Bracts and bracteoles 2.5–4 cm, acute-acuminate, connate for about half length, softly tomentose, margin entire, forming a tubiform involucre. Hypanthium tubiform, 6–9 cm long, c. 1 cm diameter, glabrous. Sepals oblong, 2–4.5 × 1–1.5 cm, subobtuse, aristulate below apex. Petals oblong, almost equal to the sepals, obtuse, pink. Corona reduced to a purple ± ring with a few tubercles or crenulations; operculum an inward curved membrane at the base of hypanthium. Androgynophore 8–12 cm; stamens 5; filaments 1.5–2 cm long, dilated. Ovary ellipsoid or oblong, 10–12 mm long, pubescent; styles 3, 10–12 mm long. Fruit ellipsoid or oblong-ovoid, 7–12 cm long, softly pubescent. Seeds many, ellipsoid or broadly obovate, 6 × 5 mm, reticulate, dark-brown to black.

Distr. A native of Western tropical America, introduced in many tropical countries and naturalized in some of them.

Ecol. Along margins and clearings in montane forests, 1,500–2,000 m; fairly common; fls & frts Oct–Dec.

Vern. Banana passion fruit (E).

Note. It is cultivated for ornamental purposes and for its fruits.

Specimens Examined. NUWARA ELIYA DISTRICT: Kandapola Forest Reserve, along loop road, 22 may 1967, *Amaratunga 1296* (PDA), 19 Oct 1994, *Wadhwa, Weerasooriya & Samarasinghe 430* (K, PDA); Kandapola Forest Reserve, along forest road, 19 June 1972, *Maxwell & Jayasuriya 875* (PDA); Horton Plains, trail to Kirigalpotta peak, 29 June 1973, *Nowicke & Jayasuriya 243* (K); Queen's Cottage Garden, Nuwara Eliya, c. 2,000 m, 1 Apr 1985, *Koyama, Mii Sumithraarachchi, Strudwick & Hanashiro*

16024 (PDA); Horton Plains, along Ohiya road, 2,100 m, 29 March 1968, *Koyama 13533* (K); Kandy-Nuwara Eliya road, mile post 47, 19 Oct 1994, *Wadhwa, Weerasooriya & Samarasinghe 368* (PDA); Mahacoodagala, 21 Oct 1994, *Wadhwa, Weerasooriya & Samarasinghe 438* (K, PDA); Horton Plains, 8 Sept 1969, *Kostermans 23023A* (K), 20 Oct 1994, *Wadhwa, Weerasooriya & Samarasinghe 449* (K, PDA); Nuwara Eliya, Prime Minister's Lodge, 20 Nov 1994, *Wadhwa, Weerasooriya & Samarasinghe 482* (K, PDA).

2. ADENIA

Forssk., Fl. Aegypt.-Arab. 77. 1775; King, J. Asiat. Soc. Bengal 71 (2): 1, 51. 1903; Harms in Pflanzenfam. 3, 6a: 83. 1894; ed. 2, 21: 488. 1925; Chakravarty, Bull. Bot. Soc. Bengal 3 (1): 64. 1951; G. Cusset, Fl. Camb. Laos & Vietnam 5: 132. 1967; de Wilde, Adansonia ser. 2, 10: 111. 1970; de Wilde, Meded. Landbouwhogeschool Wageningen 71 (18): 1–281. 1971; de Wilde in Fl. Males. ser. 1, 7: 417. 1972. Type species: *A. venenata* Forssk.

Modecca [Rheede, Hort. Mal. 8: tt. 20–23. 1688] Lam., Enc. 4: 208. 1797; Blume, Bijdr. 15: 938. 1826; DC., Prod. 3: 336. 1828; Roxb., Fl. Ind. 3: 132. 1832; Thw., Enum. Pl. Zeyl. 128. 1859; Benth. & Hook. f., Gen. Pl. 1: 813. 1867; Masters in Hook. f., Fl. Br. Ind. 2: 601. 1879; Trimen, Handb. Fl. Ceylon 2: 240. 1894. Type species: Not designated.

Herbaceous to more or less woody perennial climbers with tendrils, sometimes erect herbs or shrublets without tendrils. Leaves alternate, simple, entire or lobed, or palmately divided; glands (0–) 1–2 at the base of the leaf blade, at or near the apex of the petiole, and with or without glands elsewhere on the lower surface or margin of the blade. Stipules minute, reniform or narrowly triangular. Tendrils axillary. Inflorescences axillary, cymose, the middle flower(s) often replaced by tendrils. Bracts and bracteoles minute, ± subulate. Flowers unisexual, rarely hermaphrodite, campanulate or urceolate to tubular, mostly greenish to yellowish, glabrous; stipe articulate at base. Hypanthium saucer-shaped, cupular or tubular. Sepals (4–) 5 (–6), free or partially connate into a calyx tube, imbricate, persistent. Petals (4–) 5 (–6), free, included in the calyx, sometimes adnate to the calyx tube, mostly fimbriate or laciniate. Corona annular or consisting of 5 cap-shaped parts or of a row of filamentous hairs or absent. Disk-glands 5, strap-shaped or capitate, inserted at or near base of hypanthium opposite the sepals, or absent. Male flowers: stamens (4–) 5 (–6), hypogynous or perigynous, free or partially connate into a tube; anthers basifixed, linear, often apiculate, 2-celled, opening introrsely to laterally; ovary minute. Female flowers: usually smaller than the male flowers, with smaller petals. Staminodes ± subulate. Ovary superior, subsessile, globose to oblong, 1-celled, with 3 (–5) parietal placentas; ovules usually numerous, anatropous; styles 3 (–5), free or partially united;

stigmas mostly subglobose, laciniate, densely woolly-papillate. Fruit a stip-itate 3 (–5)-valved capsule; pericarp coriaceous to rather fleshy, greenish to yellow or bright red. Seeds ± compressed, with crustaceous pitted testa, in a membranous to pulpy aril; embryo large; cotyledons foliaceous.

de Wilde has divided the genus into 6 sections: 5 sections in tropical and S. Africa, of which 2 extend to Asia, and one section confined to Australia. The genus has about 93 species and is represented by 2 species in Sri Lanka.

KEY TO THE SPECIES

1 Flowers very small; sepals free or nearly so; sepals and petals inserted at or about the same level; anthers short; filaments combined in a tube; gland at base of leaf blade 1 . . . . 1. A. wightiana
1 Flowers rather large; sepals partially connate in a calyx tube, extending wholly or partially above the insertion of petals; calyx lobes and petals not inserted at the same level; anthers long; filament forming a cup below; glands at base of leaf blade 2 . . . . 2. A. hondala

1. **Adenia wightiana** (Wall. ex Wight & Arn.) Engler, Bot. Jahrb. Syst. 14: 376. 1892; Gamble, Fl. Pres. Madras 535. 1919; Harms in Pflanzen-fam ed. 2, 21: 482. 1925; Alston in Trimen, Handb. Fl. Ceylon 6: 132. 1931; Chakravarty, Bull. Bot. Soc. Bengal 3 (1): 65. 1951; de Wilde, Meded. Land-bouwhogeschool Wageningen 71 (18): 79. t. 8. 1971; de Wilde in Polhill, Fl. Trop. E. Africa—Passifloraceae 27. 1975.

Modecca wightiana Wall. ex Wight & Arn., Prod. 353. 1834; Wight, Ic. Pl. Ind. Or. t. 179. 1839; Thw., Enum. Pl. Zeyl. 128. 1859; Masters in Hook. f., Fl. Br. Ind. 2: 601. 1879; Trimen, Handb. Fl. Ceylon 2: 240. 1894. Type: India, Madras, *Wallich 6764* (K-WALL, holotype).

subsp. **wightiana**

Climbers to ± 7 m, growing from a tuberous rootstock. Leaves herba-ceous, brownish-green above, greyish-green beneath, broadly ovate to subtri-angular, or oblong-lanceolate, entire to deeply 3–5-lobed, 2–12 × 1.5–11 cm, base acute to slightly cordate or hastate, apex acute to obtuse, margin entire, 3–5-nerved from the base; lobes broadly rounded to lanceolate, obtuse; peti-oles 0.5–5.5 cm long. Gland at base of leaf blade 1, margin of the blade with-out gland. Stipules triangular, c. 1 mm long, entire. Peduncle 1–15 cm. Inflo-rescence distinctly cincinnal, up to 30-flowered in male plants, 2–6-flowered in female plants; peduncle often ± spiral. Sterile tendrils simple, up to 12 cm. Bracts & bracteoles narrowly triangular to lanceolate, acute. Plants generally dioecious. Male flowers campanulate, 3–6 mm long (including stipe). Calyx tube absent. Sepals ovate, 1–2 mm long, obtuse to subacute. Petals ovate to elliptic-oblong, 1–1.5 × 0.5 mm, subentire, 1–3-nerved. Hypanthium cup-shaped, 1 × 2–2.5 mm. Stamens monadelphous; filaments 1.5–2.5 mm long, connate into a narrow tube, inserted at the base of hypanthium; anther c.

1 mm, basifixed, obtuse. Corona consisting of minute threads; disk glands very minute. Vestigial ovary c. 1 mm. Female flowers broadly campanulate, fleshy, 2.5–6 mm long including stipe. Calyx tube 0. Sepals subovate, obtuse, 1–2 mm long, entire. Petals obovate-oblong. 1–1.5 mm long, obtuse. Hypanthium cup- to saucer-shaped, ± fleshy, 1–2 × 1.5–2 mm. Staminodes to 2 mm, free, inserted at the base of the hypanthium. Corona consisting of minute filaments; disk minute. Gynophore c. 0.5 mm. Ovary ellipsoid to subglobose, 1.5–2.5 mm long; styles ± free, c. 0.5 mm; stigmas subreniform, papillate. Fruits 1–2 per inflorescence, subglobose to ovoid, 1.5–3 × 1.2–2 cm, excluding 1–3 mm long gynophore; pericarp chartaceous to coriaceous, bright red. Seeds 10–25 per capsule, subovate, 5–6 mm long, reticulately pitted.

Distr. Sri Lanka and South India.

Ecol. In scrub forests of dry region at low elevations, rather rare; fls & frts Aug–Feb.

Note. de Wilde in Blumea 17: 179. 1969, has described another subsp., *Adenia wightiana* (Wall. ex Wight & Arn.) Engler spp. *africana* de Wilde, from E. Africa, which differs in having margin of leaf-blade mostly shallowly sinuate, remotely dentate, marginal glands present; stipules broadly reniform, coarsely dissected; inflorescences mostly with 3 tendrils and sterile tendril mostly 3-fid.

Specimens Examined. JAFFNA DISTRICT: Pallavarayankaddu-Mankulum Road, 13 Nov 1970, *Kundu & Balakrishnan 643* (PDA); Chunnavil, Feb 1890, *s. coll. s.n.* (PDA). PUTTALAM DISTRICT: Puttalam, Nov 1881, *Ferguson s.n.* (PDA), Aug 1848, *Thwaites C.P. 1621* (BM, CAL, PDA). Karativoe Island, Kalpitiya, Aug 1883, *s. coll. s.n.* (PDA); LOCALITY UNKNOWN: *Thwaites C.P. 1623* (K); *Mrs Walker s.n.* (K); *Walker s.n.* (K).

2. Adenia hondala (Gaertn.) de Wilde, Blumea 15: 265. 1967; de Wilde, Meded. Landbouwhogeschool Wageningen 71 (18): 137. 1971; Sriniv. in Nair & Henry, Fl. Tamilnadu 1: 169. 1983; Saldanha, Fl. Karnataka 1: 278. 1984.

Granadilla hondala Gaertn., Fruct. 2: 480 t. 180. f. 10. 1791. *Type: Hermann s.n.* (L, apparently lost).

Modecca palmata Lam., Enc. 4: 209. 1797; Wight & Arn., Prod. 1: 353. 1834; Wight, Ic. Pl. Ind. Or. t. 201. 1839; Thw., Enum. Pl. Zeyl. 128. 1859; Masters in Hook. f., Fl. Br. Ind. 2: 603. 1879; Trimen, Handb. Fl. Ceylon 2: 241. 1894. Type: Rheede, Hort. Mal. 8: 45 & tt. 20, 21, 22. 1688.

Adenia palmata (Lam.) Engler, Bot. Jahrb. Syst. 14: 375. 1891; Gamble, Fl. Pres. Madras 525. 1919; Chakravarty, Bull. Bot. Soc. Bengal 3 (1): 69. 1951.

Climber, up to 20 m; stem woody in the older parts; bark smooth or scaly, whitish or greyish-green. Leaves ovate-oblong to suborbicular in outline, (6–) 8–20 × (3.5–) 9–20 cm, entire to deeply 3–5-palmately lobed, 3–5- nerved from the base, with 2 glands just at the base of the blade and another 2 glands above them towards base of the middle lobe, base cordate, lobes oblong-lanceolate, apex acute-acuminate, middle one constricted at the base, up to 16 cm long and 4–7 cm broad, lateral lobes bulging at the base, margin entire; petioles 3–9 cm long, sulcate. Stipules narrowly triangular to subulate, 3–4 mm, persistent. Peduncle 2–6 cm long. Inflorescences cymose, up to 25-flowered in male plants, 1–7-flowered in female plants; sterile tendril to 18 cm long. Bracts and bracteoles narrow-triangular, acute, c. 1.5–2 mm long. Male flower: campanulate to broadly tubiform, including stipe 15–28 × 8–12 mm, pedicellate; pedicels 5–15 mm. Hypanthium 5–saccate, 2–4 mm. Calyx tube 4–11 mm long; lobes triangular or ovate, subacute, 4–7 mm long, entire. Petals linear-oblong, 7–12 × 1–1.5 mm, 3–5-nerved, fimbriate, arising from the base of the calyx tube. Stamens 5; filaments subulate, 3–4 mm long, connate below, inserted at the base of the hypanthium; anthers linear-oblong, 4.5–7 mm long. Corona filiform; filaments 1–1.5 mm long, arranged in a ring on the calyx tube. Vestigial ovary including gynophore c. 1.5 mm. Female flower: campanulate, including stipe 11–16 × 1–1.5 mm, pedicellate; pedicels 3–9 mm long. Hypanthium 5–saccate, 2–3 mm. Calyx tube 5–8 mm; lobes triangular, 4–7 mm, subacute, entire. Petals linear, acute, 5–8 × c. 1 mm, 1–3-nerved, sparsely fimbriate, inserted at the same level as the corona. Staminodes 2–4 mm, connate for c. 0.5 mm, inserted at the base of the hypanthium. Corona hairs 1–1.5 mm long, sometimes membranous at the base. Ovary ellipsoid to subglobose, 3 × 2.5 mm; style arms connate for about 1 mm, 2.5–3 mm long; stigmas subglobular, woolly-papillate. Fruit globose to ellipsoid, 3.5–5 × 2.5–4 cm, yellow, orange or scarlet; pericarp coriaceous, sometimes spotted. Seeds suborbicular to reniform, 7–8 mm long, with a funicular beak at the base, strongly pitted.

Distr. Sri Lanka and South India.

Ecol. In dry and wet regions, along forest edges, up to 500 m, common; fls & frts Sept–Mar.

Vern. Hondala, Potahonda (S).

Specimens Examined. ANURADHAPURA DISTRICT: Medawa-chchiya, 80°29′ E, 8°32′ N, 22 Jan 1972, *Jayasuriya, Balakrishnan, Dassanayake & Balasubramaniam 572* (PDA); Mihintale—Anuradhapura, 27 March 1963, *Amaratunga 556* (PDA), MATALE DISTRICT: 6 miles

E. of Naula along Elahera Road, 5 Oct 1971, *Jayasuriya 294* (K).
COLOMBO DISTRICT: Wage Forest, 150 m, 28 Aug 1971, *Dassanayake 496* (K); Mirigama, Kandalama, 10 Oct 1974, *Kostermans 25581* (K).
KANDY DISTRICT: Hantane, *Gardner 491* (K, BM); Wariagala, 26 June 1927, *Alston s.n.* (PDA); Near Madugoda, 8 Nov 1931, *Simpson 8793* (BM); Rattota-Illukkumbura Road, *Clayton & Weerasooriya 6183* (K, PDA).
RATNAPURA DISTRICT: Sinharaja Forest, near Wathurawa Nature Trail, 4 Nov 1994, *Wadhwa, Weerasooriya & Samarasinghe 466* (K, PDA); Sinharaja Forest, 5 Nov 1994, *Wadhwa, Weerasooriya & Samarasinghe 469* (PDA); Kurulugala, Bulutota Pass, 1,000 m, 9 Oct 1977, *Nooteboom 3318* (PDA); Kuttapitiya, Pelmadulla, 450 m, 28 Sept 1953, *Worthington 6435* (K); Dehenella, Gallela-Wewelwatte, 4 March 1987, *Jayasuriya & Adachi 17144* (PDA); Sinharaja Forest, near Martin's Bungalow, 4 Nov 1994, *Wadhwa, Weerasooriya & Samarsinghe 467* (K, PDA), Along Halmandiya Trail, 445 m, 24 Sept 1987, *Jayasuriya & Balasubramaniam 3929* (PDA), Halmandiya dola, 21 Aug 1988, *Jayasuriya & Balasubramaniam 4425* (PDA); Sinharaja Forest, 21 Dec 1994, *Weerasooriya, Fernando & Ekanayake 164* (K, PDA), near Forest office, 5 Nov 1994, *Wadhwa, Weerasooriya & Samarasinghe 472* (K, PDA). HAMBANTOTA DISTRICT: Ruhuna National Park, Block II, E. of Menik Ganga, 100 m, 1 Oct 1967, *Mueller-Dombois, Comanor & Cooray 67100103* (PDA), Block I, at Andunoruwa wewa, 27 Feb 1968, *Mueller-Dombois 68022702* (PDA). LOCALITY UNKNOWN: *Thwaites C.P. 1627* (BM, K, PDA); *Col. Walker 176* (K); *Mrs. Walker s.n.* (K).

PHYTOLACCACEAE
(by D. Philcox*)

R.Br. in Tuckey, Narr. Exped. Congo 454. 1818 ("Phytolaceae"); H.Walter in Pflanzenr. 4, 83: 1–154. 1909. Type genus: *Phytolacca* L.

Herbs or shrubs (or trees, not in Ceylon). Leaves alternate, simple, entire; stipulate or not. Flowers hermaphrodite or unisexual, actinomorphic or zygomorphic, in terminal, axillary or lateral racemes. Sepals 4–5, free, imbricate in bud, equal or unequal, coloured during and often after anthesis, persistent. Stamens 3-many in 1–2 series, inserted usually on disk, arranged regularly or irregularly, with those of only or outer series alternating with sepals; filaments slender, free; anthers dorsi- or basifixed, dehiscing longitudinally. Ovary superior, of 1 or more carpels; carpels laterally connate or free. Styles equalling carpels, short, free or absent. Ovules solitary in each carpel. Fruit of 1 or more carpels, juicy or dry. Seed erect.

Family of about 20 genera, mostly tropical and mainly from America; represented in Ceylon by 3 genera.

KEY TO THE GENERA

1 Carpels 2-many; sepals 5; styles 5–10; fruit a juicy, 5–10-seeded, red or black berry ......
.... **1. Phytolacca**
1 Carpels 1; sepals 4; style 1 or lacking; fruit 1-seeded
  2 Flowers actinomorphic; stamens 4; ovary glabrous; style distinct; adult inflorescence up to 10 cm long, erect; fruit a juicy berry .... **2. Rivina**
  2 Flowers zygomorphic
    3 Inflorescence 4–10 cm long in flower, up to 20 cm in fruit, flowers spreading, not appressed to axis; sepals 4, 3 united to midway, 1 free, becoming brightly coloured; stamens 4; fruit not spinose .... **3. Hilleria**
    3 Inflorescence 10–40 cm long, flowers spaced 0.7–1.5 cm apart, appressed to axis; sepals 4, free, green; stamens 6–8; fruit dry, spinose .... **4. Petiveria**

1. PHYTOLACCA

L., Sp. Pl. 441. 1753; L., Gen Pl. ed. 5: 521. 1754. Lectotype Species: *Phytolacca americana* L. (vide Britton & Brown, Ill. Fl. N. U.S. ed. 2, 2: 26. 1913).

* Royal Botanic Gardens, Kew.

Erect or scandent herbs which at times may appear shrubby or, more rarely, trees; glabrous except for pubescent or papillose inflorescence axis; stems at times angular. Stipules absent. Leaves membranous. Flowers many, short- to very short-pedicellate, in dense, long, erect, sometimes spiciform racemes; sepals 5, coloured, spreading during anthesis, afterwards subreflexed, often dark-red in fruit; stamens c. 6–30, in 1 or 2 series, inner or only series inserted on margin of disk, outer series on undersurface; filaments filiform-subulate; anthers dorsifixed, bilobed at both ends. Carpels 5–10(–16), whorled, laterally connate into a depressed-globose ovary; styles short, subulate, erect or recurved. Fruit baccate, depressed-globose, longitudinally 5–10-furrowed. Seeds laterally-compressed, ovoid, shiny black.

Genus of about 25 species from tropical, subtropical and warm temperate areas of the world.

KEY TO THE SPECIES

1 Herb up to 1 m tall, sometimes woody; stamens 7–9 .... **1. P. octandra**
1 Tree 7–10 m tall; stamens 20–30 .... **2. P. dioica**

1. Phytolacca octandra L., Sp. Pl. ed. 2: 631. 1762; Thw., Enum. Pl. Zeyl. 250. 1861; Trimen, Handb. Fl. Ceylon 3: 410. 1895; Alston in Trimen, Handb. Fl. Ceylon 6: 240. 1931. Type: from Mexico.

Shrubby herb up to 1 m tall; stem somewhat angular. Leaves: lamina 5–13 × 2–4.5 cm, elliptic-lanceolate to narrowly ovate-lanceolate, acute to acuminate, glabrous; petiole 1–3 cm long. Inflorescence terminal or pseudolateral, 6–10 cm, or possibly more, long when mature, rhachis subdensely minutely papillose to almost smooth. Flowers white to cream; pedicels 1–2 mm long at anthesis, becoming 2–3 mm long in fruit; sepals 2.5–3 mm long during anthesis, oblong, acute to obtuse, pinkish-green, enlarging to 3–4 mm long when in fruit; all flowers staminate, stamens 7–9, more usually 8, inserted on outer margin of disk. Fruit 5–7 mm diameter, black, with 8 (7–9) longitudinal furrows, glabrous. Seeds 2.5 × 2 mm, ovoid.

D i s t r. Native of tropical America from Mexico to Columbia, naturalized elsewhere in the tropics.

E c o l. Roadsides and a weed of cultivation, originally escaping from garden cultivation.

S p e c i m e n s E x a m i n e d. KANDY DISTRICT: Kandy to Hewaheta road, road marker 30/5, 19 Feb 1974, *Sumithraarachchi 117* (PDA). RATNAPURA DISTRICT: road across Adam's Peak Wilderness from Bogawanthalawa to Pinnawela, 16 Nov 1973, *Sohmer & Waas 8753* (PDA); Pinnawela, 30 June 1971, *Balakrishnan 563* (PDA). NUWARA ELIYA DISTRICT: Nuwara Eliya, close to lake, c. 200 m, 25 Jan 1977, *Cramer 4806*

(K, PDA); between km 68 and 69, Kandy to Nuwara Eliya road, 1900 m, 1 Mar 1994, *Philcox et al. 10689* (K, PDA); Hakgala, 1700 m, 14 July 1972, *Cramer 3815* (K); base of ascent to Pidurutalagala, 1 Mar 1978, *Cramer 5189* (K); Moon Plains, 1525 m, 12 July 1978, *Meijer 1897* (K); Kandapola Road, Moon Plains, c. 1830 m, 6 Mar 1970, *Balakrishnan 345* (PDA); Hawa Eliya, 15 Sept 1920, *Silva s.n.* (PDA); Rahatungoda, January 1859, *Thwaites s.n.* (PDA); Kandapola, 9 Sept 1942, *s. coll. s.n.* (PDA); near Ragala, 18 Sept 1931, *Simpson 8697* (BM).

2. Phytolacca dioica L., Sp. Pl. 632. 1753. Type: from South America.

Pircunia dioica (L.) Moq. in DC., Prod. 13, 2: 30. 1849.

Tree, erect, 7–10 m tall, dioecious; branches glabrous, spreading. Leaves long-petiolate: lamina 4.5–8(–14) × 2.25–4.5(–8.5) cm, broadly elliptic to oblong-ovate, rounded-obtuse at base with sides at times unequal, apex shortly acuminate-mucronate, mucro c. 3.5 mm long, recurved, glabrous, 5–8 (–10) pairs of lateral veins, prominent beneath with midrib markedly so; petiole 1.5–4 (–5) cm long. Inflorescence terminal or axillary, shortly pedunculate, 5–12 cm long, rhachis shortly pedunculate, flexuous, pubescent, sublaxiflorous. Flowers sublaxly arranged, spreading, becoming subreflexed at maturity; bracts 1–1.5 mm long, narrowly acute to subulate, glabrous; pedicels 6–7 mm long, puberulous, thickened towards apex; calyx lobes 3 × 2 mm, orbicular-ovate, apex obtuse, shortly laciniate; stamens 20–30, unequal; filaments about 5 mm long, filiform; anthers 1 mm long, oblong. Carpels 7–10, joined, forming fruit 7–10 mm diameter, ripening black.

Distr. Native of tropical South America. Naturalized in parts of southern India and elsewhere in the Old World tropics.

Specimens Examined. BADULLA DISTRICT: Pitaratmalie Estate, above Haputale, 1700 m, 8 May 1969, *Kostermans 23406* (K).

2. RIVINA

L., Sp. Pl. 121. 1753; L., Gen. Pl. ed. 5: 57. 1754, as "Rivinia". Type Species: *Rivina humilis* L.

Erect herbs, often woody at base. Leaves exstipulate, long petiolate, ovate-oblong or ovate-lanceolate, long-acuminate, membranous. Flowers actinomorphic, hermaphrodite, in terminal or pseudoterminal racemes, pedicellate with pedicels slender, minutely bracteolate at or above middle. Sepals 4, subequal, patent or subreflexed during anthesis, white or pinkish, somewhat accrescent, becoming green. Stamens 4, alternating with and shorter than sepals; anthers dorsifixed, bifid at both ends. Ovary 1-carpellate, globose, glabrous to ellipsoid; style solitary, short, with peltate stigma, recurved after anthesis. Fruit globose, 1-seeded berry with thin pericarp, juicy. Seed glabrous or shortly pubescent.

Genus of 1 (or 3) tropical American species (vide infra).

Rivina humilis L., Sp. Pl. 121. 1753; Trimen, Handb. Fl. Ceylon 3: 410. 1895; Alston in Trimen, Handb. Fl. Ceylon 6: 240. 1931. Type from the West Indies.

Herb, erect, slender, 30–60 cm, or much more, tall, much branched, often woody at or towards base; stem patent-pubescent to subglabrous above. Leaves long petiolate, somewhat distant: lamina 4–12 × 1.5–5 cm, ovate-lanceolate or ovate-oblong, acute to long-acuminate, base very obtuse, rounded to subcordate, glabrous to subglabrous above, similar beneath except shortly pubescent on midrib; petiole (1–) 1.5–4 cm long, very slender, patent-pubescent. Racemes many-flowered, 3.5–5(–8) cm long, up to 11 cm long in fruit, erect or erect-spreading, sparsely patent-short-pubescent to subglabrous. Pedicels very slender, spreading, 2–3 mm long during anthesis, 3–4 mm long afterwards; sepals 2–2.5 mm long during anthesis, white or pinkish, 3–3.5 mm long in fruit. Berry 3.5–4 mm diameter, bright red. Seeds spherical, c. 2–2.5 mm diameter, short-pubescent.

Distr. Native of tropical America; naturalized widely throughout the tropics elsewhere, including Ceylon.

Ecol. Roadsides and waste-places, in full sun or shade.

Note. The three entities comprising this genus in the view of some authors, are so close that the present author considers the limits, if any exist, are better left to the decision of a monographer.

Specimens Examined. MATALE DISTRICT: Dambulla, 100–200 m 3 May 1978, *Soejarto 4899* (K); Dambulla Hill, 16 May 1931, *Simpson 8123* (BM). POLONNARUWA DISTRICT: Manampitiya, 66 m, 25 Mar 1971, *Dassanayake 352* (K). KANDY DISTRICT: Kandy, S of lake, 500 m, 30 July 1970, *Dassanayake 164* (K, PDA); Peradeniya, 23 Feb 1964, *Amaratunga 773* (PDA). Alawatugoda, 27 May 1967, *Amaratunga 1306* (PDA); Milapitiya to Haragama, 16 June 1965, *Amaratunga 883* (PDA); Alawatugoda, 27 May 1967, *Amaratunga 1306* (PDA). BADULLA DISTRICT: Nikapatha, 26 Apr 1974, *Waas 479* (PDA). RATNAPURA DISTRICT: Allerton Tea Estate, Rakwana, 457 m, *Hepper et al., 4563* (K, PDA). MONERAGALA DISTRICT: c. 1 km E of Wellawaya, 27 Jun 1976, *Meijer 177* (K, PDA); along Highway A–4 near bridge 154/6 on way to Arugam Bay, c. 175 m, 20 Jan 1971, *Theobald & Krahulik 2867* (PDA). HAMBANTOTA DISTRICT: Menik Ganga, c. 1 km above Yala Bungalow, 5 Jan 1969, *Fosberg et al. 51061* (PDA). WITHOUT LOCALITY: Fraser 53 (BM); *Walker s.n.* (K).

3. HILLERIA

Vell. Conc., Fl. Flum. 47. 1825, Atlas 1, tab. 122. 1835.
Mohlana Mart., Nov. Gen. & Sp. 3: 170, tab. 290. 1832.

Herbs, often appearing shrubby. Flowers terminal- or in axillary race-mose inflorescences, zygomorphic, hermaphrodite. Sepals 4, with 3 united to about midway, 1 free, becoming markedly 3-veined, and at times brightly coloured. Stamens 4; anthers dorsifixed. Ovary 1-carpellate; style very short to equalling ovary. Fruit subdiscoid, 1-seeded. Seed similar shape, black.

Genus of 4 species from South America, with *H. latifolia* widespread and occurring apparently native in Africa, Madagascar and the Mascarene Islands.

Hilleria latifolia (Lam.) H. Walter in Pflanzenr. IV, 83: 81, fig. 25. 1909.

Rivina latifolia Lam., Enc. 6: 215. 1804. Type from Mauritius.
Mohlana nemoralis Mart., Nov. Gen. & Sp. 3: 171, tab. 290. 1832; Trimen,
    J. Bot. 23: 173. 1885. Type from Brazil.

Shrubby herb to 1.5 m or more tall; branches bristly pubescent when young, otherwise glabrous. Leaves long-petiolate: lamina 7–15 (–20) × 3–7 (–9) cm, long- or short-acuminate, broadly cuneate into slightly unequal-sided, somewhat decurrent base, pubescent only on lateral veins, especially beneath; 4–8 pairs of lateral veins; petiole 1.25–3.5 cm long, lightly pubescent. Racemes 4–10 cm long in flower, lengthening to c. 20 cm in fruit, many-flowered, pubescent on ridges of rhachis. Flowers pedicellate, pedicels c. 1.5–2 mm long, increasing to c. 4 mm long in fruit. Sepals 1.5–2 mm long, becoming 3–4 mm long in fruit, accrescent, becoming yellow to red in fruit. Stamens 4, slightly shorter than sepals. Style very short to lacking; stigma capitate. Fruit 2–3 mm diameter, drying dark-red or purple.

Distr. Native of tropical South America and purportedly native in West Africa, Angola and South Africa (Natal); in East Africa, Madagascar and the Mascarene Islands. Elsewhere in the Old World tropics, including Ceylon, this is considered to be an introduction or an adventive

Ecol. Rain forests and wet riverine areas.

Note. The specimen cited below and in the Kew herbarium, bears the name *Mohlana nemoralis* and a note by Trimen as follows: "I am at a loss to understand how this reached this wild district of Ceylon. We have never had it in the Gardens". Without doubt it is correctly identified, but is not represented in the Peradeniya herbarium.

Specimens Examined. BADULLA DISTRICT: Uma ("Ooma") Oya, in 1880, *Trimen s.n.* (K).

4. PETIVERIA

L., Sp. Pl. 342. 1753; L., Gen. Pl. ed. 5: 417. 1754. Type Species: *Petiveria alliacea* L.

Erect undershrubs, stipulate; stipules subaxillary, minute. Leaves reportedly smelling of onions when bruised. Flowers small, hermaphrodite, zygomorphic. Inflorescence terminal and axillary racemose, erect or nodding at top. Pedicels bibracteolate. Sepals 4, linear-oblong, spreading during anthesis, yellowish, becoming erect, accrescent, green. Stamens 4–8 at base of, and shorter than, sepals; anthers linear, dorsifixed, deeply bifid apex, shallowly 2-lobed base. Ovary oblong, 1-carpellate, pubescent with 4–6 bristles at top, 1-celled. Style absent; stigma lateral, consisting of a number of long hairs. Fruit narrowly cuneiform, dry, compressed, indehiscent, with 4 reflexed, long, appressed, spines, exserted from enclosing sepals. Seeds linear.

Genus of 1 or 2 tropical and warm American species.

Petiveria alliacea L., Sp. Pl. 342. 1753; Alston in Trimen, Handb. Fl. Ceylon 6: 240. 1931. Type: from Jamaica.

Erect, up to 1.5 m tall; stems thin, hairy above. Leaves: lamina 6–13 × 2–5 cm, oblong-elliptic to subobovate, apex acute, acuminate, obtuse or rounded, base acute, margin slightly undulate, pubescent above on main veins, glabrous beneath; petiole 0.5–1.5 cm long. Racemes spiciform, up to 25 cm long, simple or branched near base, often nodding at apex, flowers lax. Bracts c. 1.75 mm long, ovate, acuminate. Pedicels 0.5–1 mm long; sepals 3–4 mm long during anthesis, up to 6 mm long afterwards. Stamens 6–8, erect, shorter than sepals. Fruit 6–8 mm long, appressed to inflorescence axis, spines c. 3 mm long, yellowish.

Distr. Native of the warmer parts of America, occurring as a spasmodic weed elsewhere in the tropics.

Ecol. Waste ground; not yet permanently established anywhere in Ceylon.

Specimens Examined. KEGALLE DISTRICT: Bulatkohupitiya, 10 Dec 1968, *Amaratunga 1683* (PDA). KANDY DISTRICT: junction of Lady Blake Road and Kandy to Peradeniya road, c. 500 m, 26 Apr 1978, *Soejarto 4834* (K); Peradeniya, 21 July 1963, *Amaratunga 652* (PDA), W of Botany building, University, 507 m, 16 May 1967, *Comanor 323* (K, PDA), Galaha Road, University campus, 20 Oct 1970, *Balakrishnan 396* (PDA).

PLANTAGINACEAE
(by M.D. Dassanayake*)

Juss., Gen. Pl. 89. 1789 ('Plantagines'). Type genus: *Plantago* L.

Annual or perennial herbs with short stems. Leaves usually spirally arranged, simple, parallel-veined, often with sheathing bases, exstipulate. Flowers small, regular, bisexual, more or less strongly protogynous, in heads or spikes in the axils of broad bracts, wind-pollinated. Calyx gamosepalous, with 4, rarely 3, frequently unequal, mostly imbricate, membranous segments persistent in the fruit. Corolla gamopetalous, with 4, rarely 3, dry, membranous, imbricate lobes spreading or reflexed at anthesis. Stamens 4, rarely fewer, often subequal, inserted in corolla tube at different levels, alternating with corolla lobes; filaments long, filiform, induplicate in the bud, strongly exserted at anthesis. Anthers versatile, introrse, longitudinally dehiscent, often more or less heart-shaped, with the connective produced beyond the anthers to form a process. Pollen powdery. Ovary superior, syncarpous, with 2 carpels, 2-locular (rarely 1-locular), each loculus with 1- many anatropous to hemianatropous ovules with a single integument, usually on axile placenta. Style simple, filiform; stigma 2-lobed. Fruit a circumscissile capsule or an achene or a nut. Seeds usually peltate. Embryo mostly straight, sometimes curved; endosperm fleshy, translucent, copious.

3 genera, with about 255 species, almost cosmopolitan in distribution.

PLANTAGO

L., Sp. Pl. 112. 1753; L., Gen. Pl. ed. 5. 52. 1754. Lectotype species: *P. major* L.

Herbs or rarely subshrubs. Leaves spirally arranged, mostly radical. Inflorescence a head or spike. Calyx segments equal or subequal, imbricate. Corolla tube about the same length as the calyx tube or longer, segments equal. Stamens 4. Ovary mostly 2-locular, with 1-many ovules in each loculus. Capsule exserted from the persistent calyx, dehiscing transversely across the middle or close to the base, the upper part separating as a lid. Seeds 2-many, angular, subglobose or compressed, one face flattened or concave

---

* Flora of Ceylon Project, Peradeniya.

and attached to axis. Testa thin, of some species mucilaginous when wet. Embryo straight or curved.

About 250 species, cosmopolitan, mostly in the Mediterranean region, the Himalayas, southwestern N. America and the mountains of S. America.

KEY TO THE SPECIES

1 Leaves elliptic, capsule with 8-many seeds .... ₁. P. erosa
1 Leaves lanceolate, capsule with 2 seeds .... 2. P. lanceolata

1. Plantago erosa Wall. in Roxb., Fl. Ind. ed. Carey 1: 423. 1820; Decne. in DC., Prod. 13(1): 696. 1852; Pilger in Pflanzenr. iv. 269. Heft 102: 60. 1956. Type: from India.

Plantago major Auct.: Thw., Enum. Pl. Zeyl. 245. 1861; Hook. f., fl. Br. Ind. 4: 705. 1885, p.p., non L. 1753.
Plantago major var. *asiatica* sensu Trimen, Handb. Fl. Ceylon 3: 389. 1895, non Decne. 1852.
Plantago asiatica sensu Alston in Trimen, Handb. Fl. Ceylon 6: 237. 1931, non L. 1753.

Perennial herb. Stem erect, to about 10 cm long. Leaves 5–15, 3–11 × 1.5–7 cm, elliptic, broadly elliptic or ovate, with obtuse apex and more or less broadly cuneate base, with mostly 3 main veins diverging at base and meeting at tip, and 2 marginal veins; glabrous or sparsely hairy, when young densely villous; margin entire or more often dentate to varying extents, rarely lobed; veins prominent beneath; petiole channelled, 2–8 cm long, villous, sheathing at base, densely hairy when young. Inflorescence a spike; peduncle axillary, erect or ascending, more or less arcuate, furrowed, nearly glabrous, 6–30 cm long; spike 4–20 cm long, with flowers more or less distant towards the base. Bracts 1–2 mm long, elliptic, green, pale and membranous at margins, glabrous or ciliolate with short hairs, keeled. Calyx lobes 2–2.5 mm long, ovate, subequal, green, pale and membranous at margins, glabrous, with keel extending to tip. Corolla 3–3.5 mm long, lobes lanceolate, narrowly ovate or ovate, acute, reflexed between the sepals; corolla tube oblong-ovoid. Filaments c. 5 mm long; anthers c. 1 mm long, white. Ovary with many ovules in each loculus. Capsule ovoid, attenuate at apex and truncate, glabrous, purplish, c. 5 mm long, dehiscing by a circular line c. 2 mm from base. Seeds mostly many, ovoid-oblong or angular, black-brown, 1–1.5 mm long.

Distr. Also in Nilgiri Hills in Peninsular India, Himalayas, China.

Ecol. Montane and upper montane zone. Open places, roadsides, paths, cultivated ground; occasionally at lower elevations, as at Pussellawa. Flowering almost throughout the year.

Note. A very variable plant, in size, leaf-shape, etc.

Specimens Examined. NUWARA ELIYA DISTRICT: Horton Plains, 19 Sept 1969, *Van Beusekom 1461* (PDA); Horton Plains, 17 May 1968, *Cooray & Wirawan 68051710R* (PDA); Perawela, 26 Oct 1970, *Cramer 3240* (PDA); Nuwara Eliya, 19 May 1971, *Jayasuriya et al. 174* (PDA), *Gardner* in *C.P. 2246* (PDA); Ambewela Road, 19 March 1906, *Willis s.n.* (PDA); Pidurutalagala, April 1893, *Nock s.n.* (PDA); Dimbula, in 1890, *Smith s.n.* (PDA); Hakgala, Jan 1888, *s. coll. s.n.* (PDA), *Dassanayake 105* (PDA); Hakgala jungle, Oct 1906, *A.M. Silva s.n.* (PDA).

2. Plantago lanceolata L., Sp. Pl. 113. 1753; Decne. in DC., Prod. 13(1): 714. 1852; Trimen, Handb. Fl. Ceylon 3: 389. 1895; Alston in Trimen, Handb. Fl. Ceylon 6: 237. 1931. Type: From Europe.

Perennial herb. Stem erect, stout, to c. 10 cm long, covered with leaf bases and long silky hairs. Leaves 10–20, 5–12 × 1.5–3 cm, narrowly lanceolate or lanceolate, rarely elliptic, attenuate towards apex and base, acute, with 3–5 veins, mostly pilose, particularly along the veins; margin entire, sometimes remotely denticulate; petiole 2–6 cm long, sheathing at base. Inflorescence a spike; peduncle axillary, erect or ascending, more or less deeply 5-furrowed, pilose, 6–50 cm long; spike 1–4.5 cm long, dense, cylindric or conical to more or less globose. Bracts 3–5 mm long, broadly ovate, caudate, thin and membranous, pale brownish, glabrous. Calyx 3–3.5 mm long, lobes ovate, obtuse or acute at apex, glabrous or villous at upper edge, pale, purplish coloured at apex, membranous, with dark green keel; two anterior calyx lobes connate, forming a structure bilobed at the apex, with two keels ending just short of the apex. Corolla tube 2–3 mm long, c. 0.6 mm broad, lobes c. 2 mm long, ovate, acuminate, pale brown, membranous, apical half dark purplish brown along mid-vein; filaments 5–6 mm long; anthers pale yellow, 2–3 mm long. Style c. 8 mm long; ovules 2. Capsule c. 5 mm long, ellipsoid; seeds 2, hemispherical, with one side concave, finely ribbed, brown-black.

Distr. The 'ribwort plantain', native to countries in Europe and north and central Asia; introduced and established throughout most of the world.

Ecol. Upper montane zone. In grassy places, edges of forests, roadsides, weed in cultivated fields. Common around Nuwara Eliya. Flowering more or less throughout the year.

Specimens Examined. NUWARA ELIYA DISTRICT: Nuwara Eliya, *s. coll. C.P. 2247* (PDA), May 1971, *Jayasuriya 175* (PDA); Ambewela, *Muller-Dombois & Cooray 68011212* (PDA); Moon Plains, *Dassanayake 366* (PDA); Pidurutalagala, May 1971, *Jayasuriya 187* (PDA).

PODOSTEMACEAE
(by D. Philcox[*])

Rich. ex Agardh, Aphor. Bot. 125. 1822 ("Podostemeae"). Type genus: *Podostemum* Michaux.

Perennial, aquatic herbs, usually submerged in running water; stems thallus-like, closely attached to rocks, with flattened or subcylindric fronds. Leaves small, usually fasciculate, deciduous or fragile and breaking away below apex. Flowers bisexual, small, solitary, consisting of short spathe, with perianth of 3 lobes or segments, or 0. Stamens 1–3 (usually 2); filaments connate to above midway. Ovary of 2–3 cells; ovules numerous; styles 2 or 3. Capsule dehiscing into 2 or 3 valves.

Family of some 50 genera and about 275 tropical species, mostly American and Asian.

In 1977, Cusset & Cusset (Bull. Mus. Hist. Nat. (Paris) ser. 4, 10: 149–177) divided the hitherto single family Podostemaceae into two, the other being Tristichaceae. Of our material the Tristichaceae would only include *Dalzellia*, but for reasons of convenience in this Flora, it has been decided to retain Podostemaceae in its original state to encompass all four of our genera. However, in a later study based on her concept of Podostemaceae, Colette Cusset (l. c. 14: 13–54. 1992) published a revision of the Asian genera and species, and in this treatment her taxonomy and nomenclature have been followed, and where deemed prudent, a number of the Ceylon (Sri Lankan) localities have been clarified.

KEY TO THE GENERA

1 Capsule indehiscent containing 2 seeds; ovary bilocular with one loculus aborted and sterile, the other with 2 ovules .... **1. Farmeria**
1 Capsule dehiscent; ovary bi- or trilocular, with all loculi fertile and with more than 2 ovules in each loculus
  2 Ovary trilocular .... **4. Dalzellia**
  2 Ovary bilocular
    3 Ovary bilocular of 2 unequal loculi; capsule with 8 stout ribs or 6 stout and 2 winged; loculi somewhat unequal with one valve persistent after anthesis and the other caducous; numerous seeds in each loculus .... **2. Zeylanidium**

* Royal Botanic Gardens, Kew.

3 Ovary bilocular of 2 equal loculi; capsule with 8 ribs; both valves persistent after anthesis; more than 20 seeds in each loculus .... **3. Polypleurum**

1. FARMERIA

Willis ex Hook. f. in Trimen, Handb. Fl. Ceylon 5: 386. 1900; Willis, Ann. Roy. Bot. Gard. Peradeniya 1 (3): 246. 1902. Type species: *Farmeria metzgerioides* (Trimen) Willis.

Genus distinguished by the bilocular ovary with one loculus aborted with a stout placenta bearing two ovules. The fruit is sessile, globose, smooth, indehiscent, subtended by 2 bracteoles.

Monotypic genus, endemic to Ceylon.

Farmeria metzgerioides (Trimen) Willis ex Hook. f. in Trimen, Handb. Fl. Ceylon 5: 386. 1900; Willis, Ann. Roy. Bot. Gard. Peradeniya 1 (3): 247. 1902, 1 (4): 397. 1902.

Podostemon metzgerioides Trimen, Handb. Fl. Ceylon 3: 419. 1895. Type: Ceylon: Mahaweli river at Hakkinda [near Gannoruwa], N of Peradeniya, February, *Trimen s.n.* (K, holotype).

Thallus narrow, flat, ribbonlike, creeping over rocks to which it is solidly attached, marginally bearing very short, 1-flowered shoots, with 3–4 obtuse, linear to linear-spathulate leaves up to 10 mm long, and terminated by one curved flower towards the apex. Spathe dehiscent by splitting at apex on upper surface. Flowers sessile, not exserted, with only stamen and stigmas emerging from spathe. Tepals 2, as long as ovary, very slender. Stamen 1, emerging from spathe, with very long filament and triangular anthers with cells divergent from base. Ovary bilocular, very asymmetric, with one loculus aborted, other containing 2 ovules carried on swollen placenta. Stigmas linear or subulate. Capsule obliquely globose, sessile, indehiscent, with thin membranous walls, containing 2 large seeds which germinate within capsule.

Distr. Endemic.

Ecol. Running rivers and cataracts.

Specimens Examined. KANDY DISTRICT: Hakkinda, on rocks in Mahaweli River, below Peradeniya, Feb 1894, *Farmer s.n.* (K, PDA), 12 Mar 1899, *Willis s.n.* (K, PDA), 3 Feb 1993, *Philcox et al. 10440* (PDA); Haragama, 10 Jan 1929, *Alston 2378* (PDA).

2. ZEYLANIDIUM

(Tulasne) Engler, Pflanzenfam. ed. 2, 18a: 61. 1928. Type species: *Zeylanidium olivaceum* (Gardner) Engler.

Thallus ribbonlike or cylindric, bearing acaulous shoots or well developed stems. Leaves distichous, occasionally reduced to scale or sheath after fall of leaf-blade. Spathes terminal or subterminal, solitary or grouped together. Flowers arranged within spathe. Tepals 2, surrounding androecium formed of 1–2 stamens, androphore longer than filaments. Ovary bilocular with 2 unequal loculi, sessile; stigmas short, linear; placentas entirely covered by anatropous ovules. Capsule with 8 substantially large ribs, dehiscent into 2 unequal valves, smaller caducous. Seeds numerous, flattened-ellipsoid, covered with longitudinal, sinuate ribs, or smooth.

Genus of 4 species from India, Burma and Ceylon.

KEY TO THE SPECIES

1 Sterile, acaulous shoots borne on the thallus; flowers solitary; stamens 2
  2 Leafy thallus bearing shoots over whole surface; fertile shoots prostrate, sterile shoots erect, 2–3 cm long . . . . 1. Z. olivaceum
  2 Thallus narrow, branched in regular fashion with shoots arising in the angles of branching . . . . 2. Z. lichenoides
1 Plant with developed stem, erect, longer than 2–3 cm; stamens 2; capsule 8-ribbed, not winged . . . . 3. Z. subulatum

1. Zeylanidium olivaceum (Gardner) Engler in Pflanzenfam., ed. 2, 18a: 61. 1928.

Podostemon olivaceum Gardner, Calcutta J. Nat. Hist. 7: 181. 1847; Wedd. in DC., Prod. 17: 75. 1873; Trimen, Handb. Fl. Ceylon 3: 41. 1895; Hook. f., Fl. Br. Ind. 5: 66. 1896; Trimen, Handb. Fl. Ceylon 5: 386. 1900. Type: Ceylon: *Gardner s.n.* (K, holotype).

Hydrobryum olivaceum (Gardner) Tulasne var. *zeylanicum* Willis, Ann. Roy. Bot. Gard. Peradeniya 1 (3): 240. 1902. Syntypes: Ceylon: *Thwaites C.P. 3065* (PDA); Mahaweli Ganga, Ramboda, *Gardner s.n.* (K); Talawakele, *Willis s.n.* (K), *Macmillan s.n.* (K); Maskeliya, *Harvey s.n.* in *C.P. 2989* (K, PDA), *Trimen s.n.* (K).

Leafy thallus bearing very small fertile branches on surface, arranged in concentric fashion, and sterile shoots forming simple, long stem of 2–3 cm long, terminating in very dense tuft of filaments, 2–5 cm long. Fertile shoots composed of 4–6 basally arranged linear leaves, disappearing gradually, then followed by 6, densely imbricate, distichous, boat-shaped, bract-like leaves. Spathe ovoid, dehiscing by ventral slit. Pedicel lengthening to extrude flower from spathe after anthesis. Flower erect. Tepals 2, narrowly triangular, 1.5 mm long, inserted at base of androecium formed of 2 stamens; androphore longer than filaments; anthers 1.2×0.8 mm. Stigmas conical, entire or sometimes slightly divided. Capsule 8-ribbed, dehiscent into 2 unequal valves, smaller caducous.

Distr. Ceylon and southern India.

Ecol. Running rivers and cataracts.

Specimens Examined. KANDY DISTRICT: on rocks in Mahaweli River, Hakkinda, 3 Feb 1993, *Philcox et al. 10438* (PDA); Palagalla, Jan 1834, *s. coll.* in *C.P. 3065*, p.p. (PDA); Peradeniya, *s. coll.* in *C.P. 3065* (K, PDA); Maskeliya, 2 Mar 1883, *s. coll. s.n.* (PDA); Adam's Peak, 1200 m, Sept, *van Beusekom & van Beusekom 1586* (PDA); Ambagamuwa, Feb 1846, *Gardner s.n.* (K, P). NUWARA ELIYA DISTRICT: Ramboda, *Harvey s.n.* (E), Jan 1847, *Gardner s.n.* (K); Pundalu Oya, Nov 1888, *Green s.n.* (K, PDA). LOCALITY UNKNOWN: *Gardner s.n.* (K).

2. Zeylanidium lichenoides (Kurz) Engler, Pflanzenfam., ed. 2, 18a: 61. 1928.

Hydrobryum lichenoides Kurz, J. Asiat. Soc. Bengal 42: 103. 1873; Willis, Ann. Roy. Bot. Gard. Peradeniya 1 (3): 242. 1902. Type: from Burma.

Thallus narrow, several mm wide, regularly branched with small shoots lying in angles of branching. Leaves 4–5, thick, distichous, boat-shaped, narrowly imbricate. Spathe ovoid, dehiscing by longitudinal, ventral slit. Pedicel lengthening to extrude flower from spathe after anthesis. Flower erect. Tepals 2, linear. Androecium formed of 2 stamens. Ovary sessile, obtuse, with 2 conical stigmas at apex. Capsule 8-ribbed, dehiscent into 2 unequal valves, smaller caducous.

Distr. India, Burma and Ceylon.

Ecol. Running rivers and cataracts.

Specimens Examined. NUWARA ELIYA DISTRICT: Kehel Ganga, Dickoya, Mar 1899, *Macmillan s.n.* (K).

3. Zeylanidium subulatum (Gardner) C. Cusset, Bull. Mus. Hist. Nat. (Paris) ser. 4, 14: 32. 1992.

Podostemon subulatum Gardner, Calcutta J. Nat. Hist. 7: 184. 1847; Hook. f., Fl. Br. Ind. 5: 65. 1886; Trimen, Handb. Fl. Ceylon 3: 418. 1895; Willis, Ann. Roy. Bot. Gard. Peradeniya 1 (3): 229 & (4): 328.1902. Type: Ceylon, Mahawelle [Mahaweli] Ganga, Holnicut, Ambagamuwa, Feb 1846, *Gardner s.n.* (K, P).
Podostemon subulatum var. *mavaeliae* Willis, Ann. Roy. Bot. Gard. Peradeniya 1 (3): 230. 1902. Syntypes: Hakkinda, *Trimen s.n.* (K); Holnicut, *Gardner s.n.* (K); Raxawa (or Dolosbage), Feb 1855, *s. coll.* in *C.P. 3366*, p.p. (PDA), s. loc., *Thwaites s.n.* (K); s. loc., *Willis s.n.* (K).

Thallus cylindric, stems simple or more frequently branched, 1–2 cm long. Leaves distichous, 3 mm long, linear, broadly sheathed at base, blade of lower caducous, leaving base of sheath to form scale. Flowers in axils of leaves or terminal; bracts similar to leaves but blade not caducous. Spathe ovoid, dehiscing irregularly at summit at anthesis. Pedicel enlarging to 5–7 mm long after anthesis. Tepals 2, filiform, 2–5 mm long. Stamens 2, longer than ovary, androphore 3 times longer than free portion of filaments. Ovary sessile with 2 linear, acute stigmas at apex. Capsule with 8 ribs, cells unequal with smaller caducous.

Distr. Ceylon and south and east India.

Ecol. Running rivers, falls and cascades.

Specimens Examined. KANDY DISTRICT: Without exact locality, Mar, *Borgensen s.n.* (C); Raxawa (or Dolosbage), Feb 1855, *s. coll.* in *C.P. 3366*, p.p. (K, PDA); Mahaweli River, *Thwaites* in *C.P. 3088* (K), Feb, *Wight s.n.* (C, E); Gannoruwa, Dec, *Balakrishnan 1137* (PDA); Hakkinda, 24 Feb 1928, *Alston 1937* (PDA), 21 Sep 1928, *de Silva 91* (PDA), 8 Jan 1899, *Willis s.n.* (PDA); Holnicut, Ambagamuwa, Feb 1846, *Gardner s.n.* (K, holotype; P, isotype); Halloluwa, Mar, *Jacobsen 23/21* (C). LOCALITY UNKNOWN: *Ward s.n.* (E).

3. POLYPLEURUM

(Taylor ex Tulasne) Warming, Kongel Danske Vidensk. Selsk. Naturvidensk. Math. Afh. ser. 3, 11: 464. 1901. Type species: *Polypleurum wallichii* (R. Br.) Warming.

Thallus foliaceous or ribbon-like and attached to rocks, bearing small acaulous shoots or long cylindrical or flattened thalli flowing with the current and bearing short, endogenous shoots. Leaves distichous, entire, linear, blade usually caducous, the broad sheath marcescent. Spathe solitary, ovoid, enclosing erect flower, dehiscent at apex. Pedicel exserted at anthesis. Tepals 2. Stamens 1 or 2, on long androphore; anthers small, oblong. Ovary ellipsoid, sessile or subsessile, both cells equal; placenta thick, bilobed, bearing numerous ovules over whole surface. Capsule bilocular, 8-ribbed, ribs slender or thick, dehiscent into 2 equal valves, not caducous. Seeds more than 20 in each loculus.

Genus of 7 species from India, Ceylon and Thailand.

KEY TO THE SPECIES

1 Flowering stems bearing 2 leaves
  2 Thallus cylindrical reaching up to 60 cm long; stigmas filiform .... 1. P. elongatum
  2 Thallus ribbon-like, 10–20 cm long; stigmas subconical .... 2. P. stylosum
1 Flowering stems bearing 4 leaves .... 1. P. elongatum

1. Polypleurum elongatum (Gardner) J.B. Hall, Kew Bull. 26: 131. 1971.

Podostemon elongatum Gardner, Calcutta J. Nat. Hist. 7: 188. 1847; Hook. f., Fl. Br. Ind. 5: 65. 1886; Trimen, Handb. Fl. Ceylon 3: 417. 1895. Type: Ceylon, s. loc., *Gardner s.n.* (K, holotype; P, isotype).

Dicraeia elongata (Gardner) Tulasne 'Dicraea', Ann. Sci. Nat. Bot., ser. 3, 11: 101. 1849; Wight, Ic. Pl. Ind. Or. 5: 33, t. 1917, I. 1852; Wedd. in DC., Prod. 17: 70. 1873; Willis, Ann. Roy. Bot. Gard. Peradeniya 1(3): 219. 1902.

Free-floating thalli cylindric, 50–60 cm long, filamentous, bearing distichous, alternately arranged, secondary sessile shoots, fertile towards base, sterile above. Main leaves linear, 2–5 mm long, very slender, fragile, shed rather rapidly, next leaves 2–4 in number, distichous, well developed with sheath clasping stem, blade caducous, sheath eventually withering. Flower solitary, terminal. Pedicel up to 7 mm long at anthesis. Tepals 2, filiform, 2 mm long. Stamens 2, exceeding ovary. Ovary sessile; stigmas 1 mm long, filiform. Capsule with 8 slender, but well defined ribs.

Distr. Endemic.

Ecol. Running rivers, cataracts and waterfalls.

Specimens Examined. KEGALLE DISTRICT: Kelani River at Kitulgala, 22 Feb 1899, *Macmillan s.n.* (K, PDA), 4 Mar 1882, *s. coll. s.n.* (PDA). KANDY DISTRICT: Hakkinda, Dec 1898, *Willis s.n.* (K), 25 Feb 1928, *Alston 1938* (PDA), Feb, *Alston 1988* (C), 21 Sep 1928, *de Silva 89* (K, PDA), 8 Jan 1899, *Willis s.n.* (PDA), 20 Jan 1900, *Willis s.n.* (PDA); Mahaweli River, *Thwaites* in *C.P. 2259* p.p. (E, K, P), 450 m, 27 Feb 1979, *Kostermans 27371* (K); near Gannoruwa, 22 Dec 1971, *Balakrishnan & Dassanayake 1138* (PDA), *1139* (K, PDA), *Balakrishnan 1140* (PDA); Feb 1846, Ambagamuwa, *Gardner s.n.* (K, P); near Peradeniya, Mar, *Giesenhagen s.n.* (M); Peradeniya, Jan 1854, *Gardner s.n.* (K), in *C.P. 2259*, p.p. (PDA). LOCALITY UNKNOWN: Mar, *Boergesen s.n.* (C), *Farmer s.n.* (M), *Goebel s.n.* (M), *Wight s.n.* (E, P).

2. Polypleurum stylosum (Wight) J. B. Hall, Kew Bull. 26: 131. 1971.

Dicraeia stylosa Wight 'Dicraea', Ic. Pl. Ind. Or. 5: 33, t. 1917, II. 1852; Wedd. in DC., Prod. 17: 70. 1873; Willis, Ann. Roy. Bot. Gard. Peradeniya 1(3): 225. 1902. Type from India.

Podostemon stylosum (Wight) Benth. in Benth. & Hook. f., Gen. Pl. 3: 112. 1880; Hook. f., Fl. Br. Ind. 5: 64. 1886.

Dicraeia stylosa var. *fucoides* Willis, Ann. Roy. Bot. Gard. Peradeniya 1(3): 226. 1902. Syntypes: Mahaweli River, Hakkinda, *Willis s.n.* (K) and from India.

Dicraeia stylosa var. *laciniata* Willis, Ann. Roy. Bot. Gard. Peradeniya 1(3): 117. 1902. Syntypes: Mahaweli River, Hakkinda, *Thwaites s.n.* (PDA); *Trimen s.n.* (PDA); *Willis s.n.* (K).

Thallus attached to rocks for some centimetres, then free-flowing, or frequently attached by few haptera or small hairs, less often creeping, ribbonlike, much branched, from 10–40 cm long, 5–6 mm wide, rarely more. Flowers solitary, arranged somewhat densely on upper surface of lower part of thallus, 3–5 mm apart on margin, surrounded by 2 enlarged, wider than long, leaf-bases (blade linear, caducous), concealing the spathe before flowering. Pedicel 6–20 mm long, long exserted. Tepals 2, 2 mm long, filiform. Stamens 2, exceeding ovary; filaments very short. Ovary sessile; stigmas 2, 0.8–1 mm long, subconical. Capsule 8-ribbed, ellipsoid. Seeds c. 0.2×0.12 mm, smooth.

Distr. India and Ceylon.

Ecol. Running rivers, cataracts and waterfalls.

Specimens Examined. KANDY DISTRICT: Kandy, Mar, *Boergesen s.n.* (C); Mahaweli River, 19 Feb 1893, *Willis s.n.* (K, PDA), 11 Aug 1898, *Willis s.n.* (K), *Goebel 1899* (M); Hakkinda, 8 Jan 1899, *Willis s.n.* (K), 21 Sep 1928, *de Silva 90* (PDA), 24 Feb 1928, *de Silva s.n.* (PDA), 28 Feb 1928, *de Silva 92* (PDA), 19 Feb 1899, *s. coll. s.n.* (PDA); Gannoruwa, Dec, *Balakrishnan 1141* (PDA); foot of Primrose Hill, Kandy, 490 m, 5 Feb 1977, *Wheeler 12911* (PDA); Lady Blake's Drive, Kandy, 24 Sep 1977, *Meijer 1467* (K). LOCALITY UNKNOWN: *Farmer s.n.* (M).

4. DALZELLIA

Wight, Ic. Pl. Ind. Or. 5 (2): 34. 1852. Type species: *Dalzellia ceylanica* (Gardner) Wight.

Thallus flattened-frondose, branched with dilated, flabelliform lobes. Leaves linear, fasciculate on upper surface of thallus. Flowers many, solitary at apices of lobes, pedicellate. Pedicels arising from terminal leafy cupules. Perianth segments 3, slightly imbricate. Stigmas 3. Ovary 3-celled; ovules numerous. Styles 3. Capsule 3-valved.

Genus of 4 species from Ceylon, Thailand and Indo-China.

Dalzellia ceylanica (Gardner) Wight, Ic. Pl. Ind. Or. 5 (2): 35. 1852, 'zeylanica'.

Tristicha ceylanica Gardner, Calcutta J. Nat. Hist. 7: 177. 1846. Type: Ceylon, Mahaweli river, March 1846, *Gardner s.n.* (K, holotype).
Lawia zeylanica (Gardner) Tulasne, Ann. Sci. Nat. Bot., ser. 3, 11: 112. 1849.
Tulasnea ceylanica (Gardner) Wight, Ic. Pl. Ind. Or. 5 (2): t. 1919. 1852, pro syn.

Lawia zeylanica var. *parkiniana* Willis, Ann. Roy. Bot. Gard. Peradeniya 1(3): 215. 1902. Type: Ceylon, *Parkin s.n.* (K, holotype).

Lawia zeylanica var. *gardneriana* Willis, Ann. Roy. Bot. Gard. Peradeniya 1(3): 214. 1902, nom. illegit.

Fronds of thallus very thin, closely attached to rocks, much branched, widespreading in dense areas up to 30 cm or more across; thalli 0.5–0.75 mm thick, hard, siliceous, deep crimson to dark green. Leaves very small, narrowly linear-oblong, obtuse, in small stellate tufts. Spathe densely covered with short, soft, spinelike processes. Pedicel twice length of flowers, 2–6 mm long at anthesis. Perianth segments united at base for 2/3 their length, oblong-lanceolate, acute. Stamens equalling perianth segments. Capsule ovoid-oblong, 9-ribbed.

Distr. Southwest India and Ceylon.

Ecol. Running rivers, waterfalls and cataracts.

Specimens Examined. KEGALLE DISTRICT: Kitulgala, on rocks in Kelani River, 22 Feb 1899, *Macmillan s.n.* (K, PDA); in Galata Oya, Pussellawa, 30 Jan 1929, *Alston s.n.* (PDA). KANDY DISTRICT: Mahaweli River, *Trimen s.n.* (PDA); Hakkinda, 3 Feb 1993, *Philcox et al. 10449* (PDA), *10451* (PDA); Haragama, 10 Jan 1929, *Alston 2377* (PDA), Peradeniya, *s. coll.* in *C.P. 3089*, p.p. (PDA). MATALE DISTRICT: Matale East, *s. coll.* in *C.P. 3089*, p.p. (PDA).

POLYGALACEAE (Continued)

(by B.M. Wadhwa[*])

R. Brown in Flinders, Voy. Terra Austr. 2: 542. 1842 ('Polygaleae'). Type genus: *Polygala* L.

Two genera (*Polygala* L. and *Salomonia* Lour.) of this family covering 12 species have already been published in this Rev. Handb. Fl. Ceylon, volume 6: 301–317. 1988.

The genus *Xanthophyllum* Roxb., has often been treated under a separate family Xanthophyllaceae (Cronquist, Integr. Syst. 763. 1981), based on incomplete knowledge of the genus. The anatomical evidence of wood supports a close affinity with Polygalaceae (Bridgwater & Baas, IAWA Bull. Leiden n.s. 3: 115–125. 1982). The studies on the morphology of ovules, fruits and seeds (Verkerke, Blumea 29: 409–421, 1984) and also foliar anatomy (Dickson, J. Linn. Soc. Bot. 67: 103–115. 1973) do not present arguments to split *Xanthophyllum* from Polygalaceae. The present treatment follows Meijden's monograph of the genus (in Leiden Bot. ser. 7:1–159. 1982), which included it in Polygalaceae.

Herbs, shrubs, lianas or trees. Leaves simple, entire, usually spirally arranged, sometimes alternate, sometimes scale-like or absent. Stipules absent. Inflorescence usually a raceme, unbranched, extra-axillary and/or terminal, sometimes thyrsoid. Bracts and bracteoles present, rarely absent. Flowers bisexual, zygomorphic. Sepals 5, free and quincuncial, or abaxial 2 connate, subequal or the lateral ones larger and then often wing-like and petaloid. Petals 3 or 5, free or variously united, usually adnate to the base of staminal tube, often unequal, with lower petal often keel-like and frequently pouched and lobed. Stamens usually 8, filaments usually ± connate except the upper stamens and often adnate to the petals; anthers basifixed, 1–2-celled, opening by a single, often oblique pore. Ovary superior, sessile, sometimes stipitate, usually 2-locular, but occasionally 1-, 5-, 7- or 8-locular; ovules 1 per cell and subapical, or 4-more in a 1-locular, bicarpellate ovary (as in *Xanthophyllum*), anatropous; style simple, often dilated at apex. Fruit various, generally a 2-celled, 2-seeded loculicidal capsule, or a 1-seeded drupe or samara, sometimes a berry.

[*] Royal Botanic Gardens, Kew.

About 15 genera and over 1,000 species, widespread in temperate and tropical regions, especially in S. America and S. Africa; 3 genera and 13 species in Sri Lanka.

KEY TO THE GENERA (REVISED)

1 Herbs, sometimes woody at base
  2 Inflorescence a raceme; sepals unequal, lateral sepals wing-like, larger than the 3 outer sepals, petaloid, ± as long as the petals; lower petal (keel) apically bearded; stamens 8; fruit laterally dehiscent, margins entire; seeds always with aril or caruncle .... **1. Polygala**
  2 Inflorescence a spike; sepals ± equal, not petaloid, much shorter than the petals; lower petal (keel) apically not bearded; stamens 4–6; fruit laterally dehiscent, margins dentate; seeds without aril or caruncle .... **2. Salomonia**
1 Shrubs, trees or climbers
   3 Petals 3 or with an additional pair of much reduced ones; lateral sepals wing-like, petaloid and at least twice as large as the other sepals; ovary and fruit 2-locular with 1 ovule in each locule; fruit a loculicidally dehiscent capsule .... **1. Polygala**
   3 Petals 5; lateral sepals not petaloid, and less than twice as large as the other sepals; ovary and fruit 1-locular, with 4 or more ovules; fruit an indehiscent drupe .... **Xanthophyllum**

XANTHOPHYLLUM

Roxb., Pl. Corom. 3: 81. 1820 ('1819'), nom. cons.; Thw., Enum. Pl. Zeyl. 23. 1858; Benth. & Hook. f., Gen. Pl. 1: 139. 1867; Benn. in Hook. f., Fl. Br. Ind. 1: 208. 1872; Trimen, Handb. Fl. Ceylon 1: 84. 1893; Chodat, Bull. Herb. Boiss. 4: 254. 1896; Hutch., Gen. Fl. Pl. 2: 339. 1967; Meijden, Leiden Bot. ser. 7: 1–159. 1982. *Lectotype species*: *X. flavescens* Roxb. (typ. cons.).

Banisterodes (L., Fl. Zeyl. 192. 1747 '*Bannisterioides*') Kuntze, Rev. Gen. Pl. 1: 45. 1891, nom. rej.

Trees or shrubs; branches terete, sometimes with cap-like nodal glands. Leaves alternate, stalked, nearly always with glands beneath, mostly near base of blade; exstipulate, coriaceous. Inflorescence axillary, branched or unbranched, panicle or racemelike. Flowers solitary or in lower half 3 together. Sepals 5, free, usually slightly subequal, caducous after flowering. Petals 5, free, usually unequal, glabrous; lower petal keeled, usually boat-shaped, clawed, apically without crest or appendage. Stamens mostly 8; 4 epipetalous, 2 placed at the base of keel and partially adnate to it, 2 alternipetalous and opposite the lateral sepals; filaments free or connate basally; anthers tetrasporangiate, opening introrsely with slits. Disk annular, hypogynous. Ovary usually shortly stipitate, syncarpous, bicarpellary, 1-locular or sometimes imperfectly ± 2-locular; style terminal; stigma slightly bilobed or sometimes peltate; ovules 2-seriate, 4 or (6–) 8–20. Fruit indehiscent, globose or subglobose, c. 1–15 cm, usually with a hard pericarp. Seed(s) 1 or 4–20, glabrous, estrophiolate; embryo large, flat; albumen copious to nearly absent.

About 94 species in tropical S.E. Asia, northwards to S. China and Hainan, throughout Malesia to Australia; represented by 1 species in Sri Lanka.

Xanthophyllum zeylanicum Meijden, Leiden Bot. ser. 7: 131. 1982. *Type*: *Herb. Hermann*, vol. 2, folio 5, left hand plant *'Bannisterioides'* (BM, holotype; L, isotype: 139 *'Boromus'*).

Xanthophyllum flavescens auct. (non. Roxb., 1820): Thwaites, Enum. Pl. Zeyl. 23. 1858; Trimen, Handb. Fl. Ceylon 1: 84. 1893.
Xanthophyllum geminiflorum auct. (non Alston, 1931): Alston in Trimen, Handb. Fl. Ceylon 6: 16. 1931, excl. type; Ooststr., Blumea Suppl. 1: 206. 1937.

Evergreen trees, 15–20 m high. Axillary buds in basal part of a shoot 2, the upper bud 2–4 mm, supra-axillary, base slightly thickened. Leaves elliptic or oblong-lanceolate, 6–16 × 2.5–6.5 cm, apex acute-acuminate, subacute at base, entire to subundulate along margin, coriaceous, upper side yellowish-brown, lower greenish to greenish-yellow, sometimes faintly waxy; main lateral nerves 5–7 pairs, with short intermediaries, not forming an intramarginal nerve; a few perforated glands scattered but mostly near midrib or at the axils of nerves; petioles (4.5–) 5.5–8 mm long, often with 2 tiny glands halfway, rugose. Inflorescences unbranched racemes sometimes with a side branch at base only, shorter than the leaves; axes angular, sparingly minutely hairy. Bracts alternate. Flowers all solitary, 10–14 mm in diam.; pedicels 3–6 mm long, subglabrous to shortly appressedly hairy. Sepals subequal, elliptic, densely pubescent, sometimes with tiny glandular spots; outer sepals 1.3–2.5 × 1.2–2.2 mm; inner sepals 2.2–3 × 1.5–3.2 mm, brownish-yellow. Petals subequal, narrowly oblong, yellowish when dry, the longest one (8.5–) 9–11.5 mm long; carina or keel appressedly hairy outside; other petals glabrous to sparsely hairy at apex outside. Stamens 8, 7–9 mm long, strigose-ciliate at base; anthers 0.5 mm long, opening introrsely with splits. Disk annular, hypogynous. Ovary globose to subglobose, shortly stipitate, half patent to appressedly hairy; ovules 4; styles covered with brown hairs; stigma bilobed. Fruit globose, c. 1.5–2.0 cm in diam., smooth, dull, dark green, brownish to grey, appressedly hairy apically; pericarp thin and leathery or thicker and hard.

Distr. Endemic.

Ecol. In evergreen forests in wet zone, to 600 m, fairly common; fls & frts Nov–May.

Vern. Palala (S).

Note. The name *Xanthophyllum geminiflorum* (Dennst.) Alston is superfluous under Art. 52. 1 of International Code of Botanical Nomenclature (Tokyo ·Code, 1994) because Alston refers to both *X. flavescens* Roxb. and *X. virens* Roxb.

This species comes close to *Xanthophyllum virens* Roxb., but in the latter species the leaf-blade is glaucous-papillose on the lower side, and the leaf-base is attenuate to cuneate; in the latter the carina (keel) is auriculate, which is not the case in the former.

Plate No. 135 in PDA based on *C.P. 1244* belongs to this species.

Specimens Examined. COLOMBO DISTRICT: Colombo, *Beckett 957* (BM), July 1884, *Ferguson s.n.* (PDA); Labugama, 18 March 1975, *Waas 1235* (PDA); R.B. Garden, Gampaha, 17 Aug 1963, *Amaratunga 677* (PDA). KALUTARA DISTRICT: Yagirala, alt. low, 29 June 1975, *Waas 1298* (PDA); Near Moragala, Pasdoon Korale, Mar 1887, *s. coll. s.n.* (PDA); Swamp forest near Honaka, Bulathsinhala, 7 Oct 1979, *Kostermans 27867* (PDA); Molkawa, July 1988, *Jayasuriya & Balasubramaniam 4412* (PDA); Beraliya, 17 May 1928, *Alston 2361* (K, PDA); Road Pelawatte-Moragala, 9/3 culvert, 150 m, 29 Jan 1979, *Gunatilleke 296* (PDA); Kalutara, *Macrae 214* (BM, K). RATNAPURA DISTRICT: Udakarawita Kanda, 450 m, 1 May 1970, *Balakrishnan 321* (K, PDA); Sinharaja Forest, Weddagalle Entrance, 200 m 26 March 1979, *Kostermans 27453* (PDA); Derangala other state forest, 29 June 1992, *Jayasuriya 6496* (PDA); Karawita Kanda, 30 Sept 1973, *Waas 26* (PDA); Ratnapura-Kalwana Road, mile 151, 6 Aug 1971, *Meijer 1091* (PDA); Between Kudawe & Sinharaja, 28 May 1988, *Jayasuriya & Balasubramaniam 4364* (PDA); Ratnapura, behind Rest House, 5 Oct 1977, *Nooteboom 3265* (PDA); Ratnapura, Mar 1853, *Thwaites C.P. 1244* (BM, PDA); Karawita Proposed Reserve, Kalawana Range, 23 May 1990, *Jayasuriya & Balasubramaniam 5112 & 5114* (PDA). GALLE DISTRICT: Godakanda, near Hiniduma, 200 m, 6 May 1979, *Kostermans 27614* (K, PDA), 100–120 m, 27 Sept 1977, *Huber 325* (PDA); Kanneliya, 13 Apr 1974, *Waas & Peeris 565* (PDA); Halwitigala, 30 June 1975, *Waas 1315* (PDA); Hiyare Reservoir, 20 Nov 1971, *Balakrishnan 991* (K, PDA); Kanneliya Forest, near Hiniduma, 200 m, 10 June 1979, *Kostermans 27575* (PDA); Kandewattegoda Proposed Reserve, Galle Range, 29 June 1990, *Jayasuriya & Balasubramaniam 5346* (PDA); Hiniduma, 23 Sept 1977, *Nooteboom 3180* (PDA); Goda Kanda, E. of Hiniduma, 80°20′ E, 6°18′ N, 11 May 1972, *Jayasuriya, Cramer & Balakrishnan 799* (K, PDA); Pituwala, SE of Elpitya, 180 m, 26 & 27 Oct 1977, *Huber 310, 331* (PDA); Thawalama, Banks of Gin Ganga, 30 June 1975, *Waas 1332* (K, PDA). MATARA DISTRICT: Akuressa Forest Range, Dediyagala Forest Reserve, 14 Dec 1991, *Jayasuriya 6001* (PDA); Kadugalakanda Forest Reserve, 2 July 1990, *Jayasuriya & Balasubramaniam 5393* (PDA). LOCALITY UNKNOWN: in 1847, *Gardner 59* (K); *Col. Walker s.n.* (K); Herb. Hort. Soc. 1839, *Mackenzie s.n.* (K); in 1800, *Jonville s.n.* (BM); *Macrae s.n.* (BM); *Moon s.n.* (BM).

PORTULACACEAE

(by B.M. Wadhwa* and A. Weerasooriya**)

Juss., Gen. Pl. 312. 1789 ("Portulaceae"). Type genus: *Portulaca* L.

Annual or perennial herbs or shrubs, mostly branched, erect or creeping, some woody at base, or with a tuberous main root occasionally rooting at the nodes. Leaves simple, alternate, spirally arranged or opposite, subsessile, entire, occasionally with axillary hairs or scales. Flowers in axillary and/or terminal clusters (capitula) or in corymbose cymes or thyrses, or dichasia, rarely solitary, bisexual, actinomorphic, 4–5-merous, bracteate or not. Bracts leaf-like or membranous, bracteoles hairy or scarious. Sepals 2 (4–8 in some extra-Sri Lankan species), boat-shaped, deltoid to obovate, imbricate, carinate or not, deciduous, at base shortly connate and confluent with petals and stamens. Petals 4–6 (–8), mostly obovate, subequal, shortly connate, imbricate, fugaceous or marcescent, variously coloured. Stamens (1–) 3-many in one or more whorls; filaments basally shortly connate; anthers 2- or 4-celled, dorsifixed, dehiscing longitudinally. Ovary superior or semi-inferior, 1- celled; ovules 4-many on free central placenta, campylotropus; style apically with 3–5 (–18) arms. Capsules globose, conical or ovoid, dehiscence circumscissile with operculum or valvular, occasionally surrounded by the persistent calyx. Seeds mostly numerous, reniform to ± globular, smooth or minutely tubercled, mostly with a caruncle.

Cosmopolitan, 15 genera with c. 200 species; 2 genera and 8 species in Sri Lanka. Most of the species occur as adventives or weeds in waste places, or cultivated as ornamental or food plants.

KEY TO THE GENERA

1 Ovary semi-inferior; capsule dehiscing circumscissile with an operculum; seeds usually tuberculate; leaves less than 2 cm long .... **1. Portulaca**
1 Ovary superior; capsule valvular or irregularly dehiscent; seeds usually smooth; leaves more than 6 cm long .... **2. Talinum**

1. PORTULACA

L., Sp. Pl. 445. 1753; Benth. & Hook. f., Gen. Pl. 1: 156. 1862; Dyer in Hook. f., Fl. Br. Ind. 1: 246. 1874; Trimen, Handb. Fl. Ceylon 1: 89. 1893;

---

* Royal Botanic Gardens, Kew.
** Flora of Ceylon Project, Peradeniya.

Pax & Hoffm. in Pflanzenfam ed. 2, 16C: 246. 1934; Poelln. in Fedde, Repert. 37: 240. 1934; Geesink, Blumea 17(2): 283. 1969; Geesink in Fl. Males. ser. 1, 7: 126. 1971. Type species: *P. oleracea* L.

Annual or perennial, erect or creeping and rooting at nodes, much branched, often succulent herbs. Leaves spirally arranged or opposite, mostly subsessile, linear to orbicular, with axillary hairs or scales. Flowers bisexual, actinomorphic, solitary or in 2–30-flowered terminal capitula; receptacle infundibular, mostly with hairs or scales in the axils of the bracts. Sepals 2, boat-shaped, deltoid to obovate, shortly connate and confluent with petals and stamens, keeled or hooded, persistent or caducous. Petals 4–6 (–8), mostly obovate, free or subconnate at base, marcescent. Stamens 4-many in one whorl. Filaments for 1/4th length connate; anthers 2–4-celled, dorsifixed, dehiscing longitudinally. Ovary semi-inferior, 1-locular; styles with 2–18 arms. Capsules with a caducous, circumscissile operculum. Seeds many; reniform, tuberculate, rarely smooth.

About 40 species, cosmopolitan, mainly tropical and subtropical; 6 species in Sri Lanka.

KEY TO THE SPECIES

1 All leaves opposite; hairs and scales infra- and inter-petiolar; bracteoles absent
  2 Nodes with whorls of membranous scales concealing almost all leaves; internodes very short; flowers 2–7 in clusters; stamens 8–10 ....:..... **1. P. wightiana**
  2 Nodes with a ring of setose hairs; leaves distinctly visible; flowers solitary; branchlets rooting and fluffy at nodes .... **2. P. quadrifida**
1 At least middle cauline leaves spirally arranged; hairs only axillary; bracteoles membranous
  3 Stem nodes unappendaged, glabrous or axillary hairs inconspicuous; flowers 3–6 in terminal clusters; sepals distinctly carinate; leaves obovate to spathulate .... **3. P. oleracea**
  3 Stem nodes with a ring of setose hairs; sepals not carinate, but membranous, occasionally with an apical, dorsal dome-shaped spur; leaves linear to obovate, less than 4 mm wide
    4 Flowers more than 2.5 cm across; petals 12–22 mm across; foliar involucre 5–8 ......
    .... **4. P. grandiflora**
    4 Flowers to 1.25 cm across; petals c. 12 mm long
     5 Erect or suffruticose herbs; roots much branched, woody, not tuberous; leaves terete or elliptic; axillary hairs dense, c. 2 mm long; flowers yellow, stamens >20 ....
     .... **5. P. suffruticosa**
     5 Spreading herbs, with tuberous roots; leaves linear; axillary hairs short, about c. 1 mm long; flowers whitish-yellow, stamens 16 .... **6. P. tuberosa**

1. Portulaca wightiana Wall. ex Wight & Arn., Prod. 356. 1834; Dyer in Hook. f., Fl. Br. Ind. 1: 247. 1874; Trimen, Handb. Fl. Ceylon 1: 89.1893; Dunn in Gamble, Fl. Pres. Madras 1: 66. 1915; Geesink, Blumea 17(2): 290. 1969; Matthew & Britto in Matthew, Fl. Tamilnadu Carnatic 1: 83. 1983. Chandrasekaran in Nair & Henry, Fl. Tamilnadu 1: 25. 1983. Type: India, *Wallich 6845*; Herb. Wight propr. *1166* (K).

Herbs c. 8 cm high; stem much branched, prostrate; internodes very short. Leaves alternate, sessile, cordate to ovate, c. 4 × 2.7 mm, acute, margins curved, with large axillary, silvery scales covering internodes; scales deltoid to ovate, c. 4 × 2 mm, imbricate, acute, membranous, persistent. Capitula terminal, 2–7-flowered. Flowers sessile, subtended and surrounded by scales and hairs towards centre of the capitulum. Sepals c. 3 × 3.3 mm, broadly ovate-lanceolate. Petals 4(–5), elliptic, c. 2.7–1.7 mm. Stamens 10, rarely 5; filaments c. 1 mm long. Style c. 2 mm long, with 3–6 arms. Capsules c. 2 mm in diam., globose; operculum nearly 2/3 the length of the capsule, pale-yellow. Seeds numerous, granulate, 0.5 mm in diam.; testa cells ± hexangular with central tubercle.

Distr. Sri Lanka and S. India.

Ecol. On exposed gravelly soil; rare: fls & frts Oct–Dec.

Notes. Plate No. 143 in PDA belongs to this species.

Specimens Examined. MANNAR DISTRICT: Nayorpalam, Vidathal Tivu water supply turn-off road, mile post 13/9 to 14/9 on Mannar Pooneryn road, 9°00′ N, 80°03′ E, 11 Jan 1977, *R.B. & A.J. Faden 77/126* (K); Iluppaikkadavai, 10 Feb 1890, *Ferguson s.n.* (K, PDA). PUTTALAM DISTRICT: Wilpattu National Park, Galbendi Wewa, 28 Apr 1959, *Mueller-Dombois, Wirawan, Cooray & Balakrishnan 69042826* (PDA). HAMBANTOTA DISTRICT: Ruhuna National Park, Patanagala rock, 3–5 m, 3 Apr 1968, *Fosberg & Mueller-Dombois 50136* (PDA); Ruhuna (Yala) National Park, Block II near Sithulpawwa Vihara, 8 Oct 1994, *Wadhwa, Weerasooriya & Samarasinghe 416* (K, PDA).

2. Portulaca quadrifida L., Mant. 73. 1767; Thw., Enum. Pl. Zeyl. 23. 1858; Dyer in Hook. f., Fl. Br. Ind. 1: 247. 1874; Trimen, Handb. Fl. Ceylon 1: 90. 1893; Dunn in Gamble, Fl. Pres. Madras 1: 66. 1915; Geesink, Blumea 17(2): 290. 1969; Matthew & Britto in Matthew, Fl. Tamilnadu Carnatic 1: 82. 1983; Chandrasekaran in Nair & Henry, Fl. Tamilnadu 1: 25. 1983. Type: from Egypt.

Annual herb, much branched; branches up to 10 cm long, creeping, profusely rooting at the nodes; nodes with a whorl of dense, silvery white, c. 5 mm long hairs. Leaves fleshy, elliptic-cordate to ovate-lanceolate, 2–20 × 1–7 mm, acute at apex, entire, with c. 5 mm long, axillary, white hairs. Flowers terminal, c. 1 cm across, solitary on an infundibular receptacle subtended by 4 leaves at its edge and encircled by hairs. Sepals 2, oblong, c. 3 mm long. Petals 4, obovate, 5×4 mm, obtuse, yellow. Stamens 8 or 12, filaments to 3 mm long. Ovary ellipsoid, to 3 mm; half-embedded in receptacle, 1-celled; style up to 3.5 mm, with (3–) 4 (–5) arms. Capsule ± obovate, 3.5–4.5 × 3 mm; operculum to nearly 2/3 the length of capsule, shining, straw-yellow. Seeds 2-many, c. 1 mm in diam., dull black, testa cells elliptic, radially elongated, margin straight, surface either convex or tubercled.

Distr. Pantropical, except Australia and Pacific east of Samoa.

Ecol. In waste places, arable lands, on the floor of scrub forests; up to 1000 m; very common; fls & frts throughout the year.

Vern. Heengendakola (S).

Notes. Plate No. 144 in PDA belongs to this species.

Specimens Examined. MANNAR DISTRICT: Mannar causeway, Feb 1890, *s. coll. s.n.* (PDA). TRINCOMALEE DISTRICT: Trincomalee lagoon, *Thwaites C.P. 1100* (K, PDA); Marble Point Beach, 4 March 1995, *Weerasooriya & Samarasinghe 175* (K, PDA). KURUNEGALA DISTRICT: Melsiripura, 30 Apr 1965, *Amaratunga 865* (PDA). MATALE DISTRICT: Dambulla rock, 12 Oct 1966, *Amaratunga 1153* (PDA), 10 Apr 1995, *Wadhwa, Weerasooriya & Samarasinghe 578* (K, PDA); Naula-Elahera Road, off road marker 12/4, 6 Oct 1971, *Jayasuriya 320* (PDA); Sigiriya rock, 14 July 1973, *Nowicke, Fosberg, & Jayasuriya 343* (PDA). BATTICALOA DISTRICT: Batticaloa, in 1846, *Gardner s.n.* (PDA). KANDY DISTRICT: Royal Botanic Gardens, Peradeniya, 27 Oct 1927, *Alston 2008* (K, PDA), 23 Feb 1964, *Amaratunga 782* (PDA), Medicinal Plot, 4 May 1995, *Wadhwa & Weerasooriya 598* (K, PDA). BADULLA DISTRICT : Past Dunhinda Falls, 21 Oct 1994, *Wadhwa, Weerasooriya & Samarasinghe 451* (K, PDA). HAMBANTOTA DISTRICT: Ruhuna National Park, Block I, Patanagala Smithsonian Camp, 22 Oct 1969, *Cooray 68102211R* (PDA), *Wirawan 636* (PDA), Patanagala, 2–3 m, 5 Apr 1968, *Fosberg 50228* (K, PDA).

3. Portulaca oleracea L., Sp. Pl. 445. 1753; Thw., Enum. Pl. Zeyl. 23. 1858; Dyer in Hook. f., Fl. Br. Ind. 1: 246. 1874; Trimen, Handb. Fl. Ceylon 1: 89. 1893; Dunn in Gamble, Fl. Pres. Madras 1: 66. 1915; Geesink, Blumea 17(2): 292. 1969; Ramesh in Saldanha, Fl. Karnataka 1:155. 1984. Type: from Europe, Austria, Herb. Linn.

Annual erect or decumbent herb up to 35 cm. Leaves alternate or sub-opposite, obovate to spathulate, 20–40 × 15–20 mm, fleshy, glabrous, base attenuate, margin entire, apex obtuse or truncate, with inconspicuous, c. 1 mm long axillary hairs. Flowers 3–6 in capitula, rarely solitary, c. 1 cm across, with mostly 2 involucral leaves, surrounded by c. 5 × 6 mm bracteoles and inconspicuous hairs. Sepals 2, suborbicular, c. 9 mm long, carinate; carina c. 3.5 mm long and 1.5 mm, high. Petals (4–) 5, yellow, 3–9 × 7 mm, broadly obovate, apex emarginate to mucronate; basally subconnate. Stamens 8–10 (–15); filaments to c. 4 mm long; anthers small. Ovary obovoid, 1.5 mm, half enclosed by calyx, 1-locular; style up to 4.5 mm long, with 3–5 arms.

Capsule ovoid or subglobose, 4 × 3 mm; operculum to 2/3 of the length of capsule, shining, straw-yellow. Seeds numerous, reniform, c. 1 mm in diam., shining, black, granulose; testa cells stellulate with many tubercles.

Distr. Pantropical.

Ecol. A weed in waste places, seashores, waysides and arable lands throughout the country; common; fls & frts throughout the year.

Vern. Gendakola (S); Pulikkirai, Pulichchankirai (T).

Notes. Plate No. 142 in PDA belongs to this species.

Specimens Examined. JAFFNA DISTRICT: Jaffna, in 1846, *Gardner C.P. 1103* (PDA). PUTTALAM DISTRICT: Chilaw Rest House, 22 Nov 1994, *Wadhwa, Weerasooriya & Samarasinghe 391* (K, PDA). ANURADHA-PURA DISTRICT: Awukana, 10 July 1971, *Jayasuriya 238* (PDA). KURUNE-GALA DISTRICT: Kandy-Kurunegala Road, before Kurunegala, c. 300 m, 23 Apr 1978, *Seojarto & Balasubramaniam 4821* (K). POLONNARUWA DIS-TRICT: Polonnaruwa-Habarana Road, 5 Sept 1970, *Balakrishnan 349* (PDA). COLOMBO DISTRICT: Kollupitiya Station, along Railway line, 23 Sept 1994, *Wadhwa, Weerasooriya & Samarasinghe 393* (K, PDA); Negombo, 9 Oct 1968, *Amaratunga 1886* (PDA). KANDY DISTRICT: Near Kandy Lake, 7 Nov 1994, *Wadhwa, Weerasooriya & Samarasinghe 477* (K, PDA); Royal Botanic Gardens, Peradeniya, 29 March 1964, *Amaratunga 797* (PDA), on the drive of SB's Bungalow, 5 Apr 1927, *Alston 1128* (PDA). BADULLA DISTRICT: Haputale, 27 July 1927, *A. de Alwis 8* (PDA).

4. Portulaca grandiflora Hook. in Curtis, Bot. Mag. t. 2885. 1829; Matthew & Britto in Matthew, Fl. Tamilnadu Carnatic 1: 81. 1983; Rao in Sharma et al., Fl. India 3: 3. 1993. Type: from Tropical America, *Hooker s.n.* (K).

Portulaca pilosa L. ssp. *grandiflora* (Hook.) Geesink, Blumea 17: 297. 1969; Geesink in Fl. Males. ser. 1, 7: 131. 1971; Chandrasekaran in Nair & Henry, Fl. Tamilnadu 1: 24. 1983.

Herbs, up to 30 cm high; branches decumbent or ascending, puberulous. Leaves alternate to subopposite, linear-subulate, 20–30 × 2.5–3 mm, terete, often curved, ± obtuse at both ends, with c. 5 mm long axillary hairs. Flowers 2–7 in capitula, c. 2.4 cm across, flowering successively, subtended by 5–8 involucral bracts, deltoid bracteoles and 8–9 mm long hairs. Sepals 2, ovate, 8–10 mm long, with a very small apical keel. Petals 5, obovate, 10–25 × 10 mm, pink, red, orange or yellow. Stamens numerous; filaments 3–6 mm long; anthers elliptic. Styles up to 13 mm long, with 5–12 arms. Capsules globose, 5 mm in diam., operculum to 2/3 of the length of the capsule, shining, straw-yellow. Seeds shining, minute; testa cells ± stellulate, the marginal ones with a central tubercle.

Distr. Native of tropical America, widely cultivated.

Ecol. Widely cultivated as an ornamental, occurs as an escape; fls & frts throughout the year.

Vern. Sun plant, common rose moss (E).

Specimens Examined. MATALE DISTRICT: Sigiriya Rest House ground, 15 July 1973, *Nowicke, Fosberg & Jayasuriya 352* (PDA). COLOMBO DISTRICT: Kollupitiya Station, along Railway line, 23 Sept 1994, *Wadhwa, Weerasooriya & Samarasinghe 395* (K, PDA). NUWARA ELIYA DISTRICT: Boundary of the Hakgala Botanic Gardens, as an escape, 29 March 1995, *Wadhwa & Samarasinghe 519* (K, PDA).

5. Portulaca suffruticosa Wall. [Cat. n. 6842. 1832, nom. nud.] ex Wight & Arn., Prod. 356. 1834; Thw., Enum. Pl. Zeyl. 24. 1858; Dyer in Hook. f., Fl. Br. Ind. 1: 247. 1874; Trimen, Handb. Fl. Ceylon 1: 90. 1893; Dunn in Gamble, Fl. Pres. Madras 1: 66. 1915; Matthew & Britto in Matthew, Fl. Tamilnadu Carnatic 1: 82. 1983; Chandrasekaran in Nair & Henry, Fl. Tamilnadu 1: 25. 1983. Type: from India, *Wight* in Wallich *6842* (K); *Wight 1165* (K).

Portulaca pilosa L. ssp. *pilosa* Geesink, Blumea 17(2): 295. 1969, p.p.

Herb to 40 cm, with a woody, much branched rootstock and several short, spreading, prostrate branches. Leaves alternate, linear, 5–15 × 1–1.5 mm, sub-succulent, puberulous, apex subacute, margin entire, with 2–2.5 mm long axillary leaves. Flowers terminal, solitary, sometimes 2–3 in clusters, sessile, c. 1.25 mm across; foliar involucres 6–8; axillary hairs dense. Sepals 2, oblong, to 4 mm long, acuminate. Petals 4, oblong, 4–5 mm long, obtuse. Stamens 30–40. Styles thick, with 5–6 arms. Capsule subglobose or oblong, to 5 mm long; operculum to 2/3 of the length of the capsule. Seeds many, very small, shining, black, with concentric rows of minute tubercles.

Distr. India & Sri Lanka.

Ecol. Scrub forests and dry lowland areas, on poor gravelly/ sandy soil, rather common fls & frts almost throughout the year.

Notes. Plate No. 145 in PDA belongs to this species.

Specimens Examined. ANURADHAPURA DISTRICT: Mihintale Sanctuary, 55 mile marker, Culvert No. 5, 5 Oct 1973, *Sohmer 8123* (PDA); Rajangane Reservoir, 24 March 1985, *Koyama, Mii, Jayasuriya, Strudwick & Hanashiro 15982* (PDA). MATALE DISTRICT: Dambulla Rock, 12 Oct 1966, *Amaratunga 1158* (PDA), 50 m, 1 Oct 1978, *Kostermans 26855* (PDA), 30 Nov 1926, *J.M. Silva s.n.* (PDA), 20 Dec 1981, *s. coll. s.n.* (PDA); Dambulla, July 1848, *Gardner C.P. 1101* (K, PDA), 4 Feb 1928, *Alston 1968* (PDA). TRINCOMALEE DISTRICT: Palakadu, sea level, 5 Dec 1976,

Cramer 4770 (PDA). POLONNARUWA DISTRICT: Dimbulagala, 24 March 1985, *Cramer & Jayaratnam 5652* (PDA). COLOMBO DISTRICT: Colombo seashore, Jan 1881, *Ferguson s.n.* (PDA); KANDY DISTRICT: Alawatugoda, 27 May 1967, *Amaratunga 1308* (PDA). GALLE DISTRICT: Ambalangoda, 16 May 1928, *Alston 2348* (K, PDA); Koggala, Close to Airforce runway-Matara road, 22 Apr 1995, *Wadhwa, Weerasooriya & Samarasinghe 565* (K, PDA). MATARA DISTRICT: Dondra Head, 23 Feb 1981, *s. coll. s.n.* (PDA). HAMBANTOTA DISTRICT: Ruhuna National Park, Block II near Sithul Pawwa Vihara, 8 Oct 1994, *Wadhwa, Weerasooriya & Samarasinghe 417* (K, PDA), Buttawa Bungalow area, 8 m, 18 June 1967, *Comanor 376* (PDA), 18 July 1973, *Nowicke & Jayasuriya 401* (PDA), Block I, Patanagala rock outcrop area, 8 Dec 1967, *Mueller-Dombois 67120825* (PDA), near Patanagala Beach, outside Plot R-10, 31 Aug 1967, *Mueller-Dombois 67083115* (PDA); Yala, Jamburagala, 27 Apr 1973, *Cramer 4138* (K, PDA).

6. Portulaca tuberosa Roxb. [Hort. Beng. 91. 1814, nom. nud.] Fl. Ind. 2: 464. 1832; Thw., Enum. Pl. Zeyl. 401. 1864; Dyer in Hook. f., Fl. Br. Ind. 1: 247. 1874, Trimen, Handb. Fl. Ceylon 1: 90. 1893; Dunn in Gamble, Fl. Pres. Madras 1: 66. 1915; Matthew & Britto in Matthew, Fl. Tamilnadu Carnatic 1: 83. 1983. Type: India, Walajabad, *Roxburg s.n.* (CAL).

Portulaca pilosa L. ssp. *pilosa* 'race' *tuberosa* Geesink, Blumea 17: 296. 1969; Geesink in Fl. Males. ser. 1, 7: 131. 1971.

Herbs to 12 cm, with thick, fusiform tuberous rootstock 5–8 cm long; branchlets spreading, glabrous. Leaves alternate, lanceolate or linear-oblong, 6–10 × 4 mm, fleshy, apex acute, base cuneate, margin entire, with a ring of axillary c. 1–1.5 mm long hairs. Flowers terminal, solitary or paired, with an involucre of 4–8 leaves. Sepals 2, broadly ovate, 4–5 mm long. Petals 5, pale-yellow or whitish yellow, obovate, 8–10 × 5 mm, connate at base. Stamens up to 20; filaments unequal, 2–4 mm long; anthers oblong. Ovary ovoid-globose, 1-locular; style filiform, 5–6 mm long, 3–6-armed. Capsule ovoid-globose, 5 × 3 mm; operculum to 2/3 the length of capsule, shining. Seeds numerous, black, shining, minutely tubercled.

Distr. Pantropical.

Ecol. On sandy coast in dry region, fairly common; fls & frts Dec–April.

Vern. Urugenda (S).

Specimens Examined. JAFFNA DISTRICT: Elephant Pass, near Rest House on sandy shore, 14 Sept 1897, *s. coll. s.n.* (PDA). PUTTALAM DISTRICT: On sandy soil along coast, 12 Apr 1995, *Wadhwa, Weerasooriya & Samarasinghe 583* (K, PDA). COLOMBO DISTRICT: Uswetakeiyawa, 6 Sept 1962, *Amaratunga 343* (PDA); Colombo, 24 July 1914, *Petch s.n.* (K).

BADULLA DISTRICT: Ooma Oya, June 1881, *s. coll. s.n.* (PDA). GALLE DISTRICT: Bentota, May 1859, *s. coll. C.P. 3638* (PDA). HAMBANTOTA DISTRICT: Ruhuna National Park, Block I, at Gonalabba, 27 Oct 1968, *Cooray & Balakrishnan 68102701* (PDA).

2. TALINUM

Adans., Fam. Pl. 2: 245, 609. 1763, nom. cons.; Benth. & Hook. f., Gen. Pl. 1: 157. 1862; Alston in Trimen, Handb. Fl. Ceylon 6: 19. 1931; Poelln. in Fedde, Repert. 35: 1. 1934; Dandy, Taxon 18: 464. 1969; Geesink in Fl. Males. ser. 1, 7: 123. 1971. Type species: *T. triangulare* (Jacq.) Willd. (= *Portulaca triangularis* Jacq.), typ. cons.

Perennial herbs or subshrubs. Leaves alternate, spirally arranged, sometimes the lowermost opposite, linear to obovate, sessile to petiolate. Flowers in terminal corymbiform thyrsi or racemiform or paniculiform inflorescences, seldom axillary or solitary. Sepals free or shortly connate, mostly caducous. Petals mostly 5, caducous, rarely persistent. Stamens 5-many. Ovary superior; style mostly with 3 arms. Fruit mostly globose, usually 3-valved, sometimes irregularly dehiscent. Seeds many, tuberculate or ribbed or smooth, shining, with caruncle.

About 50 species, natives of S. & C. America and S. Africa, with a few species now pantropical; represented by 2 species in Sri Lanka.

KEY TO THE SPECIES

1 Flowering stem cylindrical; branches without lateral basal buds; leaves acute; stamens not more than 15 .... 1. T. paniculatum
1 Flowering stem triangular; branches with 2 lateral basal buds; leaves obtuse, emarginate; stamens more than 15, (up to 40) .... 2. T. triangulare

1. Talinum paniculatum (Jacq.) Gaertn., Fruct. 2: 219. t. 128. 1791; Alston in Trimen, Handb. Fl. Ceylon 6: 19. 1931; A.C. Smith, Bull. Torrey Bot. Club 70: 537. 1943; Geesink in Fl. Males. ser. 1, 7: 124. 1971.

Portulaca paniculatum Jacq., Enum. Pl. Carib. 22. 1760. Type: from Jamaica?
Portulaca patens L., Mant. 242. 1771. Type: from America.
Talinum patens (L.) Willd., Sp. Pl. 2: 863. 1799.

Erect subshrub, to 30 cm. Leaves elliptic to obovate, up to 8.5 × 3.5 cm, acute, entire. Axillary buds with 2 subulate cataphylls. Inflorescence terminal, thyrsi up to 22 × 12 cm, with up to 8 dichasia, each with up to 25 flowers. Bracts and bracteoles subulate. Sepals suborbicular, c. 1.1 mm long, acute. Petals (4–) 5 (–6), obovate, 3.5–1.5 mm, pink, emarginate. Stamens 4 (–15); filaments c. 2.5 mm long. Ovary globose; style c. 1.5 mm long, 3-fid. Fruit globose, c. 3 mm in diam., yellow or pink-yellow, 3-valved. Seeds ovoid with a notch at base, 1.2 mm in diam., black, testa cells radially elongated, tubercled, with small pits between the cells.

Distr. Pantropical weed, native of America; naturalized in Malesia and Sri Lanka.

Ecol. Wayside waste places, rare; fls & frts Dec–May.

Vern. Gasniviti, Rataniviti (S).

Specimens Examined. COLOMBO DISTRICT: Colombo, in 1884, *Ferguson s.n.* (PDA). KANDY DISTRICT: Kandy, by lakeside, 21 March 1928, *Alston 2215* (PDA). RATNAPURA DISTRICT: Road to Rassagala, 18 March 1995, *Weerasooriya & Samarasinghe 176* (K, PDA).

2. Talinum triangulare (Jacq.) Willd., Sp. Pl. 2: 862. 1799; Geesink in Fl. Males. ser. 1. 7: 124. 1971; Rao in Sharma et al., Fl. India 3: 10. 1993.

Portulaca triangularis Jacq., Enum. Pl. Carib. 22. 1760. Type: from the West Indies.
Portulaca racemosa L., Sp. Pl. ed. 2. 640. 1762. Type: from America.

Erect perennial herb, to 80 cm. Leaves elliptic to obovate, up to 15 × 3.5 cm, apex obtuse or rotund, emarginate; margin entire; fleshy, glossy above, pinnately nerved. Axillary buds with 2 small cataphylls, caducous, in dried specimens with a scar. Inflorescence terminal, corymboid, up to 15 cm across, the axes sharply triangular, with 8–24 flowers. Bracts and bracteoles subulate. Sepals deltoid, 4.5 × 3.5 mm, acuminate. Petals 5, obovate, 4–10 × 2–4 mm, emarginate, pink. Stamens 20–35; filaments to 5 mm long; anthers c. 1 mm long. Ovary superior, globose; style c. 3 mm long, 3-fid. Fruits c. 5 mm in diam., 2–3-valved, yellow. Seeds numerous, c. 1.2 mm diam.; testa cells radially elongate, smooth, tuberculate at the edges.

Distr. Native of Tropical America, now a pantropical weed.

Ecol. Weed along roadsides, waste places, forest edges; c. 800 m; fairly common; fls & frts almost throughout the year.

Notes. Commonly used as a vegetable; easily propagated by cuttings.

Specimens Examined. MATALE DISTRICT: Rattota-Illukkumbura Road, 23 Nov 1994, *Wadhwa, Weerasooriya & Samarasinghe 497* (K, PDA). KANDY DISTRICT: Govt. Stock Garden, 17 Jan 1922, *C. Drieberg s.n.* (PDA); Peradeniya, cultivated, 9 Aug 1928, *F.W. de Silva 106* (K, PDA); Peradeniya, Vegetable Garden, 9 Aug 1922, *J.M. Silva 257* (K, PDA). BADULLA DISTRICT: Badulla-Mahiyangana Road, 23 Oct 1994, *Wadhwa, Weerasooriya & Samarasinghe 461* (K, PDA). HAMBANTOTA DISTRICT: Ruhuna National Park, Buttawa Bungalow, 19 March 1970, *Cooray 70031905* (PDA).

RANUNCULACEAE
(by M.D. Dassanayake*)

Juss., Gen. Pl. 231. 1789. Type genus: *Ranunculus* L.

Rhizomatous herbs, sometimes aquatic, woody climbers or rarely low shrubs. Leaves spirally arranged or opposite, simple, lobed or compound; stipules minute or absent. Flowers in cymose inflorescences, rarely solitary, usually bisexual, regular or irregular, hypogynous, with a more or less elongated receptacle. Perianth of 5–8 or more petaloid or sepaloid tepals, rarely a true calyx and corolla, spirally arranged or in whorls, often caducous. Petals often with basal nectaries. Stamens usually many, free, mostly spirally arranged and centripetal. Filaments filiform. Anthers 2-lobed, extrorse or introrse, dehiscing longitudinally. Ovary with mostly several carpels, free and spirally arranged with several to many marginal ovules, or a single basal ovule. Ovules anatropous, sometimes hemitropous, bitegmic or unitegmic. Style well developed. Fruit usually of separate follicles, achenes or berries. Embryo small, endosperm copious.

c. 58 genera and 1750 species, widespread, mostly in northern temperate regions.

5 genera in Sri Lanka.

KEY TO THE GENERA

1 Climbers. Leaves opposite
  2 Leaves trifoliolate, terminal pinna modified into a tendril .... **1. Naravelia**
  2 Leaves simple or if compound pinnae not modified into tendrils .... **2. Clematis**
1 Erect herbs. Leaves spirally arranged
  3 Flowers yellow, with sepals and petals .... **3. Ranunculus**
  3 Flowers not yellow, with perianth of one type of segment
    4 Leaves pinnate. Flowers c. 1 cm across, without an involucre .... **4. Thalictrum**
    4 Leaves lobed or dissected. Flowers c. 4 cm across, with an involucre .... **5. Anemone**

1. NARAVELIA

Adans., Fam. Pl. 2: 460. 1763 ("Naravel") (orth. cons.). Type species: *N. zeylanica* (L.) DC. (*Atragene zeylanica* L.) (typ. cons.).

* Flora of Ceylon Project, Peradeniya.

Atragene L., Sp. Pl. 542. 1753; L., Gen. Pl. ed. 5. 241. 1754. Lectotype species: *A. alpina* L.

Woody climbers. Leaves opposite, trifoliolate, exstipulate, terminal pinna transformed into a generally trifid tendril. Pinnae entire. Flowers many, in axillary and terminal panicles, regular, bisexual. Bracts and bracteoles present. Tepals 4–5, sepaloid, valvate, caducous. Staminodes 10–14, petaloid, linear to club-shaped, elongating after fall of perianth. Stamens many. Filaments short, flattened. Anthers linear; connectives prolonged beyond the anthers. Carpels many. Ovules unitegmic. Achenes with a short stalk, linear, sometimes twisted at the apex, terminated by the elongated feathery style.

7 species, in the Indo-Malesian region and southern China.

Naravelia zeylanica (L.) DC., Syst. Nat. 167. 1817; Trimen, Handb. Fl. Ceylon 1: 2. 1893; Alston, Kandy Fl. 3, fig. 13. 1938.

Atragene zeylanica L., Sp. Pl. 542. 1753. Type: From Sri Lanka, *Hermann Herbarium 226* (BM).

Climbing shrub. Young parts pubescent. Leaves trifoliolate, the central pinna usually transformed into a 8–15 cm long tendril recurved sharply and trifurcate at the end, with rigid, sharp-pointed and hooked branches. Pinnae 6–15 × 2–9 cm, ovate, acute at the apex, more or less truncate, rounded and often unequal at the base, entire or with a few deep cuts in the margin. Lamina glabrous and dark green above, pale and densely pubescent with simple hairs beneath, thin, with 5 primary veins prominent beneath. Petiole 5–8 cm, broadened at the base, with appressed hairs on the upper surface proximally. Petiolules 1–2 cm long, thick, purplish, channelled above. Inflorescence an axillary panicle. Peduncle to 7 cm long, purplish below, silky pubescent. Bracts to c. 1 cm long, narrowly oblanceolate, pubescent. Pedicel 1.5–3 cm long, with 2 opposite bracteoles c. 1 mm long, c. 5 mm above the base. Tepals 5–8 × 3–5 mm, ovate, concave, brownish green, purplish below, velvety, caducous early. Staminodes 12, in 4 groups of 3 each in the bud, 8–12 mm long, subulate, rounded at the apex, green, spreading, later more or less decurved, caducous. Stamens many, erect, 3–5 mm long, pale yellowish green. Filaments 1–2 mm long, as broad as the anther. Anthers 2–3 mm long, with broad connective prolonged c. 1 mm beyond the anther to form a conical projection. Carpels covered with long erect silky hairs. Style 1–2 mm long, flattened and bent outwards at the apex. Stigma simple, on upper surface of reflexed end of style, papillose. Achenes 15–20, linear, 2–3 mm long, subsessile, slightly pubescent, with the style now 4–5 cm long, spirally twisted, feathery, with smooth hairs.

Distr. Also in India, north to the Himalayas.

Ecol. Moist lowlands, rather rare. Flowering September to January. Flowers fragrant.

Vern. Narawel (S).

Specimens Examined. ANURADHAPURA DISTRICT: Medawachchiya, 22 Jan 1972, *Jayasuriya 571* (PDA). BATTICALOA DISTRICT: Kalkudah, 11 Feb 1972, *Jayasuriya 689* (PDA). COLOMBO DISTRICT: Sept 1856, *s. coll. C.P. 1009*, p.p. (PDA). BADULLA DISTRICT: Lunugala, Jan 1888, *s. coll. s.n.* (PDA). GALLE DISTRICT: Galle, Dec 1853, *Gardner in C.P. 1009*, p.p. (PDA); Hinuduma, 20 Feb 1923, *Livera s.n.* (PDA); Bona Vista, 20 Oct 1972, *Balakrishnan 967* (PDA). LOCALITY UNKNOWN: *Alston 16* (PDA).

2. CLEMATIS

L., Sp. Pl. 543. 1753; L., Gen. Pl. ed. 5. 242. 1754. Lectotype species: *Clematis vitalba* L.

Mostly woody climbers, rarely erect shrubs or herbs. Leaves opposite, simply pinnate or ternately compound. Petiole and rachis often twining. Flowers in many- to few-flowered axillary and terminal panicles or solitary, regular, bisexual. Tepals 4–8, valvate, petaloid. Stamens many. Filaments filiform. Anthers short. Carpels many, each with one pendulous untiegmic ovule. Style pilose. Achenes in a head, ovoid, sessile or stalked, with a persistent, usually elongated, plumose or naked style.

c. 230 species, cosmopolitan, chiefly temperate.

KEY TO THE SPECIES

1 Leaves simple or trifoliolate. Flowers more than 3 cm across, with maroon tepals . . . .
. . . . **1. C. smilacifolia**
1 Leaves pinnate with many pinnae. Flowers to 1.5 cm across, with white tepals . . . .
. . . . **2. C. gouriana**

1. Clematis smilacifolia Wall., Asiat. Res. 13: 402. 1820; Trimen, Handb. Fl. Ceylon 1: 1. 1893. Type: From India.

Woody climber, with glabrous branches. Leaves simple or trifoliolate, lamina 6–12 × 3–8.5 cm, ovate, acuminate, subcordate to obtuse at the base, thick, chartaceous, entire or remotely and irregularly serrate, glabrous on both sides, dark green above; veins distinct, with 5–9 primary veins from the base, purple. Petiole prehensile. Flowers in terminal and axillary dichasial panicles. Peduncle 3–5 cm. Bracts 2, linear, 1–2 cm long. Pedicel 1–3.5 cm long. Perianth segments 10–15 × 4–6 mm, oblong, velvety outside, glabrous within, spreading, later reflexed, brownish yellow outside, dark purple within, with 7–10 veins. Stamens 50–60, white, the outer longer. Filaments 3–6 mm long. Anthers 2–3 mm long; connective produced beyond the anthers for 1.5–2 mm. Carpels c. 40, c. 7 mm long. Style filiform. Achene with style 5–7 cm long, pubescent.

Distr. Also in India and Malesia to the Philippines.

Ecol. Lower montane zone, rare. Flowering January.

Vern. Narawel (S) (same as *Naravelia zeylanica*).

Note. No collection since 1858.

Specimens Examined. KANDY DISTRICT: Alagalla, in 1853, and Hantane, in 1858, *s. coll. C.P. 1858* (PDA).

2. Clematis gouriana Roxb. [Hort. Beng. 43. 1814, nom. nud.] ex DC., Syst. Nat. 1: 138. 1817; Trimen, Handb. Fl. Ceylon 1: 2. 1893. Type: from India.

Woody climber, with wide-spreading branches. Young parts pubescent. Leaves pinnate or bipinnate, to c. 16 cm long. Pinnae usually unequal, 3.7–7 × 1–4 cm, ovate, acuminate at apex, obtuse or cordate at base, margin usually serrate. Lamina glabrous above, with 3–5 primary veins. Veins distinct and hairy beneath. Petiole 1.5–6 cm long, petiolule c. 1 cm long, of terminal pinna c. 2 cm. Inflorescence a many-flowered panicle. Peduncle to c. 7.5 cm long. Pedicel 1–2 cm long. Perianth 5–7 × 2–4 mm, greenish white, pubescent. Stamens 30–35. Filaments 3–4 mm long. Connective not produced beyond anther. Carpels 10–15. Style 2–3 mm long. Achene 3–4 mm long, with style to c. 5 cm long, pubescent.

Distr. Also in India, Burma, Malaysia, Java, Philippines.

Ecol. Montane zone, rare. Flowering December to February.

Note. No collection since 1906.

Specimens Examined. NUWARA ELIYA DISTRICT: Hakgala, "jungle above laboratory", March 1906, *Willis s.n.* (PDA): "Nuwara Eliya, Dimboola and Ramboda", *Gardner in C.P. 680* (PDA).

3. RANUNCULUS

L., Sp. Pl. 548. 1753; L., Gen. Pl. ed. 5. 243. 1754. Lectotype species: *R. auricomus* L.

Perennial herbs. Leaves radical or cauline, spirally arranged, entire, lobed or dissected, with sheathing base, exstipulate. Flowers in many-flowered inflorescences or solitary, regular, bisexual. Sepals 5, imbricate, spreading or reflexed at anthesis, caducous. Petals 5–8 or more, spreading, yellow or white, with a basal nectariferous pocket with or without a scale. Stamens many. Carpels many, each with one unitegmic, erect ovule. Achenes many, in a head or spike, usually with glabrous styles.

c. 250 species, cosmopolitan; particularly in temperate and cold countries, and tropical mountains.

KEY TO THE SPECIES

1 Leaves undivided .... **1. R. sagittifolius**
1 Leaves much divided .... **2. R. wallichianus**

1. Ranunculus sagittifolius Hook., Ic. Pl. t. 173. 1873; Trimen, Handb. Fl. Ceylon 1: 4. 1893; Bond, Wild Flowers of Ceylon Hills 2. 1953. Type: Hook., Ic. Pl. t. 173. 1873, from a Sri Lankan specimen.

Perennial herb, to c. 90 cm tall. Rhizome horizontal or ascending, pale, c. 1 cm thick, hairy, clothed with withered leaves and the fibrous remnants of decayed leaves. Roots fleshy, from the lower surface of rhizome, 3–4 mm thick. Leaves mostly radical, sheathing at the base; lamina 3–11 × 2.5–6.5 cm, cordate-oblong, obtuse or acute at apex, strongly sagittate with rounded lobes at the base, thick, crenate at margin, glabrous, dark green and glossy above, paler and pubescent beneath along veins; primary veins 7–9, depressed on upper surface. Petiole to c. 35 cm long, densely long-pubescent. Cauline leaves few, distant, to c. 8 cm long, oblong or linear, entire or incised-dentate, with sheathing base. Inflorescence to c. 90 cm tall, branched. Flowers 2.2–3 cm across. Pedicel to c. 5 cm at anthesis, elongating later to c. 7 cm, terete. Sepals 4–6 × 3–4 mm, broadly ovate, green, yellow at edges, glabrous, deeply boat-shaped, caducous. Petals 6–15 × 5–10 mm, orbicular to obovate, spreading, overlapping forming a cup, bright yellow and glossy, greenish yellow at base, glabrous, each petal with a nectary pit within near base, covered by a scale. Filaments c. 3 mm long, yellow. Anthers c. 1 mm long, yellow. Carpels in a globose head, smooth, green, each keeled at the back, with a short yellow beak. Achenes smooth, glabrous, 1.5–2.5 mm long, in a more or less globose head on an axis c. 5 mm long; beak c. 1 mm long.

Distr. Endemic.

Ecol. Upper montane zone. In wet or marshy places, among grass. Common. Flowering April to October.

Specimens Examined. NUWARA ELIYA DISTRICT: Ambewela, 26 March 1906, *Willis s.n.* (PDA); Mahagastota, 3 June 1949, *Niles s.n.* (PDA); Nuwara Eliya, June 1880, *s. coll. s.n.* (PDA), 15 June 1922, *J.M. Silva s.n.* (PDA); Horton Plains, 29 Jan 1974, *Sumithraarachchi 849* (PDA), 28 April 1970, *Cramer 2951* (PDA), 11 March 1969, *Robyns 6963* (PDA), 27 March 1968, *Fosberg 49987* (PDA), Sept 1890, *s. coll. s.n.* (PDA). 26 Jan 1906, *Willis s.n.* (PDA), 27 April 1932, *Simpson 9522* (PDA), 22 April 1968, *Mueller-Dombois 68042201* (PDA), 17 May 1968, *Cooray 68051722*

(PDA), 9 July 1967, *Mueller-Dombois et al., 67070902* (PDA), 13 Sept 1967, *Mueller-Dombois et al. 67091322* (PDA). BADULLA DISTRICT: Pattipola, 10 March 1969, *Hoogland 11526* (PDA).

2. Ranunculus wallichianus Wight & Arn., Prod. 1: 4. 1834; Trimen, Handb. Fl. Ceylon 1: 4. 1893; Bond, Wild Flowers of Ceylon Hills 4. 1953. Type: From India.

Perennial stoloniferous herb to c. 25 cm tall. Stem to c. 2 cm long. Stolons c. 3 mm thick, pubescent, rooting at nodes. Leaves radical; lamina 1–8 × 1–9 cm, ternatisect or trifoliolate; segments tripartite, deeply cut or toothed at ends, long silky.pubescent on both surfaces, pale green. Petiole to c. 40 cm long, green, purplish at base, terete or channelled above, pubescent, basal 1–5 cm with a membranous sheath enclosing younger leaves and shoot apex. Flowers solitary, 0.5–1 cm across. Peduncle pubescent, to c. 5 cm long at anthesis. Sepals 2–3 mm long, with long white hairs, yellow-green, reflexed. Petals 4–6 × 2–3 mm, broadly ovate, greenish yellow outside, pale yellow within, lower part greenish. Filaments c. 2 mm long. Anthers c. 1 mm long, yellow. Achenes c. 10, each 3–4 mm long, glabrous, compressed, sides tuberculate or smooth, with a hooked beak 1–2 mm long.

Distr. Also Nilgiri Hills in South India.

Ecol. Upper montane zone, common. Grows gregariously in damp shady places with some exposure, such as edges of jungle paths. Flowering April to October.

Specimens Examined. NUWARA ELIYA DISTRICT: Horton Plains, April 1856, *s. coll. C.P. 1012*, p.p. (PDA), 20 May 1911, *J.M. Silva s.n.* (PDA). 28 March 1968, *Fosberg et al. 50025, 50082* (PDA), 29 March 1968, *Fosberg 50113* (PDA), 24 April 1970, *Gould 13560* (PDA), 11 May 1970, *Gould et al. 13841* (PDA), 6 Dec 1972, *Trivengadum et al., 98* (PDA); Nuwara Eliya, 17 Sept 1969, *Van Beusekom 1421* (PDA), 27 Feb 1970, *Cramer 2863* (PDA); Pidurutalagala, 20 March 1971, *Robyns 7303* (PDA), BADULLA DISTRICT: Haputale, *s. coll. C.P. 1012*, p.p. (PDA); Bandarawela, Craig Estate, 25 March 1906, *Willis s.n.* (PDA).

4. THALICTRUM

L., Sp. Pl. 545. 1753; L., Gen. Pl. ed. 5. 242. 1754. Lectotype species: *T. aquilegifolium* L.

Perennial herbs. Leaves spirally arranged, compound, with sheathing bases. Flowers in terminal and subterminal racemes or panicles, small, regular, bisexual or some male only. Tepals 3–5, imbricate in the bud, green or petaloid, caducous. Stamens 3-many, often with conspicuous linear anthers. Carpels 1-many, each with a solitary, pendulous, mostly bitegmic, ovule.

Style short. Achenes sessile or stipitate, in a head, sometimes with the style persistent and forming a beak.

c. 85 species, in northern temperate regions, tropical South America, tropical and southern Africa.

Thalictrum javanicum Blume, Bijdr. 1: 2. 1825; Trimen, Handb. Fl. Ceylon 1: 3. 1893. Type: from Java.

Herb 40–120 cm tall. Stem erect, c. 5 mm thick, smooth, striate, green, often zig-zag at the nodes, clothed at the base with persistent sheaths of former leaves. Roots fibrous. Leaves mostly tri- or quadri-pinnate. Basal leaves including petioles c. 18 cm long. Cauline leaves to c. 55 cm long including petiole, the upper cauline leaves progressively smaller. Petiole to c. 13 cm long, striate, at base with amplexicaul membranous sheath. Pinnules mostly 3 or 5, the terminal largest. Petiolules filiform, sparsely pubescent, of terminal pinnule 5–11 mm long, of lateral pinnules 2–4 mm. Terminal pinnule 10–20 × 10–20 mm, lateral pinnules 6–12 × 6–12 mm. Lamina oblate, orbicular or very broadly ovate, apiculate, rounded to cordate or truncate at the base, lobed with 3–7 lobes, each with a marginal tooth at the middle, thin, dark green above, glaucous and sparsely pubescent beneath; veins clearly marked beneath, the midrib and secondary veins ending in marginal teeth (hydathodes). Inflorescence an axillary or terminal, lax, many-flowered panicle. Peduncle 2–6 cm long. Bracts 2–3 mm long, filiform. Pedicel 4–10 mm long. Tepals 4 (5), caducous, 3–5 × 1–2 mm, ovate, white. Stamens many, erect. Filaments 3–4 mm long, white, filiform, swollen at the end. Anthers to c. 0.5 mm long, yellow-brown, narrow, linear. Carpels (4–) 5–6 (–7), green, ribbed. Stigma c. 1 mm long, conical, papillose on the two sides, often bent. Achenes 4–5 mm long, strongly ribbed, with a hooked beak 2–3 mm long.

Distr. Also in Central India, Himalayas, Malesia.

Ecol. Upper montane zone, rare. Flowering January to April.

Specimens Examined. NUWARA ELIYA DISTRICT: Hills around Nuwara Eliya, *Van Beusekom 1431* (PDA); Nuwara Eliya and Dimbula, April 1852 and Feb 1857, *s. coll. C.P. 2574*, p.p. (PDA); Hakgala Gardens, Fernery, 12 Aug 1929, *F.W. de Silva s.n.* (PDA). LOCALITY UNKNOWN: *s. coll. C.P. 2574*, p.p. (PDA); in 1836, *s. coll. s.n.*, Herb. Wight (PDA).

5. ANEMONE

L., Sp. Pl. 538. 1753; L., Gen. Pl. ed. 5. 241. 1754. Lectotype species: *A. coronaria* L.

Perennial rhizomatous herbs. Leaves radical, more or less lobed or dissected. Flowers large, solitary or in a simple or branched cymose inflorescence, with an involucre or 3 green leaves. Tepals mostly 5–8, imbricate, petaloid. Stamens many, the outer sometimes petaloid. Carpels many, each with a solitary pendulous unitegmic ovule. Achenes clustered in a head, with bent or straight, naked or bearded, styles.

c. 120 species, cosmopolitan, mostly in the northern temperate regions.

Anemone rivularis Buch.-Ham. in DC., Syst. Nat. 1: 211. 1817, non Wall. 1831; Trimen, Handb. Fl. Ceylon 1: 3. 1893. Type: From India.

Herb to c. 90 cm tall. Rhizome woody, to c. 1 cm thick, clothed at the base with the fibrous remains of former leaves. Whole plant more or less pubescent. Lamina of radical leaves 2.5–11 × 3.5–15 cm, tripartite, almost trifoliolate, segments very deeply dissected, serrate. Petiole to c. 35 cm long. Inflorescence trichotomous, with erect branches. Involucral leaves 2, opposite, c. 14 cm long, with linear segments to c. 1.5 cm broad. Bracts smaller, opposite, below branches of inflorescence. Peduncle to c. 40 cm long. Flowers c. 4 cm across. Pedicel 4–17 cm long. Tepals (5–)7 (–8), 12–18 × 4–8 mm, slightly silky outside, white or bluish. Stamens c. 5 mm long; filaments c. 4 mm long. Achenes 5–7 mm long, smooth, beaked with the curved style.

D i s t r. Also in the higher mountains of the Indian peninsula and the Himalayas.

E c o l. Upper montane zone, rare. Flowering in April.

N o t e. No recent collections.

S p e c i m e n s   E x a m i n e d. NUWARA ELIYA DISTRICT: Nuwara Eliya, April 1856, s. coll. *C.P. 1011*, p.p. (PDA); Ambewela, March 1922, A. de Alwis s.n. (PDA). LOCALITY UNKNOWN: s. coll. *C.P. 1011*, p.p. (PDA).

RHAMNACEAE

(by B.M. Wadhwa*)

Juss., Gen. Pl. 376. 1789 ("Rhamni"). Type genus: *Rhamnus* L.

Trees or shrubs, erect or climbing, unarmed or armed with stipular spines or ramal thorns, sometimes bearing tendrils. Leaves simple, alternate, opposite or subopposite, coriaceous, 3–5-nerved from the base or penninerved; stipules usually present, small, deciduous or modified into spines. Inflorescence usually in axillary cymes, rarely in clusters or umbels, sometimes in panicles or spikes. Flowers small, regular, bisexual or polygamous-dioecious. Calyx lobes 5 or 4, triangular-ovate, valvate, mostly keeled within; calyx-tube campanulate or turbinate, ± united with the ovary or the disk, mostly persistent. Petals 5 or 4, sometimes 0, obovate or spathulate, cucullate, often clawed at base, inserted on the throat of calyx tube and usually shorter than the calyx lobes. Disk thick, fleshy, filling the calyx tube or thin, membranous and lining it, glabrous or tomentose, entire or crenulate, rarely pitted. Stamens 5 or 4 and opposite to the petals, often embraced by petals; filaments filiform, rarely flat; anthers small, 2-celled, dehiscing longitudinally. Ovary sessile, free or immersed in the disk, (1)2–3(4)-loculed; ovule usually 1 per cell, anatropous; style short, simple or shortly divided into as many lobes as locules; stigma usually capitate. Fruit superior or inferior, a 3-celled capsule or a drupe, dry or fleshy, with one or more pyrenes, sometimes winged. Seeds solitary in each cell, ovoid, compressed with copious endosperm and large, straight embryo.

51 genera and 900 species widely represented, but mostly in the tropical and subtropical regions of the world. 7 genera and 14 species in Sri Lanka.

KEY TO THE GENERA

1 Branches often tendrilled; fruit inferior; capsule of 3-winged, indehiscent mericarps crowned with the persistent calyx .... **1. Gouania**
1 Branches without tendrils; fruit superior or partly enclosed in the calyx-tube, either exalate or with single apical wing of enlarged style
  2 Branchlets with recurved, sharp stipular spines; leaves with 3–5-nerves from the base ....
  .... **2. Ziziphus**
  2 Branchlets without stipular spines; straight or recurved ramal thorns may be present

* Royal Botanic Gardens, Kew.

3 Disk thin, lining the calyx tube, flowers pedicelled .... **3. Rhamnus**
3 Disk thick, fleshy, filling the calyx-tube; flowers pedicelled or sessile
  4 Flowers distinctly pedicelled; polygamous or bisexual
    5 Flowers polygamous, in axillary, many-flowered cymes; fruit a drupe, eventually splitting into 3 cocci .... **4. Colubrina**
    5 Flowers bisexual; fruit samaroid, 1-seeded indehiscent nut, with single apical wing of enlarged style .... **5. Ventilago**
  4 Flowers sessile, bisexual
    6 Thorns recurved; flowers in axillary fascicles or condensed umbellate clusters ......
      .... **6. Scutia**
    6 Thorns straight or absent; flowers in terminal or axillary panicles ...... **7. Sageretia**

1. GOUANIA

Jacq., Select. 263. 1763; DC., Prod. 2: 40. 1825; Benth. & Hook. f., Gen. Pl. 1: 385. 1862. Type species: *G. glabra* Jacq.

Unarmed climbing shrubs or lianes; branches often circinnately tendrillar. Leaves alternate, petiolate, ovate or ovate-oblong, crenate-serrate, 3–5-nerved; stipules free, deciduous. Inflorescences terminal or axillary spikes or racemes, sometimes in panicles, rhachis often cirrhose. Flowers polygamous, epigynous, 5-merous. Calyx subinfundibuliform; lobes 5, ovate-triangular, valvate, glabrous or pubescent. Petals 5, ovate-oblong, cucullate, shortly clawed at base, inserted below the margin of the disk. Stamens 5, enclosed in petals; anthers oval, dehiscing longitudinally. Disk massive, filling the calyx-tube, glabrous or hairy, 5-lobed; lobes alternating with the stamens or opposite to calyx-lobes. Ovary connate to bottom of calyx-tube, 3-celled, immersed in the disk, usually urceolate; style 3-cleft; stigma papillose. Schizocarpic capsules inferior, trigonous, 3-winged, coriaceous, crowned by the persistent calyx-lobes, splitting into 3 cocci through the wings, leaving a slender axis. Seeds 3, obovate, plano-convex; testa shining, horny; albumen thin, embryo erect.

About 70 species; 35 in America, 14 in Asia, 15 in Madagascar and other islands of the Indian Ocean, 2 in Australia, 2 in Africa and one in Sri Lanka.

Gouania microcarpa DC., Prod. 2: 40. 1825; Thw., Enum. Pl. Zeyl. 75. 1858; Lawson in Hook. f., Fl. Br. Ind. 1: 643. 1875; Trimen, Handb. Fl. Ceylon 1: 286. 1893; Bhand. & Bhans. in Fasc. Fl. India 20: 40. f. 7. 1990. Type: India, *Wallich* n. *4271* (K; BM, isotype).

Shrubs, climbing by axillary or extra-axillary tendrils; branchlets and tendrils with stripes of antrorse, dense, brownish hairs. Leaves elliptic-ovate or orbicular, 2.8–9.2 × 1.2–5.5 cm, acute at apex, rounded at base, distantly and minutely serrate along margin with teeth tipped by a deciduous glandular mucro, glabrous, secondary reticulation extremely fine, main veins with fine brown hairs on ventral side; petioles 0.5–1.6 cm long, brown-villous; stipules subulate, 2.5–6 × 0.5–1.5 mm, pubescent. Flowers in racemes of 4–8-flowered clusters, white, 4–5 mm in diam., polygamous, male sessile, females on short woolly pedicels; bracteoles conspicuous, lanceolate, 2–3 mm long. Calyx lobes deltoid, 1.5–2.0 × 1.0–1.5 mm, glabrous within, densely pubescent outside, keeled within to middle. Petals spathulate, 1.5 × 2.0 mm. Disk orbicular, villous; lobes rotund, emarginate. Stamens 1 mm long; filament filiform. Style 3-cleft to 2/3 of its length, arms hairy. Capsule trigonous, 3-winged, 8–15 × 7–12 mm, at first pubescent, eventually glabrescent, shining, green; wings parchment-like, crowned by persistent calyx. Seeds ovate, 5 × 4 mm, cordate at base.

Distr. India, Burma, Malaysia, Indonesia and Sri Lanka.

Ecol. Submontane forests, along water channels, up to 1400 m; fls & frts June–Dec.

Notes. On the specimen *C.P. 1238* in PDA, two localities—Hantane and "Hunguran", which may be Hanguranketa, July 1851—*Wight* are given in pencil. Plate No. 440 (habit & fruit only) in PDA belongs to this species.

Specimens Examined. KANDY DISTRICT: East of Hunnasgiriya, near road mark 24/16, 600 m, 8 Aug 1975, *Jayasuriya & Austin 2260* (K, PDA); Kandy-Mahiyangana road, 30/10 mark, 2 Aug 1975, *Sumithraarachchi & D.F. & S.K. Austin 972* (PDA); Hillside, 2 miles east of Hunnasgiriya, 1,000 m, 15 April 1973, *Stone 11130* (PDA). BADULLA DISTRICT: Namunukula, ± 1,400 m, 22 Sept 1976, *Cramer 4725* (K, PDA); Moratota, 81°01′E, 6°56′ N, 1,050 m, 15 Dec 1972, *Jayasuriya & Tirvengadum 1002* (PDA); Along Badulla road, close to waterfalls, 1,100 m, 15 Dec 1972, *Tirvengadum & Jayasuriya 150* (PDA); Near Lunugala, Jan 1888, *s. coll s.n.* (PDA). RATNAPURA DISTRICT: Warangula, Adam's Peak, on way to Sitagangula, ± 1,000 m, 8 Dec 1977, *Cramer 5017* (K). LOCALITY UNKNOWN: *s. coll. C.P. 1238* (BM, PDA, CAL); in Dec, *Macrae 761* (BM); *Col. Walker s.n.* (K, PDA).

2. ZIZIPHUS

Mill., Gard. Dict. ed. 4, 1754; Benth. & Hook. f., Gen. Pl. 1: 373. 1862. Lectotype species: *Z. jujuba* Mill., Gard. Dict. ed. 8, 1768. (=*Rhamnus ziziphus* L., 1753).

Trees or shrubs, rarely decumbent, mostly armed with single or paired stipular spines, one usually long and straight, the other short, recurved. Leaves alternate or opposite, entire or crenate-serrate, coriaceous, 3–5-nerved, usually leathery and oblique at base, petiolate. Inflorescences axillary or terminal cymes. Flowers 5-merous, usually bisexual, rarely polygamous, pedicellate.

Calyx 5-fid or saucer-shaped; lobes keeled. Petals 5, sometimes 0, cucullate, incurved or deflexed. Stamens included or excluded, inserted below the disk. Disk pitted or flat, 5 or 10-lobed, rarely entire, mostly lining the calyx-tube. Ovary 2–4-loculed, immersed in the disk and adnate to its base; style 2–(3)–4 partite, free; stigma small, papillose. Fruit drupaceous, fleshy and succulent, globose or oblong; pyrenes 1–4, rugose or tuberculate, 1-seeded in each stone. Seeds plano-convex, smooth with scanty endosperm.

About 135 species in temperate and tropical parts of the world, chiefly in Asia, tropical America, Africa, Mediterranean region, Australia and Indo-Malaya; 6 species in Sri Lanka.

KEY TO THE SPECIES

1 Flowers apetalous, in pedunculate cymes forming long terminal panicles ..... **1. Z. rugosa**
1 Flowers petalous, in pedunculate or sessile cymes or cymose fascicles
  2 Leaves glabrous on both surfaces, with sparse to dense hairs along nerves on the lower surface
    3 Leaves ovate-lanceolate, densely hairy on nerves beneath .... **2. Z. lucida**
    3 Leaves elliptic-oblong, sparsely hairy on nerves beneath .... **3. Z. napeca**
  2 Leaves usually glabrous above, hairy to silky tomentose beneath
    4 Flowers in distinctly pedunculate cymes; peduncles up to 1.5 cm long ....
    .... **4. Z. xylopyrus**
    4 Flowers in sessile or shortly pedunculate cymes or cymose fascicles
      5 Leaves obliquely ovate-lanceolate, minutely denticulate, base oblique. ....
      .... **5. Z. oenoplia**
      5 Leaves ovate-oblong or semi-orbicular, serrulate, base ± rounded. ....
      .... **6. Z. mauritiana**

1. Ziziphus rugosa Lam., Enc. 3: 319. 1789; Wight & Arn., Prod. 162. 1834; Wight, Ic. Pl. Ind. Or. 2: t. 339. 1840–1843; Thw., Enum. Pl. Zeyl. 73. 1858; Lawson in Hook. f., Fl. Br. Ind. 1: 636. 1875; Trimen, Handb. Fl. Ceylon 1: 282. 1893; Gamble, Fl. Pres. Madras 1: 221. 1918; Bhand. & Bhans. in Fasc. Fl. India 20:108. 1990. Type: from India (E. India).

Small trees or straggling, evergreen armed shrubs, 3–6 m high; young branches fulvous tomentose; spines usually solitary, recurved, tomentose except the tip. Leaves broadly elliptic or orbicular-rotund, 5–10.5 × 4–6 cm, apex rotund, mucronate, base oblique, glabrous above, fulvous-tomentose beneath, margin serrate, teeth with callous points, basally 3-nerved; petioles 0.5–1.0 cm long, tomentose. Inflorescence of pedunculate cymes forming terminal panicles, up to 30 cm long, tomentose; peduncles to 6 cm long, densely tomentose. Flowers yellowish-green, 4–5 mm across; pedicels 4–6 mm long, densely tomentose. Calyx lobes 5, ovate, c. 2 mm long, acute, pubescent without. Petals 0. Stamens 1–2 mm long; anther lobes broadly ovate. Disk 5-lobed, glabrous. Ovary ovoid, 2-celled, tomentose; style 2-cleft from below the middle, curved; stigma 2-fid. Drupes pyriform to globose, 6–12 × 9–10 mm, fleshy, white when ripe. Seeds 2 (–1), obovoid, compressed, black.

Distr. Pakistan, India, Burma and Sri Lanka.

Ecol. In forests in moist low country, c. 500 m; fls, & frts Dec–Apr.

Vern. Maha-eraminiya (S); Churai (T).

Notes. Bhandari & Bhansali (l.c.) have made 3 varieties on the basis of leaves glabrous or fulvous—tomentose beneath. Since these variations can be seen on the same plant, their views are not accepted here.

On the specimens *C.P. 2719* in PDA, 3 localities, viz. Hantane, Ratnapura and Ambagamuwa, are annotated in pencil.

Specimens Examined. KANDY DISTRICT: Kandy, in 1919, *Moon s.n.* (BM), Udawattakela Sanctuary, 500 m, 22 Jan 1975, *Jayasuriya 1881* (K, PDA). KALUTARA DISTRICT: Kalutara, *s. coll. s.n.* (K); Jeep track to Morapitiya, 11 Jan 1988, *Jayasuriya 4259* (PDA). RATNAPURA DISTRICT: Parakaduwa, Ratnapura, *Jayasuriya 2692* (PDA); Sinharaja forest, ascent along Halmandiyadola, 12 Jan 1988, *Jayasuriya et al. 4330* (PDA). LOCALITY UNKNOWN: *Thwaites C.P. 2719* (K, BM, PDA), *Col. Walker 331* (K).

2. Ziziphus lucida Moon [Cat. 17.1824, nom. nud.] ex Thw., Enum. Pl. Zeyl. 74. 1858; Lawson in Hook. f., Fl. Br. Ind. 1: 635.1875; Suesseng. in Pflanzenfam. ed. 2, 20d: 127. 1953. Type: Sri Lanka, *C.P. 1241* (PDA).

Ziziphus linnaei Lawson in Hook. f., Fl. Br. Ind. 1: 635. 1875. Type: Sri Lanka, *Col. Walker s.n.* (K).

Ziziphus napeca var. *lucida* (Thw.) Trimen, Handb. Fl. Ceylon 1: 281. 1893.

Erect or climbing armed shrubs; branches round, glabrous; young branches brownish-tomentose, more markedly at nodes; spines solitary, very short, recurved. Leaves alternate, ovate-lanceolate, 4.0–7.0 × 1.8–2.8 cm, apex acute-subacuminate, base unequal, glabrous above, hairy on veins beneath, margin crenate-serrate with a minute bristle on each tooth, basally 3-nerved, secondary veins from the outer 2 primary veins diverging towards margin at 30°–40° angle; petioles 4–7 mm long, brown pubescent. Inflorescence in pedunculate, axillary lax cymes, 3–4 cm long, brownish tomentose; peduncles c. 5 mm long, densely tomentose. Flowers 3–5 mm across, 5-merous; pedicels c. 2–3 mm long, tomentose. Calyx lobes ovate, 1.5–2 mm long, acute, tomentose without, keeled within. Petals 5, obovoid, 1–1.5 mm long, cucullate, shortly clawed. Stamens 5, filaments 1–1.5 mm long. Disk faintly 5-angled, glabrous, not pitted. Ovary 3–4-locular; style 3-cleft from below the middle. Drupes globose, 8–10 mm diam., fleshy, smooth, black when ripe; stone very hard, rugulose.

Distr. Endemic.

Ecol. In drier regions, at lower altitude; fls July–Nov; frts Mar.

Vern. Eraminiya (S).

Notes. Trimen (l.c.) has mentioned Matale (Moon's material) and Kurunegala, as two localities. On the specimen *C.P. 1241* kept in PDA, Kurunegala, July 1854, *Gardner* and Kalutara, *Moon* are given in pencil.

A plate bearing No. 432 in PDA and based on the specimen *C.P. 1241* (habit & fruit) belongs to this species.

Specimens Examined. KURUNEGALA DISTRICT: Hettipola, 27 March 1970, *Amaratunga 2042* (PDA); Erankeliya Eraniya forest, 17 Nov 1992, *Wadhwa 304* (PDA). LOCALITY UNKNOWN: Herb. *Linn. 262. 30* (BM); *Col. Walker s.n.* (K, PDA), *Macrae 603* (BM), *s. coll. C.P. 1241* (BM, CAL, K, isotypes).

3. Ziziphus napeca (L.) Willd., Sp. Pl. 1 (2): 1104. 1798; Lawson in Hook. f., Fl. Br. Ind. 1: 635. 1875; Trimen, Handb. Fl. Ceylon 1: 281. 1893, excl. var. *lucida*; Suesseng. in Pflanzenfam. ed. 2, 20d: 127. 1953.

Rhamus napeca L., Sp. Pl. 194. 1753. Type: Sri Lanka, Herb. *Hermann* 3: 43. n. *87* (BM).

Profusely armed, straggling shrub; young branches, inflorescence and petioles rusty hispid; spines solitary or in pairs, recurved, 3–4 mm long, stout, hairy. Leaves elliptic-oblong, 2.8–8.5 × 1.5–4.6 cm, apex shortly acuminate, base obliquely subrotund, finely crenate-serrate, glabrous on both surfaces except sparse hairs on main nerves, basally 3-nerved; petioles 3–4 mm long, hairy. Flowers in axillary, pedunculate cymes; peduncles c. 8 mm long, brownish hairy. Flowers 4–5 mm across, 5-merous; pedicels 1.5–2 mm long, tomentose. Calyx lobes triangular, c. 2 mm long, acute, hairy without. Petals 1.5 mm long, cucullate, shortly clawed. Stamens as long as the petals. Disk 5-lobed, not pitted. Ovary 2-celled; styles 2, as long as the ovary, recurved. Fruit not seen.

Dist. Endemic.

Ecol. In moist and dry forests, at lower elevations; rare.

Vern. Yakeraminiya (S).

Notes. The species should be looked for carefully in future plant survey trips and conserved in Botanical Gardens.

Specimens Examined. PUTTALAM DISTRICT: Puttalam-Kurunegala road, Anamaduwa Teak Plantation, 10 Km mark, 16 Nov 1992, *Wadhwa 295* (PDA). ANURADHAPURA DISTRICT: Konwewa forest, Anuradhapura, 16 Nov 1992, *Wadhwa 289* (PDA). KURUNEGALA DISTRICT: Erankeliya Eraniya forest, 17 Nov 1992, *Wadhwa 327* (PDA). KEGALLE

DISTRICT: Bible rock, ± 650 m, 19 March 1974, *Sumithraarachchi & Fernando 157* (PDA). RATNAPURA DISTRICT: Sinharaja forest, Tundole ela, 25 Nov 1992, *Wadhwa & Weerasooriya 337* (PDA). LOCALITY UN-KNOWN: Herb. *Hermann* 3: 43. n. *87* (Type, BM).

4. Ziziphus xylopyus (Retz.) Willd., Sp. Pl. 1(2): 1104. 1798; Roxb., Fl. Ind. 1: 611. 1832; Wight & Arn., Prod. 162. 1834; Thw., Enum. Pl. Zeyl. 74. 1858; Lawson in Hook. f., Fl. Br. Ind. 1: 634. 1875; Trimen, Handb. Fl. Ceylon 1: 282. 1893; Gamble, Fl. Pres. Madras 1: 220. 1918; Worthington, Ceylon Trees 137. 1959; Bhand. & Bhans. in Fasc. Fl. India 20: 112. f. 26. 1990.

Rhamnus xylopyrus Retz., Obs. 2: 11. 1781. Type: from India.

Small trees or large straggling shrubs, 7.5–12 m high; young branches rusty tomentose; spines in pairs, one straight and the other curved; internodes up to 3.5 cm long, swollen at the leaf scars. Leaves broadly elliptic or ovate-oblong, base obliquely subcordate, apex obtuse with mucro, glabrous above, hairy beneath, serrate, each tooth with a deciduous glandular mucro, basally 3–4 nerved, primary nerves prominent, secondary nerves diverging towards margin at 40°–50° angles; petioles c. 1.2 cm long, fulvous tomentose. Inflorescences dense axillary, dichotomous cymes; peduncles up to 1.5 cm long. Flowers 4–6 mm across, yellowish-green, densely pubescent; pedicels 3–4 mm long, tomentose. Calyx-lobes 5, triangular, 2.0–2.5 mm long, keeled up to the middle, pubescent without, glabrous within. Petals 5, 1.5–2 mm long, spathulate, yellowish-green. Stamens equal to petals. Disk 10-lobed, rarely 5-lobed, pitted, glabrous. Ovary 3-celled, globose, almost hidden by the disk, ovule solitary; style 3-cleft up to 1/2 or 2/3 of length. Drupes globose, 1.6–2.5 cm in diam., woody, slightly rugose; stone hard. Seeds 3, oblong, compressed, black.

Distr. India, Nepal and Sri Lanka.

Ecol. In drier regions, mostly at lower altitude, common; fls Apr–Aug, frts May–Jan.

Vern. Kakuru (S); Nariilantai (T).

Specimens Examined. JAFFNA DISTRICT: Near Elephant Pass, 14 Sept 1897, *s. coll. s.n.* (PDA). VAVUNIYA DISTRICT: Without exact locality, July 1890, *A.H. Brown s.n.* (PDA). KURUNEGALA DISTRICT: Kurunegala, *Gardner C.P. 1243* (BM, K, PDA). POLONNARUWA DISTRICT: Polonnaruwa Sacred area, S. of Pabulu Vehera, 61 m, 26 Aug 1971, *Ripley 449* (PDA), section 4c, 61 m, 26 May 1969, *Ripley 136* (PDA), Polonnaruwa Sacred area, 61 m, 31 Aug 1971, *Dittus 71083103* (K), Polonnaruwa Sacred area, east of Siva Devale 2, 26 Aug 1971, *Ripley 450* (PDA); base of Gunners quoin, alt. low, Jan 1980, *Kostermans 28011* (K), 3 Jan 1973, *Tirvengadum, Cramer & Balasubramaniam 236* (PDA);

Manampitiya, 3 Aug 1959, *Worthington 538* (BM). MATALE DISTRICT: Dambulla, 18 Jan 1945, *Worthington 1669* A (K). BATTICALOA DISTRICT: Bibile-Batticaloa road, 50/71 mark, 150 m, 17 June 1953, *Worthington 6311* (K). BADULLA DISTRICT: Bibile, 43.5 mile, 175 m, 22 Feb 1951, *Worthington 5118* (BM); between Bibile and Nilgala, Jan 1888, *s. coll. s.n.* (PDA); Bulupitiya-Bibile, 210 m, 23 April 1950, *Worthington 4667* (BM); Pethiyagoda, 25 Sept 1976, *Waas 866* (K, PDA), between Uraniya and Ekiriyankumbure, 81° 09′ E, 7° 13′ N, alt. low, 28 Oct 1971, *Jayasuriya 379* (K, PDA); Arawakumbura, S. of Bibile, 6 March 1987, *Jayasuriya & Adachi 3806* (PDA); Gurukumbura, near Dambana, 150–180 m, 17 Oct 1977, *Huber 466* (PDA). RATNAPURA DISTRICT: Balangoda Kallote, c. 10 mile point, 480 m, 28 Oct 1947, *Worthington 3304* (K). MONERAGALA DISTRICT: Serawa, between Pitakumbura and Bulupitiya, alt. low, 1 May 1975, *Jayasuriya 1945* (K, PDA); Sellakaoya Sanctuary, Galoya National Park, alt. low, 5 May 1975, *Jayasuriya 2076* (K, PDA).

5. Ziziphus oenoplia (L.) Mill., Gard. Dict. ed. 8, 1768; Roxb., Fl. Ind. 1: 611. 1832; Wight & Arn., Prod. 163. 1834; Thw., Enum. Pl. Zeyl. 74. 1858; Lawson in Hook. f., Fl. Br. Ind. 1: 634. 1875; Trimen, Handb. Fl. Ceylon 1: 280. 1893; Gamble, Fl. Pres. Madras 2: 220. 1918; Bhand. & Bhans. in Fasc. Fl. India 20:103. 1990.

Rhamnus oenoplia L., Sp. Pl. 194. 1753. Type: Sri Lanka, Herb. Linn. 262. 33.

Straggling or climbing shrubs, 3–5 m high; branchlets strigose, densely rusty tomentose; spines solitary or in pairs, usually recurved, 3.5–6 mm long, rusty tomentose at base. Leaves obliquely ovate-lanceolate, 2.4–6.5 × 1.5–3 cm, glabrous or pubescent above, golden-brownish appressed-hairy below, base obliquely subacute, apex acute—acuminate, margin minutely denticulate, basally 3–4 nerved; petioles 3–6 mm long, brown tomentose. Inflorescences axillary, shortly pedunculate cymes, 20–25-flowered; peduncles c. 3 mm long, brown-tomentose. Calyx lobes triangular or ovate, 1.5–2 mm long, glabrous within, brownish tomentose without. Petals obovate, c. 1 mm long, emarginate, green. Stamens as long as the petals. Disk 10-lobed, grooved, glabrous; style 2-fid from above the middle. Drupes globose or ovoid, 3–5 mm in diam., 1–2-celled, shining and black when ripe, with acrid pulp. Seeds 1–2, ovoid.

Distr. India, Pakistan, Malaysia, Australia and Sri Lanka.

Ecol. In forests of dry region, common; fls Oct–Mar, frts Feb–Apr.

Vern. Hineraminiya (S); Perilantai, Churai (T).

Notes. The treatment of Bhandari & Bhansali (l.c.) as 4 varieties has not been accepted in this work.

Specimens Examined. PUTTALAM DISTRICT: Marai villu, 30 Dec 1968, *Fosberg, Mueller-Dombois, Wirawan, Cooray & Balakrishnan 50875* (PDA); Chilaw, in 1880, *Trimen s.n.* (PDA); Palavi, airport, Puttalam-Kalpitiya road, abundant, 15 Nov 1992, *Wadhwa 281* (PDA); Puttalam-Kurunegala road, Anamaduwa Teak Plantation, 10 km mark, 16 Nov 1992, *Wadhwa 295* (PDA). ANURADHA-PURA DISTRICT: Ritigala, 60 m, 18 March 1973, *Bernardi 14314* (PDA); Ritigala nature Reserve forest, 14 Nov 1992, *Wadhwa 260* (PDA); Anuradhapura, March 1883, *s. coll. s.n.* (PDA); Anuradhapura-Puttalam road, 15 Nov 1992, *Wadhwa 262* (PDA); Konwewa forest, Anuradhapura, 16 Nov 1992, *Wadhwa 294* (PDA). TRINCOMALEE DISTRICT: Trincomalee, Apr 1854, *Gardner C.P. 1240* (BM, PDA). KURUNEGALA DISTRICT: Ibbagamuwa, 2 Feb 1968, *Amaratunga 1563* (PDA); Batalagoda tank, 5 Feb 1973, *Townsend 73/37* (K, PDA); Sundapola forest, *Chatterjee 459* (CAL). MATALE DISTRICT: Matale-Dambulla road, before mile mark 44, 215 m, 12 Jan 1968, *Comanor 719* (K); Dambulla, 18 Jan 1945, *Worthington 1669* B (K); Dambulla Rock, 150–300 m, 23 Jan 1971, *Robyns 7213* (K, PDA); Dambulla, IFS Popham Arboretum, *Cramer 6519* (PDA); along shores of Sigiriya wewa, 11 Nov 1973, *Sohmer 8646* (PDA). POLONNARUWA DISTRICT: Polonnaruwa sacred area, 61 m, 4 Sept 1971, *Dittus 71090401* (PDA); NW shore of Minneriya tank, near Kedawatte Bendimulla, 80°52′ E, 8°04′ N, 90 m, 27 Feb 1977, *B. & K. Bremer 918* (PDA). COLOMBO DISTRICT: Delgoda, 17 Feb 1970, *Amaratunga 2006* (PDA); Millakande, 20 Dec 1969, *Reid 2364* (PDA); Mirigama, 27 Feb 1967, *Amaratunga 1253* (PDA). KEGALLE DISTRICT: Kotteyakumbara, 22 Aug 1966, *Amaratunga 1146* (PDA). KANDY DISTRICT: Haragama, 80°43′ E, 7°16′ N, 500 m, 14 Dec 1971, *Jayasuriya & Balasubramaniam 442* (K, PDA), 19 Jan 1974, *Sumithraarachchi, H.N. & A.L. Moldenke & Jayasuriya 13* (K, PDA); c. 1 mile W. of Hunnasgiriya near mile post 20/11, 750 m, 14 Nov 1974, *Davidse & Jayasuriya 8377* (PDA); trail to Nitre cave area from St. Martin's Estate, near base of Dumbanagala Spur, 14 Nov 1975, *Sohmer & Jayasuriya 10688* (PDA). BADULLA DISTRICT: Mahiyangana, 150 m, 25 April 1973, *B.C. Stone 11,158* (PDA). AMPARAI DISTRICT: Lahugala Reserve, 20 m, 15 Dec 1975, *Bernardi 16046* (PDA). RATNAPURA DISTRICT: Belihuloya, 600 m, 10 May 1969, *Kostermans 23458* (K). MONERAGALA DISTRICT: Yala, main Park road, mile mark 7, 6–10 m, 10 Dec 1967, *Comanor 657* (K, PDA); Ruhuna National Park, 10 m, 9 March 1973, *Bernardi 14199* (PDA), before gate, 5 m, 26 Jan 1968, *Comanor 858* (K, PDA), 23 March 1970, *Cooray 70032309* (K, PDA), Block 1, at Bambuwa, 17 Nov 1969, *Cooray 69111725* (K, PDA), Rakinawala, 22 Nov 1968, *Cooray 68102213* (K, PDA). GALLE DISTRICT: Hindalgoda-Hiniduma forest

path, 30 Nov 1992, *Wadhwa & Weerasooriya 368* (PDA). HAMBANTOTA
DISTRICT: Bundala Sanctuary, 10 m, 8 March 1973, *Bernardi 14170* (PDA).
LOCALITY UNKNOWN: *Col. Walker 440, 362* (K); *Thompson s.n.* (K);
Fraser 13 (BM); *Macrae 758* (BM).

6. Ziziphus mauritiana Lam., Enc. 3: 319. 1789; Bhand. & Bhans. in Fasc.
Fl. India 20: 99. 1990. Type: Sri Lanka, *Hermann* 3: 14. n. *89* (BM).

Rhamnus jujuba L., Sp. Pl. 194. 1753. Type: from India.
Ziziphus jujuba (L.) Gaertn., Fruct. 1: 203. 1788 (non Mill., 1768); Lam.,
Enc. 3: 318. 1789; Roxb., Fl. Ind. 1: 608. 1832; Wight, Ic. Pl. Ind. Or.
1: t. 99. 1838; Thw., Enum. Pl. Zeyl. 74. 1858; Lawson in Hook. f., Fl.
Br. Ind. 1: 632. 1875; Trimen, Handb. Fl. Ceylon 1: 280. 1893; Gamble,
Fl. Pres. Madras 1: 219. 1918.

var. **mauritiana**

Large armed shrubs or trees, 5–10 (–12) m high; branches spreading
or drooping; bark dark-grey with vertical cracks, young branches rusty tomentose; stipular spines solitary or in pairs, straight or one of them recurved. Leaves alternate, ovate-oblong or elliptic or semiorbicular, 2.0–9.0
× 1.5–5.0 cm, rounded at both ends or very slightly oblique at base, often
mucronate at apex, densely white or brown silky-tomentose beneath, margin
serrulate or entire, basally 3-nerved; petioles 5–12 mm long, tomentose. Inflorescences axillary, sessile cymes or fascicles of 10–12 flowers, tomentose.
Flowers greenish-yellow, 4–6 mm across; pedicels 2–4.5 mm in flowers, and
up to 8 mm long in fruits, tomentose. Calyx lobes ovate, 1.5–2 mm long,
acute, glabrous within, tomentose without; calyx-tube campanulate. Petals
1–1.5 mm long, spathulate, reflexed, greenish-yellow. Stamens as long as
the petals. Disk 10-grooved, fleshy. Ovary 2-locular, glabrous, ovule 1; style
short, 2-cleft from the middle; stigma lobes 2, curved. Drupes ± globose,
1–1.5 cm in diam., (1–)2-celled, fleshy, glabrous, orange or red when ripe;
kernel irregularly furrowed, with hard, stony shell. Seeds compressed.

Distr. Pakistan, India, Afghanistan, China, Australia, Tropical Africa
and Sri Lanka.

Ecol. In forests of dry zone, common; fls Nov–May; frts Mar–Sept.

Vern. Mahadebara, Dabara, Masan (S); Ilantai (T).

Notes. On specimen *C.P. 1242* in PDA, two localities, viz. Anuradhapura, *Gardner* and Trincomalee, are recorded in pencil.

Specimens Examined. JAFFNA DISTRICT: Near Ponneryn, 5 m,
17 March 1973, *Bernardi 14283* (K, PDA); Elephant Pass, scrub jungle,
11 March 1932, *Simpson 9252* (BM). MANNAR DISTRICT: Madhu road
to Mannar island, alt. low, 17 July 1974, *Kostermans 25299* (K, PDA).

PUTTALAM DISTRICT: Puttalam, 12 Sept 1981, *Balasubramaniam 2621* (K), Rest House grounds, 15 Nov 1992, *Wadhwa 272* (PDA). ANURAD-HAPURA DISTRICT: Maduruodai, 2.5 miles S. of Marai villu, 30 Dec 1968, *Fosberg, Mueller-Dombois, Wirawan, Cooray & Balakrishnan 50868* (K, PDA); Panwewa Estate, Nochchiyagama, 26 May 1974, *Sumithraarachchi & Sumithraarachchi 330* (K, PDA); Wilpattu National Park, between Sengapattu and Kattankandal Kulam in Plot W 16, 6 July 1969, *Wirawan, Cooray & Balakrishnan 966* (PDA). TRINCOMALEE DISTRICT: Shellbay, Trincomalee, 3 m, 1 Jan 1944, *Worthington 1358* (K). POLONNARUWA DISTRICT: Polonnaruwa sacred area, section 4C, 61 m, 18 Dec 1970, *Ripley 324* (PDA). COLOMBO DISTRICT: Negombo-Jaela, roadside, 9 Oct 1969, *Amaratunga 1875* (PDA); Walpitamulla, 9 Oct 1969, *Amaratunga 1876* (PDA); Katuwa, 24 miles on Colombo-Negombo road, 7 March 1970, *Amaratunga 2043* (PDA). HAMBANTOTA DISTRICT: Ruhuna National Park, Andunoruwa wewa, 3–5 m, 3 April 1968, *Fosberg & Mueller-Dombois 50138* (K, BM, PDA), *Fosberg 50139* (PDA), Block 1, main Yala road, 4–8 m, 28 March 1968, *Comanor 1163* (K, PDA); 1 mile E. of Hambantota on A2, alt. sea level, 19 Nov 1973, *Sohmer, Waas & Eliezer 8863* (BM, PDA); Hambantota, 9 April 1985, *Jayasuriya & Balasubramaniam 3296* (CAL, PDA), near Hambantota salterns, 26 Nov 1974, *Sumithraarachchi 573* (K, PDA); Tissamaharama-Kataragama road, 4 Oct 1969, *Cooray 69090415* (K, PDA). LOCALITY UNKNOWN: *s. coll. C.P. 1242* (BM, PDA); *Macrae 757* (BM).

3. RHAMNUS

L., Sp. Pl. 193. 1753; L., Gen. Pl. ed. 5, 89. 1754; Benth. & Hook. f., Gen. Pl. 1: 377. 1862; Grub., Fl. Syst. Pl. Vasc. 8: 286. 1949. Type: *R. catharticus* L.

Small trees or shrubs, thorny or unarmed. Leaves mostly alternate, rarely subopposite, petiolate, entire or crenate-dentate, penninerved; stipules minute, free, deciduous or persistent. Inflorescences axillary, solitary or in fascicles, rarely simple or branched racemes. Flowers 5- or 4-merous, bisexual or polygamous-dioecious, small, sessile or pedicellate. Calyx 4- or 5-lobed, triangular-ovate, erect or spreading; calyx-tube urceolate. Petals 4 or 5 or 0, inserted at the margin of the calyx-tube, cucullate or flat embracing the stamens, narrowly clawed. Stamens 4–5; filaments very short; anthers oblong, dehiscing longitudinally. Disk thin, lining the calyx-tube. Ovary 3 or 4-celled, ovoid; styles 3–4-cleft about half the length; stigmas small, obtuse, papillose. Drupes with 2–4 free, 1-seeded stones, globose, encircled at base by persistent calyx tube; pyrenes horny or cartilaginous, indehiscent or dehiscing inwardly. Seeds oblong or globose, smooth or furrowed on the back; albumen fleshy; cotyledons with recurved margins or flat.

150 species in Europe, Asia and America, often rare in the tropics; 2 in Sri Lanka.

KEY TO THE SPECIES

1 Flowers 4-merous, apetalous, in fascicles of 2–4 .... **1. R. arnottianus**
1 Flowers 5-merous, petalous, in fascicles of 10–20 .... **2. R. wightii**

1. Rhamnus arnottianus Gardner ex Thw., Enum. Pl. Zeyl. 74. 1858; Lawson in Hook. f., Fl. Br. Ind. 1: 638. 1875; Trimen, Handb. Fl. Ceylon 1: 283. 1893; Grub., Fl. Syst. Pl. Vasc. 8: 293. 1949. Type: Sri Lanka, Nuwara Eliya District, Horton Plains, 2135 m, July 1846, *Thwaites C.P. 201* (K, holotype; BM, CAL, HAK, PDA, isotypes).

Tall, unarmed shrubs, up to 3 m high; young branches puberulous. Leaves ovate-lanceolate, 2.5–7 × 1.7–2.5 cm, coriaceous, glabrous, base ± obtuse, apex acute or subacuminate, margin closely serrate, teeth with fine, minute glands; lateral nerves slightly raised below, secondary nerves converging at the margin; petioles 4–10 mm long; stipules linear, c. 5 mm long, caducous, with minute, persistent glands within. Flowers 2–4 in axillary fascicles, 4-merous; pedicellate, red in colour; pedicels slender, longer than the petioles. Calyx lobes 4, deltoid, very acute, pilose; calyx-tube urceolate. Petals 0. Stamens 4; filaments short, dehiscing longitudinally. Disk thin, filling the calyx-tube. Ovary 3–4-celled, ovoid; style branches 2–3. Drupes 3-celled, pyriform-globose, 4–6 mm across, apiculate, on a persistent but not adherent calyx-tube, when young green, on maturity dark-red or purple-black; pyrenes 1-seeded, cartilaginous. Seeds obovate, membranous, furrowed on the back.

Distr. Endemic.

Ecol. Along forest edges in wet zone, up to 2400 m, uncommon; fls Mar–May, frts Apr–July.

Note. A plate bearing No. 435 (habit only) in PDA belongs to this species.

Specimens Examined. NUWARA ELIYA DISTRICT: Nuwara Eliya, 1830 m, *Gardner 182* (K, BM); Horton Plains, 2400 m, 12 Dec 1971, *Balakrishnan 405* (K, PDA), 2100 m, 21 Jan 1945, *Worthington 1684* (K), 29 Jan 1980, *Rasiah 83* (PDU); Pidurutalagala below peak, 2700 m, 21 May 1971, *Jayasuriya 190* (K, PDA), ("Pedrotalagala"), 2285 m, Apr 1899, *Gamble 27599* (K), ("Pedrotalaga"), Apr 1846, *Thwaites C.P. 636* (K), to the summit, 21 April 1906, *A.M. Smith s.n.* (PDA), top, 21 April 1906, *Willis s.n.* (PDA); Hakgala peak trail, above Hakgala Botanic Garden, 1675–2170 m, 12 July 1968, *Meijer 1870* (K). LOCALITY UNKNOWN: *Col. Walker s.n.* (K).

2. Rhamnus wightii Wight & Arn., Prod. 1: 164. 1834; Wight, Ic. Pl. Ind. Or. t. 159. 1839; Thw., Enum. Pl. Zeyl. 74. 1858; Bedd., Fl. Sylv. 70. 1871; Lawson in Hook. f., Fl. Br. Ind. 1: 639. 1875; Trimen, Handb. Fl. Ceylon 1: 283. 1893; Gamble, Fl. Pres. Madras 1: 22. 1918; Grub., Fl. Syst. Pl. Vasc. 8: 297. 1949; Bhand. & Bhans. in Fasc. Fl. India 20: 59. 1990. Type: India: Peninsular India, Courtallum, *Wight 507* (K; BM, PDA, isotypes).

Ceanothus wightianus Wall., Cat. 4264. 1828, nom. nud.

Large shrubs or small trees, c. 3.5 m high, unarmed; stem grooved; younger parts puberulous; nodes enlarging around the leaf scars. Leaves alternate, rarely subopposite, 2.4–10.2 × 1/3–4.1 cm, subcoriaceous, ovate-oblong or elliptic-ovate, acuminate, ± rounded at base, closely serrate, teeth with glandular tip, lateral nerves 5–8 pairs, merging at the margins; petioles 7–15 mm long, puberulous; stipules 3 mm long, subulate, pubescent, caducous. Flowers 10–20 in axillary fascicles, yellowish-green, 5-merous, pedicellate; pedicels 2–4 mm long, accrescent; calyx lobes 5, deltoid, up to 4 mm long, thickened at apex; Calyx-tube campanulate, 2 mm in diam., minutely pubescent without. Petals spathulate or obovate, c. 2 mm long, cucullate. Stamens nearly as long as petals, pale-brown. Disk saucer-shaped, glabrous. Ovary in male flowers rudimentary, in others 3-carpellary, glabrous, superior, almost immersed in the disk but free; style short, 3-cleft up to 1/3 of the length, branches diverging. Drupes 3-celled, obovoid or globose, 4–5 mm in diam., supported at base by the adnate, accrescent calyx-tube, tipped with remains of style, purplish-red when ripe; stones 3, 1-seeded. Seeds c. 4 mm long, with deep furrow on the back; cotyledons flat, foliaceous, enclosed in fleshy, thin albumen.

Distr. Sri Lanka, India, Nepal.

Ecol. In wet evergreen montane forests, 1400–2200 m; fls Apr–Nov, frts May–Dec.

Specimens Examined. BADULLA DISTRICT: Ohiya, 1650 m, 2 Sept 1978, *Huber 885* (PDA); Jungle above Thotulagala Estate, 29 June 1975, *D.B. & D. Sumithraarachchi 918* (PDA). NUWARA ELIYA DISTRICT: Ambewela, 2 Feb 1927, *Alston 1038* (PDA); Pattipola, 1900 m, 9–10 Sept 1981, *Rajah 25, 59* (PDU); Nuwara Eliya, *Gardner 183* (BM, K, PDA), 9 Sept 1981, *Sivalingam 60* (PDU), *Ranasinghe 25* (PDU), *Fernando 71* (PDU), *Caldera 20* (PDU), *Amaratunga 1295* (PDA); Elephant Plains, in 1851, *Thwaites C.P. 2525* (BM, CAL, HAK, K, PDA); Sita Eliya near Hakgala, 28 April 1932, *Simpson 9602* (BM); Hakgala, *Amaratunga 420* (PDA), 23 May 1911, *J.M. Silva s.n.* (PDA), Apr 1920, *Alston s.n.* (PDA); 15 Oct 1992, *Wadhwa & Weerasooriya 99* (PDA); Jungle above the laboratory, Hakgala, 17 March 1906, *A.M. Smith s.n.* (PDA), Ohiya, ± 1400 m, 15 May 1975, *Cramer 4475*

(PDA); Elk Plains, between Nuwara Eliya and Kande Ela Reservoir, 1730 m, 17 May 1968, *Mueller-Dombois 6805* (PDA); Jungle edge, Sita Eliya, Hakgala, 6 April 1906, *A.M. Smith s.n.* (PDA); Horton Plains, trail behind Farr Inn, 2130 m, 23 April 1969, *Hladik 739* (PDA); Along forest road, Kandapola Forest Reserve, 19 June 1972, *Maxwell & Jayasuriya 873* (PDA); Nuwara Eliya to Hakgala road, c. 1650 m, *C.F. & R.J. van Beusekom 1438* (PDA); Mount Galaha, Hakagala, near river, 9 Dec 1975, *Bernardi 15846* (PDA). LOCALITY UNKNOWN: *Col. Walker s.n.* (K, PDA); *Col. Walker 103* (K).

4. COLUBRINA

Rich. ex Brongn., Ann. Sci. Nat. Bot. 10: 368. 1827, nom. cons.; Benth. ex Hook. f., Gen. Pl. 1: 379. 1862; Johnston, Brittonia 23: 7. 1971. Type: *C. ferruginosa* Brongn. (=*Rhamnus colubrina* Jacq.), typ. cons.

Erect shrubs or small trees, usually unarmed, sometimes armed; young stems usually pubescent. Leaves alternate, petiolate, penninerved or triplinerved from the base, margin entire or crenate-serrate; stipules small, caducous. Flowers small, polygamous, 5-merous, in axillary, shortly peduncled cymes, with a calyx cup at the rim of which are borne the sepal lobes, petals and the androecium. Calyx deeply 5-lobed; lobes valvate in buds, deltoid, spreading, pubescent without, deciduous; calyx tube in male flowers cupular and in bisexual obconical. Petals yellowish-green, slightly shorter than the calyx, oblanceolate-spathulate, cucullate, convolute, shortly clawed, deciduous. Stamens 5, about as long as the petals; anthers clasped by petals. Disk fleshy, nectariferous, flat or slightly saucer-shaped, nearly filling the calyx-tube, accrescent along with the cup and adnate to lower fifth to half of the fruit. Ovary in male flowers reduced with a short stunted style and in bisexual half-inferior, immersed in the disk, 3-celled, with one ovule in each cell; style slender, 3-cleft; stigmas 3, small, obtuse. Fruits nearly globose, slightly 3-lobed, dehiscing loculicidally into 3 cocci, at the base with the adnate, accrescent remains of the cup and disk; exocarp thin. Seeds obovoid, compressed, trigonous, usually lustrous, dark-brown; albumen fleshy, thin.

31 species in the warmer parts of the world, mostly distributed in America, Hawaii, Madagascar, Australia, and S.E. Asia (Malay Peninsula, Java, Borneo), India and Sri Lanka; 1 in Sri Lanka.

Colubrina asiatica (L.) Brongn., Ánn. Sci. Nat. Bot. 10: 369. 1827; Thw., Enum. Pl. Zeyl. 75. 1858; Lawson in Hook. f., Fl. Br. Ind. 1: 642. 1875; Trimen, Handb. Fl. Ceylon 1: 285. 1893; Gamble, Fl. Pres. Madras 1: 224. 1918; Parkinson, For. Fl. Andaman 131. 1923; Johnston, Brittonia 23: 46. 1971; Bhand. & Bhans. in Fasc. Fl. India 20: 32. 1990.

var. **asiatica** Johnston, Brittonia 23: 47. 1971.

Ceanothus asiaticus L., Sp. Pl. 196. 1753. Type: Sri Lanka, Herb. *Hermann*, 2.11. n. *98*. (BM, Lectotype, designated by Johnston, l.c.).
Rhamnus asiaticus (L.) Lam. ex Poir., Enc. 4: 474. 1796.

Straggling unarmed shrubs, 2–6 m high; branches slender and often somewhat zig-zag; the internodes 0.5–4.5 cm long, covered with sparse, appressed, golden silky hairs. Leaves alternate, ovate, 2.9–10.6 × 1.8–6.5 cm, membranous, glaucous on both surfaces, at base broadly rounded to often shallowly cordate, acuminate at apex, finely crenulate-serrate, each tooth with a minute, dark glandular mucro, basally 3-nerved with 2–3 pairs of lateral nerves; petiole 0.6–1.9 cm long; stipules deltoid, c.1 mm long. Inflorescence a short, peduncled cyme of a few lower bisexual and several upper male flowers; peduncles 2–3 mm long. Flowers yellowish-green, fragrant; pedicels greatly varying in length, in flowers 2–3 mm long and in fruits 5–15 mm long. Calyx lobes 5, deltoid, 2.5–3 mm long, acute; calyx-tube glabrescent. Petals c. 2 mm long, emarginate, with a very small claw. Stamens 1.5 mm long; anthers dorsifixed. Disk 5-lobed, orange yellow, when young with some silky hairs, later glabrate. Ovary globose, glabrous; style in male flowers stunted, in bisexual flowers up to 1.5 mm long, 3-fid halfway down; stigmas truncate. Fruit nearly globose, 7–8 mm long, slightly depressed at top, dehiscing longitudinally into 3 cocci; cocci thin-walled, occasionally adherent to receptacle after dehiscence. Seeds obovoid, 3.5–4 × 3–3.5 mm, slightly emarginate at base, brown.

Distr. Sri Lanka, India, Burma, Indonesia, Malay Peninsula (Java, Borneo), Australia, E. Africa, Madagascar, America, Hawaii, Cuba, Jamaica and Carribean Islands.

Ecol. Along coast, especially on calciferous beaches forming a wind break, common; fls Jan–June; frts Mar–July.

Vern. Telhiriya (S); Mayirmanikkam (T).

Notes. Johnston (l.c.) has recognized 2 varieties under this species on the basis of hairs or tomentum on the young internodes, breadth of leafblade or lamina, narrowly or broadly rounded base of the leaf, glabrous to tomentose ventral leaf surface and on the length of the seeds. Sri Lankan plants belong to var. *asiatica* only.

On the specimen *C.P. 1239* in PDA, two localities, viz., Trincomalee and Kurunegala, are given, which are also cited by Trimen (l.c.).

Specimens Examined. JAFFNA DISTRICT: Near Karaitivu, 12 Oct 1975, *Bernardi 15325* (PDA); Karaitivu Island, Aug 1883, *s. coll. s.n.* (PDA); Jaffna, Feb 1890, *s. coll. s.n.* (PDA). VAVUNIYA DISTRICT: Lagoon near Alampil, sea level, 7 July 1974, *Meijer 764* (K, PDA). PUTTALAM DISTRICT: Chilaw, 30 May 1931, *Simpson 8152* (BM), 10 Nov 1971, *Kundu & Balakrishnan 547* (PDA); Kalpitiya, Palavi, 5 Oct 1962, *Amaratunga 337*

(PDA). TRINCOMALEE DISTRICT: Trincomalee town, sea level, 27 March 1977, *Cramer 4917* (K, PDA), 29 Jan 1786, Herb. *Rottler s.n.* (K), 20 May 1932, *Simpson 9686* (BM). BATTICALOA DISTRICT: Batticaloa-Kalmunai Road, 28 June 1931, *Simpson 8283* (BM), in 1862, *Gardner C.P. 1239* (BM, CAL, PDA). KANDY DISTRICT: Kandy, in 1819, *Moon s.n.* (BM). HAMBANTOTA DISTRICT: Ruhuna National Park, Block I, near Buttawa lagoon, South side, plot R 27, 8 March 1969, *Mueller-Dombois, Wirawan, Cooray & Balakrishnan 69030804* (K, PDA). LOCALITY UNKNOWN: *Forster s.n.* (K); *Macrae 416* (K, BM); *Duke s.n.* (K).

5. VENTILAGO

Gaertn., Fruct. 1: 223. t. 49, f. 2. 1788; DC., Prod. 2: 38. 1825; Benth. & Hook. f., Gen. Pl. 1: 375. 1862; Suesseng. in Pflanzenfam. ed. 2, 20d:151. 1953. Type: *Ventilago madraspatana* Gaertn.

Climbing woody or scandent shrubs; branches slender or stout, puberulous, striate. Leaves alternate, simple, coriaceous, petiolate; leaf-blade orbicular to elliptic-ovate or oblong-elliptic, glabrous to variously hairy, base asymmetrically obtuse or attenuate, margin entire or crenate-serrate or dentate, apex acute to acuminate; penninerved, secondary nerves ascending and converging along the margin; stipules very small. Flowers bisexual, in axillary or terminal panicles or umbellate cymes. Sepals 5, acute, keeled within, puberulous outside; calyx-tube obconical or flat and saucer-like. Petals 5, alternate with sepals, obcordate, truncate, membranous, cucullate, clawed at base. Stamens 5, opposite to and enclosed by petals, slightly adnate to their bases; filaments filiform; anthers 2-celled, dorsifixed. Disk thick, filling the calyx-tube, 5-lobed, adnate to the lower half of ovary. Ovary more or less sunk in the disk, 2-celled; style very short, hairy at base, with 2 short stigmatic lobes. Fruit a samaroid, 1-seeded, globose nut, prolonged above by the style enlargement into a linear or linear-oblong, coriaceous wing, the tip crowned by the remains of the stigma; wing puberulous to tomentose; mesial line on the wing single or double, the seed chamber thinly double-walled; the fruit at the base enclosed by calyx-tube or saucer-like calyx only adnate to the base. Seeds globose to subglobose, exalbuminous; cotyledons thick, fleshy.

About 40 species, occurring in the Indo-Malayan region, Burma, Polynesian Islands, Australia and Madagascar; 2 in Sri Lanka.

KEY TO THE SPECIES

1 Branchlets slender, pale; leaves ovate or ovate-lanceolate; panicles minutely grey pubescent; nut with grey-puberulous wing .... **1. V. madraspatana**

1 Branchlets stout, dark; leaves oblong-lanceolate; panicles golden or brown-villous; nut with brown tomentose wing .... **2. V. gamblei**

1. Ventilago madraspatana Gaertn., Fruct. 1: 223. t. 49, f. 2. 1788; Roxb., Pl. Corom. t. 76. 1798; Wight, Ic. Pl. Ind. Or. 1: t. 163. 1839; Thw., Enum. Pl. Zeyl. 74. 1858; Lawson in Hook. f., Fl. Br. Ind. 1: 631. 1875; Trimen, Handb. Fl. Ceylon 1: 279. 1893; Gamble, Fl. Pres. Madras 1: 218. 1918; Banerjee & Mukerjee, Indian Forester 96: 207. 1970; Matthew, Fl. Tamilnadu Carnatic 1: 269. 1983 "*maderaspatana*"; Bhand. & Bhans. in Fasc. Fl. India 20: 85. 1990. Lectotype: Rumphius t. 2 in Herb. Amboinense, 1747.

var. **madraspatana**

Large, much branched, woody climber, young branches grey-pubescent, older dark grey, glabrous. Leaves alternate, elliptic-ovate or ovate-lanceolate, 1.7–8.5 × 1.4–3.2 cm, entire or crenate, apex acute or sub-acuminate, base generally rounded or sometimes slightly oblique, glabrous; lateral nerves 4–8 pairs, ascending and converging near the margin; petioles up to 8 mm long, glabrous; stipules very small, lanceolate, pubescent. Inflorescences axillary and terminal panicles, minutely grey-pubescent, occasionally with leafy bracts. Flowers numerous, 3–5 mm across, yellowish-green; pedicels 3 mm long, pubescent. Calyx lobes 5, 1.5–2.5 mm long, spreading; calyx-tube obconical, pubescent. Petals 5, obcordate, truncate, shorter than the calyx-lobes. Stamens 5, as long as the petals; anthers oval. Disk glabrous. Ovary 2-celled, pubescent; style arms divergent. Fruit a samaroid globose nut, prolonged above into an elliptic-oblong wing and at the base adnate to the saucer-shaped, 5–7 mm diam. calyx-tube; wing 2.2–4.0 × 0.6–0.8 cm, grey puberulous; mesial line on the wing single. Seed globose, 4 × 3 mm, thin-walled, brown, exalbuminous; cotyledon fleshy, thick.

Distr. Sri Lanka, India, Burma, Indonesia and Malaysia.

Ecol. In the drier hilly forests, ascending up to 700 m, common; fls Jan–July, frts May–Sept.

Vern. Yakadawel (S); Vempadam (T).

Notes. Santapau (Kew Bull. 34. 1949) has made two varieties on the basis of apex of fruit wing deeply bifid or not. In Sri Lankan material, the fruit wing is not bifid and belongs to var. *madraspatana* only.

On sheet *C.P. 1236* in PDA, Dambulla, Anuradhapura (*Gardner*) and Guruwala (July 1851) are recorded in pencil.

Specimens Examined. VAVUNIYA DISTRICT: c. 2 miles SW of Nedunken, along road to Puliyankulam near mile post 21/3, 95 m, 5 May 1974, *Davidse & Sumithraarachchi 9077* (PDA); Vavuniya-Kebithigollewa road, 24 June 1975, *D.B. & D. Sumithraarachchi 825* (PDA). ANURAD-HAPURA DISTRICT: Ritigala, 13 July 1970, *Cramer 3031* (PDA), west of Reserve, 13 July 1970, *Meijer 332* (K, PDA), ± 600 m, 29 May 1973, *Cramer*

4146 (PDA); Andiyagala, 10 July 1965, *Amaratunga 932* (PDA), 600 m, 18 March 1973, *Bernardi 14307* (K, PDA), 17 Oct 1970, *Kundu & Balakrishnan 327* (PDA); Anuradhapura to Mannar road, alt. low, 27 May 1975, *Kostermans 24878* (K), 11 Jan 1974, *Waas 334* (K, PDA); between Wilpattu National Park and Anuradhapura, 27 Sept 1969, *C.F. & R.J. van Beusekom 1637* (PDA); Kekirawa, 15 Aug 1885, *s. coll s.n.* (PDA). POLONNARUWA DISTRICT: Attanakadawala-Jamburawala road, 17 Oct 1906, *Jayasuriya 326* (K, PDA); Polonnaruwa sacred area, 7 Jan 1970, *Fosberg 51878* (PDA), 25 June 1970, *Meijer & Balakrishnan 110* (PDA), Northside in section 3A, 61 m, 10 Jan 1970, *Ripley 298* (PDA), 60 m, 4 Dec 1969, *Hladik 113* (PDA), 60 m, 11 Oct 1977, *Huber 432* (PDA); Welikanda-Kandakaduwa road, 8 June 1974, *Waas 610* (K, PDA). TRINCOMALEE DISTRICT: Monkey bridge, Trincomalee, 24 May 1975, *Waas 1264* (K, PDA); Trincomalee to Kuchchaveli road, alt. low, 16 May 1973, *Kostermans 24802* (BM, K, PDA). MATALE DISTRICT: At Nityandagala along Matale-Illukumbura road, 3 July 1974, *Sumithraarachchi 399* (PDA); near Bambuwa, 80°34' E, 7°45' N, 18 June 1975, *D.B. & D. Sumithraarachchi 728* (K, PDA); Beligama, 18 July 1942, *Worthington 1292* (K, PDA); Dambulla, *Amaratunga 1366* (PDA). KANDY DISTRICT: Lady Hortons, 15 May 1928, *J.M. de Silva s.n.* (PDA); Guruoya, 80°47' E, 7°17' N, 400 m, 30 April 1975, *Jayasuriya 1191* (K); Kandy-Mahiyangane road, 16/14 mark, 9 Sept 1974, *Trivengadum & Waas 438* (K, PDA); Medamahanuwara, 9 Sept 1974, *Waas & Trivengadum 786* (K, PDA); Hunnasgiriya, 15 Nov 1967, *Amaratunga 1481* (PDA), 600 m, 6 June 1971, *Kostermans 24425 A* (K, PDA); Hantane, 650 m, *Gardner 181* (K, BM); Uma Oya-Teldeniya road, 300 m, 28 July 1974, *Kostermans 25268* (K); Halambe, 26 June 1931, *Simpson 8242* (BM); Uma Oya, in 1880, *s. coll. s.n.* (PDA). AMPARAI DISTRICT: Lahugala Reserve, 15 Dec 1975, *Bernardi 16049* (PDA). NUWARA ELIYA DISTRICT: Siyambalagastenna, 7 Sept 1926, *Alston s.n.* (PDA). MONERAGALA DISTRICT: Wellawaya-Moneragala road, 1 Aug 1969, *Cooray 69080102* (PDA); c. 1 mile E. of Wellawaya, 27 June 1970, *Meijer 180* (PDA); Kataragama Peak, 450 m, 28 Oct 1975, *Bernardi 15505* (PDA); Bibile, 25 Oct 1925, *J.M. de Silva s.n.* (PDA). MATARA DISTRICT: Beraliya proposed reserve, 10 Jan 1992, *Jayasuriya 6057* (PDA). LOCALITY UNKNOWN: *Thwaites C.P. 1236* (BM, CAL, K, PDA); *Macrae 380* (BM); *Col. Walker s.n.* (K, PDA); 24 Sept 1978, *Kostermans 24978* (PDA).

2. Ventilago gamblei Susseng. in Pflanzenfam. ed. 2, 20d: 152. 1953; Banerjee & Mukerjee, Indian Forester 96: 209. 1970; Ramachand. & Nair, Fl. Cann. 117. 1988; Bhand. & Bhans. in Fasc. Fl. India 20: 84. 1990. Lectotype: India, Malabar, Kannoth, *C.A. Barber s.n.* (MH).

Ventilago lanceolata Gamble, Bull. Misc. Inform. 134. 1916 (non Merrill, 1915); Gamble, Fl. Pres. Madras 1: 218. 1918; Alston in Trimen, Handb. Fl. Ceylon 6: 49. 1931.

Climbing shrubs, up to 15 m high; young branches puberulous, older dark, glabrous. Leaves alternate, 5.2–11.2 × 1.8–5.1 cm, oblong-lanceolate, coriaceous, strongly crenate, young ones puberulous abaxially, base obtuse-subrotundate, unequal, apex acuminate or mucronate; lateral nerves 6–7 pairs converging near the margins; petioles 6–10 mm long, puberulous. Inflorescences axillary panicles, golden or brown-villous interspersed with short glandular hairs. Flowers 3–4 mm across; bracts caducous; bracteoles subglomerulate-linear. Calyx lobes 3 mm long, triangular, keeled within at apex, rusty pubescent without. Petals obcordate, 1–2 mm long, alternating with the calyx-lobes, cucullate. Stamens 5, nearly equal to petals in size; anther connectives apiculate, recurved. Disk glabrous. Ovary two-celled, immersed in the disk, glabrous; style arms 2, short, divergent. Fruit a samaroid globose nut, 3.5–4 mm in diam., with apical wing, 5.4 × 1.2 cm, and saucer-shaped calyx-tube adnate to the base; wing flat, mucronate, with double median line, rusty brown-tomentose. Seeds globose, 2.5–3.0 × 2–2.5 mm, blackish-grey.

Distr. Sri Lanka and India.

Ecol. In forests of dry zone and lower wet zone, not common; fls & frts Dec–July.

Specimens Examined. ANURADHAPURA DISTRICT: Ritigala Strict Natural Reserve, slopes North of Weweltenna Plain, 27 May 1974, *Jayasuriya 2167* (PDA). KURUNEGALA DISTRICT: Badagama forest reserve, 7°30′ N, 80°23′ E, low alt. 12 Jan 1972, *Jayasuriya & Balasubramaniam 517* (PDA). KANDY DISTRICT: Hantane, 600 m, *Gardner 180* (K, BM), *Walker 169* (K). RATNAPURA DISTRICT: Sinharaja forest, near MAB Laboratory, 26 Nov 1992, *Wadhwa & Weerasooriya 345* (PDA); Sinharajagala, 28 Nov 1992, *Wadhwa & Weerasooriya 365* (PDA). GALLE DISTRICT: Kanneliya, 14 July 1991, *Jayasuriya 6509* (PDA). LOCALITY UNKNOWN: 24 July 1928, *Alston s.n.* (PDA); *s. coll. C.P. 1236 A* (PDA); *s. coll. 8138 A* (PDU).

6. SCUTIA

(Comm. ex DC.) Brongn., Ann. Sci. Nat. Bot. 10: 362. 1827, nom. cons.; Benth. ex Hook. f., Gen. Pl. 1: 379. 1862.

Ceanothus L. sect. *Scutia* Comm. ex DC., Prod. 2: 29. 1825. Type: *S. circumscissa* (L. f.) Radlkofer (=*Rhamnus circumscissus* L. f.).

Small trees or shrubs, rarely scandent, glabrous or nearly glabrous, unarmed or armed with recurved thorns; branchlets usually angular. Leaves opposite or subopposite, ovate or obovate, often variable, obscurely dentate, rounded at base and at apex, but usually mucronate, penninerved; petiole short; stipules triangular, usually deciduous. Flowers 5-merous, bisexual, either in axillary cymes or in sessile, cymose, few-flowered axillary fascicles or solitary in the axils. Calyx lobes deltoid, more or less spreading; calyx-tube turbinate or hemispherical. Petals deeply obcordate or bilobed, cucullate or flat, much shorter than the sepals. Stamens equalling the petals; anthers dorsifixed, ovate, 2-celled. Disk lining the calyx-tube, rather thin. Ovary sunk in the disk, 2–3-celled; style very short, slightly lobed. Drupes subglobose, dry or subfleshy with 2–3 seeds enclosed in free endocarpous stones, irregularly dehiscent.

Nine species, distributed in tropical America, South Africa, Indo-China, India and Sri Lanka; one in Sri Lanka.

Scutia myrtina (Burm. f.) Kurz, J. Asiat. Soc. Bengal 44: 168. 1875; Gamble, Fl. Pres. Madras 1: 223. 1918; Johnston, Bull. Torrey Bot. Club 101(2): 65. 1974; Bhand. & Bhans. in Fasc. Fl. India 20: 71. 1990.

Rhamnus myrtinus Burm. f., Fl. Ind. 60. 1768 *"myrtina"*. Neotype: India, *BSI s.n.* (MH), designated by Johnston, l.c.

Rhamnus circumscissus L. f., Suppl. 152. 1781 *"Circumscissa"*; Roxb., Fl. Ind. ed. Carey 1: 603. 1832. Type: India, Herb. Linn. 262. 34.

Rhamnus lucida Roxb., Fl. Ind. 2: 353. 1824. Type: Mauritius, *Roxburgh s.n.* (BM).

Scutia indica Brongn., Mem. Fam. Rham. 56. t. 4. 1826, nom. superfl.; Wight & Arn., Prod. 1:165. 1834; Wight, Ill. Ind. Bot. t. 73. 1840; Thw., Enum. Pl. Zeyl. 75. 1858; Lawson in Hook. f., Fl. Br. Ind. 1: 640. 1875; Trimen, Handb. Fl. Ceylon 1: 284. 1893.

Scutia myrtina var. *emarginata* Bhand., & Bhans. in Fasc. Fl. India 20: 71. 1990. Type: India, *Wagh 5478* (BLAT).

Ceanothus zeylanicus Heyne ex Roth, Nov. Sp. Pl. 153. 1821. Type: India, *Heyne s.n.* (K).

Straggling or scandent shrubs, rarely small trees, 2–5 (–10) m tall; young herbage and inflorescences puberulent; branchlets angulate, decussate, usually armed with recurved, axillary, solitary, often 2 per node, 2.5–6 mm long thorns. Leaves coriaceous, glabrous, elliptic to ovate or orbicular, 1.5–5 × 1.5–2 cm, at base rounded to cuneate, at apex acute to obtuse, or retuse to emarginate and always mucronulate, margin entire or often in the distal two-thirds with a few indistinct crenulae, quite often after drying much wrinkled and in-rolled; petioles 4–8 mm long; stipules deltoid, 2–4 × 2.0–2.5 mm. Inflorescences much condensed cymes, often 2–16-flowered fascicles, never more than one flower per cyme maturing into fruit; peduncle 2–6 mm long.

Flowers 2.0–2.5 mm diam; pale-white or yellowish-green; bracts subulate, 1 mm long, glabrous, sometimes puberulous at apex. Calyx lobes deltoid, 1.5–2 mm long, erect, glabrous. Petals unguiculate, c. 1 mm long, shortly clawed, deeply emarginate to bilobed. Stamens as long as the petals; filaments inserted on calyx-tube, filiform. Disk lining the calyx-tube, thin and inconspicuous. Ovary 2-locular, glabrous; style up to 1 mm long, shortly bilobed. Drupes 7–8 mm long, glabrous, pale when young, purplish or bluish-black when mature, readily separating into 2 free indehiscent stones.

Distr. India, Burma, Mauritius and Sri Lanka.

Ecol. In dry lowlands, besides lagoons and brackish water; common; rare at higher elevations; fls Mar–Aug., frts June–Oct.

Vern. Tuvadi (T).

Notes. The species is variable in the size, shape and denticulation of the leaves and the recognition of taxa on these characters is not practical. Hence the treatment of 2 varieties on these characters by Bhandari & Bhandali (l.c.) is not accepted.

Specimens Examined. JAFFNA DISTRICT: Pooneryn, 5 m, 17 March 1973, *Bernardi 14290* (K, PDA); Karativu, 12 Oct 1975, *Bernardi 15332* (PDA). MANNAR DISTRICT: Akattikulam, 4 miles S. of Madhu road, alt. sea level, 10 July 1971, *Meijer 789* (PDA); Madhu road, alt. sea level, 23 March 1973, *Cramer 4087* (K, PDA). VAVUNIYA DISTRICT: Road Vavuniya to Jaffna, 21 June 1973, *Kostermans 25115* (K). ANURADHAPURA DISTRICT: Wilpattu National Park, between Sengapaduvillu and Kattankandal Kulam, 6 July 1969, *Wirawan 965* (K); Horowupotana, 28 March 1963, *Amaratunga 545* (PDA); Anuradhapura, Mar 1883, *Trimen s.n.* (PDA); Ritigala Hill base jungle, 6 April 1974, *Sumithraarachchi & Jayasuriya 224* (PDA). TRINCOMALEE DISTRICT: Uppuveli, alt. sea level, 28 March 1977, *Cramer 4937* (K, PDA). MATALE DISTRICT: Near Sigiriya tank, fringing forest, 18 June 1932, *Simpson 9817* (BM); Dambulla; 4 miles SE of Kandalama tank, 17 Sept 1974, *Sumithraarachchi 454* (PDA); Naula to Dambulla road, at 36/4 mark, 14 July 1973, *Nowicke, Fosberg & Jayasuriya 328* (PDA). POLONNARUWA DISTRICT: Polonnaruwa Sacred area, 61 m, 11 Oct 1971, *Dittus 71101101, 71090705* (PDA); on way to Somawathiya, at Sungawila, 28 May 1974, *Sumithraarachchi 351* (PDA). KANDY DISTRICT: Hantane, 600 m, *Gardner 184* (BM, K). BADULLA DISTRICT: Welimada, 1800 m, 21 March 1971, *Balakrishnan 442* (K, PDA). HAMBANTOTA DISTRICT: Ruhuna National Park, Block I, behind Yala camp, Plot R 26, 27 Feb 1968, *Cooray 68022704* (PDA), Andunoruwa Wewa, 30 March 1968, *Mueller-Dombois 68093003* (PDA), Block II, Plot R 14, 30 Aug 1967, *Mueller-Dombois & Comanor 67083003* (PDA), Block 1, N. of Buttawa, Plot R

27, 10 Dec 1967, *Mueller-Dombois & Cooray 67121030* (PDA), near Yala Bungalow, Plot R 1, 26 Aug 1967, *Mueller-Dombois 67082602* (PDA), W. of Talgasmankada, 6 Sept 1992, *Jayasuriya 6686, 6692* (PDA); Block III, 5 miles N. of Kataragama to Vaddengewadiya, 23 Oct 1968, *Wirawan 641* (K, PDA), Udawalawe area, 17 Oct 1971, *Balakrishnan & Jayasuriya 889* (K); Palatupana, alt. low, 10 April 1985, *Jayasuriya & Balasubramaniam 3302* (PDA). LOCALITY UNKNOWN: in 1854, *Thwaites C.P. 1233* (BM, CAL, PDA); *Col. Walker 22* (K).

7. SAGERETIA

Brongn., Ann. Sci. Nat. 'Bot. 10: 359. t. 13, f. 2. 1827; Benth. & Hook. f., Gen. Pl. 1: 379. 1862. Type: not designated. (The lectotype *S. theezans* (L.) Brongn. (=*Rhammus theezans* L.) as indicated by Suessenguth, 1953 is the type of *Ampeloplis* Rafin., 1938).

Deciduous or evergreen, erect or sometimes scandent or scrambling shrubs, armed or unarmed. Leaves opposite or subopposite, entire or serrate, serratures tipped with deciduous glandular mucro, penninerved, glossy and glabrous above, stellate-pubescent along nerves below; stipules minute, caducous. Inflorescences terminal or axillary panicles up to 15 cm; peduncle to 1 cm; bracts linear, up to 1.5 mm. Flowers small, bisexual, 5-merous. Calyx-tube urceolate or hemispherical, persistent; lobes 5, deltoid or ovate-triangular, acute, keeled within, pubescent without, thickened towards apex and margin. Petals cucullate, shortly clawed. Disk cupular, 5-lobed, fleshy, nectariferous. Stamens equalling the petals; filaments filiform; anthers 2-celled, dehiscing longitudinally. Ovary superior, immersed in the disk but free, glabrous, 2–3-loculed; style short, thick, 2–(3)-grooved; stigma capitate. Drupe 3-celled, globular or obovate, red, surrounded at base by the adnate, accrescent calyx; pyrenes 3, coriaceous, indehiscent. Seeds 2 or 3, smooth, ovoid or oblong; testa crustaceous.

35 species, distributed in Asia minor, Somalia, South America, Pakistan, India and Sri Lanka; 1 in Sri Lanka.

Sageretia hamosa (Wall.) Brongn., Mem. Fam. Rhamn. 53. 1826; Brogn., Ann. Sci. Nat. Bot. 10: 360. 1827; Royle, Ill. Bot. Himal. 5:169. 1835, sphalm. *"ramosa"*; Lawson in Hook. f., Fl. Br. Ind. 1: 641. 1875; Gamble, Fl. Pres. Madras 1: 223. 1918; Ramchand. & Nair, Fl. Cannannore 99. 1988; Bhand. & Bhans. in Fasc. Fl. India 20: 63. f. 14. 1990.

Zizyphus hamosa Wall. in Roxb., Fl. Ind. 2: 369.1824. Type: Nepal, *Wallich n. 4253 A* (K–W; CAL, isotype).
Sageretia costata Miq., Fl. Ind. Bat. 1: 645. 1855; Trimen, Handb. Fl. Ceylon 1: 284. 1893. Type: Java, Herb. *Horsfield.*

Berchemia parviflora sensu Thw., Enum. Pl. Zeyl. 74. 1858.

Sageretia affinis sensu Thw., Enum. Pl. Zeyl. 410. 1864.

Large straggling shrubs; branches divaricate, lateral ones often modified into straight, blunt, reflexed thorns, young branches brown-pubescent. Leaves subopposite, elliptic-oblong, or lanceolate, 9.1–13.8 × 4.4–5.8 cm, long-acuminate at apex, rounded at base, glabrous, coriaceous, finely serrulate, lateral nerves 6–8 pairs, prominent beneath; petiole 0.8–1.3 cm long, channelled above; stipules 1–3 mm long, linear, slightly pubescent, caducous. Flowers sessile in terminal and axillary, pubescent panicles not longer than leaves, 3.0–4.5 mm across, whitish-pink; bracts minute, lanceolate. Calyx lobes triangular, 1.2–2.5 × 1.0–1.5 mm, acute, brown-pubescent. Petals oblong, 1.5 × 2.0 mm, obtuse, somewhat emarginate at apex, margin inrolled, hiding the stamens. Stamens 2.0 mm long; anthers dorsifixed. Disk entire or slightly 5-lobed, glabrous. Ovary urceolate; style very short; stigma 3-grooved, convex. Drupes ovoid to obovoid, 1.0 × 0.8 cm, on persistent calyx, smooth, red.

Distr. India, Nepal, Taiwan and Sri Lanka.

Ecol. Upper montane zone, along edges of evergreen forest, common; fls Apr–Oct, frts Nov–Jan.

Notes. The original type is heterogenous; *4253 A* is the type of this species, and *4253 B* is the type of S. *Wallichii* Bhand. & Bhans. (l.c. 70. 1990).

Plate No. 438 in PDA, based on *C.P. 2477* (habit & dissected flowers) and named as *Sageretia costata* Miq., represents this species.

Specimens Examined. NUWARA ELIYA DISTRICT: Maturata, in 1851, *s. coll. C.P. 2477* (BM, HAK, K, PDA), Haputale, 1500 m, 18 Apr 1969, *Kostermans 23201* (K); Hakgala, Aug 1885, *W.F. s.n.* (PDA); Jungle S. of laboratory, Hakgala, *Willis s.n.* (PDA); Jungle below garden, Hakgala, *A.M. Silva s.n.* (HAK, PDA).

STEMONACEAE
(by B.M. Wadhwa* and A. Weerasooriya**)

Engler in Pflanzenfam. 2(5): 8. 1887. Type genus: *Stemona* Lour.

Climbing, trailing or erect perennial herbs, mostly with fasciculate tubers or with a short rhizome. Leaves alternate or opposite, sometimes verticillate, papyraceous when dry, ovate or broadly ovate, subcaudately acuminate, margin entire; nerves basal or lateral, curved, shallowly depressed above, secondary venations finely trabeculate; petiole at base pulvinate or slightly sheathing (as in *Stichoneuron*). Inflorescence axillary, usually a peduncled cincinnus, appearing to be a short raceme, rarely 1-flowered. Flowers bracteate, actinomorphic, bisexual. Perianth lobes 4, biseriate, subequal, free, rarely connate at the base, valvate, petaloid, outcurved at anthesis, persistent; pedicels articulated. Stamens 4, epipetalous; filaments short, adnate to base of petals, at base mutually free or shortly connate; anthers dorsifixed or basifixed, 2-celled, longitudinally dehiscent, connective either without or sometimes produced above into a long appendage, the thecae themselves in addition often protruding into a common sterile appendix, thus forming a crown-like structure over the stigma (as in *Stemona*). Ovary superior or semi-inferior, small, 1-locular; ovules few to many, basally or apically attached, anatropous or hemianatropous; style absent; stigma inconspicuous, papillose. Fruit a 2-valved, compressed capsule. Seeds few to many, broadly ellipsoid or ovoid-oblong, longitudinally grooved, beaked, albuminous.

About 3 genera and 25 species distributed in Tropical Asia to Japan and through Malesia to northern Australia and southeast United States; represented by 1 genus and 1 species in Sri Lanka.

STEMONA

Lour., Fl. Cochinch. 1: 404. 1790; Benth. & Hook. f., Gen. Pl. 3: 747. 1880; Hook. f., Fl. Br. Ind. 6: 298. 1892; Trimen, Handb. Fl. Ceylon 4: 281. 1898; Prain, J. Asiat. Soc. Bengal 73: 39. 1904; J. Sm., Bull. Jard. Bot. Buitenzorg III, 6: 73. 1924; Telford in Fl. Australia 46: 177. 1986. Type species: *S. tuberosa* Lour.

* Royal Botanic Gardens, Kew.
** Flora of Ceylon Project, Peradeniya.

Trailing or erect herbs or weak climbers, mostly with perennial tubers. Leaves alternate, opposite or verticillate (as in species of China & Japan), lamina with curved basal nerves; petioles not sheathing. Inflorescence peduncled or sessile, raceme-like, rarely 1-flowered. Flowers with 4 tepals, biseriate, lanceolate, spreading at anthesis, the inner two tepals slightly wider than the outer ones. Stamens sub-hypogynous; filaments very short, ± connate in a ring, anthers with thecae dorsally attached to the base of a long petaloid outgrowth of connectives, the thecae apically with or without subulate appendix. Ovary superior, free, compressed; stigma small, sessile. Capsule 2-valved, compressed, few to many-seeded. Seeds basally inserted, ovoid or oblong, grooved, beaked; funicle bearded.

About 20 species distributed in Japan and continental Asia; represented by 1 species in Sri Lanka.

Stemona curtisii Hook. f., Fl. Br. Ind. 6: 298. 1892; Curtis, Bot. Mag. 48: t. 7254. 1892; Duyfjes in Fl. Males. ser. 1, 11(2): 403. 1993. Type: Penang, *Curtis 1522* (K, holotype).

Roxburghia gloriosoides Roxb. var. *minor* Thw., Enum. Pl. Zeyl. 432. 1864.
   Type: Sri Lanka, Trincomalee, Sept. 1862, *Glenie* in *C.P. 3775* (PDA).
Stemona tuberosa Lour. var. *minor* (Thw.) Trimen, Syst. Cat. 94. 1885.
Stemona minor (Thw.) Hook. f., Fl. Br. Ind. 6: 298. 1892, p.p.; Trimen,
   Handb. Fl. Ceylon 4: 281. 1898. Type: Sri Lanka, Trincomalee, Sept
   1862, *Glenie C.P. 3775* (PDA, lectotype; K, isolectotype).
Stemona tuberosa Auct. (non Lour., 1790); Ridley, Mat. Fl. Malay Penins.
   2: 86. 1907.

Weak climber, up to 2 m; stem glabrous; roots tuberous in fascicle. Leaves alternate, rarely opposite, broadly ovate, 3.5–10 × 2.0–5.5 cm; base very often cordate, sometimes shallowly cordate, apex acuminate, margin entire; nerves 9–11 (–13); petioles 3.0–4.5 cm long. Inflorescence 2-many flowered; peduncle 2.5–6 cm long. Bracts subulate, 6–10 mm long. Flowers pendent, bisexual, actinomorphic; pedicels 8–10 mm long. Tepals 4, in 2 rows, pink to dark-brownish red, lanceolate, 1.5–2.2 cm long. Stamens 13–18 mm long; anthers 7–10 mm long, thecae with tepaloid outgrowth of the connectives, separated by smooth, projecting ridge up to 1.5 mm high; additional appendix of the thecae absent. Fruit ovoid-oblong, c. 2.5–3.0 × 1.5 cm, shortly beaked, 2–6-seeded. Seeds ± oblong, 1.5–2.0 cm long, shortly beaked, enveloped at the base by an aril consisting of hollow, finger-like appendages.

Distr. Thailand, Malay Peninsula and Sri Lanka.

Ecol. In scrub forest along coast, on sandy/limestone soil, rare; fls & fts July–Aug.

Notes. There is no recent collection of this·species. A thorough search from forests along the coast may be made to locate the species.

Specimens Examined. TRINCOMALEE DISTRICT: Trincomalee, Sept 1862, *Glenie* in *C.P. 3775* (PDA, Lectotype; K, isolectotype); Kanniya, 23 Aug 1925, *A. de Alwis 3* (PDA).

THEACEAE
(by B.M. Wadhwa[*])

D. Don, Prod. Fl. Nepal. 224. 1825. Type genus: *Thea* L.

Evergreen shrubs or trees, sometimes lianes. Leaves simple, usually spirally arranged, sometimes alternate, often serrate, mostly coriaceous; stipules 0. Flowers usually large, solitary and axillary, sometimes in terminal racemes or panicles, regular, bisexual, sometimes unisexual or polygamous, often bibracteolate. Sepals (4) 5 (–7), imbricate, free or basally connate, sometimes persistent. Petals (4) 5 (-many), imbricate, often slightly basally connate, sometimes not clearly distinguishable from sepals. Stamens 5–15-numerous, free or basally connate in a ring or in 5 bundles, opposite and adnate to petals, often basally nectariferous; anthers basifixed or versatile, longitudinally dehiscent. Ovary superior, sometimes semi-inferior, mostly 3–5-celled; styles as many as the cells, free or basally united; ovules 2-many in each cell, rarely 1 in each cell; placentation axile. Fruit a loculicidally dehiscent capsule or a juicy, indehiscent berry. Seeds few or many, winged or wingless; endosperm scanty or 0; embryo large, straight or curved.

Approximately 25 genera and about 520 species in the tropical and warm temperate areas of the world; represented by 5 genera and 12 species in Sri Lanka.

KEY TO THE GENERA

1 Fruit a woody capsule, loculicidally dehiscent; flowers bisexual; anthers versatile
  2 Capsules cylindrical; seeds expanded at the top into a wing .... **1. Gordonia**
  2 Capsules subglobose; seeds rounded or plano-convex, wingless .... **2. Camellia**
1 Fruit a berry; flowers bisexual, polygamous or unisexual; anthers basifixed
   3 Flowers at most 1 cm diam., 1–6 together in leaf-axil, unisexual; anthers glabrous; leaves up to 12 cm long .... **3. Eurya**
   3 Flowers more than 1 cm diam., 1–2 in leaf-axil, bisexual or polygamous; anthers pilose or glabrous; leaves up to 18 cm long
    4 Anthers pilose; flowers always bisexual; fruits many-seeded .... **4. Adinandra**
    4 Anthers glabrous; flowers bisexual or polygamous; fruits 1–4-seeded ....
     .... **5. Ternstroemia**

[*] Royal Botanic Gardens, Kew.

1. GORDONIA

Ellis, Philos. Trans. 60: 518, t.11. 1771, nom. cons.; Thw., Enum. Pl. Zeyl. 40: 1858; Benth & Hook. f., Gen. Pl. 1: 186. 1862 (incl. *Laplacea*); Dyer in Hook. f., Fl. Br. Ind. 1: 290. 1874; Trimen, Handb. Fl. Ceylon 1: 110. 1893; Burkill, J. Straits Branch Roy. Asiat. Soc. 76: 133. 1917; Melchior in Pflanzenfam. ed. 2, 21: 136, f. 63. 1925 (incl. *Laplacea*); Keng, Gard. Bull. Singapore 33: 308. 1980. Type species: *G. lasianthus* (L.) Ellis (= *Hypericum lasianthus* L.).

Lasianthus Adans., Fam. Pl. 2: 398. 1763 (*non* Jack, 1823, nec Zucc. ex DC., 1836). Type species: *Hypericum lasianthus* L. (=*Gordonia lasianthus* (L.) Ellis).

Laplacea H.B.K., Nov. Gen. Sp. 5: 207 t. 461. 1822. Type species: *L. speciosa* H.B.K.

Polyspora Sweet ex G. Don., Gen. Hist. 1: 564, 574. 1831. Type species: *P. axillaris* (Roxb. ex Ker-Gawl.) Sweet.

Carria Gardn., Calcutta J. Nat. Hist. 7: 6. 1847. Type species: *C. speciosa* Gardner.

Dipterosperma Griff., Notul. 4: 564. 1854, non Hassk., 1842. Type species not designated.

Small to medium-sized trees, rarely shrubs. Leaves simple, coriaceous, alternate, spirally or distichously arranged, entire or serrate. Flowers rather large, bisexual, axillary, solitary or 2–3 together in a cluster, often sub-sessile; bracteoles 2–3 (5) borne on the pedicels caducous. Sepals 5–6, unequal, inner ones often petal-like. Petals 5–6, rarely 9–10, sub-equal, usually slightly connate at the base. Stamens numerous, in 3–4 whorls, sometimes united below in 1 or 5 bundles (1 or 5-adelphous), adnate to the base of the petals; anthers versatile, usually antrorse. Ovary 3–5(–6)-celled; ovules 2–8 in each cell, on axile placentation in 2 vertical rows; styles mostly 5, sometimes 3 or 6, fused to varying extent proximally, sometimes free. Fruit a woody capsule, ovoid-cylindric, bluntly angulate, dehiscing loculicidally up to a persistent central columella. Seeds usually 2–5 in each cell, compressed with a thin membranous, oblique, unilaterally attached wing; testa woody, embryo oblong, straight or oblique.

About 70 species, distributed in S.E. Asia and America. About 40 species are found from India, Sri Lanka, Burma, Thailand, Indo-China, S. China, Taiwan and southwards to Malaysia, Philippine Islands and New Guinea and 30 species are distributed in Central, North and South America, and the West Indies; 4 species in Sri Lanka.

Gordonia axillaris (Ker-Gawl.) D. Dietr. has been introduced in Hakgala Botanic Garden for its beautiful, large, white flowers.

KEY TO THE SPECIES

1 Flowers deep crimson or red; leaves strongly revolute, deep green and shining above . . . . .
. . . . **1. G. speciosa**
1 Flowers white or pink with reddish tinge; leaves flat or slightly revolute in the lower part,
  neither deep green nor shining above
  2 Young branches and leaves pilose or hairy; leaves lanceolate or oblong-lanceolate, margin
    flat; flowers white . . . . **2. G. ceylanica**
  2 Young branches and leaves glabrous; leaves elliptic or elliptic-oblong, margin flat or slightly
    revolute in the lower part; flowers white or pink with reddish-tinge
    3 Flowers white; leaves elliptic, margin slightly revolute in the lower part . . . .
    . . . . **3. G. elliptica**
    3 Flowers pink with reddish tinge; leaves elliptic-oblong or oblanceolate, margin flat, not
      revolute . . . . **4. G. dassanayakei**

1. Gordonia speciosa (Gardn.) Choisy, Mem. Soc. Phys. Geneve 14: 142
(Mem. Ternstr. 52) 1855; Thw., Enum. Pl. Zeyl. 40. 1858; Dyer in Hook. f.,
Fl. Br. Ind. 1: 292. 1874; Trimen, Handb. Fl. Ceylon 1: 111. 1893; Burkill,
J. Straits Branch Roy. Asiat. Soc. 76: 156. 1917; Melchior in Pflanzenfam.
ed. 2, 21: 137. 1925; Worthington, Ceylon Trees 45. 1959.

Carria speciosa Gardn., Calcutta J. Nat. Hist. 7: 7. 1847. Type: Sri Lanka:
  "Rambodde", *Gardner 94* ((K).

Medium-sized trees, up to 15 m; branches terete, glabrous, branchlets
with leaves crowded at their extremities; bark grey, fissured longitudinally.
Leaves coriaceous, glabrous, almost sessile, broadly elliptic or oblong-oval,
8.6–16.5 × 3.6–7 cm, very obtuse and retuse at the apex, margin often
strongly revolute, entire, green and shining above, somewhat glaucous beneath; midrib strong, channelled above, veins obsolete. Flowers bisexual, axillary and terminal, subsessile, campanulate, very large, 9.0–10.8 cm diam.,
deep red or crimson; pedicels short, thick, c. 3–4 mm long, bearing a few
bracts. Bracts oblong-ovate or obovate, glabrous. Calyx free, persistent; lobes
3–5, rotund or broadly oval, 1.5–1.7 × 1.4–1.5 cm, deeply emarginate, concave, coriaceous, margin scarious, ciliate. Petals 5, broadly obovate, 4.5–5.2
× 2.8–3.4 cm, emarginate, imbricate, connate at base, silky tomentose outside. Stamens numerous, in 5 bundles (or 5-adelphous) attached to the base of
the petals and nearly half as long as the petals; filaments stout; anthers ovate,
2-celled, extrorse, connectives broad, fleshy. Ovary superior, 5-celled, silky
hairy, with 4–6 ovules in each cell, attached to the inner angle in 2-rows,
pendulous; styles ± as long as the stamens, stout; stigma 5-lobed. Capsules
woody, oblong, 5-angled, 5-celled, 3.8–4.6 cm long, apiculate, surrounded by
persistent sepals, dehiscing loculicidally by 5-valves. Seeds 3–5 in each cell,
compressed, with a thin, membranous, unilaterally attached wing; embryo
straight, exalbuminous.

Distr. Endemic.

Ecol. In montane forest areas of wet zone, 1270–1650 m, uncommon; fls & frts Oct–Feb.

Vern. Miriheeriya (S).

Notes. Plate No. 177 in PDA, based on specimen *C.P. 2403*, belongs to this species.

This species with large, beautiful, deep crimson coloured flowers, formerly occurred gregariously in the forest above Ramboda, but due to extensive clearings, it has become rare and uncommon. This needs to be conserved by declaring the Peak Wilderness area above Moray Estate as a 'Gene Sanctuary' for this species.

Specimens Examined. KANDY DISTRICT: Bogawantalawa, ± 1525 m, 15 Oct 1955, *Worthington 6778* (K), 1420 m, 7 Feb 1940, *Worthington 785* (K); Devonford Estate, Upper Dikoya, 1570 m, 19 Oct 1947, *Worthington 3236* (BM, K); Galboda, 1235 m, 27 Jan 1954, *Worthington 6548* (K); Near Fishing Hut, Moray Estate, 17 Oct 1992, *Wadhwa & Weerasooriya 136* (K, PDA), 16 May 1971, *Kostermans 24135* (BM, K, PDA); Moray Estate, Maskeliya, 3 Feb 1971, *Balakrishnan 592* (K, PDA); Batulu Oya Stream, near Fishing Hut, edge of Peak Sanctuary, upper Moray Estate, 1270 m, 25 Oct 1978, *Fosberg 57990* (K); Adam's Peak Wilderness bordering Moray Estate, 80°31' E, 6°48' N, 150 m, 27 March 1974, *Jayasuriya & Sumithraarachchi 1563* (K, PDA); Peak Wilderness, Meriyakota, above Fairlawn Estate, 6°45' N, 80°37' E, 1630 m, 16 Aug 1984, *Jayasuriya, Balasubramaniam, Greller, S. & N. Gunatilleke 2835* (PDA); Peak Wilderness above Devonford and Maratenne Estates, 6°45' N, 80°39' E, 1650 m, 15 Aug 1984, *Jayasuriya, Balasubramaniam, Greller, S. & N. Gunatilleke 2843* (PDA); Peak Wilderness Forest, 18 Oct 1992, *Wadhwa & Weerasooriya 171* (K, PDA); Forest behind Gartmore Estate, Maskeliya, about 1700 m, 16 Jan 1971, *Theobald, Krahulik & Balakrishnan 3862* (PDA). NUWARA ELIYA DISTRICT: Ramboda, 1525 m, *Gardner 94* (K, holotype; BM, isotype); Ramboda, *Thwaites C.P. 2403* (BM, CAL, HAK, K, PDA); Nuwara Eliya, Oct 1845, *Thomson s.n.* (K). RATNAPURA DISTRICT: Gethampana, trail to Adam's Peak from Eratne, 1360 m, 2 March 1985, *Jayasuriya & N. Gunatilleke 3180* (PDA); Below Seethagangula, S. slopes of Peak Wilderness Sanctuary approach to Carney, 1575 m, 28 Feb 1985, *Jayasuriya & N. Gunatilleke 3353* (PDA); Bogawantlawa Proposed Reserve, 4 Nov 1994, *Jayasuriya 8482* (PDA).

2. Gordonia ceylanica Wight, Ill. Ind. Bot. 1: 99. 1840; Thw., Enum. Pl. Zeyl. 40. 1858; Dyer in Hook. f., Fl. Br. Ind. 1: 291. 1874 *'zeylanica'*; Trimen, Handb. Fl. Ceylon 1: 110. 1893 (excl. var. *elliptica*); Burkill, J. Straits Branch

Roy. Asiat. Soc. 76: 156. 1917 '*zeylanica*'; Worthington, Ceylon Trees 46. 1959 '*zeylanica*'. Holotype: Sri Lanka, *Col. Walker 25* (K).

Medium-sized trees, 12–15 m, tall; branchlets pilose; bark with vertical concave ridges. Leaves rather coriaceous, lanceolate or oblong-lanceolate, 5.8–9.2 × 2.2–2.8 cm long, tapering at both ends; apex acute, usually with a retuse, glandular point, margin usually entire, sometimes serrate towards the apex; midrib strongly marked beneath, channelled above, veins obsolete; young leaves garnet-brown, shining, pilose at the back, at least on midrib even in some mature leaves; petioles thick, 2–3 mm long, ciliate or pilose in young leaves. Flowers axillary, solitary on short pedicels, 3.2–4.2 cm diam., white; pedicels thick, about 4 mm long, pilose-hairy. Sepals orbicular, 7 × 8 mm, smooth, retuse, margins ciliate, 2 inner sepals somewhat larger. Petals 5, obcordate or rotundate with a broad claw, 1.5–2.1 × 1.0–1.5 cm, pubescent externally towards the base. Stamens numerous, free; filaments stout, fusiform; anthers oblong, versatile, introrse. Ovary 3–5-celled, silky-hairy; ovules 4–6 in each cell in 2 rows; stigma 5-lobed, clavate. Capsules oblong-ovoid, 2.5–4.0 cm long, apiculate, 3–5-angled, covered with silky hairs when young, loculicidally dehiscent, surrounded at the base with persistent sepals. Seeds flattened, pale-brown, winged.

Distr. Endemic.

Ecol. In montane forests of wet zone, 900–2100 m.

Vern. Mirihiriya, Mihiriya (S).

Specimens Examined. MATALE DISTRICT: Midlands, 1100 m, 23 July 1974, *Jayasuriya & Bandaranayake 1763* (PDA); Kandanuwara, Matale E, 1065 m, 7 Nov 1946, *Worthington 2398A* (BM). KANDY DISTRICT: Mirisketiya, 1000 m, 19 Oct 1977, *Nooteboom 3413* (PDA); Hunnasgiriya, 1000 m, 23 Oct 1980, *Werner 120* (PDA); Dolosbage, 975–1125 m, 9 Dec 1946, *Worthington 1862* (K), *Worthington 1895* (BM), Rangala, Corbet's Gap, 1220 m, 27 Feb 1948, *Worthington 3586* (BM), Corbet's Gap, Kobonilagala, 1220 m, 9 Sept 1951, *Worthington 5485* (BM). BADULLA DISTRICT: Namunukula, 1345 m, 5 Dec 1992, *Wadhwa & Weerasooriya 383* (PDA, K). RATNAPURA DISTRICT: Forest above Longford Division of Hayes Group, 6°23′ N, 80°40′ E, 1035 m, 28 Aug 1984, *Jayasuriya, Balasubramaniam & Greller 2888* (PDA); Kadumeriya Proposed Reserve, 23 Jan 1994, *Jayasuriya 7945* (PDA); Forest above Nagrak Tea Estate Bungalow near Galagama Falls, 2150 m, 28 Dec 1970, *Theobald & Krahulik 2841* (PDA); Trail to Gonagala above Longford Division of Hayes Group, 6°23′ N, 80°39′ E, 1125 m, 28 Aug 1984, *Jayasuriya, Balasubramaniam & Greller 2942, 2943* (PDA); Tangamale Plains, Morningside, Sinharaja, 13 Jan 1993, *Jayasuriya 7059* (PDA); Rakwana, Aigburth Estate, 990 m, 17 Sept 1946, *Worthington 2170* (K). NUWARA ELIYA DISTRICT: Above Hakgala

Botanic Garden, 11 Feb 1970, *Cooray 70021106R* (K, PDA); Edge of forest along Hakgala Botanic Garden, 15 Oct 1992, *Wadhwa & Weerasooriya 105* (K, PDA); Pedro Estate, Nuwara Eliya, 1950 m, 14 March 1956, *Worthington 6813* (K); Near 52 mile mark, culvert No. 5, North side of Nuwara Eliya-Hakgala road, 31 Oct 1973, *Sohmer, Jayasuriya & Eleizer 8448* (PDA); Trail to Mt. Pidurutalagala, about 1/3 of the way to summit, ± 2000 m, 31 Oct 1973, *Sohmer, Jayasuriya & Eleizer 8430* (PDA); Pidurutalagala, 2135–2525 m, 14 July 1978, *Meijer 1915* (K), 1800–2400 m, 1 Oct 1977, *Nooteboom 3254* (PDA); Jungle path, Pidurutalagala, 21 Apr 1906, *A.M. Silva s.n.* (PDA, HAK); Elk Plains near Ambawela, 1800 m, Apr 1969, *Kostermans s.n.* (K); Hakgala-Nuwara Eliya road near Botanic Garden, 30 June 1973, *Nowicke & Jayasuriya 276* (K, PDA); Kandapola road, near 6th km post, 17 Oct 1992, *Wadhwa & Weerasooriya 132* (K, PDA); Pattipola, 1891 m, 5 Sept 1981, *Sittambalam 76* (PDU); Lower side of Nuwara Eliya road, Hakgala, 12 May 1906, *A.M. Silva s.n.* (HAK, PDA); Mooloya, 1585 m, 5 Feb 1975, *Waas 1140* (PDA); Ramboda area above Frotoft Estate, 2000 m, 21 June 1973, *Kostermans 25099* (PDA); Horton Plains-Diyagama, 1950 m, 28 Jan 1945, *Worthington s.n.* (BM); Pattipola-Ambewela Proposed Reserve, 1 Nov 1994, *Jayasuriya 8447, 8453* (PDA); Kikiliyamana Proposed Reserve, Nuwara Eliya Range, 9 Sept 1994, *Jayasuriya 8291* (PDA); Mahacoodugala, near the bridge at the Estate, 17 Oct 1992, *Wadhwa & Weerasooriya 134* (K, PDA); Mahacoodugala, 21 Oct 1994, *Wadhwa & Weerasooriya 3941* (PDA); Nuwara Eliya, Mar 1846, *Thwaites C.P. 788* (Type of *G. ceylanica* var. *lanceolata* Thw. (BM, CAL, HAK, K); Kandapola-Sita Eliya Forest Reserve, 7 Sept 1994, *Jayasuriya & Karunaratne 8252* (PDA); Walawe Basin approach, Walaboda, 2 March 1994, *Jayasuriya 8052* (PDA). MATARA DISTRICT: Ensalwatte-Kurulugala Division, Deniyaya forest range, 27 June 1992, *Jayasuriya 6444* (PDA). LOCALITY UNKNOWN: Aug 1936, *Willis 118* (CAL); *Col. Walker 25* (K, holotype).

3. Gordonia elliptica Gardner, Calcutta J. Nat. Hist. 7: 448. 1847; Dyer in Hook. f., Fl. Br. Ind. 1: 291. 1874. Type: Sri Lanka, Elephant Plains, 1830 m, *Gardner 93* (K).

Gordonia ceylanica Wight var. *elliptica* (Gardner) Thw., Enum. Pl. Zeyl. 40. 1858; Trimen, Handb. Fl. Ceylon 1: 111. 1893.

Large trees, 20–25 m tall; branchlets glabrous, with leaves crowded at their extremities; bark (dry) pale brown, fissured, about 2 mm thick; living bark red, about 10 mm thick, yellowish within. Leaves ± sessile, somewhat coriaceous, elliptic, 5.3–11.1 × 3.2–4.8 cm, ± obtuse or rounded at both ends, retuse at the apex, margin entire, slightly revolute in the lower portion when dry, upper surface reticulately veined; midrib strongly marked beneath, channelled above. Flowers axillary, solitary, sessile, 4.2–6.5 cm across, white. Sepals 5; lobes orbicular, 8 × 9 mm; emarginate, puberulous without, margin ciliate. Petals 5, obcordate, 2.2–3.2 × 1.6–1.8 cm, connate at the base, thick, puberulous externally. Stamens numerous, in 3-whorls, free; filaments stout, pilose hairy; anthers oblong, versatile, connectives broad. Ovary 5-celled, silky-pilose; ovules 4–6 in each cell in 2 rows; styles somewhat fused; stigmas 5-lobed. Capsule oblong, 5-angled, 2.8–4.5 cm long, 5-valved, loculicidally dehiscent, pilose, with persistent sepals at the base. Seeds not seen.

Distr. Endemic.

Ecol. In montane forests of wet zone, 800–1830 m, occasional; fls July–Oct, frts Oct–Apr.

Notes. Plate No. 176 in PDA, based on specimen *C.P. 2417*, belongs to this species.

Specimens Examined. MATALE DISTRICT: Rattota-Illukkumbura Road, 23 Nov 1994, *Wadhwa, Weerasooriya & Samarasinghe 501* (K, PDA), Kandanuwara, Matale E, 1005 m, 18 Aug 1953, *Worthington 6372* (K), 1065 m, 7 Nov 1946, *Worthington 2398B* (BM). KANDY DISTRICT: Hunnasgiriya hill, above Madulkele Estate, 13 Nov 1975, *Sohmer & Jayasuriya 10625* (K); Hills E. of Hoolankande Estate, N. of Madulkele, 1370 m, 20 Aug 1978, *Huber 797* (PDA); Knuckles, Madulkele area, 800 m, 12 June 1973, *Kostermans 25066* (PDA); Corbet's Gap, 1220 m, 27 Feb 1946, *Worthington 3598* (K), 13 Aug 1970, *Meijer & Dassanayake 661* (K, PDA); Mountain Ridge between Knuckles and Ritigala No. 2 Peaks, E. of Bambrella, 1500 m, 9 Apr 1974, *Davidse 8279* (K, PDA); Knuckles, 1400 m, 22 July 1946, *Worthington 1950* (BM). NUWARA ELIYA DISTRICT: Elephant Plains, *Thwaites C.P. 2417* (BM, CAL, HAK, K); 830 m, *Gardner 93* (K, holotype; BM, isotype); Nuwara Eliya, *Thomson s.n.* (K); Ramboda, *Thwaites C.P. 2403*, p.p. (K); Mahacoodagala, 26 Sept 1994, *Jayasuriya 8349* (PDA). BADULLA DISTRICT: Namunukula Kanda, 25 Apr 1971, *Waas 471* (K, PDA), ± 2000 m, 6 May 1976, *Waas 1614* (K, PDA); Haputale, 1585 m, Sept 1890, *s. coll. s.n.* (PDA); Totulagalla Estate, above Haputale, 1600 m, 19 Apr 1969, *Kostermans 3230* (K); Haputale, Thangamalai Forest near Adisham, ± 1200 m, 21 Nov 1994, *Wadhwa, Weerasooriya & Samarasinghe 500* (K, PDA).

4. Gordonia dassanayakei Wadhwa et Weerasooriya, sp. nov.—**Fig. 1.**

Species *G. ellipticae* Gardner foliis juvenibus et ramulisque glabris affinis, sed foliis elliptico-oblongis vel oblanceolatis, venationae reticulata supra non valde conspicua, marginibus ubique planis; floribus breviter pedicellatis, petalis roseis ad extremitates rubentibus differt.

Medium-sized trees, 12–15 m tall; branchlets glabrous; bole 20.0 cm diam.; bark (dry) cracked, soft, thin, brown, live pale-red with paler spots.

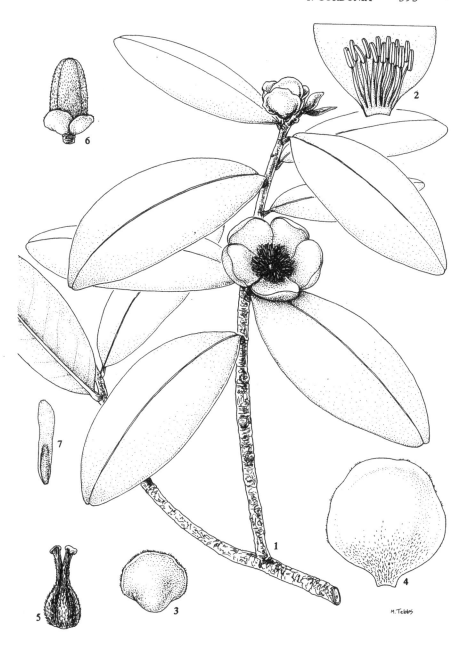

Fig. 1. *Gordonia dassanayakei* Wadhwa et Weerasooriya sp. nov. 1. Habit (×2/3); 2. Stamens with part of a petal (×2); 3. Sepal (×2); 4. Petal (×2); 5. Ovary (×2); 6. Fruit (×2/3); 7. Seed (×2).

Leaves coriaceous, elliptic-oblong or oblanceolate, 7.1–16.4 × 2.8–5.4 cm, apex obtuse or subacute, retuse, margin entire, flat; midrib strongly marked beneath, channelled above; young leaves almost glabrous; petioles 2–3 mm long. Flowers solitary, axillary, short- pedicellate, 4.6–5.4 cm diam., pink with reddish tinge; pedicels thick, stout, c. 3 mm long. Sepals 5, orbicular, 0.8 × 1.2 cm, somewhat pubescent without, margins scarious, ciliate, persistent. Petals 5, thick, suborbicular or obovate, 2.3–2.7 × 1.4–1.6 cm, pink, reddish tinged near extremities, silky pubescent at base externally. Stamens numerous, 5-adelphous, 1.2 cm long, adnate to the base of the petals; filaments thick, hairy; anthers oblong, versatile, introrse. Ovary globose, 5–celled, slightly 5-angled, covered with golden hairs; ovules 4–6 in each cell; style thick, columnar, 2–3 mm long, slightly branched near the tip. Fruit cylindric-oblong, 5-angled, 3 cm long, loculicidally dehiscent, minute silky hairy, with persistent sepals at the base. Seeds 3–5 in each cell, 1.2 cm long (incl. wing), compressed with thin, membranous oblique wing.

This species comes close to *G. elliptica* Gardner in having glabrous young leaves and branchlets but differs in its elliptic-oblong or oblanceolate leaves, without well marked reticulate venation on upper surface, margin flat throughout; flowers short-pedicellate, pink with reddish tinge at extremities.

This species has been named after Prof. M.D. Dassanayake for his contributions to the Botany of Sri Lanka.

Distr. Endemic.

Ecol. In submontane forests of the wet zone, 1300–1650 m, rare; fls Aug-Nov, frts Oct-Dec.

Specimens Examined. BADULLA DISTRICT: Tothulagalle Estate above Haputale, 1600 m, 7 May 1969, *Kostermans 23386* (BM, holotype; K, PDA, isotypes); Namunukula Hill forest, 1340 m, 5 Dec 1992, *Wadhwa & Weerasooriya 380* (K, PDA), 16 Oct 1992, *Wadhwa & Weerasooriya 126* (K, PDA); Adisham, Haputale, 1615 m, 11 Aug 1951, *Worthington 5396* (K). NUWARA ELIYA DISTRICT: Forest around bridge near Mahacoodagala Estate, 80° 50′ E, 7° 03′ N, 29 Oct 1975, *Sohmer & Sumithraarachchi 10147* (K, PDA), 21 Oct 1994, *Wadhwa & Weerasooriya 395* (K, PDA), near bridge, 16 Oct 1992, *Wadhwa & Weerasooriya 133* (K, PDA, paratype). KANDY DISTRICT: West slopes of Knuckles above Bambrella, 1330 m, 22 Nov 1977, *Huber 685* (PDA).

2. CAMELLIA

L., Sp. Pl. 698. 1753; L., Gen. Pl. ed. 5, 311. 1754 emend. Sweet, Hort. Suburb. Londin. 157. 1818; Benth. & Hook. f., Gen. Pl. 1: 187. 1862; Sealy, Rev. Gen. Camellia 14. 1958. Type: *C. japonica* L.

Thea L., Sp. Pl. 515. 1753. *T. sinensis* L.

Evergreen trees or shrubs. Leaves alternate, chartaceous or coriaceous. Flowers solitary or 2–3 in a cluster, axillary or subterminal; shortly pedicellate, bisexual. Sepals 5–6, imbricate, often gradually passing from bracts to sepals, persistent. Petals mostly 5–7, slightly connate at the base. Stamens numerous, free or outer ones connate below and adnate to the corolla; anthers versatile. Ovary mostly 3–5 locular, with 4–6 ovules in each loculus; styles free or fused at base. Fruit a woody, subglobose capsule dehiscing loculicidally. Seeds rounded or plano-convex.

About 82 species in the world, mostly in Indo-Malayan region, China and Japan; represented in Sri Lanka by 1 species.

Camellia kissi Wall., Asiat. Res. 13: 429. 1820; Wall., Pl. Asiat. Rar. 3: 36. t. 256. 1832; Brandis, Ind. Trees 61. 1906; Sealy, Rev. Gen. Camellia 197. f. 91 A-K. 1958. Type: Nepal, *Wallich 977* (K).

Camellia druperifera sensu Seeman, Trans. Linn. Soc. London 22: 344. 1859; Dyer in Hook. f., Fl. Br. Ind. 1: 293. 1874, non Lour., 1790.

Shrub or tree up to 12 m; young branches and shoots pale-grey or brown, pubescent to villose, sometimes glabrous. Leaves elliptic-ovate or narrowly oblong-elliptic, 5.5–9.0 × 1.7–3.5 cm, acuminate, base cuneate or rounded, margin finely serrate for one-quarter to two-thirds from the apex, thinly coriaceous or coriaceous, upper surface bright green and glabrous, lower surface pubescent at first; petioles 3–7 mm long, densely pubescent or villose. Flowers bracteolate, fragrant, 1 or 2 at the ends of the shoots and branches, often also in the axils of the uppermost leaves; pedicels 2–6 mm, erect, densely scarred. Bracteoles and sepals ± indistinguishable, caducous, suborbicular, outermost 1.25–2 mm long, inner 9 mm long, very strongly concave, brown, margin membranous, subglabrous to pubescent. Petals caducous, obovate-cuneate, 0.8–1.2 × 0.6–1.0 cm, emarginate to retuse. Stamens a little shorter than the petals, glabrous; outer filaments slightly united at the base; anthers yellow. Ovary ovoid, about 4 mm long, densely silky tomentose; styles 5–6 mm long, divided for half their length or more into 3 arms, glabrous or villose to the point of division and often to the apex. Capsule subglobose, 1.5–2.5 × 1.2–2.2 cm, villose, 3-locular, but usually 1-locular by abortion, with usually 1-seed in each loculus.

D i s t r. India (Sikkim & N.E. India), Nepal, Bhutan, Burma, China, Indo-China and Sri Lanka.

E c o l. Along forest edges in montane zone, 2120 m, rare, fls and frts Oct–Nov.

Camellia sinensis (L.) Kuntze, the tea plant, is extensively cultivated for conventional and fermented tea. Sri Lanka along with India, Java and China are bulk tea producing countries. There are two varieties in cultivation, the typical var. *sinensis* with leaves 5–9 × 2–3 cm, obtuse or with short rounded point, and var. *assamica* (Masters) Kitamura having large leaves 8–14 × 3.5–5.5 cm, ± acuminate at apex.

Camellia japonica L. is cultivated in gardens for its foliage and beautiful flowers.

3. EURYA

Thunb., Nov. Gen. Pl. 67. 1783; Thw., Enum. Pl. Zeyl. 41. 1858; Benth. & Hook. f., Gen. Pl. 1: 183. 1862; Dyer in Hook. f., Fl. Br. Ind. 1: 284. 1874; Trimen, Handb. Fl. Ceylon 1: 109. 1893; Melchior in Pflanzenfam. ed. 2, 21: 146. 1925; Kobuski, Ann. Missouri Bot. Gard. 25: 299. 1938; De Wit, Bull. Jard. Bot. Buitenzorg ser. 3, 17(3): 332. 1947. Type species: *E. japonica* Thunb.

Shrubs or small trees, evergreen, with spreading, often feathered branches. Leaves membranous or coriaceous, usually distichous, glabrous, often crenate-serrate. Flowers generally small, axillary, mostly dioecious, solitary or in small clusters or fascicles of 2–4, shortly pedicellate. Bracteoles 2, persistent. Sepals 5, free, imbricate, persistent. Petals 5 (7), mostly connate at the base. Stamens in male flowers 5–15, sometimes more than 20, adnate to the base of the petals; filaments glabrous; anthers basifixed, glabrous, 2-loculate, sometimes multi-chambered, connective mostly apiculate; in female flowers stamens rudimentary. Ovary in female flowers 3–5-locular, with several ovules in each loculus; styles 3–5, free or united, in male flowers ovary rudimentary. Fruit a berry, indehiscent, many-seeded. Seeds usually small, angled or pitted; albumen fleshy.

About 70 species in the tropical and warm temperate areas of S.E. Asia, Indian Archipelago and Pacific Islands; 4 species in Sri Lanka.

KEY TO THE SPECIES

1 Terminal leaf-buds and young branches glabrous; young branches 2-angled or winged . . . . .
. . . . 1. **E. nitida**
1 Terminal leaf-buds and young branches pubescent or pilose; young branches terete
  2 Leaves membranous or nearly membranous, narrowly oblong-lanceolate, short- to long-acuminate, margin flat, not revolute; veins inconspicuous on the upper surface . . . .
. . . . 2. **E. acuminata**
  2 Leaves coriaceous, thick, broadly elliptic, oval-oblong or obovate-elliptic, bluntly acuminate, margin revolute; veins on upper surface ± impressed
    3 Leaves small, usually 4 cm long, sometimes about 6 cm long, obovate-elliptic or oval-elliptic, bluntly acuminate, margin rigidly revolute; pedicels glabrous; stamens 9–12 . . . .
. . . . 3. **E. chinensis**
    3 Leaves large, 6–12 cm long, broadly elliptic or oval-oblong, margin somewhat revolute in the lower half; pedicels and bracteoles hairy; stamens ± 15 . . . . 4. **E. ceylanica**

1. Eurya nitida Korth. in Temminck, Verh. Naturf. Gesch. Bot. 3: 115. t. 17. 1840; Blume, Ann. Mus. Bot. Lugduno-Batavum 2: 111. 1856; Rehder, J. Arnold Arbor. 15: 99. 1934; Kobuski, Ann. Missouri Bot. Gard. 25: 310. 1938; De Wit, Bull. Jard. Bot. Buitenzorg ser. 3, 17(3): 357. 1947; Keng in Fl. Thailand 2(2): 156. 1972; Chitra in Nair & Henry, Fl. Tamilnadu 1: 29. 1983. Type: Borneo, *Korthals s.n.* (K).

var. **nitida**

Eurya wightiana Wall. [Cat. n. 3662. 1829, nom. nud.] ex Wight, Ill. Ind. Bot. 1: t. 38. 1840. Type: India, *Wallich 3662* (K).
Eurya fasciculata Wall. [Cat. n. 4399. 1830, nom. nud.] ex Vesque, Bull. Soc. Bot. France 42: 153. 1895. Type: Bangladesh: Sylhet, *Wallich 4399* (K).
Eurya elliptica Gardner, Calcutta J. Nat. Hist. 7: 443. 1847. Type: Sri Lanka: Kandy District, Adam's Peak, Mar 1846, *s. coll. C.P. 787* (K).
Eurya japonica Thunb. var. *thunbergii* Thw., Enum. Pl. Zeyl. 41. 1858; Dyer in Hook. f., Fl. Br. Ind. 1: 284. 1874; Trimen, Handb. Fl. Ceylon 1: 109. 1893. Type: Sri Lanka, *Thwaites C.P. 777* (K, holotype; BM, isotype).
Eurya japonica Thunb. var. *nitida* (Korth.) Dyer in Hook. f., Fl. Br. Ind. 1: 284. 1874.
Eurya japonica Auct.: Gamble, Fl. Pres. Madras 1: 79. 1915; Balak., Fl. Jowai 1: 93. 1981; Matthew, Fl. Tamilnadu Carnatic 1: 97. 1983, non Thunb., 1784.

Shrubs 2–5 m high; young ultimate branches and leaf-buds entirely glabrous; branches angulate or winged. Leaves chartaceous, glabrous, often ovate, sometimes obovate or elliptic-ovate, 2.8–7.0 (–10.8) × 1.4–2.5 (–4.4) cm, apex acute or broadly acuminate, base cuneate or slightly attenuate, margin sharply serrate in the upper two-thirds; veins prominent beneath; petioles 1–3 mm long. Flowers axillary, solitary or 2–3 in clusters, polygamous, c. 2.5 mm across, white; pedicels slender, 1–2 mm long, glabrous, reflexed in fruits. Bracts 2 or more, ovate, slightly less than half as long as the sepals, glabrous. Sepals 5, subequal, broadly ovate or rotundate, 1.5–2 mm long and wide, persistent, imbricate. Petals 5, obovate, 4.0 × 2–2.5 mm, connate at base, imbricate, apex emarginate, margin entire or crenulate. Stamens in male flowers 10–15, often connate; filaments about 1.5 mm long; anthers oblong, 2-celled, basifixed, longitudinally dehiscent, connectives mucronulate, in female flowers rudimentary or absent. Ovary in female flowers ovoid, about 4 mm diam., 3-locular, glabrous; ovules many in each cell on axile placentae; styles 3, 2 mm long, connate to half their length; stigmas 3, reflexed, pistillode in male flowers rudimentary, ovoid, c. 1.5 mm long. Berries subglobose or ovoid, 3–4 mm diam., 3-celled, with persistent style, many-seeded. Seeds angular, ovate, compressed.

Distr. India, Sri Lanka, Indo-China, Malaysia, Philippine Islands, S. and S.W. China and Hainan.

Ecol. Along edges of submontane forests in wet zone, 425–2100 m, fairly common; fls and frts Sept–Mar.

Vern. Neya Dassa (S).

Notes. Two other varieties, viz. var. *aurescens* (Rehder & Wilson) Kobuski from China and var. *venosa* De Wit from Malaysia, are reported under this species.

Specimens Examined. MATALE DISTRICT: Campbells Lane, 1220 m, 18 Jan 1975, *Waas 958, 962* (K); Pittawella, 1200 m, 1 Dec 1971, *Jayasuriya, Dassanayake & Balasubramaniam 440* (PDA). KEGALLE DISTRICT: Alagalla, Dekanda, 2135 m, 10 Oct 1929, *Worthington 594* (K). KANDY DISTRICT: Hunnasgiriya, 1370 m, 19 Jan 1975, *Waas 973* (PDA); Laxapana-Maskeliya Road, 850 m, 13 Sept 1977, *Nooteboom & Huber 3108* (PDA); Moray Estate, along banks of stream, 15 Nov 1983, *Sohmer & Waas 8726* (PDA); Norton Bridge, 880 m, 1 May 1947, *Worthington 2719* (K); Ginigathena, 425 m, 11 Sept 1946, *Worthington 2080* (K); Kandy, 17 Feb 1819, *Moon s.n.* (BM); Road to Maskeliya, 900 m, June 1971, *Kostermans 24525* (K, PDA); upper part of Moray Estate, Maskeliya, 1500 m, *Kostermans 25408* (K, PDA); Hunnasgiriya, 1370 m, 19 Jan 1975, *Waas 973* (K); Adam's Peak, Mar 1846, *s. coll. C.P. 787* (Type of *E. elliptica* Gardn., K); Balangoda-Bogawantalawa, 1200 m, 7 Feb 1940, *Worthington 776* (K). RATNAPURA DISTRICT: Below Seethagangula, S. slope of Peak Wilderness Sanctuary from Carney, 1575 m, 28 Feb 1985, *Jayasuriya & Gunatilleke 3351* (PDA). NUWARA ELIYA DISTRICT: About 2 miles E. of Pundaluoya, road to Nuwara Eliya, 1300 m, 26 Dec 1970, *Theobald & Krahulik 2817* (PDA); Kandapola-Sita Eliya Forest Reserve, 7–8 Sept 1994, *Jayasuriya 8230, 8265* (PDA). LOCALITY UNKNOWN: Thwaites *C.P. 777* (Holotype of *E. japonica* Thunb. var. *thunbergii* Thw. K; BM, isotype); in 1836, *Herb. R. Wight 586* (K).

2. Eurya acuminata DC., in Mem. Soc. Phys. Geneve 1: 418 (Mem. Fam. Ternstroem. 26) 1822; DC., Prod. 1: 525. 1824; Wall., Cat. n. 1464. 1828; Royle, Ill. Bot. Himal. 1: 127. 1834; 2: t. 24, f.1. 1839; Blume, Ann. Mus. Bot. Lugduno Batavum 2: 117. 1856; Dyer in Hook. f., Fl. Br. Ind. 1: 285. 1874 (excl. var. *wallichiana*); Trimen, Handb. Fl. Ceylon 1: 110. 1893; Melchior in Pflanzenfam. ed. 2, 21: 147. 1925; Kobuski, Ann. Missouri Bot. Gard. 25: 321. 1938; De Wit, Bull. Jard. Bot. Buitenzorg ser. 3, 17(3): 336. f.1. 1947; Balak., Fl. Jowai 1: 93. 1981; Grierson in Grierson & Long, Fl. Bhutan 1(2): 363. 1984. Type: Nepal, *Wallich 1464* (K; PDA, isotype).

Eurya angustifolia Wall. [Cat. n. 1465. 1828, nom. nud.] ex Walp., Repert. 1: 370. 1842. Holotype: India, *Wallich 1465* (BM).

Eurya lucida Wall. [Cat. n. 1462. 1828. nom. nud.] ex Blume, Ann. Mus. Bot. Lugduno Batavum 2: 118. 1856. Type: Burma: Martaban, *Wallich 1462* (K, holotype; BM, isotype).

Eurya euprista Korth. in Temminck, Verh. Naturf. Gesch. Bot. 3: 113. 1840. Type: from Sumatra.

Eurya membranacea Gardner, Calcutta J. Nat. Hist. 7: 444. 1847. Type: Sri Lanka: Nuwara Eliya, Elephant Plains, 1830 m, *Gardner 92* (K, holotype; BM, isotype).

Eurya japonica Thunb. var. *acuminata* (DC.) Thw., Enum. Pl. Zeyl. 41. 1858.

Eurya acuminata DC. var. *euprista* (Korth.) Dyer in Hook. f., Fl. Br. Ind. 1: 285. 1874.

Shrubs or small trees, 10–12 m tall; young branchlets usually terete, lanate to pubescent, usually whip-like, terminal leaf-buds hairy. Leaves narrow oblong-lanceolate, 4.5–10.2 × 1.2–3 cm, usually membranous, occasionally coriaceous, glabrous or minutely appressed puberulous beneath, apex short-to long-acuminate, acumen ± blunt, base cuneate, margin in the upper half finely serrate, in the lower ± entire; midrib softly hairy beneath, lateral nerves anastomosing forming prominent loops near the margins; petioles slender, 1–3.5 mm long, hairy. Flowers 1–3 in axillary clusters, ± 3 mm diam., white, fragrant; pedicels 3–4 mm long, glabrous or minutely puberulous, curved. Bracteoles ovate, pubescent to hairy, nearly half as long as the sepals. Sepals broadly elliptic, 1.5–2 × 1.5 mm, glabrous. Petals obovate, 3–4.5 × 2–2.5 mm. Stamens in male flowers c. 20; filaments slender; anthers 2-celled. Ovary in female flowers ovoid to globose, 1.5–2 mm diam., 3–5-celled, glabrous; styles 3–5, connate below, about 2 mm long. Berries globose, ± 4 mm diam., with persistent style at the top, and supported by persistent sepals below, bluish-black. Seeds trigonous, brown, shining.

Distr. India, Nepal, Sri Lanka, Thailand, Bhutan, S.W. China, Formosa and W. Malaysia.

Ecol. Along edges of montane forests in wet zone, 610–1830 m, common; fls and frts Feb–Nov.

Notes. The slender twigs spread horizontally and often have a feather-like appearance given by the distichous branches and leaves. The ultimate branchlets are very thin and whip-like. The flowers are placed on curved pedicels along the lower surface of the branches.

Specimens Examined. KANDY DISTRICT: Corbets Gap, *Worthington 5020* (K); Nawalapitiya, *Worthington 374* (K); Madugoda, 5/7 mile beyond Rest House, ± 640 m, 1 Sept 1944, *Worthington 1599* (K, BM); Kandy, in 1819, *Moon s.n.* (BM), 15 March 1819, *Moon 532* (BM).

BADULLA DISTRICT: Top of Namunukula, 25 Apr 1907, *J.M. Silva s.n.* (PDA); Kukuluwa Viharekande, 16 Feb 1974, *Wass 266* (PDA). KALUTARA DISTRICT: Delgoda Reserve, Pitakele off Morapitiya, 11 Jan 1988, *Jayasuriya, Balasubramaniam & Wijesinghe 4297* (PDA); Trail Kudawa-Pitakele, Sinharaja forest, 22 Aug 1988, *Jayasuriya 4428* (PDA); Hadigalla, 7 Nov 1974, *Waas 915* (PDA). RATNAPURA DISTRICT: Sinharaja Forest, Trail from Pitakele, 11 Jan 1988, *Jayasuriya 4297* (PDA); Tundoleela, Sinharaja forest, 25 Nov 1992, *Wadhwa & Weerasooriya 334* (K, PDA); Sinharaja forest, Weddagale Entrance, 200 m, 26 March 1979, *Kostermans 27454* (PDA); Bulutota Pass from Rakwana, ± 750 m, 29 June 1972, *Hepper, Maxwell & Fernando 4564* (PDA); Bulutota Pass on A12, from Bend No. 9 to Sinharaja forest, 29 June 1972, *Maxwell, Hepper & Fernando 980* (PDA); Trail on descent from Nagrak Tea Estate, near Galagama Falls to Belihuloya, 1200 m, 29 Dec 1970, *Theobald & Krahulik 2844* (PDA); Longford Div. of Hayes Group, below Gangala Cardamon Plantation, 6°23′ N, 80°40′ E, 1050 m, 14 March 1985, *Jayasuriya & Balasubramaniam 3377* (PDA); Trail from Kudawa, Pitakele, Sinharaja, 22 Aug 1988, *Jayasuriya & Balasubramanian 4428* (PDA). NUWARA ELIYA DISTRICT: Forest near Perawella, Hakgala, 13 Oct 1992, *Wadhwa & Weerasooriya 54* (K, PDA); Mulhalkele, 1525 m, 21 Apr 1939, *Worthington 142* (BM); Galagama, Feb 1846, *Thwaites C.P. 785* (K); Elephant Plains, 1830 m, *Gardner 92* (*Holotype* of *E. membranacea* Gardn., K; BM, isotype). GALLE DISTRICT: Sinharaja forest above Beverley Estate, 800 m, 15 March 1985, *Jayasuriya & Balasubramaniam 3238* (PDA); Kanneliya Forest Reserve, road to Dediyagala, 3 Oct 1985, *Jayasuriya 3413* (PDA); Hills N. of Nelluwa, 240 m, 29 June 1977, *Huber 353* (PDA). MATARA DISTRICT: Morawaka Kanda, ± 500 m, 15 Feb 1976, *Waas 1496* (PDA); Sinharaja, near Ensalwatta, 8 Nov 1975, *Sohmer & Waas 10434* (PDA); Deniya Forest Range, 27 Jan 1992, *Jayasuriya 6079* (PDA). LOCALITY UNKNOWN: *Col. Walker s.n.* (K); *Thwaites C.P. 1078* (BM, K, PDA); *Macrae s.n.* (BM).

3. Eurya chinensis R. Br. in Abel, Narr. Jour. China 379. t. 1818; DC., Prod. 1: 525. 1824; Champion, Hooker's J. Bot. Kew Gard. Misc. 3: 307. 1851. Champion, Trans. Linn. Soc. London 21: 113. 1855; Dyer in Hook. f., Fl. Br. Ind. 1: 285. 1874; Kobuski, Ann. Missouri Bot. Gard. 25: 319. 1938. Type: China, *Abel s.n.* (K).

Eurya parvifolia Gardner, Calcutta J. Nat. Hist. 7: 445. 1847. Type: Sri Lanka: Nuwara Eliya, Feb 1846, *Col. Walker C.P. 784* (K, holotype; BM, HAK, isotypes).
Eurya japonica Thunb. var. *parvifolia* (Gardner) Thw., Enum. Pl. Zeyl. 41. 1858.

Eurya japonica Thunb. var. *chinensis* (R. Br.) Trimen, Handb. Fl. Ceylon 1: 109. 1893.

Small shrubs, 1–4 m high; branchlets terete, sometimes 2-edged, covered with black, patent hairs, leaf-buds hairy. Leaves coriaceous, obovate-elliptic or oval-elliptic, 1.8–4.2 × 1.2–1.6 cm, bluntly acuminate at apex, cuneate at base, glabrous except midrib slightly hairy beneath, margin serrate, revolute; petioles c. 2 mm long, hairy. Flowers small, axillary, bracteolate, subsessile, 3–5 in clusters, 2.5–3 mm across, white; pedicels c. 1.5 mm long, glabrous. Bracteoles 2, ovate, obtuse, usually glabrous, sometimes slightly hairy without. Sepals 5, orbicular, emarginate, unequal, 2 outer ones smaller, 1–1.5 mm wide, imbricate, glabrous. Petals 5, obovate, 2–2.5 × 1.5–2 mm, connate at base. Stamens in male flowers 9–12, slightly united with petals at the base; filaments filiform, slightly longer than the anthers, glabrous; anthers ovate–cordate, 2-celled, basifixed, longitudinally dehiscent, in female flowers rudimentary. Ovary in female flowers globose, glabrous, 3-celled; ovules many in each cell; style 1, subulate; stigmas 3, filiform, shorter than the style, in male flowers absent. Berries ovoid or globose, 3 × 2 mm, 3-celled, crowned with persistent style and stigmas and subtended below with persistent sepals, black; many-seeded.

Distr. China, Formosa and Sri Lanka.

Ecol. Along edges of montane forests in wet zone, rare; fls & frts Jan–May.

Specimens Examined. BADULLA DISTRICT: Ohiya, 1700 m, 2 Sept 1978, *Huber 884* (PDA). NUWARA ELIYA DISTRICT: Horton Plains, 2000 m, 7 Apr 1969, *Kostermans 23060* (K); Horton Plains on hill top, 8 May 1970, *Cooray 70050802 R* (K, PDA); Hakgala, Apr 1921, *Alston s.n.* (HAK, PDA); Behind Rest House, Horton Plains, 10 May 1970, *Cooray 70051001 R* (K, PDA); Horton Plains, 2135 m, 2 Apr 1945, *Worthington 1720* (BM); Hakgala, 24 Aug 1926, *J.M. Silva 155* (HAK); Horton Plains on Ohiya road, 1.5 miles from Rest House, 2175 m, 9 July 1967, *Mueller-Dombois & Comanor 67070935* (PDA); Nuwara Eliya from Hakgala Botanic Garden, 1800–2000 m, 30 July 1977, *Nooteboom 3240* (PDA); 1 mile from A7 on Ambawela road, 1860 m, 11 Feb 1968, *Comanor 987* (PDA); Along trail from Big World's End near Galagama Falls, c. 2300 m, 27 Dec 1970, *Theobald & Krahulik 2832* (PDA); Trail from Small World's End to Big World's End, Horton Plains, c. 2300 m, 28 Dec 1970, *Theobald & Krahulik 2840* (PDA); Hakgala Botanic Garden, Block G-56, 15 Oct 1992, *Wadhwa & Weerasooriya 98* (K, PDA); Forest edge along Hakgala Botanic Gardens, 15 Oct 1992, *Wadhwa & Weerasooriya 108* (K, PDA); Horton Plains, forest edge on road to Baker's Falls, 14 Oct 1992, *Wadhwa & Weerasooriya 81* (K, PDA); Hakgala, 11 Feb 1970, *Cooray 700211102 R* (PDA); Hakgala Natural

Reserve, 1800 m, 18 Apr 1973, *Stone 11244* (PDA); Cardamom Forest, Hermitage Div., 1400 m, 28 June 1980, *Werner 78* (PDA); Nuwara Eliya, Feb 1846, *Col. Walker* in *C.P. 784* (K, holotype of *E. parvifolia* Gardner; BM, HAK, isotypes). LOCALITY UNKNOWN: 1800 m, *Col. Walker 82* (K); Kikiliyamana Proposed Reserve, 9 Sept 1994, *Jayasuriya 8284* (PDA); Mahacoodagala Proposed Reserve, 26 Sept 1994, *Jayasuriya 8335* (PDA); *Maxwell s.n.* (K, mounted along with *C.P. 784*).

4. Eurya ceylanica Wight, Ill. Ind. Bot. 1: 98. 1838; Gardner, Calcutta J. Nat. Hist. 7: 444. 1847; Dyer in Hook. f., Fl. Br. Ind. 1: 285. 1874; Kobuski, Ann. Missouri Bot. Gard. 25: 320. 1938. Type: Sri Lanka, *Col. Walker s.n.* (not seen).

Eurya japonica Thunb. var. *chinensis* Thw., Enum. Pl. Zeyl. 41. 1858. Type: Sri Lanka: Nuwara Eliya Dist, Pidurutalagala ('Pedrotalagala'), *Thwaites C.P. 2600* (K, holotype; BM, HAK, isotypes).
Eurya japonica Thunb. var. *ceylanica* (Wight) Trimen, Handb. Fl. Ceylon 1: 109. 1893 'zeylanica'.
Eurya chinensis R. Br. subsp. *ceylanica* (Wight) De Wit, Bull. Jard. Bot. Buitenzorg ser. 3, 17(3): 345. 1947.

Shrubs, 4–5 m high; branches terete; extreme branches and leaf–buds hairy. Leaves coriaceous, thick, broadly elliptic or oval-oblong, 3.5–11.2 × 1.5–4.2 cm, glabrous except midrib beneath, obtusely acuminate at apex, tip retuse, cuneate at base, margin mucronate-serrate for upper three quarters, somewhat revolute; midrib channelled above, raised and hairy beneath; lateral veins impressed on upper surface, pronounced and upraised on the lower surface; petioles 2–3 mm long, hairy. Flowers axillary, small, 2–4 in clusters, 3–4 mm across; pedicels about 2 mm long, glabrous or slightly pubescent. Bracteoles 2, ovate, nearly half as long as the sepals, hairy. Sepals 5, rotundate, about 1.5 mm wide, emarginate, glabrous, persistent. Petals 5, obovate, 3 × 2 mm. Stamens 15 in male flowers; filaments slender, small; anthers oblong, 2-celled, obtuse. Ovary in female flowers ± globose, 2.5–3 mm diam., 3-locular, with many ovules in each cell; styles 3, ± united; stigmas 3, reflexed. Berries ovoid or globose, 3 × 2 mm, 3-celled, many-seeded, pubescent or short-hairy, crowned with short, persistent styles and stigmas and subtended below by persistent calyx.

Distr. Endemic.

Ecol. In montane forests in wet region, common; fls & frts throughout the year.

Notes. Plate No. 174, in PDA, based on specimen *C.P. 2600*, belongs to this species.

Specimens Examined. KANDY DISTRICT: Nawalapitiya, 14 May 1970, *Balakrishnan 339* (K, PDA); Hantane, *Worthington 265* (K); Kandy-Patiagama road, 945 m, 8 Feb 1945, *Worthington 1765* (K); Deltota, 875 m, *Worthington 6626* (K, PDA); Deanstone Estate Forest Reserve, 915 m, 2 Feb 1975, *Waas 1064* (K, CAL); Kellie Estate, Dolosbage, 975 m, 11 June 1946, *Worthington 1881* (BM); Catherine Estate near Dolosbage, 1010 m, 25 Aug 1978, *Huber 807* (PDA); Nallathanie, 15 March 1975, *Waas & W.D.A. Gunatilleke 1182* (PDA); Loolkandura Estate, Kondagala Div., S. of Deltota, 7°09′ N, 80°42′ E, 1325 m, 25 Oct 1984, *Jayasuriya & Gunatilleke 3424* (PDA); Fishing Hut area, Moray Estate, S.E. base of Adam's Peak, 1400 m, 21 Nov 1974, *Davidse & Sumithraarachchi 2694* (PDA); N.E. Knuckles Mts., Rattota-Illukkumbara road, Midlands Cardamom Estate, 1500 m, 8 Nov 1974, *Kostermans 27950, 27951* (K); 1200 m, 22 Oct 1980, *Werner 109* (PDA); Rangala Forest along Rangala-Corbet's Gap road, 14 Nov 1975, *Sohmer & Jayasuriya 10648* (PDA); Corbet's Gap-Rangala, 1220 m, 9 Sept 1951, *Worthington 5474* (K). BADULLA DISTRICT: Deyanegalla Spur of Namunukula Kande, 1790 m, 7 Sept 1978, *Huber 915* (PDA); Namunukula top, 16 Oct 1992, *Wadhwa & Weerasooriya 130* (K, PDA). NUWARA ELIYA DISTRICT: Elk Plains near Ambawela, 1800 m, Apr 1969, *Kostermans s.n.* (K); Nuwara Eliya, Mile 51, culvert 7, 1860 m, 20 June 1970, *Meijer 38* (K, PDA); Nuwara Eliya-surrounding hills, c. 1850 m, 17 Sept 1969, *C.F. & R.J. van Beusekom 1397* (PDA); Ramboda Pass, 42/5 mile mark, 1765 m, 28 July 1947, *Worthington 2898* (K); Hakgala, above Botanic Garden, 11 Feb 1970, *Cooray 70021104 R* (K, PDA); Kirgalpotha Summit, 2375 m, 26 Jan 1940, *Worthington 1728* (K); Pidurutalagala, 1830–1980 m, 21 June 1970, *Meijer 51* (K, PDA), 22 Sept 1974, *Waas 838* (PDA), 2135–2525 m, 14 July 1978, *Meijer 1914* (K), *Thwaites C.P. 2600 (Type* of *E. japonica* Thunb. var. *chinensis* Thw. K, holotype; BM, HAK, isotypes); Hakgala Peak Trail, above Botanic Garden, 1675–2175 m, 12 July 1978, *Meijer 1808* (K); Horton Plains, behind Rest House, *Cooray 70051105 R* (K); Horton Plains, halfway on hill, 8 May 1970, *Cooray 70050803 R* (K, PDA); Nuwara Eliya, 1830 m, *Gardner 91* (K, BM), Oct 1845, *Thomson s.n.* (K, CAL); Pedro proposed reserve, 27 Sept 1994, *Jayasuriya 8352* (PDA). RATNAPURA DISTRICT: Carney from Adam's Peak, 1500 m, 7 Dec 1975, *Bernardi 15773* (PDA); Below Seethagangula, S. slope of Peak Wilderness Sanctuary from Carney, 1575 m, 28 Feb 1985, *Jayasuriya & Gunatilleke 3485* (PDA); Indikatupana, S. slope of Peak Wilderness Sanctuary above Palabaddala, 1375 m, 1 March 1985, *Jayasuriya & Gunatilleke 3168* (PDA). LOCALITY UNKNOWN: in 1836, Herb. R. *Wight 585* (K); *s. coll. 12* (K).

4. ADINANDRA

Jack, Malay. Misc. 2(7): 50. 1822; Jack in Hook., Companion Bot. Mag. 1: 153. 1835; Korth. in Temminck, Verh. Naturf. Gesch. Bot. 103. 1841; Choisy, Mem. Soc. Phys. Hist. Nat. Geneve 1: 111 (Mem. Ternstr. 23) 1855; Benth. & Hook. f., Gen. Pl. 1: 182. 1862; Dyer in Hook. f., Fl. Br. Ind. 1: 281. 1874; Trimen, Handb. Fl. Ceylon 1: 108. 1893; Melchior in Pflanzenfam. ed. 2, 21: 143. 1925; Kobuski, J. Arnold Arbor. 28(1): 1. 1947. Type species: *A. dumosa* Jack.

Sarosanthera Korth. in Temminck, Verh. Nat. Gesch. Bot. 103. 1841; Thw., Enum. Pl. Zeyl. 41. 1858. Type: *S. excelsa* Korth.

Trees or shrubs, evergreen, alternately branched. Leaves alternate, usually petiolate, coriaceous, occasionally chartaceous, entire-crenulate or glandular-denticulate to crenate-serrate. Flowers axillary, solitary or in pairs, bisexual, often silky outside; peduncles and flowers recurved, seldom erect, 2-bracteolate at the apex; bracteoles opposite or alternate, caducous or persistent. Sepals 5, unequal, with inner ones larger, imbricate, thick, concave, persistent. Petals 5, connate at the base, glabrous or sericeous on the outer surface, imbricate. Stamens numerous (15–60), 1–5-adelphous, adnate to the base of the corolla; filaments usually united, pubescent or glabrous; anthers basifixed, extrorse, hispid; connectives apiculate. Ovary superior, 3–5-locular, pubescent or glabrous; ovules numerous in each loculus, affixed with prominent placentae; style simple, entire, rarely 3–5-fid, persistent; stigma usually single, entire, sometimes 3–5-lobed. Fruits baccate, indehiscent. Seeds few to many, small; albumen fleshy.

About 70 species in the world distributed in subtropical and warm temperate areas of East and S.E. Asia, India, Bangladesh, Sri Lanka, Malaysia, Philippines Islands, New Guinea and W. Africa; represented by 1 species in Sri Lanka.

Adinandra lasiopetala (Wight) Choisy, Mem. Soc. Phys. Geneve 1: 112 (Mem. Ternstr. 24) 1855; Dyer in Hook. f., Fl. Br. Ind. 1: 283. 1874; Szyszylowicz in Pflanzenfam. 3(6): 189. 1893; Trimen, Handb. Fl. Ceylon 1: 108. t. 9. 1893; Melchior in Pflanzenfam. ed. 2, 21: 144. 1925; Kobuski, J. Arnold Arbor. 28(1): 54. 1947; Worthington, Ceylon Trees 44. 1959.

Cleyera lasiopetala Wight, Ill. Ind. Bot. 1: 99. 1840. Type: Sri Lanka, *Col. Walker 243* (K).
Eurya lasiopetala (Wight) Gardner, Calcutta J. Nat. Hist. 7: 446. 1847.
Sarosanthera lasiopetala (Wight) Thw., Enum. Pl. Zeyl. 41. 1858.

Small, slender trees, 12–15 m tall; young parts minutely hairy; branches terete, greyish-brown, glabrous. Leaves coriaceous, oblong-lanceolate or oblanceolate, 4.7–13.4 × 1.5–3.4 cm, glabrous above, minutely pubescent beneath, apex obtuse or subacute, margin finely serrulate, conspicuously recurved especially along the lower half of the leaf; petioles semi-terete, flat on the upper surface, 4–6 mm long. Flowers axillary, solitary, c. 2 cm across, white; pedicels somewhat recurved, up to 1.5 cm long, very stout, appressed-pubescent, thickened above. Bracteoles 2, unequal, opposite, immediately beneath the flowers, persistent, triangular or sub-rotund, 3–5 mm long, finely appressed-pubescent. Calyx lobes 5, imbricate, thickened, unequal, ± rotund, 10 × 8 mm, the outer lobes finely appressed-pubescent, the inner ones sericeous on the central part of outer surface, with membranous or scarious margins. Corolla lobes 5, oblong, 2.0 × 1.0 cm, obtuse, covered on the back with dense sericeous pubescence. Stamens 18–25, monadelphous, adherent to the base of the petals; filaments unequal, sparsely hairy; anthers sparsely pubescent. Ovary 3-celled, glabrous, attenuate at the apex into a filiform, glabrous style, which on maturing of the fruit becomes 3-fid, 8 mm long; stigma 3-fid. Fruit globose, 1.2 × 1.0 cm, 3-celled, apiculate, many-seeded, surrounded by persistent sepals. Seeds kidney-shaped, small, black, shining, punctate.

Distr. Endemic.

Ecol. In montane forests of wet zone, 1220–2300 m, rather common; fls & frts Feb-June.

Vern. Rutumihiriya (S).

Notes. On one of the herbarium sheets of *C.P. 689* in PDA, 3 localities, viz. Horton Plains, Jan & Feb 1846, *Gardner*; Maturata, July 1851 and Dambulla, Apr 1854, are annotated, while on the sheet of *C.P. 689* in HAK, Horton Plains as locality has been annotated.

Plate No. 173 in PDA, based on specimen No. *C.P. 775*, belongs to this species.

Specimens Examined. KANDY DISTRICT: Near Fishing Hut, Moray Estate, 1500 m, 8 Feb 1973, *Jayasuriya, Burtt & Townsend 1119* (K, PDA); Batulu Oya Stream near Fishing Hut, edge of Peak Sanctuary, 1370 m, 25 Oct 1978, *Fosberg 57987* (K); Loolkandura Estate, S. of Deltota, 80°42′ E, 7°09′ N, 1325 m, 25 Oct 1984, *Jayasuriya & Gunatilleke 3425* (PDA); Maskeliya-Balangoda, Bogwantalawa-Maratenna, 1800 m, 13 Dec 1975, *Bernardi 15977* (K, PDA); Above Moray Estate, 1000–1600 m, 15 Oct 1977, *Nooteboom 3373* (PDA); Maskeliya, above Moray Estate, 1350–1700 m, 14 Sept 1977, *Nooteboom & Huber 3137* (PDA). NUWARA ELIYA DISTRICT: World's End trail, Horton Plains, 2400 m, 3 Nov 1971, *Balakrishnan 1041, 1043* (PDA); Pattipola, ± 2000 m, 5 Dec 1972, *Cramer 3943* (PDA); Horton Plains, forest beside Pattipola road, ± 2200 m, 26 Jan 1977, *Cramer 4811* (PDA); Road up to Farr Inn from Pattipola, 1 Nov 1973,

Sohmer, Jayasuriya & Eleizer 8535 (PDA); Trail from Small World's End to Big World's End, Horton Plains, c. 2300 m, 28 Dec 1970, *Theobald & Krahulik 2838* (PDA); Peak Wilderness forest, 18 Oct 1992, *Wadhwa & Weerasooriya 164* (K, PDA); Horton forest reserve, 2100 m, 17 Nov 1977, *Huber 640* (PDA); Nuwara Eliya, 1830 m, 20 June 1953, *Worthington 6327* (K, PDA); Hakgala forest, 1830 m, 30 Oct 1952, *Worthington 6096* (BM, K, PDA); Mooloya, 1585 m, 5 Feb 1975, *Waas 1133* (K, PDA), *Waas 1140* (PDA); W. of Hewaheta, forest above Hope Estate, 7°06′ N, 80°44′ E, 1650 m, 23 Aug 1984, *Jayasuriya, Balasubramaniam & Greller 2855* (CAL, PDA); Pattipola road, halfway up slope to Horton Plains, 2000 m, 7 Oct 1967, *Comanor 459* (K, PDA); 5 Dec 1972, *Tirvengadun & Cramer 92* (K, PDA); Horton Plains, 2300 m, 10 Apr 1969, *Kostermans 23107* (K, BM), *Gardner C.P. 689* (HAK, PDA), Road to World's End, 11 May 1970, *Cooray 70051101 R* (K, PDA); MAB Reserve, road to Horton Plains, 2000 m, 15 July 1978, *Meijer 1977* (K); Hakgala Hills, *Beddome 460* (BM); Hakgala Peak trail, above Hakgala Botanic Garden, 1585–2170 m, 12 July 1978, *Meijer 1844* (K); Great Western Range above Lindula, 2000 m, 23 Apr 1974, *Kostermans 24616* (K, PDA); Rambode, *s. coll. s.n.* (K); *Thwaites 783* (K), 1830 m, in 1847, *Gardner 95* (K); Kandapola-Sita Eliya Forest Reserve, 7 Sept 1994, *Jayasuriya 8233* (PDA). LOCALITY UNKNOWN: *Thwaites C.P.775* (BM, CAL, K), *C.P.776* (K); *Col. Walker 243* (K, holotype; PDA, isotype).

5. TERNSTROEMIA

Mutis ex L. f., Suppl. Pl. 39. 1781, nom. cons.; Benth. & Hook., f., Gen. Pl. 1: 182. 1862; Dyer in Hook. f., Fl. Br. Ind. 1: 280. 1874; Trimen, Handb. Fl. Ceylon 1: 107. 1893; Melchior in Pflanzenfam. ed. 2, 21: 140. 1925. Type species: *T. meridionalis* Mutis ex L. f.

Evergreen trees or shrubs, glabrous, usually dioecious; branchlets often ± whorled. Leaves coriaceous, often subverticillate or spirally arranged, petiolate, margin entire to crenate-serrate. Flowers solitary, axillary or extra-axillary, bisexual or unisexual, 2-bracteolate. Sepals 5, imbricate, persistent, the margin sometimes glandular-serrate. Petals 5, imbricate, connate at base. Stamens 20-many, in 1-several series; filaments connate below; anthers basifixed, glabrous, connectives usually slightly apiculate at apex, sometimes emarginate or truncate. Ovary superior, usually 2-celled, sometimes 1 or 3-celled; ovules few, pendulous; style simple or 2–3-lobed; stigma simple, disk-shaped or lobed. Fruits berry-like, indehiscent or irregularly dehiscent from apex. Seeds usually 2 in each locule, arillate; endosperm fleshy.

Pantropical genus of c. 85 species, distributed in Central and South America, Sri Lanka, India, Nepal, Bhutan, China, Japan, Taiwan and Malaysia; 1

species in Queensland and 2 species in S.W. Africa; represented in Sri Lanka by 2 species.

KEY TO THE SPECIES

1 Leaves oblanceolate or elliptic, apex obtuse or shortly and obtusely pointed, base cuneate, margin entire, flat, not revolute .... **1. T. gymnanthera**
1 Leaves obovate or spathulate, apex obtuse, emarginate, base attenuate, margin crenate-serrate in the upper half, revolute .... **2. T. emarginata**

1. Ternstroemia gymnanthera (Wight & Arn.) Bedd., Fl. Sylv. t. 91. 1871; Sprague, J. Bot. 61: 18. 1923; Ohwi, Fl. Jap. 629. 1965; Keng, Fl. Thailand 2(2): 154. 1972; Whitmore in Hara & Williams, Enum. Fl. Pl. Nep. 2: 65. 1974; Grierson in Grierson & Long, Fl. Bhutan 1(2): 364. 1984.

Cleyera gymnanthera Wight & Arn., Prod. 1: 87. 1834; Wight, Ic. Pl. Ind. Or. t. 47. 1838; Thw., Enum. Pl. Zeyl. 40. 1858. Type: Peninsular India, *Wight 300* (K).

Ternstroemia japonica Auct.: Sieb. et Zucc., Fl. Jap. 1: 148. t. 81. 1841; Dyer in Hook. f., Fl. Br. Ind. 1: 280. 1874; Trimen, Handb. Fl. Ceylon 1: 107. 1893; Gamble, Fl. Pres. Madras 1: 78. 1915; Melchior in Pflanzenfam. ed. 2, 21: 141. 1925; Worthington, Ceylon Trees 43. 1959; Chitra in Nair & Henry, Fl. Tamilnadu 1: 30. 1983, non Thunb., 1794.

Small to medium-sized trees, 5–12 m tall; branches thick, glabrous, rusty-grey. Leaves alternate, often closely approximate near the top of the branches, coriaceous, oblanceolate, or elliptic, 4.5–9.5 × 1.5–4.5 cm, apex obtuse or shortly and obtusely pointed or subacute, base cuneate, margin entire, flat; midrib slightly raised beneath; petioles 0.5–1.5 cm long. Flowers axillary, 0.8–1.2 cm across, deflexed, fragrant; pedicels 1.0–2.2 cm long, slightly 2–ridged, bearing 2 minute, ovate bracteoles near the apex. Sepals 5, broadly elliptic, 4–6 × 3–4 mm, leathery, obtuse-subacute at apex. Petals obovate or oblong, 6–8 × 5–6 mm, obtuse or rounded at apex, cuneate at base, yellow or pale-yellow, fleshy. Stamens many, connate at base, 3-seriate in male flowers, 1-seriate in bisexual flowers; filaments nearly as long as the anthers. Ovary ovoid, 2-locular; style 2-fid; stigma 2-lobed. Fruit subglobose, 1.5–2.0 × 1.0–1.5 cm, tipped with persistent style, surrounded at base with persistent bracteoles and sepals, fleshy, brown.

Distr. India, Bangladesh, China, Formosa, Japan, Sri Lanka and Malaysia.

Ecol. In montane forests in wet zone, 1250–2000 m, common; fls & frts April–Oct.

Vern. Penamihiriya, Rattota, Rattiya (S).

Notes. Plate No. 171 in PDA belongs to this species.

Specimens Examined. KANDY DISTRICT: Trail to Adam's Peak from Moray Estate, 26 Oct 1975, *Sohmer & Sumithraarachchi 9894* (K, PDA); Upper part of Moray Estate, path to Adam's Peak, 1800–2000 m, 16–22 May 1971, *Kostermans 24141, 24209, 24212, 24235* (BM, K, PDA); Trail to Adam's Peak from Moray Estate, Peak Wilderness, ± 2000 m, 23 June 1976, *Waas 1708* (PDA); Rajamallay, Maskeliya, ± 1540 m, 2 July 1974, *Cramer 4282* (K); Fishing Hut area, margin of Moray Estate, S.E. base of Adam's Peak, 1400 m, 21 Nov 1974, *Davidse & Sumithraarachchi 8692* (PDA); Maskeliya near Tenna, 1700–1750 m, 11 Dec 1975, *Bernardi 15915* (K, PDA). BADULLA DISTRICT: Namunukula, ± 2000 m, 22 Sept 1976, *Cramer 4739* (PDA); Haputale, 1585 m, 20 July 1951, *Worthington 5348* (K, PDA); Deyangalla, northern spur of Namunakula Kande, 1800 m, 7 Sept 1978, *Huber 918* (PDA). RATNAPURA DISTRICT: Adam's Peak trail, N.E. of Carney, 900–1350 m, 23 Nov 1974, *Davidse & Sumithraarachchi 8732* (PDA); Balangoda, 3 miles above Tea Estate, 1220 m, 21 July 1970, *Meijer 440* (K); Mandagalaoya forest, ± 700 m, 22 June 1976, *Waas 1692* (PDA). NUWARA ELIYA DISTRICT: Ramboda, 1525 m, in 1847, *Gardner 96* (BM, CAL, K); Pidurutalagala, 1800–1950 m, 21 June 1970, *Meijer 68* (PDA); Nuwara Eliya, Jan 1846, *Thwaites C.P. 778 (779)* (HAK, K); Nuwara Eliya, mile 51, culvert 7, 1890 m, 20 June 1970, *Meijer 41* (K), 1905 m, 17 June 1946, *Worthington 1933* (K); Great Western Range near Lindula, 2000 m, 23 Apr 1973, *Kostermans 24620* (BM, K, PDA); Hakgala jungle, 18 May 1971, *Kostermans 24188* (K); Bogawantalawa Proposed Reserve, 3 Nov 1992, *Jayasuriya 8472* (PDA).

2. Ternstroemia emarginata (Gardner) Choisy, Mem. Ternstr. 14. 1855; Dyer in Hook. f., Fl. Br. Ind. 1: 281. 1874; Trimen, Handb. Fl. Ceylon 1: 108. 1893; Melchior in Pflanzenfam. ed. 2, 21: 142. 1925.

Cleyera emarginata Gardner, Calcutta J. Nat. Hist. 7: 447. 1847; Thw., Enum. Pl. Zeyl. 1858. Type: Sri Lanka, Feb 1846, *Thwaites C.P. 782* (PDA).

Shrubs, 3–4 m high; branches dichotomous, terete, glabrous. Leaves alternate, crowded at the ends of the branchlets, coriaceous, glabrous, obovate or spathulate, 2.4–5.2 × 1.2–2.4 cm, dark green and shining above, pale green beneath, apex obtuse, emarginate, base attenuate, margin crenate-serrate from the middle upwards with a distinct gland in the serrature, revolute; midrib slightly raised beneath, channelled above; petioles 4–6 mm long, flat, purple. Flowers axillary, solitary, 1.2–1.8 cm across; pedicels terete, glabrous, 2.0–2.4 cm long, bearing 2 minute, ovate bracteoles. Sepals 5, broadly elliptic, concave, 4–5 × 3–3.5 mm, subacute at apex, imbricate. Petals 5, obovate or orbicular, concave, 8–9 × 5–6 mm, pale yellow, outer ones tinged with purple. Stamens numerous, in several series, connate at base; filaments shorter than the anthers; anthers linear-oblong, 2-celled, introrse, dehiscing longitudinally, separated by broad connectives. Ovary conical, glabrous, 2-celled with 2 collateral ovules in each cell; style deeply 2-fid; stigma capitate or subreniform. Fruit ovoid, 1.7–2.2 × 0.9–1.0 cm, tipped with persistent style, and surrounded at base with persistent calyx, 2-seeded, brownish-green.

Distr. Endemic.

Ecol. In secondary montane forests in wet zone, 1700–2400 m, fairly common; fls & frts Feb-Oct.

Notes. Plate No. 172, based on specimen *C.P. 782* in PDA, belongs to this species.

Though Choisy (l.c.) and Dyer (l.c.) have pointed out that this species is barely distinguishable from *T. cuneifolia* Gardner from Brazil, this differs from the latter in its leaves without punctate glands on the lower surface, margin crenate-serrate in upper part only; calyx-lobes ± equal and without glandular-ciliate margins.

Specimens Examined. KANDY DISTRICT: Adam's Peak, c. 1700 m, 21 Sept 1969, *C.F. & R.J. van Beusekom 1356* (BM); Adam's Peak Wilderness above Moray Estate, 1820 m, *Jayasuriya, Burtt & Townsend 1108* (K, PDA); Upper part of Moray Estate along Maskeliya Oya, 1600 m, 16 May 1971, *Kostermans 24157* (K, PDA), Batulu Oya stream near Fishing Hut, edge of Peak Sanctuary, 1370 m, 25 Oct 1978, *Fosberg 58012* (K); Peak Wilderness, ± 2000 m, 23 June 1976, *Waas 1710* (K); Near Fishing Hut, upper part of Moray Estate, 1700 m, 16 May 1971, *Kostermans 24156* (BM, PDA), 17 Oct 1971, *Balakrishnan, Dassanayake & Balasubramaniam 925* (K, PDA); 22 Nov 1974, *Sumithraarachchi & Davidse 569* (PDA), 27 March 1974, *Sumithraarachchi & Jayasuriya 174* (PDA), 16 May 1971, *Kostermans 24123* (PDA); Rajamalay, Peak Wilderness, 2 Oct 1973, *Waas 287* (PDA); Maskeliya, above Moray Estate, 1350–1700 m, 14 Sept 1977, *Nooteboom & Huber 3133* (PDA); Trail to Adam's Peak from Moray Estate, 15 Nov 1973, *Sohmer & Waas 8702* (PDA); Trail to Adam's Peak, above Moray Estate, 1900 m, 13 Nov 1978, *Kostermans 27031* (PDA); Road to Fishing Hut above Moray Estate, 1500 m, 14 Nov 1978, *Kostermans 27044* (PDA); Adam's Peak, *s. coll. s.n.* (K). NUWARA ELIYA DISTRICT: Horton Plains, Feb 1846, *Thwaites C.P. 782* (PDA, holotype; BM, K, isotypes); Baker's Falls, Horton Plains, ± 2400 m, 14 May 1975, *Cramer 4456* (K, PDA); Near Galagama Falls, along trail from Big World's End, c. 2300 m, 27 Dec 1970, *Theobald & Krahulik 2834* (PDA); Horton Plains, 11 May 1970, *Cooray 70051108 R* (PDA). RATNAPURA DISTRICT: Near Galagama Falls, 80°46′ E, 6°46′ N, 1980 m, 5 March 1977, *B. & K. Bremer 966* (PDA). LOCALITY UNKNOWN: *s. coll. 1662* (K).

TILIACEAE   (Continued)

(by B.M. Wadhwa* and A. Weerasooriya**)

Six genera and 42 species of the family Tiliaceae have already been published in this Revised Handb. Fl. Ceylon, volume 7: 402–437. 1991.

The genus *Muntingia* L. is represented by its species *M. calabura* L. which, introduced in this country some time ago, has now become naturalised. This genus has been treated differently by different taxonomists. Long & Lakela, Fl. Trop. Florida 587. 1971, Mabberley, Plant Book 383. 1987 and Murti in Sharma et al., Fl. India 3: 570. 1993, have treated this genus in the Elaeocarpaceae. Cronquist, Integr. Syst. 349–350. 1981 has treated it under the Flacourtiaceae, while DC., Prod. 1: 514. 1824, Benth. & Hook. f., Gen. Pl. 1: 236. 1867, Hutchinson, Gen. Fl. Pl. 2: 485. 1967 and Brummitt, Vasc. Pl. Fam. & Gen. 680–681. 1992 have treated it under the Tiliaceae; this is followed in the present work.

KEY TO THE GENERA (REVISED)

1 Sepals free to the base or nearly so
  2 Stamens inserted on the torus adjacent to the petals; petals nude and not pitted within the base
    3 Fruit dehiscent, a capsule, opening lengthwise or by lateral slits .... **3. Corchorus**
    3 Fruit indehiscent, a berry .... **Muntingia**
  2 Stamens inserted on the ± elevated torus bearing the gynoecium; petals pitted or glandular within the base
    4 Trees and shrubs; fruits neither echinate, nor setose nor tuberculate, sometimes winged
      5 Inflorescences of umbellate cymes or flowers solitary .... **1. Grewia**
      5 Inflorescences paniculate, the ultimate ramifications cymose, the cymes usually 3-flowered
         .... **2. Microcos**
    4 Mostly herbs; fruits echinate, setose or tuberculate .... **6. Triumfetta**
1 Sepals united into a 3–5-lobed campanulate calyx; anthers short, mostly globose or didymous; carpels united, fruits sometimes winged
      6 Flowers without staminodes; capsules winged, largely 6 (–8)-alate ..... **4. Berrya**
      6 Flowers with 5 staminodes; capsules not winged, obovoid, conspicuously 5-angulate-lobate, verrucose, densely lepidote .... **5. Diplodiscus**

* Royal Botanic Gardens, Kew.
** Flora of Ceylon Project, Peradeniya.

MUNTINGIA

L., Sp. Pl. 509. 1753; L., Gen. Pl. ed. 5, 225. 1754; Benth. & Hook. f.,
Gen. Pl. 1: 236. 1867; Hutchinson, Gen. Fl. Pl. 2: 485. 1967. Type species:
M. calabura L.

Small trees or shrubs. Leaves alternate, oblong-lanceolate, acuminate,
base very oblique, irregularly or coarsely serrate, green above, densely grey
or white stellate pubescent below. Flowers pedicellate, single, axillary or
supra axillary, 2–3 flower-stalks arising together, white, medium-sized. Sepals
mostly 5, long acuminate. Petals 5, obovate. Stamens numerous, free, inserted
on an annular disk; anthers rounded. Ovary densely surrounded by glandular hairs, 5–7-locular; ovules numerous; stigmas sessile, thick, sulcate-lobed.
Fruit a berry, ovoid-globose, smooth, indehiscent. Seeds numerous, small,
immersed in sweet pulp.

About 3 species in Tropical America, now widespread in Tropical Asia;
represented by 1 species in Sri Lanka.

Muntingia calabura L., Sp. Pl. 509. 1753; Corner, Wayside Trees Malaya
1: 644. 1940; Murti in Sharma et al., Fl. India 3: 570. 1993. Type: from
Jamaica.

Evergreen small trees or large shrub, 3.5–8 m tall, with dense, spreading crown; the branches drooping; bark smooth, pale-brownish grey, tough
fibrous; branchlets densely villous, glandular pubescent. Leaves alternate,
oblong-lanceolate or narrowly ovate, 5–10 × 1.5–4 cm, obliquely subcordate
at base, acuminate at apex, serrate, chartaceous, glandular hairy above, woolly
pubescent beneath; veins 3–5 pairs on either side of midrib; petioles 5–6 mm
long. Stipule 1, c. 5 mm long, filiform, hairy. Inflorescences sessile, usually
supra-axillary, flower-stalks arising together, with 3 filiform, small bracts at
base. Flowers 1.5–2.5 cm in diam., white; pedicels c. 2–2.5 cm long. Sepals 5,
lanceolate, 1.5 cm long, caudate-acuminate, valvate, shortly connate at base,
densely pubescent on both surfaces. Petals 5, thin, obovate or suborbicular,
shortly clawed, ± as long as sepals, imbricate. Intrastaminal disk annular,
bearing a ring of hairs on exterior margin. Stamens many, c. 1 cm long; anthers elliptic, dorsifixed, versatile, longitudinally dehiscent. Ovary superior,
ellipsoid or subglobose, 5–6 mm long, 5-carpellary, syncarpous, 5-locular;
ovules numerous; styles absent; stigmas knob-like, 5-ridged. Fruit a berry,
subglobular, 1–1.5 cm in diam., appearing imperfectly many-locular, red or
yellow with ridged stigmas at top and withered stamens at base; pulp juicy,
sweet. Seeds numerous, obovoid or ellipsoid, minute.

Distr. A native of Jamaica, now naturalized in other parts of Tropical
America and some Tropical Asian countries.

Ecol. Introduced and naturalised in dry and submontane areas along forest edges, rare; fls & frts almost throughout the year.

Uses. Cultivated for its sweet, juicy fruits.

Vern. Jam (S); Jamaican cherry (E).

Specimens Examined. TRINCOMALEE DISTRICT: Trincomalee, planted, 1 Feb 1944, *Worthington 1372* (K). MATALE DISTRICT: Open forest, road around Sigiriya Rock, 15 July 1973, *Nowicke, Fosberg & Jayasuriya 358* (K, PDA). POLONNARUWA DISTRICT: Polonnaruwa Sacred Area, 61 m, 7 Oct 1971, *Dittus 1100702* (PDA), Sector 4A, 61 m, 26 May 1969, *Ripley 140* (PDA). COLOMBO DISTRICT: Ratmalana, cultivated, 27 Nov 1970, *Amaratunga 2144* (PDA). KANDY DISTRICT: Highway A5, Peradeniya-Nuwara Eliya Road, 10 miles from Peradeniya, 520 m, cultivated, *Mueller-Dombois 67052801* (PDA); c. 1 mile NE of Teldeniya, c. 600 m, cultivated, 8 Feb 1969, *Robyns 6913* (K, PDA); Kandy, 640 m, 5 Jan 1952, *Worthington 5594* (K). BADULLA DISTRICT: Bandarawala, cultivated, (introduced from Singapore), 3 Sept 1917, *s. coll. s.n.* (PDA). RATNAPURA DISTRICT: Ratnapura, Road to Rest House, 180 m, 20 March 1968, *Comanor 1130* (K); Near Kuruwita, mile post 95/5, 18 Apr 1995, *Wadhwa, Weerasooriya & Samarasinghe 522* (K, PDA). MATARA DISTRICT: Tangalle, 29 Apr 1924, *Lewis s.n.* (PDA). LOCALITY UNKNOWN: *s. coll. s.n.* (PDA).

VIOLACEAE

(by B.M. Wadhwa* and A. Weerasooriya**)

Batsch, Tab. Affin. Regni Veg. 57. 1802 ("Violariae"). Type genus: *Viola* L.

Annual or perennial herbs, undershrubs or shrubs, rarely small trees or lianas. Leaves alternate, rarely subopposite to opposite, simple, entire or toothed, stipulate. Flowers usually bisexual, rarely dioecious, actinomorphic or zygomorphic, hypogynous or slightly perigynous, solitary or in axillary or terminal racemes, or spikes or panicles, often bracteolate; pedicels often articulated. Sepals 5, free or slightly connate at base, imbricate, often ciliate, persistent. Petals 5, free, unequal, the lowermost (anterior) often longer than others and differently shaped, gibbous or spurred. Stamens 5, mostly hypogynous; filaments free or connate into a tube, alternating with the petals, closely connivent around pistil; anthers 2-celled, basifixed, introrse, connectives with apical appendage, dehiscence by longitudinal slits. Ovary sessile, subglobose, superior, 1-locular, generally 3 (–5)-carpellary, placentae parietal with 1–2-many anatropous ovules on each; style simple, mostly sigmoid or thickened above; stigma various, usually truncate, lobed, beaked or simple. Fruit a 3-valved capsule dehiscing loculicidally, rarely indehiscent and baccate or nut-like. Seeds numerous, smooth or rough, rarely tomentose, mostly sessile, often arillate, sometimes with funicular outgrowth; embryo straight; endosperm copious to moderate, fleshy.

A family of about 20 genera and 800 species; cosmopolitian, in tropical and temperate regions; represented by 3 genera and 8 species in Sri Lanka.

KEY TO THE GENERA

1 Herbs, sometimes with woody base; flowers zygomorphic, solitary in the leaf axils; abaxial stamens often spurred at base
  2 Leaves with long petioles; pedicels not jointed; sepals pouched at the base; petals equal or subequal; seeds smooth .... **1. Viola**
  2 Leaves subsessile; pedicels jointed; sepals not pouched at the base; petals always unequal; seeds longitudinally ribbed .... **2. Hybanthus**
1 Shrubs or trees; flowers actinomorphic, in axillary fascicles or racemes; abaxial stamens not spurred .... **3. Rinorea**

* Royal Botanic Gardens, Kew.
** Flora of Ceylon Project, Peradeniya.

1. VIOLA

L., Sp. Pl. 933. 1753; L., Gen. Pl. ed. 5, 402. 1754; Ging. in DC., Prod. 1: 291. 1824; Benth. & Hook. f., Gen. Pl. 1: 117. 1862; Hook. f. & Thoms. in Hook. f., Fl. Br. Ind. 1: 182. 1872; Trimen, Handb. Fl. Ceylon 1: 65. 1893; Hutchinson, Gen. Fl. Pl. 2: 334. 1967; Jacobs & D.M. Moore in Fl. Males. ser. 1, 7: 199. 1971; S.P. Banerjee and Pramanik in Sharma et al., Fl. India 2: 351. 1993. Type species: *V. odorata* L.

Annual or perennial herbs, often suffruticose, rarely shrubs; rhizomes present or absent; stem mostly present. Leaves entire to pinnatisect, ovate-triangular or reniform, cordate, serrate or crenate; petioles sometimes winged. Stipules persistent, free or adnate to the petiole, lanceolate-ovate, entire, dentate or fimbriate. Flowers zygomorphic, solitary or in pairs on long, axillary, non-articulate, bracteolate peduncles; bracteoles 2, opposite or subopposite. Sepals 5, persistent. Petals 5, erect or spreading, lateral larger than anterior, lowermost spurred. Anthers 2-celled, subsessile, connate, encircling the ovary, each with a small triangular appendage; filaments short and distinct, connectives of two lateral stamens often produced into spurs within the spur of corolla. Ovary sessile; style variable, erect or curved, often geniculate at base, filiform to clavate; stigma variable, truncate or obtuse, lobed or triangular, straight or beaked. Fruit a 3-valved capsule, dehiscing loculicidally. Seeds rounded-ovate, smooth, shiny.

About 500 species, cosmopolitan, chiefly distributed in the temperate regions throughout the world; 3 in Sri Lanka, mostly in montane or submontane areas, 1000–2500 m.

KEY TO THE SPECIES

1 Plants without stems or stolons; stipules ± adnate to petioles; leaves variable; linear-lanceolate to triangular-hastate or triangular-ovate .... **3. V. betonicifolia**
1 Plants with decumbent or ascending stems or stolons; stipules free, not adnate to petioles
  2 Stipules subentire to fimbriate; style subclavate distally; stigma with 2 laterally patent lobes; leaves ± glabrous; capsule linear-oblong .... **1. V. hamiltoniana**
  2 Stipules usually toothed or dentate, rarely subentire; style subclavate, subtruncate, and shortly beaked at apex; leaves pubescent; capsule ellipsoid-subglobose .... **2. V. pilosa**

1. Viola hamiltoniana D.Don, Prod. Fl. Nepal. 206. 1825 (Feb 1825); W. Becker, Beih. Bot. Centralbl. 40(2): 170. 1923; Hara in Hara & Williams, Enum. Fl. Pl. Nepal 2: 47. 1979; S.P. Banerjee & Pramanik in Sharma et al., Fl. India 2: 363. 1993. Lectotype: *Buchanan-Hamilton s.n.* (BM).

Viola arcuata Blume, Bijdr. 58. 1825 (June-Dec 1825); Alston in Trimen, Handb. Fl. Ceylon 6: 13. 1931; Jacobs & D.M. Moore in Fl. Males. ser.1, 7: 205. 1971; Chithra in Nair & Henry, Fl. Tamilnadu 1: 16. 1983. Type: Java, *Blume s.n.* (L, holotype).

Viola distans Wall. (Cat. n. 4022. 1831, nom. nud.), Trans. Med. Phys. Soc. Calcutta 7: 227. 1835; Hook. f. & Thoms. in Hook. f., Fl. Br. Ind. 1: 183. 1872; Trimen, Handb. Fl. Ceylon 1: 66. 1893: Dunn in Gamble, Fl. Pres. Madras 1: 48. 1915. Type: India, *Wallich 4022.*

Viola wightiana Wall. ex Wight & Arn. var. *glabra* Thw., Enum. Pl. Zeyl. 20. 1858. Type: Sri Lanka, *C.P. 153,* p.p.

Perennial herbs; rhizome slender, vertical to oblique; stem up to 30 cm long, slender, procumbent to ascending, often rooting at lower nodes. Leaves ovate to reniform-cordate, usually as broad as long, 1.5–5 × 1.5–4.5 cm, broadly and deeply to shallowly cordate, with prominent basal lobes usually rounded, cuspidate to obtuse at apex, crenate-serrate, glabrous, rarely sparsely ciliate or pubescent on veins beneath, usually herbaceous, sometimes thicker & dark green; petioles 1–8.5 cm long, glabrous, curved upwards. Stipules lanceolate to oblong-lanceolate, 5–15 × 1–4.5 mm, acute, subentire to fimbriate, white to pale-purple. Peduncles 2–12 (15) cm long, slender, with 2 bracteoles above the middle. Flowers 1–1.5 cm across, white to light violet. Sepals ovate-lanceolate, acute, 2.5–4.5 × 1–2 mm, entire, glabrous or with scarious margin. Petals ovate-oblong, 2–4 times as long as broad, basal one shorter than the others; lateral ones slightly bearded; spur cylindrical, obtuse, 2–2.5 mm long, green. Style 1–2 mm long, geniculate at base, subclavate distally; stigma of 2 small, laterally patent lobes. Capsules oblong, 5–10 mm long, glabrous.

Distr. India, Nepal, Bhutan, Bangladesh, Myanmar, China, Malaysia, Philippines and Sri Lanka.

Ecol. Upper montane zone in wet places, rare; fls & frts April–June.

Note. Chromosome number reported: $2n = 24$ (Borgmann, Z. Bot. 17: 1–27. 1964).

Plate No. 105 in PDA based on *C.P. 153* p.p. belongs to this species.

Specimens Examined. NUWARA ELIYA DISTRICT: Nuwara Eliya, 28 Apr 1881, *s. coll. s.n.* (PDA), 1830 m, *Gardner 46* (K), 1500–2100 m, *Thwaites C.P. 153* (Acc. No. 000033 Hak, K, PDA). LOCALITY UNKNOWN: *Col. Walker s.n.* (K, PDA); 1830 m, *Col. Walker 323* (K); *Mrs. Gen. Walker s.n.* (K).

2. Viola pilosa Blume, Cat. Gew. Buitenz. 57. 1823; Hara in Hara & Williams, Enum. Fl. Pl. Nepal 3: 48. 1982; Sharma et al., Fl. India 2: 371. 1993. Type: Java, *Blume* (L, holotype).

Viola serpens Wall ex Ging. in DC., Prod. 1: 296. 1824; Wall. ex Roxb., Fl. Ind. 2: 449. 1824; Hook. f. & Thoms. in Hook. f., Fl. Br. Ind. 1: 184. 1872; Trimen, Handb. Fl. Ceylon 1: 67. 1893; Dunn in Gamble, Fl.

Pres. Madras 1: 48. 1915; Matthew & Britto in Matthew, Fl. Tamilnadu
Carnatic 1: 55. 1983. Type: Nepal, *Wallich s.n.*
Viola wightiana Wall. ex Wight & Arn., Prod. 32. 1834; Wight, Ic. Pl. Ind.
Or. t. 943. 1835. Type: India: *Wallich n. 4021* (*Viola* sp. Herb. Wight).
Viola crenata Moon, Cat. 17. 1824, nom. nud.
Viola wightiana Wall. ex Wight & Arn. var. *pubescens* Thw., Enum. Pl. Zeyl.
20. 1858. Type: Sri Lanka, *C.P. 153*, p. p.

Prostrate to subprostrate herbs; stems slender, stoloniferous, rooting at
nodes. Leaves ovate-deltoid, 1.5–8 × 1–6 cm, shallowly to deeply cordate
at base, acute to acuminate, serrate or crenate—serrate, pubescent to hirsute
on both surfaces, especially on veins, usually pale-green; petioles 2–9.5 cm
long, pubescent. Stipules free, 6–15 mm long, lanceolate, long acuminate,
subentire-dentate, pale-green, pubescent. Flowers 8–14 mm broad, white-pale
violet, with darker veins; peduncles filiform, solitary, axillary, 3–10 cm long,
pilose, especially distally. Bracteoles 2, placed above middle, lanceolate to
linear, entire, c. 5 mm long. Sepals linear-lanceolate, 4–8 × 1–2 mm, acute,
entire or denticulate, usually pilose specially near base, ciliate, appendage
c. 3 mm long, pointed, rarely rounded. Petals 2–4 times as long as broad,
basal one obovate, obtuse, sometimes slightly bearded, laterals oblanceolate,
bearded, upper one usually lightly bearded; spur c. 5 mm long, obtusely
cylindrical. Anthers distinct. Ovary ovate, 2.5 × 1.5 mm, glabrous; style
subclavate, subtruncate, 1.5–3 mm long, shortly beaked at apex, beak directed downward; stigma simple, emarginate. Capsules ellipsoid, 5–6 mm in
diam., glabrous.

Distr. Pakistan, India, Nepal, Bhutan, Myanmar, China, Thailand,
Indonesia (Sumatra and Java) and Sri Lanka.

Ecol. Montane area of wet region, descending to 1200 m, fairly common; fls & frts most of the year.

Uses. Plants are medicinally useful as a fabrifuge; flowers in lung trou-
bles; petals made in a syrup and used for infantile disorders, roots emetic
(S.P. Banerjee & Pramanik, l. c. 2: 371. 1993).

Note. Chromosome number reported: $2n = 18, 48$. (Chatterjee &
Sharma, J. Genet. 61(1): 52–63. 1973).

Plate No. 106 in PDA based on the material from Hakgala, Apr 1920,
belongs to this species.

Specimens Examined. KANDY DISTRICT: Adam's Peak, May
1891, *s. coll. s.n.* (PDA), *Thwaites C.P. 153* (K). BADULLA DISTRICT:
Namunukula, 29 Apr 1909, *Willis s.n.* (PDA). RATNAPURA DISTRICT:
Adam's Peak, 80° 30′E, 6° 48′N, 7 March 1977; *B. & K. Bremer 989*
(PDA). NUWARA ELIYA DISTRICT: Horton Plains, World's end, 30 March

1995, *Wadhwa & Samarasinghe 608* (K, PDA), near Bakers Falls, 30 March 1995, *Wadhwa & Samarasinghe 607* (K, PDA); Jungle path to Horton Plains, 25 Jan 1906, *Willis s.n.* (PDA, HAK); Horton Plains, Sept 1890, *s. coll s.n.* (PDA), ± 2400 m, 29 Feb 1976, *Cramer 4598* (PDA), 26 Jan 1977, *Cramer 4813* (K), trail to World's End, 2400 m, 11 May 1970, *Gould & Cooray 13819* (PDA), near Farr Inn, 2100 m, 18 Sept 1969, *C.F. & R.J. van Beusekom 1482* (PDA), World's End vicinity, 2350 m, 19 Apr 1973, *Stone 11276* (PDA), on way to World's End, near Baker's Falls, 20 Oct 1994, *Wadhwa, Weerasooriya & Samarasinghe 447* (K, PDA), Horton Plains, North Entrance, 2100 m, 28 March 1968, *Fosberg & Mueller-Dombois 50040* (PDA), c.1/2 way to Little World's End from Farr Inn, 23 May 1969, *Read 2012* (PDA), Along Farr Inn-Diyagama Road, 29 Jan 1974, *Sumithraarachchi & H.N. & A.L.Moldenke 61* (PDA), Diyagama Estate to Horton Plains, 27 Oct 1975, *Sohmer & Sumithraarachchi 10031* (PDA); Horton Plains to Ohiya, c. 2040 m, 6 Feb 1971, *Robyns 7145* (PDA); along trail Big World's End to Small World's End, c. 2160 m, 5 Feb 1971, *Robyns 7136* (PDA); Hakgala strict Natural Reserve, 29 March 1995, *Wadhwa & Samarasinghe 600* (K, PDA); Nuwara Eliya, Sept 1854, *Gardner C.P. 153* (HAK, PDA); Hakgala, Apr 1920, *A. de Alwis s.n.* (PDA); Ambewela, 26 March 1906, *A.M. Silva s.n.* (Acc. No. 000029, HAK); Ohiya, 2100 m, 15 March 1971, *Balakrishnan 493* (K, PDA); Pidurutalagala, 2250–2485 m, 25 Oct 1974, *Davidse & Sumithraarachchi 8011* (PDA); Pidurutalagala, 1890 m, 2 Nov 1971, *Cramer 3494* (PDA), c. 2100 m, 20 March 1971, *Robyns 7293* (K, PDA). LOCALITY UNKNOWN: *Col. Walker 157* (PDA); *Col. Walker s.n.* (PDA); in 1847, *Gardner 47* (K); in 1839, *Mackenzie s.n.* (K); *Col. Walker 2* (K); *G. Thomson s.n.* (K).

3. Viola betonicifolia Sm. in Rees, Cyclop. 37 (1): Viola n. 7. 1817; Alston in Trimen, Handb. Fl. Ceylon 6: 13. 1931; Hara in Hara & Williams, Enum. Fl. Pl. Nepal 2: 47. 1979; Jacobs & D.M. Moore in Fl. Males. 7. 202. 1971; S.P. Banerjee & Pramanik in Sharma et al., Fl. India 2: 355. 1993.

subsp. **betonicifolia** D.M.Moore in Fedde, Repert. 68: 81. 1963. Type: Australia, Botany Bay, Port Jackson, N.S. Wales, *Wight s.n.*

Viola betonicifolia Sm. ssp. *nepalensis* (Ging. ex DC.) W. Becker, Bot. Jahrb. Syst. 54 (Beibl. 120): 166. 1917.

Viola patrinii var. *nepaulensis* Ging. ex DC., Prod.1: 293. 1824. Type: Nepal, *Buchanan - Hamilton s.n.* (BM).

Viola patrinii Auct.: Thw., Enum. Pl. Zeyl. 20. 1858; Hook. f. & Thoms. in Hook. f., Fl. Br. Ind. 1: 183. 1872; Trimen, Handb. Fl. Ceylon 1: 66. 1893; Dunn in Gamble, Fl. Pres. Madras 1: 48. 1915; Matthew & Britto in Matthew, Fl. Tamilnadu Carnatic 1: 54. 1983, non DC., 1824.

Viola walkeri Wight, Ill. 1: 42. t. 18. 1840. Type: Sri Lanka, *Walker s.n.*

Viola caespitosa D. Don, Prod. Fl. Nepal. 205. 1825. Type: *Buchanan-Hamilton s.n.* (BM).

Perennial herb, 8–20 cm high, glabrous; roots slender, unbranched; stem absent. Leaves linear-lanceolate to triangular-ovate or triangular-hastate, 1.5–8.5 × 1–2.5 cm, cuneate, truncate or widely shallowly cordate at base, acute or roundish obtuse at apex, shallowly and distantly crenate, sometimes dentate on basal lobes, usually decurrent on petiole, glabrous; petioles longer than the lamina, 1.5–12 cm long, winged, glabrous. Stipules free, ovate-lanceolate, 3–8 × 1–2 mm, acuminate, short-fimbriate. Flowers 6–12 mm long, purple with darker veins; peduncles 5–15 cm long, glabrous, with 2 bracteoles at middle. Sepals ovate to ovate-lanceolate, 4–9 × 1–2 mm, acute or acuminate, entire, glabrous or ciliate, green with scarious margins. Petals oblong-ovate, 1.5–2 times as long as sepals; lateral ones usually bearded; spur cylindrical, 2–6 mm long, straight or slightly upcurved. Style 1.5–2.5 mm long, almost geniculate at base, clavate distally; stigma 3-lobed. Capsule ellipsoid to oblong, up to 1 cm long, glabrous.

Distr. Himalayas (India, Pakistan, Nepal, Bhutan), Sri Lanka, China, Japan, Indo-China, Malaysia to Australia.

Ecol. Montane to submontane areas in wet zone; 1000–2000 m, rather common; fls & frts Feb–Apr.

Uses. Plants bruised and applied to ulcers and foul sores; the flowers are said to purify blood (S.P. Banerjee & Pramanik, l. c. 2: 357. 1993).

Note. Chromosome Number reported: $2n = 48, 72$ (Moore in Fedde, Repert. 68: 84. 1963); $2n = 24$ (Miyaji, Cytologia 1: 28–58. 1929). Two more subspecies, viz. ssp. *jaunserensis* (W.Beck.) Hara (Jap. J. Bot. 49(5): 133. 1974) from India, and ssp. *nova -guinensis* D.M. Moore (Fedde, Repert. 68: 82. 1963), from the Philippines & New Guinea are reported.

Plate No. 104 in PDA based on material from Sita Eliya, May 1992 belongs to this species.

Specimens Examined. MATALE DISTRICT: Rattota-Illukkumbura road, 5 May 1995, *Wadhwa, Weerasooriya & Samarasinghe 588* (K, PDA). BADULLA DISTRICT: Passara, Jan 1888, *s. coll. s.n.* (PDA). NUWARA ELIYA DISTRICT: B.G. Hakgala, as a weed, 29 March 1995, *Wadhwa & Samarasinghe 520* (K, PDA); Pundalu Oya, 27 Apr 1895, *s. coll. s.n.* (PDA); Sita Eliya, Mar 1922, *A.de Alwis s.n.* (PDA); Between Ramboda & Nuwara Eliya, near road marker 36/A-5, 15 May 1968, *Mueller-Dombois 68051505* (PDA); Ohiya Station, 2100 m, 15 March 1971, *Balakrishnan 474* (K, PDA); Nuwara Eliya, *Thwaites C.P. 1087* (HAK, K, PDA); Albion Tea Estate, 13 March 1906, *A.M.Silva s.n.* (Acc. No. 000031-HAK). LOCALITY UNKNOWN: 1830 m, *Col. Walker s.n.* (K).

2. HYBANTHUS

Jacq., Enum. Pl. Carib. 2. 1760, nom. cons.; Melchior in Pflanzenfam. ed. 2, 21: 357. 1925; Tennant, Kew Bull. 16(3): 430. 1963; Hutchinson, Gen. Fl. Pl. 2: 334. 1967; Jacobs & D. M. Moore in Fl. Males. ser.1, 7: 194. 1971; S.P. Banarjee & Pramanik in Fasc. Fl. India 12: 2. 1983. Type: *H. havanensis* Jacq.

Ionidium Vent., Jard. Malm. sub t. 27. 1803; Thw., Enum. Pl. Zeyl. 20. 1858; Benth. & Hook. f., Gen. Pl. 1: 117. 1862; Hook. f. & Thoms. in Hook. f. Fl. Br. Ind. 1: 185. 1872; Trimen, Handb. Fl. Ceylon 1: 67. 1893. Type: *I. polygalaefolium* Vent. nom. illeg. (*Viola verticillata* Ortega, *Ionidium verticillatum* Roem. & Schult., Syst. Veg. 5: 399. 1819–1820).

Herbs or undershrubs, rarely small trees, in a few species the twigs thorny and microphyllous. Leaves alternate, rarely subopposite to opposite, often herbaceous, sometimes leathery, more or less sessile. Stipules generally small, mostly caducous. Flowers generally solitary in the leaf axils, rarely in ± reduced axillary cymes or dichasia or in raceme-like elongated monochasia, rarely in a terminal leafy panicle, bisexual, irregular, with 2 bracteoles, rarely cleistogamous; pedicels articulated. Sepals subequal, small, ± triangular, persistent, rarely the margin with deep incisions. Petals unequal, persistent, middle ones longer and falcate; posterior ones small and straight, and anterior ones extended to a lip, with a claw ± deeply saccate to shortly spurred. Filaments usually free, rarely partly connate; anthers free or ± united, 2 or rarely 4 of them gibbous, spurred or glandular at the base; connectives prolonged into a distinct membranous appendage. Ovary ovoid, with 3 placentae bearing 3–24 ovules; style clavate, incurved; stigma oblique. Capsules ovoid-globose, elastically 3-valved; valves leathery. Seeds few, ellipsoid, mostly with a small caruncle.

About 150 species in the tropical and subtropical regions of the world (America, Africa, Asia and Australia); 2 in Sri Lanka.

KEY TO THE SPECIES

1 Herbs, annual or perennial, often woody at base; leaves spreading, variable, cuneate at base, often mucronate at apex, margin subentire to crenate-serrate . . . . 1. H. enneaspermus
1 Undershrubs; leaves crowded, erect, narrowly linear, acuminate, margins entire, slightly recurved . . . . 2. H. ramosissimus

1. **Hybanthus enneaspermus** (L.) F. Muell., Fragm. Phyt. Austr. 10: 81. 1876 (excl. syn. *Ionidium ramosissimum* Thw.); Hutch. & Dalz., Fl. W. Trop. Africa 1: 97, 96, f. 29. 1927; Alston in Trimen, Handb. Fl. Ceylon 6: 13. 1931; Tennant, Kew Bull. 16:.431. 1963; Jacobs & D.M. Moore in Fl. Males. ser.1, 7: 197. 1971; Grey-Wilson, Kew Bull. 36(1): 103. t.1. 1981; Chithra

in Nair & Henry, Fl, Tamilnadu 1: 16. 1983; Matthew & Britto in Matthew, Fl. Tamilnadu Carnatic 1: 52. 1983; S.P. Banerjee & Pramanik in Sharma et al., Fl. India 2: 343. 1993.

Viola enneasperma L., Sp. Pl. 2: 937. 1753; Roxb., Fl. Ind. 2: 448. 1824. Type: Sri Lanka, *Hermann s.n.* (Hermann Herb. 1: 19, BM).

Viola suffruticosa L., Sp. Pl. 2: 937. 1753; Type: Sri Lanka, *Hermann s.n.* (Hermann Herb. 1: 41, n. 318, BM).

Ionidium enneaspermum (L.) Vent., Jard. Malm. 1. sub t. 27. 1803; Roem. & Schult., Syst. Veg. 5: 393. 1819; Ging. in DC., Prod. 1: 308. 1824; Wight & Arn., Prod. 33. 1834.

Ionidium heterophyllum Vent., Jard. Malm. in adn. sub t. 27. 1803; Cooke, Fl. Pres. Bombay 1: 52. 1901. Type: China, without locality, (Herb. d' Incarville No. 109, P-JU, holotype).

Ionidium suffruticosum (L.) Roem. & Schult., Syst. Veg. 5: 394. 1819; Ging. in DC., Prod. 1: 311. 1824; Wight, Ic. Pl. Or. t. 308. 1840; Wight, Ill. Ind. Bot. 1: t. 19. 1840; Thw., Enum. Pl. Zeyl. 20. 1858; Hook. f. & Thoms. in Hook. f., Fl. Br. Ind. 1: 185. 1872; Trimen, Handb. Fl. Ceylon 1: 67. 1893; Dunn in Gamble, Fl. Pres. Madras 1: 49. 1915.

Hybanthus suffruticosus (L.) Baill., Bot. Med. 2: 841. 1884.

Herbs, 15–60 cm high; simple or profusely branched, often woody at base; young stems angular, more or less pubescent. Leaves sessile or subsessile, very variable, linear to linear-lanceolate, elliptic-lanceolate or oblong-lanceolate, 2–4.5 × 0.3–1.2 cm, cuneate at base, apex acute to gradually acuminate, often mucronate, margin subentire to crenate-serrate or dentate-serrate, glabrous to minutely pubescent; lower leaves broader than upper ones. Stipules triangular to subulate, 1–3 mm long, sometimes fimbriate towards the base. Flowers axillary, solitary, c. 8 mm across, pink or rose-pink; pedicels slender, articulated, up to 1.6 cm long, glabrous or pubescent with a pair of subulate bracteoles above the middle of the pedicel. Sepals 5, unequal, lanceolate, 2–3.5 × 1 mm, membranous, glabrous or ciliate. Petals 5, unequal, upper 2 oblong or elliptic-oblong, symmetric, acuminate, 3–4 mm long; lateral two falcate, 4–5 mm long, lowermost one larger, suborbicular to subcordate, clawed, saccate at base. Stamens 2–3 mm long; anterior stamen with a small, recurved, fleshy appendage. Ovary subglobose, 1–1.5 mm in diam., glabrous; style suberect, thicker towards tip. Capsules subglobose, 4–5 mm in diam., glabrous. Seeds ovoid-ellipsoid, longitudinally striate.

Distr. Widely in Africa and Madagascar, tropical Asia, Australia and America.

Ecol. In low country, especially in dry region in open ground, common; fls & frts throughout the year.

Uses. The plant is considered to possess diuretic and demulcent properties. A decoction of the whole plant is taken to improve memory and vitality and as a remedy in Asthma, fever and leprosy. A shampoo made from this plant removes dandruff. (S.P. Banerjee & Pramanik, l. c. 2: 345. 1993).

Vern. Oritad Tamarai (T).

Note. On Specimen *C.P. 76* (PDA), three localities, namely Maturata (Apr 1857); Balangoda (Feb 1846), *Gardner* and Jaffna (without precise date, 1826), *Gardner* are annotated.

A very polymorphic species, very variable in habit and leaf form. Grey-Wilson (Kew Bull. 36(1): 103–110. 1981) has recognised several distinct varieties in tropical east Africa.

Plate No. 107 in PDA based on *C.P. 76* belongs to this species.

Chromosome Number reported: $n = 16$ (Gupta & Srivastava, Taxon 20: 609–614. 1971); $2n = 32$ (Sanjappa, Taxon 28: 265–279. 1979; Sarkar et al., Taxon 29: 347–367. 1980).

Specimens Examined. VAVUNIYA DISTRICT: Parayanalankulam, alt. low, in 1973, *Jayasuriya, Dassanayake & Balasubramaniam 975*, (K, PDA). ANURADHAPURA DISTRICT: Borupangoda rock, 12 Jan 1974, *Waas 349* (PDA); Wilpattu National Park, Kalivillu, 1 May 1969, *Cooray & Balakrishnan 69050108* (PDA), Maradan Maduwa, 4 miles S.E. of Kumbuk villu, 31 Dec 1968, *Fosberg, Mueller-Dombois, Wirawan, Cooray & Balakrishnan 50987* (K, PDA). TRINCOMALEE DISTRICT: Trincomalee-Habarana Road, Mile marker 153/1, 12 July 1973, *Nowicke & Jayasuriya 296* (K, PDA). BATTICALOA DISTRICT: Tiruperumdurai, bordering Airport, 5 Dec 1976, *Cramer 4787* (K, PDA). KALUTARA DISTRICT: Maggona, at base of Calvary Shrine in St. Vincent's Home, 20 May 1970, *Cramer 2977* (PDA). MONERAGALA DISTRICT: Bibile East to National Park, 5 miles from Bibile, 11 Oct 1973, *Sohmer, Jayasuriya & Eleizer 8296* (PDA). MATARA DISTRICT: Matara, on sea shore, 3 Jan 1927, *Alston 1121* (PDA); HAMBANTOTA DISTRICT: Tissamaharama, 2 Dec 1982, *s. coll. s.n.* (PDA); Ruhuna National Park, Block-I, N. of Buttawa Bungalow, 6 Apr 1968, *Mueller-Dombois & Cooray 68040608* (PDA), 26 Oct 1966, *Wirawan 669* (PDA), Rakinawewa, near Gonalabbe Lewaya, 5 Apr 1969, *Fosberg 50284* (PDA), Patanagala, 27 Nov 1969, *Cooray 69112706* (PDA), Mahaselawa, 4 Dec 1973, *Sohmer 8970* (PDA). LOCALITY UNKNOWN: *Col.Walker s.n.* (K, PDA); *s. coll. C.P. 76* (K, PDA); in 1839, *Mackenzie s.n.* (K); *Gardner 50* (K); *Mrs. Col. Walker 1854* (K).

2. Hybanthus ramosissimus (Thw.) Melchior in Pflanzenfam. ed. 2, 21: 360. 1925; Alston in Trimen, Handb. Fl. Ceylon 6: 13. 1931.

Ionidium ramosissimum Thw., Enum. Pl. Zeyl. 21. 1858; Hook. f. & Thoms. in Hook. f., Fl. Br. Ind. 1: 186. 1872; Trimen, Handb. Fl. Ceylon 1: 68. 1893. Type: Sri Lanka, Mahaweli Ganga, at Haragama, *C.P. 1084* (PDA, holotype; K, isotype).

Undershrubs, 15–20 cm high, with numerous slender, ascending or suberect branches. Leaves sessile, erect, crowded, closely imbricate, narrow linear, 2–4.5 cm long, acuminate, margin entire, slightly recurved. Flowers solitary, axillary, pink to pale violet; pedicels slender, up to 2 cm long, glabrous or pubescent, with 2 bracteoles below joint in upper half of the pedicel. Sepals subequal, ovate-lanceolate, 2–3.5 mm long, acute, ciliate. Petals unequal, upper 2 elliptic, symmetric, 3–4.5 mm long; lateral two triangular-oblong, 4–5 mm long, obtuse; lowermost 8–16 mm long, with spurred claw and obovate broad limb. Stamens c.2 mm long; anterior stamens with a recurved fleshy appendage. Ovary subglobose, glabrous; style distally thickened. Capsule 3-lobed, glabrous, 9–12-seeded. Seeds ovoid-ellipsoid, longitudinally striate.

Distr. Endemic.

Ecol. In the low country along river banks or waterchannels, very rare; fls & frts July–Aug.

Note. Plate No. 108 in PDA based on *C.P. 1084* belongs to this species.

Specimens Examined. KANDY DISTRICT: Haragama, 9 Feb 1881, *s. coll. s.n.* (PDA); Pallekele, Haragama, 26 Apr 1926, *Alston 858* (K, PDA) NUWARA ELIYA DISTRICT: Maturata, Aug 1853, *Moon C.P. 1084* (K, PDA).

3. RINOREA

Aublet, Hist. Pl. Guian. Franc. 1: 235. t. 93. 1775; Melchior in Pflanzenfam. ed. 2, 21: 349. 1925; Tennant, Kew Bull. 16(3): 409. 1963; Hutchinson, Gen. Fl. Pl. 2: 330. 1967; Jacobs, Blumea 15: 127. 1967; Jacobs & D.M. Moore in Fl. Males., ser.1, 7: 180. 1971, 12: 4. 1983. Type: *R. guianensis* Aublet.

Alsodeia Thouars, Hist. Veg. Austr. Afr. 55. t. 17–18. 1806; Thw., Enum. Pl. Zeyl. 21. 1858; Benth. & Hook. f., Gen. Pl. 1: 118. 1862; Hook. f. & Thoms. in Hook. f., Fl. Br. Ind. 1: 186. 1872; Trimen, Handb. Fl. Ceylon 1: 68. 1893. Type: *A. pauciflora* Thouars.

Scyphellandra Thw., Enum. Pl. Zeyl. 21. 1858. Type: *S. virgata* Thw.

Shrubs or small trees; twigs generally angular when young. Leaves alternate, distichous, rarely opposite or in a spiral, entire, shallowly incised or sometimes serrate; secondary nerves parallel and many, domatia occasionally present in the primary vein axils; petiole comparatively short, rarely wanting. Stipules caducous to long-persistent, often longitudinally striate. Inflorescence lateral, rarely axillary or a terminal cyme. The flowers mostly in bundles, sometimes distichous along a short rachis, sometimes in a corymb or panicle, bisexual or unisexual, sometimes plants dioicious, regular; peduncles articulated. Sepals almost equal, rigid, fimbriate. Petals isomorphic or subequal, free, sometimes very shortly clawed. Stamens inserted on the margin of an annular disk; flaments free or connate; connectives produced into a long or short, often broad membranous appendage. Ovary tricarpellary, 1-locular with 3 parietal placentae; ovules. 1–3 on each placenta; style straight, with terminal, ± 3-lobed stigma. Fruit capsular, globose, 3-valved, subtended by other floral parts. Seeds 3–6, ellipsoid, glabrous, rarely woolly, with leathery testa.

About 200 species, pantropical; in Africa, America, Indo-Malesia, (India, Sri Lanka, Myanmar, Hainan, Bangladesh, Malaysia, Indonesia) to N. Australia; 3 species in Sri Lanka.

KEY TO THE SPECIES

1 Shrubs; twigs when young angular, often puberulous with curved hairs; leaves distichous, variable, 1–3 × 0.8–1.5 cm; petioles to 5 mm long; plants dioecious, in the male flowers only a style, in the female flowers the anther cells vestigial; ovules 6 .... 3. R. virgata
1 Shrubs or small trees; leaves 6–18 × 2–9 cm; petioles to 1 cm; flowers bisexual; disk cupular; ovules 3
  2 Leaves lanceolate to oblong-lanceolate, 5–7 × 2–2.5 cm; internodes 0.3–1 cm, greenish in dried state, apex obtuse, margin crenate-serrate, or serrate .... 1. R. decora
  2 Leaves elliptic-lanceolate, 6–12 × 2–5 cm; internodes c. 1.5 cm, brownish-green in dried state; apex acute or slightly acuminate, margin finely shallowly serrate ....
  .... 2. R. bengalensis

1. **Rinorea decora** (Trimen) Melchior in Pflanzenfam. ed. 2, 21: 352. 1925; Alston in Trimen, Handb. Fl. Ceylon 6: 14. 1931.

Alsodeia decora Trimen, J. Bot. 23: 203. 1885. Type: Sri Lanka, Hewaheta, in 1868, *Thwaites C.P. 4006* (PDA).
Alsodeia obtusata Thw. mss. in Sched. in Herb. K.

Small trees, to 6 m tall; young branches finely pilose; internodes 3–10 mm long. Leaves coriaceous, lanceolate to oblong-lanceolate, often slightly trapezoid, 5–7 × 2–2.5 cm, subcordate or acute at base, apex obtuse or emarginate, margin crenate-serrate or serrate at least in the upper 1/2 to 2/3 part, greenish when dry; midrib and veins distinct, often densely pilose in the primary axils; petioles 5–8 mm long. Stipules appressed, slender, subulate, early caducous. Flowers bisexual, pedicellate, 3–5 mm across, in short contracted, axillary, racemose fascicles. Bracteoles 2, keeled, falling very soon, leaving pale scar. Sepals subequal, lanceolate, 2–2.5 × 1–1.5 mm, fleshy, ciliate, persistent. Petals subequal, oblong to strap-shaped, nearly twice as long as the sepals, thin, dark-purple-spotted, with a few hairs near the top, persistent. Stamens on a 5-lobed, fleshy disk, glabrous; filaments short, cohering into a tube; anthers oblong; appendage subapical, hooded. Ovary oblong, glabrous, 3-ovuled; style short, glabrous. Fruit ± globose, up to 1 cm in diam., apiculate, 3-valved, dehiscing longitudinally. Seeds globose, mottled.

Distr. Endemic.

Ecol. In forests of submontane zone, 600–650 m, rare; fls & frts Sept–Oct.

Specimens Examined. MATALE DISTRICT: Nitre Cave, Sept 1888, *s. coll. s.n.* (PDA). KANDY DISTRICT: Hewaheta, in 1868, 1220 m, *Thwaites C.P. 4006* (PDA).

2. Rinorea bengalensis (Wall) Kuntze, Rev. Gen. Pl. 1: 42. 1891; Jacobs, Blumea 15: 128. 1967; Jacobs & D.M. Moore in Fl. Males. ser. 1, 7: 184. 1971; Sharma et al., Fl. Karnataka 11. 1984; B.G. Singh in Saldanha, Fl. Karnataka 1: 276. 1984; S.P. Banerjee & Pramanik in Sharma et al., Fl. India 2: 348. f. 66, 1993.

Alsodeia bengalensis Wall., Trans. Med. Phys. Soc. Calcutta 7: 224. 1835; Hook. f. & Thoms. in Hook. f., Fl. Br. Ind. 1: 186. 1872; Kurz, For. Fl. Burma 1: 70. 1877. Type: Bangladesh, Sylhet, *Wallich* n. *4896* (K, holotype).
Pentaloba ceylanica Arn., Mag. Zool. Bot. 2:543. 1838. Type: Sri Lanka, *Wight* Cat. n. *268* (K).
Pentaloba bengalensis Wall. ex Arn., Mag. Zool. Bot. 2: 543. 1838. Type: same as for *Alsodeia bengalensis* Wall.
Alsodeia ceylanica (Arn.) Thw., Enum. Pl. Zeyl. 21. 1858 "*zeylanica*"; Hook. f. & Thoms. in Hook. f., Fl. Br. Ind. 1: 187. 1872; Trimen, Handb. Fl. Ceylon 1: 68. 1893; Gamble, Fl. Pres. Madras 1: 49. 1915.
Rinorea ceylanica (Arn.) Kuntze, Rev. Gen. Pl. 1: 42. 1891 "*zeylanica*"; Alston in Trimen, Handb. Fl. Ceylon 6: 14. 1931.

Shrubs or small trees, 4–8 m tall; branchlets glabrous or puberulous; internodes 1.5 cm long. Leaves coriaceous, elliptic-lanceolate, 6–12 × 2–5 cm, base acute, rarely subcordate, apex acute or slightly acuminate, margin finely crenate-serrate, glabrous except densely pilose in the primary nerve axils; lateral veins 7–12 pairs, reticulation on both sides distinct; surfaces distinctly greenish, when dry brownish; petioles up to 1 cm long. Stipules appressed, linear-lanceolate to subulate, 5–8 mm long, brown. Flowers white, small, to 4 mm across, solitary or fascicled, axillary, surrounded by small, ovate, subacute to acute, rufous bracts; peduncles up to 1 cm long, jointed at or near the base. Sepals subequal, lanceolate or broadly ovate, 2 × 1 mm, fleshy, ciliate.

Petals subequal, oblong-linear, twice as long as the sepals, acute, recurved at apex, fleshy. Stamens on a thick, 5-lobed disk, glabrous; filaments short, cohering into a tube; anthers oblong, appendage subapical, broad, hooded. Ovary oblong or subglobose, sometimes hairy, 1-locular with 3 ovules; style short, straight, glabrous. Capsules globose, up to 1 cm in diam., apiculate, 3-valved, longitudinally dehiscent. Seeds 3–4, globose, c. 5 mm in diam., smooth, glabrous.

Distr. India (S. & E. India), Bangladesh, Myanmar, Indo-China, Sri Lanka, Hainan, Malaysia, Australia and Pacific Islands.

Ecol. As an understorey plant in low, wet country rain forest, rather rare; fls & frts Mar–May.

Note. Plate No. 109 in PDA based on *C.P. 2669* belongs to this species.

Specimens Examined. RATNAPURA DISTRICT: Ratnapura, Mar 1853 & Sept 1855, *s. coll. C.P. 2669* (K, PDA). LOCALITY UNKNOWN: *Gardner s.n.* (K); *Champion s.n.* (K).

3. Rinorea virgata (Thw.) Kuntze, Rev. Gen. Pl. 1: 42. 1891; Alston in Trimen, Handb. Fl. Ceylon 6: 14. 1931; Jacobs, Blumea 15: 135. 1967; Jacobs & D.M. Moore in Fl. Males. ser.1, 7: 190. 1971.

Scyphellandra virgata Thw., Enum. Pl. Zeyl. 21. 1858. Type: Sri Lanka, *C.P. 1085* (PDA; K, isotype).
Alsodeia virgata (Thw.) Hook. f. & Thoms. in Hook. f., Fl. Br. Ind. 1: 189. 1872; Trimen, Handb. Fl. Ceylon 1: 69. 1893.

Small shrub, c. 2 m tall; twigs long, slender, often puberulous with curved hairs, when young angular. Leaves generally distichous, herbaceous to thinly coriaceous, on the same plant variable, oblong to oblong-lanceolate or rhombic, 1–3.5 (–4.5) × 1–1.5 (–2) cm, obtuse to subacute, sometimes acute, occasionally mucronate, base cuneate to rounded, margin subentire to faintly serrate; midrib slender, prominent above, with 2–6 major veins, not pilose in the primary axils, reticulation lax, irregular; surfaces sometimes with a few hairs on the midrib above, otherwise glabrous, dull, concolorous, green; petioles 1–3 mm long, hairy. Stipules small, 1–2 mm long, ± subulate. Plants dioecious; flowers white, solitary, or in small, axillary fascicles; pedicels short, 2–4 mm long, jointed near the base, puberulous. Bracts persistent, triangular. c. 1 mm long. Sepals more or less equal, triangular, 1–2 × 1 mm, subacute, thin, sometimes hairy. Petals ovate-triangular. 1.5–3 × 1–1.5 mm, glabrous, sometimes with recurved tip. Stamens nearly as long as the petals, glabrous, in the male flowers anthers sessile, with an elongate scale at the base on the outside, thecae with dorsal appendage, in the female flowers staminodes subsessile, thecae abortive. Ovary in the female flowers globose, glabrous, as long as the petals, with thick long style, in the male flowers ovary absent, only a vestigial style arising from a flat receptacle. Fruit globose, 1 cm in diam., with a long apiculus, olive green. Seeds subglobose, 2–2.5 × 1.5–2 mm, cream-coloured, with distinct raphe.

Distr. Sri Lanka, Myanmar, Thailand, Laos, S. Vietnam and Malay Penninsula.

Ecol. On dry slopes and in ravines or open low country forests, along forest edges, rather rare; fls & frts July–Sept.

Note. On Specimen No. *C.P. 1085* (PDA), two localities viz. Minneriya, July 1848, *Gardner* (Polonnaruwa Dist.), and Bintenna-Ooma oya, July 1853, *Gardner* (Badulla Dist.) are annotated.

Plate No. 111 in PDA based on *C.P. 1085* belongs to this species.

Specimens Examined. ANURADHAPURA DISTRICT: Wilpattu National Park, between Kanjuran and Kumutavillu, 26 Apr 1969, *Mueller-Dombois, Wirawan, Cooray & Balakrishnan 69042603* (K, PDA). MATALE DISTRICT: Lenadora, near Dambulla, Feb 1888, *s. coll. s.n.* (PDA). POLONNARUWA DISTRICT: Relapanawa, NE of Kawdulla wewa, Medirigiriya, 13 Sept 1991, *Jayasuriya 5720* (PDA). RATNAPURA DISTRICT: Kuragala other state Forest, 27 Nov 1993, *Jayasuriya 7727* (PDA); Kapugala other state Forest, approach from Kaltota road, turn off Rajawaka, 26 Nov 1993, *Jayasuriya 7708* (PDA); Atakalan Korale, Sept 1857, *s. coll. C.P. 1085*(K, PDA); Gallegoda, Rakwana Range, 21 March 1994, *Jayasuriya 8188* (PDA). HAMBANTOTA DISTRICT: Suandana Ara, NE of Galge, Ruhuna National Park, 23 Sept 1992, *Jayasuriya 6736* (PDA).

Printed in India

LIST OF NEW NAMES PUBLISHED IN THIS VOLUME

T - #0157 - 101024 - C0 - 229/152/24 [26] - CB - 9789054102687 - Gloss Lamination